GEOTECHNICAL SPECIAL PL

DEEP FOUNDATIONS AND GEOTECHNICAL IN SITU TESTING

PROCEEDINGS OF SESSIONS OF GEOSHANGHAI 2010

June 3–5, 2010
Shanghai, China

HOSTED BY
Tongji University
Shanghai Society of Civil Engineering, China
Chinese Institution of Soil Mechanics and Geotechnical Engineering, China

IN COOPERATION WITH
Alaska University Transportation Center, USA
ASCE Geo-Institute, USA
Deep Foundation Institute, USA
East China Architectural Design & Research Institute Company, China
Georgia Institute of Technology, USA
Nagoya Institute of Technology, Japan
Transportation Research Board (TRB), USA
The University of Newcastle, Australia
The University of Illinois at Urbana-Champaign, USA
The University of Kansas, USA
The University of Tennessee, USA
Vienna University of Natural Resources and Applied Life Sciences, Austria

EDITED BY
Robert Y. Liang
Feng Zhang
Ke Yang

ASCE

G-I GEO-INSTITUTE

Published by the American Society of Civil Engineers

Library of Congress Cataloging-in-Publication Data

GeoShanghai International Conference (2010)
Deep foundations and geotechnical in situ testing : proceedings of the GeoShanghai 2010 International Conference, June 3-5, 2010, Shanghai, China / edited by Robert Y. Liang, Feng Zhang, Ke Yang.
p. cm. -- (Geotechnical special publication ; no. 205)
Includes bibliographical references and index.
ISBN 978-0-7844-1106-3
1. Soil mechanics--Congresses. 2. Soils--Testing--Congresses. 3. Deformations (Mechanics)--Congesses. I. Liang, Robert Y. II. Zhang, Feng, 1959- III. Yang, Ke, 1978- IV. American Society of Civil Engineers. V. Title.

TA710.A1G368 2010c
624.1'5--dc22 2010012101

American Society of Civil Engineers
1801 Alexander Bell Drive
Reston, Virginia, 20191-4400

www.pubs.asce.org

ISBN 978-0-7844-1106-3
Manufactured in the United States of America.

Preface

This Geotechnical Special Publication (GSP) contains 49 technical papers in the area of deep foundations (traditional driven piles and drilled shafts as well as innovative deep foundation construction technologies) and in situ geotechnical testing and monitoring techniques. An invited keynote paper, prepared by Professor Maosong Huang of Tongji University and Mr. Weidong Wang of East China Architectural Design and Research Institute, presented recent advances in theory and analysis methods for characterizing load carrying capacity and deformation behavior of uplift piles in soft ground. The remaining 48 technical papers are grouped into four sections. The "Piled Raft System and Soil-Structure Interaction" section contains ten papers focusing on analytical study and field monitoring of piled raft foundation system and analysis techniques for soil-foundation-structure interactions. The "Deep Foundations" section contains eighteen papers on the topics related to traditional drilled shafts and driven piles. Twelve technical papers in the "Innovative Foundations" section cover a wide range of non-traditional foundation system, such as anchor piles, rammed concrete piles, SDCM piles, cast-in-place concrete pipe piles, and jacked piles, among others. Eight technical papers in the "In-Situ Testing" section present recent research findings and case studies on in-situ geotechnical testing techniques.

Each paper published in the ASCE Geotechnical Special Publication was peer reviewed in accordance with the standards of practice of the Geo-Institute of the American Society of Civil Engineers. Each paper in this GSP was evaluated by at least two anonymous, qualified, technical reviewers and selected for publication by the proceedings editors. The authors of the accepted papers have addressed the reviewers' comments to the satisfaction of the editors. All published papers are eligible for discussion in the ASCE Journal of Geotechnical and Geoenvironmental Engineering and are also eligible for ASCE awards.

We are thankful to Prof. Yongsheng Li, Chair of GeoShanghai, and Professors Maosong Huang and Imad Al Qadi, Co-chairs, and Professor Baoshan Huang and Dr. Xian Liu, General Secretaries, for their leadership in organizing this conference, and the local organizing committee for their diligent and tireless work for this conference.

The papers in this publication were presented during the GeoShanghai 2010 Conference held in Shanghai, China on June 3 to June 5, 2010. This conference was host by Tongji University, Chinese Institution of Soil Mechanics and Geotechnical Engineering, and Shanghai Society of Civil Engineering. The cooperating agencies include: ASCE Geo-Institute, Transportation Research Board (TRB) of the National Academies, East China Architectural Design and Research Institute Co., Ltd., Deep Foundation Institute, The University of Kansas, University of Illinois at Urbana-

Champaign, Vienna University of Natural Resources and Applied Life Sciences, Nagoya Institute of Technology, University of Newcastle, Alaska University Transportation Center, and University of Tennessee.

The editors wish to thank the following individuals who reviewed one or more papers for this GSP. For those reviewers whose names were inadvertently missed, we offer our sincere apologies. This list is arranged in alphabetical order, by last name:

Name	Affiliation	Name	Affiliation
Scott Anderson	Federal Highway Administration	Inmar Badwan	T.Y. Lin International
Curt Basnett	CH2M HILL	Tanner Blackburn	Hayward Baker Inc.
William Camp	S&ME, Inc.	Raymond Castelli	Parsons Brinckerhoff
Der-Wen Chang	Tamkang University	Jiun-Yih Chen	The University of Texas at Austin
Shengli Chen	University of Oklahoma	Jianye Ching	National Taiwan University
Jerry DiMaggio	Transportation Reserarch Board, NAS	Vishnu Diyaljee	GAEA Engineering Ltd
Jiwei Duan	Maryland State Highway Administration	Christopher Dumas	CH2M HILL
Raul Durand	The University of Newcastle	Roger Failmezger	In-situ Soil Testing, L.C.
Kazunori Fujisawa	Okayama University	Yingcai Han	Fluor
Takashi Hara	Gifu University	Jie Huang	California Department of Transportation
Mohammad Khasawneh	Ohio Northern University	Mamoru Kikumoto	Nagoya Institute of Technology
Shun-ichi Kobayashi	Kanazawa University	Takahiro Konda	Geo-Research Institute
Robert Kruger	CB&I	Ching Kuo	PSI
Hiroyuki Kyokawa	Nagoya Institute of Technology	Debra Laefer	University College Dublin
Peter Lai	Florida Department of Transportation	Wei Lee	National Taiwan University of Science and Technology
Robert Y. Liang	The University of Akron	Tianfei Liao	Bechtel Power Corporation
Guoming Lin	Terracon Consultants, Inc.	Horn-Da Lin	National Taiwan University of Science and Technology
Kenichi Maeda	Nagoya Institute of Technology	Kirk McIntosh	MACTEC Engineering & Consulting, Inc.
Jorge Meneses	Kleinfelder, Inc.	Anil Misra	The University of Kansas

Name	Affiliation	Name	Affiliation
Syuichi Moriguchi	Gifu University	Kentaro Nakai	Nagoya University
Masaki Nakano	Nagoya University	James Niehoff	PSI Inc
Toshihiro Noda	Nagoya University	Jamal Nusairat	E.L. Robinson Engineering
Kwabena Ofori-Awuah	KCI Technologies Inc.	Chang-Yu Ou	National Taiwan University Of Science and Technology
Suresh Parashar	CH2M HILL	Mark Petersen	Black & Veatch
Samer Rababah	Gannett Fleming, Inc.	Frank Rausche	GRL
Lance Roberts	South Dakota School of Mines	King Sampaco	CH2M HILL
Hossain Shahin	Nagoya Institute of Technology	Ehab Shatnawi	CH2M HILL
Elizabeth M. (Liz) Smith	Terracon Consultants, Inc.	Muhannad Suleiman	Lafayette College
Robert Thompson	Dan Brown and Associates, PLLC	Ching Tsai	Louisiana Department of Transportation
Wei Tu	Black & Veatch	Cumaraswamy Vipulanandan	The University of Houston
Richard Wan	University of Calgury	Yoichi Watanabe	Kumamoto University
Daping Xiao	Institute of Foundation Engineering, China Academy of Building Research	Ke Yang	CH2M HILL
Luo Yang	Fugro	Zia Zafir	Kleinfelder, Inc.
Feng Zhang	Nagoya Institute of Technology	Limin Zhang	Hong Kong University of Science and Technology
Wei Zheng	Black & Veatch		

Editors:

Robert Y. Liang, Ph.D., P.E., F. ASCE, University of Akron
Feng Zhang, Ph.D., Nagoya Institute of Technology
Ke Yang, Ph.D., P.E., M. ASCE, CH2M HILL

Contents

Section 2: Deep Foundations

Indexes

Bearing Capacity and Deformation Characteristics of Uplift Piles in Soft Ground

Maosong Huang[1] and Weidong Wang[2]

[1]Professor, Department of Geotechnical Engineering, Tongji University, Shanghai 20092, China; mshuang@tongji.edu.cn

[2] Principal Engineer, East China Architectural Design & Research Institute, Shanghai 200002, China, weidong_wang@ecadi.com

ABSTRACT: Theoretical and experimental studies on the bearing capacity and deformation characteristics of uplift piles have received much attention due to extensive use of uplift piles in underground structures in soft ground. Shanghai World Expo 500kV Underground Transformer Substation and Shanghai Hongqiao Comprehensive Transportation Hub are typical examples. Both types of uplift piles, i.e. uniform-section piles with side-grouting and enlarged-base piles, are often used in engineering practice. Despite of their extensive applications, theoretical studies have lagged far behind engineering practice. Presented in this paper was a comprehensive study on the simplified method for predicting the ultimate bearing capacity of enlarged-base piles with various pile lengths before and after deep excavation, and load-displacement responses of uniform-section piles with side-grouting based on the load transfer method. The good agreement between theoretical predictions and field pile load test results was obtained. The proposed method provides reliable estimates in a simple way for the bearing capacity and deformation characteristics of uplift piles.

INTRODUCTION

With the development of urban construction, extensive applications of uplift piles in the anti-uplift design of deeply buried underground structures in the modern cities have taken place. Shanghai World Expo 500kV Underground Transformer Substation (Fig.1) is a typical example, which is currently the biggest and deepest underground transformer substation constructed in soft ground. Generally in engineering practice, belled uplift piles without grouting and uniform-section piles with side-grouting are often used to replace the conventional uniform-section (straight-shaft) piles to improve the bearing capacity of the uplift piles. The belled uplift piles are a kind of piles with variable cross-section, the uplift bearing capacity of which can be improved conspicuously by adding a small amount of material compared to the uniform cross-section piles. Based on field measurements, the uplift bearing capacity of the piles with side-grouting can also be improved since the side-grouting improves the interface friction property between the pile and soils. These two kinds of uplift piles have been used successfully in the practice here in Shanghai, however at the present time, both theoretical and experimental researches take the dust of engineering practice obviously.

FIG.1 Shanghai World Expo 500kV underground transformer substation

Additionally, for large-scale underground projects just like the Shanghai World Expo 500kV Underground Transformer Substation, a completely new problem of the uplift piles was faced. The substation is a fully underground structure with the diameter of 130m and embedded depth of 34m, and hence all uplift piles are deeply-buried under the ground. The determination of the bearing capacity of the uplift piles becomes rather complicated due to the bearing capacity loss under deep excavation. Bearing capacity of the uplift piles would inevitably decrease due to the change of ground stress field and the normal stress of pile-soil interface under deep excavation. Due to the limitation of field conditions, it is almost impossible to obtain the bearing capacity of the uplift piles under deep excavation from field tests. Field pile tests can be performed only before the excavation, even if tested piles are constructed by double-pipe for the top section which can keep the pile separated from soils. Such a double-pipe construction technology can only neglect the side resistance of the pile from the ground to the bottom of the underground structure. The change of normal stresses of pile-soil interface caused by excavation unloading is not considered. Therefore, theoretical study in conjunction with centrifuge modeling becomes the only feasible way to solve this problem.

Usually, the failure surface of short uplift piles at an ultimate state can be assumed to be a cylindrical failure surface, an inverted cone failure surface or a curved failure surface. Balla (1961) and Meyerholf & Adams (1968) pointed out that under the ultimate uplift force, the resulting failure surface initiates tangentially to the pile surface at the tip of the pile and moves through the surrounding soil, and the inclination of the failure surface with the horizontal at the ground surface approaches 45° -φ/2. He (2001) proposed a kind of curved failure surface equation based on a power function, and all previous assumptions on the variety of failure surfaces were only special cases of He's failure mechanism. Li *et al.* (2008) generalized He's failure surface further to analyze the belled uplift short piles. The main difficulty of the failure surface based on the power function was that the parameter *N* which defines the power function was needed to be adjusted frequently to obtain satisfying results. There is no explicit formula for calculating the parameter *N* proposed in the literature so far, and the influence of the specific soil properties to the failure surface curve can not be reflected. Chattopadhyay & Pise (1986) assumed a similarly curved failure surface for straight-shaft uplift short piles and the lateral horizontal extent of the failure surface was dependent of the frictional angle of the surrounding soils, the angle of pile friction, and the slenderness ratio of the pile. However, such an analytical method has not been used to analyze the belled uplift short piles.

Due to local failure occurred near the enlarged base, the failure surfaces mentioned above are obviously not capable of capturing the failure behavior of the enlarged-base uplift long piles. The failure surface of the enlarged-base uplift long pile is similar to the local shear slip surface of the deeply buried anchor. Based on elasto-plastic finite element analyses, Li et al. (2009) proposed a local failure surface with an elliptic shape for evaluating the bearing capacity of the uplift long piles with enlarged-base using limit equilibrium method. However, until now there has been no unified approach reported yet in the literature which is able to predict the ultimate bearing capacity of the enlarged-base uplift piles with various pile lengths.

The limit equilibrium method was applied successfully in the calculation for the limit bearing capacity of the uplift piles, however, the shortcoming of the limit equilibrium method is that the deformation properties of the uplift piles under uplift load can not be analyzed. Numerical simulation method has been very popular in the analysis of load-displacement responses of the uplift piles. But calibration of soil parameters used in the numerical simulations requires much more verification works to be done before the numerical method can be applied to the practical engineering. Geol & Patra (2007), and Sun *et al.* (2008) applied the load transfer method to the predictions of the load-displacement responses of the uniform-section piles and belled piles under uplift load. The load transfer method can be used for analyzing not only the ultimate bearing capacity but also the deformation characteristics of the uplift piles, and it has been applied successfully into the uniform-section and belled uplift piles without side-grouting. However, the extension of the load transfer method to the uplift piles with side-grouting is lacked due to the difficulties in the determination of the relationship between the pile shaft friction resistance and the pile-soil relative displacement (τ-z curves) after side-grouting.

Thus, the present paper focuses on the following three aspects:

(1) Based on finite element numerical simulations and failure mechanisms proposed in the literature, a unified approach to estimating the ultimate bearing capacity of uplift piles with enlarged-base is introduced. Under the ultimate state, the failure surface consists of an elliptic local failure surface on the enlarged base and an exponential function of failure surface for the section of straight-shaft.

(2) The τ-z curves of the uplift piles with or without side-grouting at each soil layer measured in the field is normalized with the limit side resistance τ_{ult} and corresponding critical pile-soil relative displacement z_{ult}. Based on the normalized τ-z curves, i.e. τ/τ_{ult}-z/z_{ult} curves, hyperbolic τ/τ_{ult}-z/z_{ult} functions of the uplift piles are then proposed in conjunction with the load transfer method for predicting the load-displacement responses before and after side-grouting in layered soils.

(3) In order to calculate the bearing capacity of the uplift piles under deep excavation, a simplified analytical procedure using the limit equilibrium method is presented, in which the change of ground stress field is estimated by the Mindlin solution. Comparison between the simplified method and the elastic-plastic finite element method is given to verify the effectiveness of the simplified procedure.

A UNIFIED APPROACH TO ESTIMATING ULTIMATE BEARING CAPACITY OF UPLIFT PILES WITH ENLARGED BASE

Finite Element Analysis of Belled Uplift Piles

The behavior of the uplift piles with enlarged base can be assessed by numerical simulations with comparison to field test data. Finite element method was used here to analyze the behavior of the belled uplift piles regarding the development of soil plastic zone at an ultimate state. In order to verify the feasibility of the finite element method, the results of finite element simulation were compared to the field measurements of a belled uplift pile as presented below.

The length of the pile L used for the in-situ test is 50.4m, the diameter of the straight-shaft d is 0.7m, the diameter of the enlarged base d_1 is 1.4m, and the height of the enlarged base H_0 is 2.0m. Table 1 presents the soil characteristics of the ground for the in-situ test.

Fig.2 shows the comparisons between the FE simulations and field measurements, including the load-displacement curve of pile head, shaft axial force and side resistance distribution along the pile depth.

Table 1. Soil Characteristics of the Ground for the In-situ Test

Soil layer	h(m)	γ(kN/m^3)	c(kPa)	ϕ (°)	E_s(MPa)
② Silty clay	1.4	18.5	21	19	4.24
③ Mucky-silt clay	2.8	17.3	12	19.5	3.53
④ Mucky clay	7	16.9	14	13	2.37
⑤$_{1-1}$ Silty clay	9.8	17.8	15	14.5	3.49
⑤$_{1-2}$ Silty clay	3.7	18.1	15	21	4.31
⑥ Silty clay	3.3	19.4	45	16	6.98
⑦$_1$ Sandy silt	11.8	19	5	29.5	12.21
⑦ Silty clay	5.6	18.3	22	20	2.28
⑦$_2$ Fine sand	13.9	18.8	0	33	13.14

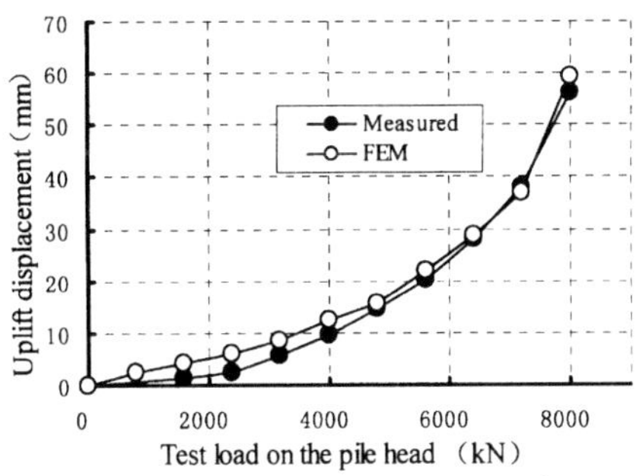

(a) Load-displacement curve of pile head

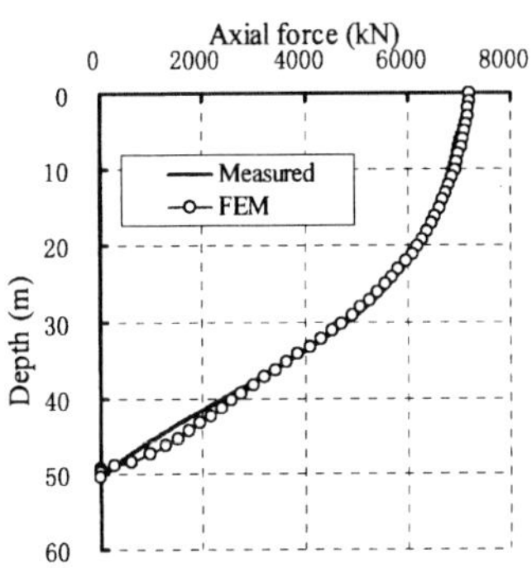

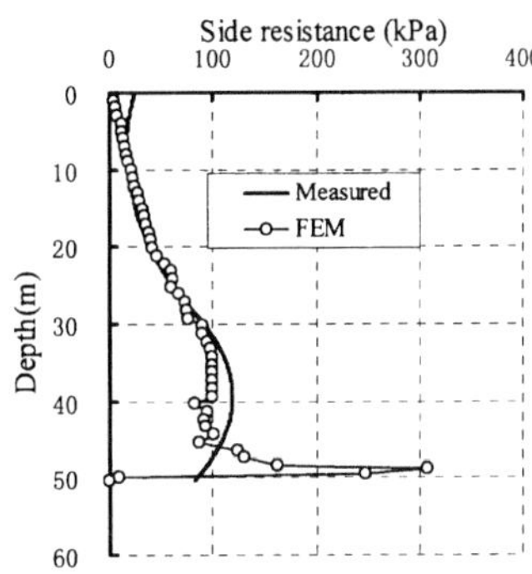

(b) Shaft axial force distribution **(c) Pile side resistance distribution**

FIG. 2. Comparison between FE simulations and measured data

Fig.2(a) shows that, when the test load on the pile head is less than 4000kN, the calculated displacement of the pile head is larger than the measured result. When the test load is larger than 4000kN, the calculated result is well in accordance with the measured results. There is little discrepancy of the shaft axial force and side resistance between the calculated results and measured data, which are shown in Figs.2(b) and 2(c). The comparisons between the FE simulations and measured results indicate that the FEM is capable of capturing the behavior of the uplift piles with enlarged base.

In what follows, the development of soil plastic zone around the belled uplift piles under the ultimate state was studied here by the FEM. The studies focus on the impact of the pile length on the shape of soil plastic zone, and a homogeneous soil layer was assumed for simplicity. Table 2 summarizes the pile and soil parameters used for the numerical simulations in regard to the development of plastic zone.

Table 2. Pile and Soil Parameters Used for FE Simulations

Cases No.	$\lambda=L/d$	d (m)	d_1 (m)	H_0 (m)
1	5	0.5	1.0	1.5
2	10	0.5	1.0	1.5
3	20	0.5	1.0	1.5
4	50	0.5	1.0	1.5
Soil Layer	γ (kN/m^3)	c (kPa)	ϕ (°)	E_s (MPa)
Layer 1	18	2	22	5

Fig.3 shows the soil plastic zone development of the belled uplift pile under ultimate states with four different slenderness ratios (length to diameter ratios) of piles λ. It can seen that when λ=5, the soil plastic zone extended from the base to the head of the pile. The plastic zones are similar with each other for the cases of λ=10, 20 and 50. An elliptic plastic zone around the enlarged base was formed, while near the pile head, a horn-shaped failure mechanism was observed.

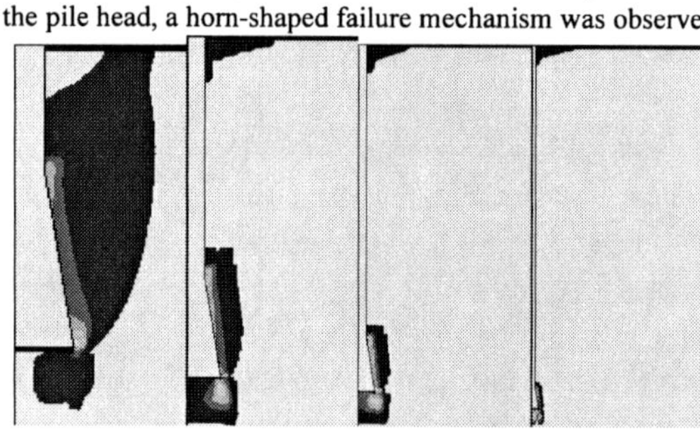

(a) λ=5 (b) λ=10 (c) λ=20 (d) λ=50

FIG. 3. Equivalent plastic strain distributions by FEM

An important parameter introduced here is the influencing height of enlarged base, H_1, which is defined as the height of the intersection of the elliptic soil plastic zone and the pile shaft (Fig.3). It is observed that in the case of λ=5, H_1 approaches to the length of the pile L, and the value of H_1 does not increase obviously with the growth of the pile length.

A Proposed Analytical Method

A proposed failure mechanism for estimating the bearing capacity of the uplift piles with enlarged base is shown in Fig.4a. Under an ultimate state, the failure surface is formed by a 1/4 elliptic local failure surface at the enlarged base, while along the straight-shaft, an exponential function of failure surface is assumed. The ultimate bearing capacity of a belled uplift pile consists of the vertical soil resistance in the failure surface, weight of soil body surrounded by the failure surface and weight of the pile body. As shown in Fig.4a, the calculation of the ultimate bearing capacity can be divided into three sections: (1) section of the enlarged base, ranged from 0 to H_0; (2) uniform section influenced by the enlarged base, ranged from H_0 to H_1; (3) uniform section without the influence of the enlarged base (identical to uniform-section pile), ranged from H_1 to L. The ultimate bearing capacity for each section is deduced respectively and details are given below.

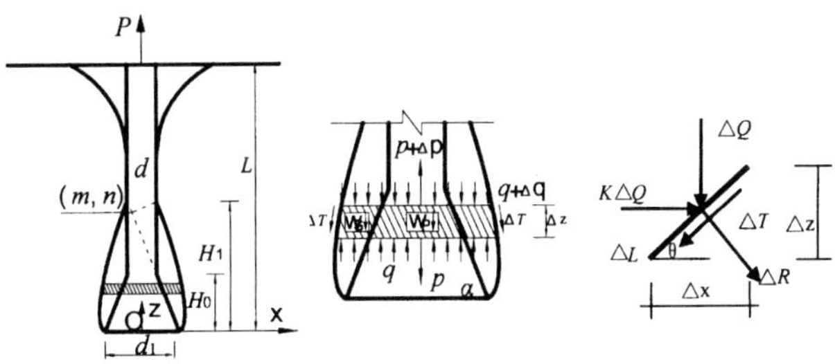

(a) A unified failure mechanism (b) Forces on the wedge (c)Forces on the failure surface

FIG. 4. Calculation model of belled uplift pile under an ultimate condition

Section of enlarged base. For the section of enlarged base, an elliptic failure surface is assumed with the vertex of semi-minor axis being the crossing point of the influencing height of enlarged base and the shaft, and the semi-major axis being the inclined surface of the enlarged base. Set (m,n) as the coordinates for the center of the elliptic surface (See Fig.4a). With the preceding assumptions, the elliptic function can be expressed as

$$\frac{[(z-n)\sin\alpha-(x-m)\cos\alpha]^2}{a^2}+\frac{[(z-n)\cos\alpha+(x-m)\sin\alpha]^2}{b^2}=1 \qquad (1)$$

in which $m=(\frac{d}{2}+\frac{d_1}{2}\tan^2\alpha-H_1\tan\alpha)/(1+\tan^2\alpha)$; $n=-(m-\frac{d_1}{2})\tan\alpha$; a is the length of the semi-major axis, $a=\frac{n}{\sin\alpha}$; b is the length of the semi-minor axis, $b=(H_1-H_0)\cos\alpha$; α is the include angle by the inclined surface of the enlarged base and the level plane.

A circular disc wedge of thickness Δz at a height z above pile tip is considered. Forces acting on the wedge are shown in Fig.4b. Considering the equilibrium of the circular disc wedge leads to

$$\frac{\partial P_1}{\partial z}=2\bar{\gamma}_s(L-z)\pi[x\cot\theta+(\frac{d_1}{2}-z\cot\alpha)\cot\alpha]+(\gamma_s-\bar{\gamma}_s)\pi[x^2-(\frac{d_1}{2}-z\cot\alpha)^2]$$
$$+2\pi x[c+\bar{\gamma}_s(L-z)(\cos\theta+K_0\sin\theta)\tan\phi]+\gamma_p\pi(\frac{d_1}{2}-z\cot\alpha)^2 \tag{2}$$

in which x can be derived from Eq. (1); $\bar{\gamma}_{si}=\sum\gamma_{si}h_i/\sum h_i$, being the weighted average unit weight of the soil layers above the considered wedge; γ_p is the unit weight of the pile; c is the cohesion of the soil; K_0 is the coefficient of lateral earth pressure; θ is the tangent angle of the failure surface (See Fig.4c).

Hence, the uplift bearing capacity for the section of the enlarged base can be obtained by integrating Eq. (2) within the limit of 0 to H_0, and it is given by

$$P_1=\int_0^{H_0}\frac{\partial P_1}{\partial z}dz \tag{3}$$

Uniform section influenced by enlarged base. For the uniform section influenced by enlarged base with the height ranged from H_0 to H_1 (See Fig. 4a), the following simplified differential equation can be obtained by considering the vertical limiting equilibrium of the disk wedge

$$\frac{\partial P_2}{\partial z}=2\pi\bar{\gamma}_s(L-z)x\cot\theta+\pi(x^2-\frac{d^2}{4})(\gamma_s-\bar{\gamma}_s)$$
$$+2\pi x[\bar{\gamma}_s(L-z)(K_0\sin\theta+\cos\theta)(\tan\phi+\cot\theta)+c]+\gamma_p\frac{\pi}{4}d^2 \tag{4}$$

Again, the uplift bearing capacity for the uniform section influenced by enlarged base can be obtained by integrating Eq. (4) within the limit of H_0 to H_1 as

$$P_2=\int_{H_0}^{H_1}\frac{\partial P_2}{\partial z}dz \tag{5}$$

Uniform section without the influence of enlarged base. Chattopahydy & Pise (1986) assumed a failure surface in an exponential function, and the slope of the curve defined as $\frac{dz}{dx}=\tan(\frac{\pi}{4}-\frac{\phi}{2})(\frac{L}{z})e^{\beta(1-\frac{z}{L})}$, in which $\beta=\frac{L}{d}\frac{50-\phi}{2\delta}$ with L being the pile length, d being the pile diameter and δ being the angle of pile friction.

Note that β is related to the properties of soils and pile-soil interface. Considering the influence of enlarged base leads to a modified expression of β as $\beta = \frac{L}{d}\frac{50-\phi}{2\delta}(1+\frac{H_1}{L})$.

Again, we consider the vertical limiting equilibrium of the wedge in the straight shaft section without the influence of enlarged base and obtain the uplift bearing capacity for this section

$$P_3 = \int_{H_1}^{L} \frac{\partial P_3}{\partial z} dz \quad (6)$$

where

$$\begin{aligned}\frac{\partial P_3}{\partial z} &= 2\bar{\gamma}_s (L-z)\pi x \cot\theta - \bar{\gamma}_s \pi [x^2 - (\frac{d}{2})^2] \\ &+ \gamma_s \pi [x^2 - (\frac{d}{2})^2] + \gamma_p \pi (\frac{d}{2})^2 + 2\pi x [c + \bar{\gamma}_s (L-z)(\cos\theta + K_0 \sin\theta)\tan\phi]\end{aligned} \quad (7)$$

As the consequence of the above, the bearing capacity of the whole uplift pile can be calculated by adding all three sections together

$$P = \int_0^{H_0} \frac{\partial P_1}{\partial z} dz + \int_{H_0}^{H_1} \frac{\partial P_2}{\partial z} dz + \int_{H_1}^{L} \frac{\partial P_3}{\partial z} dz \quad (8)$$

Determination of the influencing height H_1. The determination of the influencing height of enlarged base H_1 can be realized from measured data of field tests. Five field test cases of the belled uplift piles are summarized in Table 3, and Eq. (8) was used to back-calculate the value of H_1 for each case. As shown in Fig. 5, a simplified function for evaluating the value of H_1 was fitted which is related to the properties of the belled pile.

Table 3. Field Test Cases and Deduced Values of H_1

Case No.	1	2	3	4	5
Pile length, L (m)	18	18	27	20	20
Diameter of straight-shaft, d (m)	0.65	0.55	0.45	0.4	0.4
Diameter of enlarged base, d_1 (m)	1.3	1.1	0.8	0.8	0.8
Height of enlarged base, H_0 (m)	1.5	1.5	1.5	1.5	2.5
Ultimate bearing capacity, P_u (kN)	2660	2080	1514	978	1247
Influencing height by enlarged base, H_1(m)	4.5	4.4	2.25	3.5	7.5
$x=(L+d)/H_0$	12.43	12.37	18.3	13.6	8.16
$y= H_1/ d_1$	3.46	4	2.81	4.37	9.38

Note: Case Nos. 1, 2 and 3 - Data from Wang *et al.* (2007a); Case Nos. 4 and 5 - Data from Wang *et al.* (2007b).

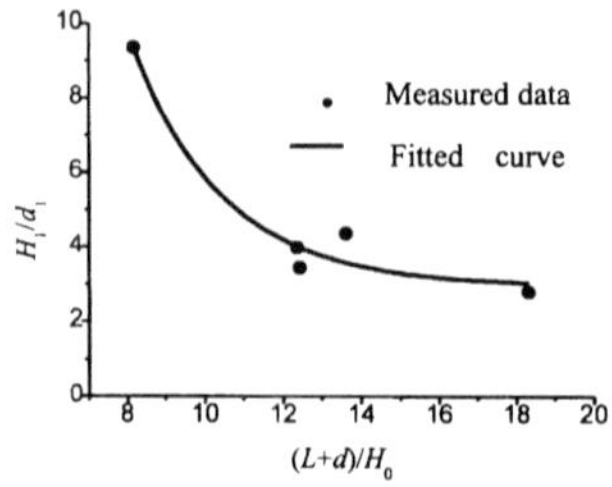

FIG. 5. Fitted curve for the relationship between H_1 and the pile sizes

It can be seen that, the ultimate uplift bearing capacity is related to the pile length, L, the diameter of the straight shaft, d, the diameter of the enlarged base, d_1, as well as the height of the enlarged base, H_0. For example, the sizes of the piles in Case 4 and Case 5 are almost identical except the value of H_0, however, the ultimate bearing capacities are different significantly. The uplift pile in Case 5 has a larger ultimate bearing capacity, whose value of H_0 is larger than the H_0 in Case 4. For the purpose of finding the relationship between H_1 and the sizes of the pile, trial calculations have been conducted and the following equation was found with satisfactory results:

$$H_1 = (2.971 + 210.644e^{-0.4284\frac{L+d}{H_0}})d_1 \tag{9}$$

Eq. (9) shows that, when the pile length L approaches infinity, H_1=2.971d_1. It is only available for the evaluation of H_1 of the belled uplift pile, whose length-diameter ratio $\lambda \geq 10$.

Validations. Table 4 shows the comparison between measured ultimate bearing capacities and predictions.

Table 4. Comparison Between Predicted Ultimate Bearing Capacities and Measurements

Case No.	L(m)	d(m)	d_1(m)	H_0(m)	Measured P_u (kN)	H_l(m)	Predicted P_u(kN)
1	18	0.65	1.3	1.5	2660	4.3	2565
2	18	0.55	1.1	1.5	2080	4.1	2020
3	27	0.45	0.8	1.5	1514	3.0	1594
4	20	0.40	0.8	1.5	978	3.47	955
5	20	0.40	0.8	1.5	1247	8.0	1278
6	22	0.40	0.8	1.5	1100	3.25	1020
7	31	0.6	1.2	1.0	>3400	2.96	3480
8	37	0.6	1.2	1.0	>3900	2.97	4490
9	50.4	0.7	1.4	2.0	>8000	3.45	8880

Note: Case Nos. 6, 7, and 8 - Data from Li *et al.* (2009).

In order to further verify the effectiveness of the proposed analytical method, four more engineering cases have been added in Table 4. It can be seen that, the predicted results have a good agreement with the measured results and the comparisons confirm that the present unified analytical model is feasible and effective.

LOAD-DISPLACEMENT RESPONSES OF STRAIGHT-SHAFT UPLIFT PILES WITH SIDE-GROUTING

In this section, the load-displacement responses of the uniform-section uplift piles before and after side-grouting are presented. An in-situ test pile was 50.4m in length and 700mm in diameter. There were several grouting pipes embedded in the pile shaft, and four grouting sections were set uniformly spaced along the pile depth. Fig.6 shows the profile of the test pile and the soil characteristics in the test areas. The vertical pullout static load test of the pile was carried out twice before and after side-grouting. The load-displacement curves on the pile head as well as the shaft axial force were measured during the static load tests.

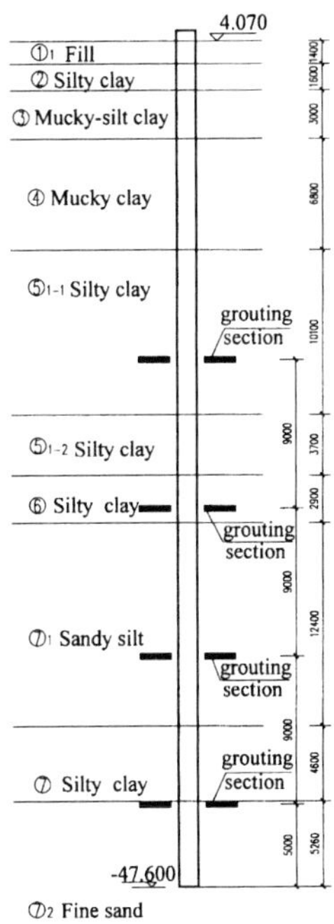

FIG. 6. Profile of the test pile

Based on the measured data of shaft axial forces and load-displacement curves of the pile head from the tests, the side resistance distribution and the relationship between the side resistance and relative displacement of pile and soil (i.e. τ-z curve) were easily obtained.

Normalization of the τ-z Curves

τ-z curves deduced from measured shaft axial forces. Fig.7 shows the measured τ-z curves of the uniform section uplift pile before and after side-grouting. It can seen that, as to the pile before side-grouting, the side resistances at all soil layers had reached the ultimate state except the layer ⑦$_2$ of fine sand, which is the bearing stratum, and for the applied test load level, the side resistance was not fully mobilized. For the pile after side-grouting, the side resistances for the layer ⑦ of silty clay and layer ⑦$_2$ of fine sand did not reach the ultimate state. For simplicity, maximum side resistances and their corresponding pile-soil relative displacements were chosen as the normalized parameters, i.e. the ultimate side resistance τ_{ult} and critical pile soil relative displacement z_{ult}, which are listed in Table 5.

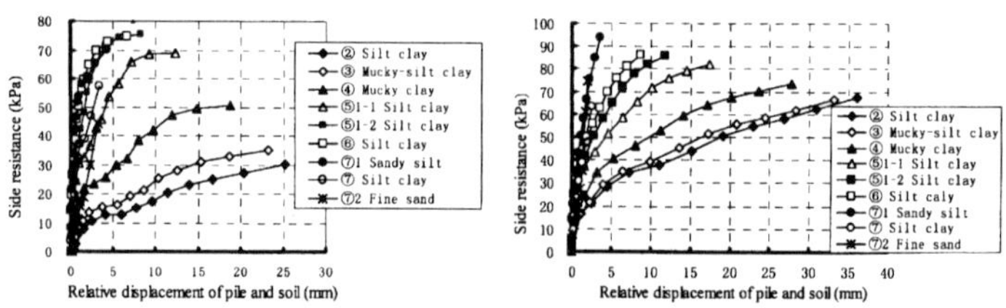

(a) Before side-grouting **(b) After side-grouting**

FIG. 7. τ-z curves of the straight-shaft uplift pile before and after side-grouting

Table 5. Normalized Parameters of the Pile before and after Grouting

Soil layer	Before side-grouting		After side-grouting	
	Measured τ_{ult} (kPa)	Measured z_{ult} (mm)	Measured τ_{ult} (kPa)	Measured z_{ult} (mm)
② Silty clay	13	6.3	27	6.23
③ Mucky-silt clay	16	6.1	32	5.44
④ Mucky clay	23	5.5	50	6.07
⑤$_{1\text{-}1}$ Silty clay	69	12.3	73	15.69

⑤$_{1-2}$ Silty clay	76	8.3	73	9.72
⑥ Silty clay	75	6.6	73	6.99
⑦$_1$ Sandy silt	70	4.3	80	3.92
⑦ Silty clay	57	3.4	63	2.48
⑦$_2$ Fine sand	44	3.4	40	2.35

Normalization of the τ-z curve before side-grouting. The normalization of the τ-z curves of the straight-shaft pile before side-grouting can be obtained by normalizing the pile movement z by z_{ult} and the shaft resistance τ by τ_{ult} for each soil layer, and the normalized curves are shown in Fig. 8a.

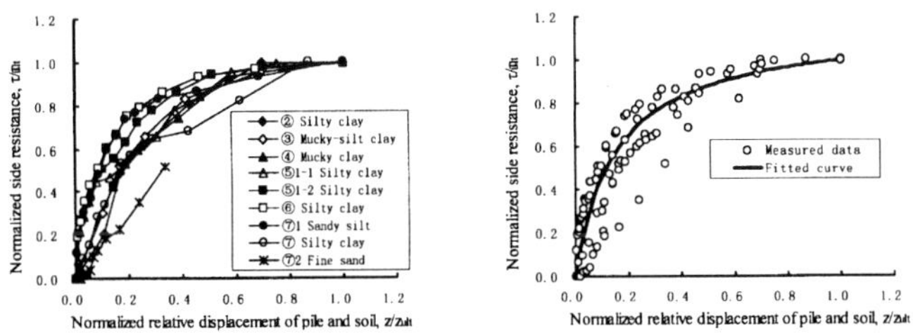

(a) Normalized τ-z curves **(b) Fitted curve of the normalized data**

FIG. 8. Normalized τ-z curves before side-grouting and the fitted curve

It can be seen from Fig. 8a, after normalization, the τ-z curves of each soil layer are located in a certain region, however, the normalization of the soil layer at the pile tip appears to be scattered such as the soil layer ⑦$_2$ of fine sand. A unified hyperbolic function can be used to fit all data points measured from all soil layers except from the layer ⑦$_2$, which is given by

$$\tau = \frac{z}{\frac{0.134 z_{ult}}{\tau_{ult}} + 0.866\frac{z}{\tau_{ult}}} \tag{10}$$

Normalization of the τ-z curve after side-grouting. For the pile after side-grouting, a similar normalization of the τ-z curves was applied. Due to the improvement of pile-soil interface resistance by side-grouting, the normalized curves of the layer ⑦ of silt clay and layer ⑦$_2$ of fine sand near the pile tip were significantly different from others as shown in Fig.9a. Again, employing a hyperbolic function to fit measured data expect for the layer ⑦ and layer ⑦$_2$, and the fitted hyperbolic function can be expressed as

$$\tau = \frac{z}{\frac{0.168 z_{ult}}{\tau_{ult}} + 0.832\frac{z}{\tau_{ult}}} \tag{11}$$

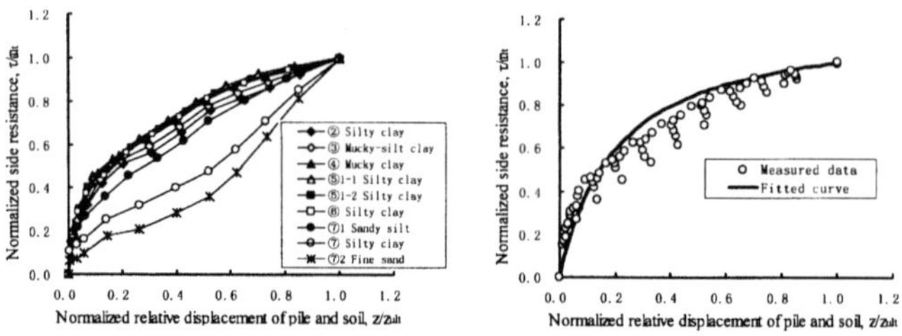

(a) Normalized τ-z curves **(b) Fitted curve of the normalized data**

FIG. 9. Normalized τ-z curves after side-grouting and the fitted curve

Load-Transfer Method and Verifications

The load-transfer method used here was introduced in details by Goel & Patra (2007) for the analysis of the uniform-section uplift piles in homogeneous soils. The method can be easily modified to be applicable in the case of layered soils. The fitted τ/τ_{ult}-z/z_{ult} curves given in Eqs. (10) and (11) were adopted in the analyses using the layered load-transfer method before and after side-grouting, and calculated results compared to the measured data are shown in Figs. 10 and 11.

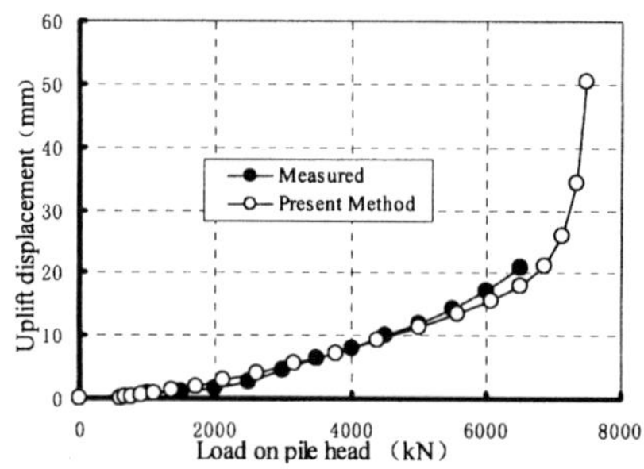

(a) Load–displacement curves at the pile head

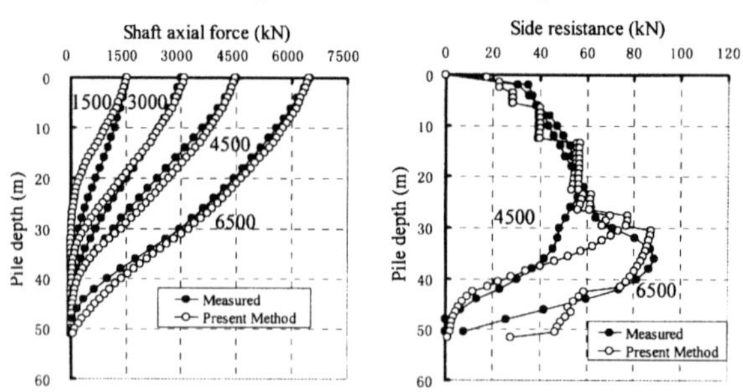

(b) Distribution of shaft axial forces **(c) Distribution of side frictions**

FIG. 10. Comparisons of calculated and measured results before side-grouting

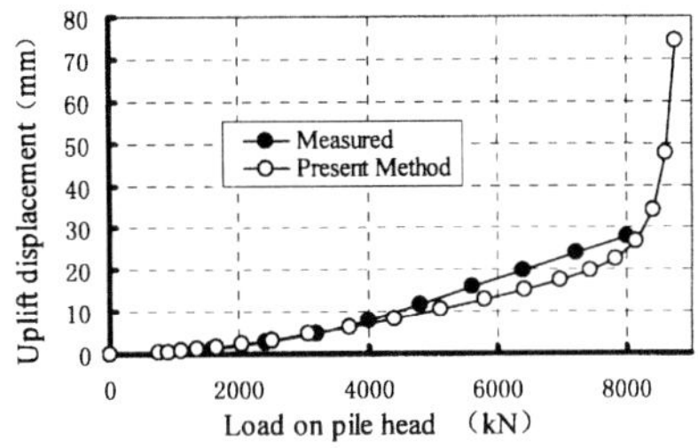

(a) Load–displacement curves at the pile head

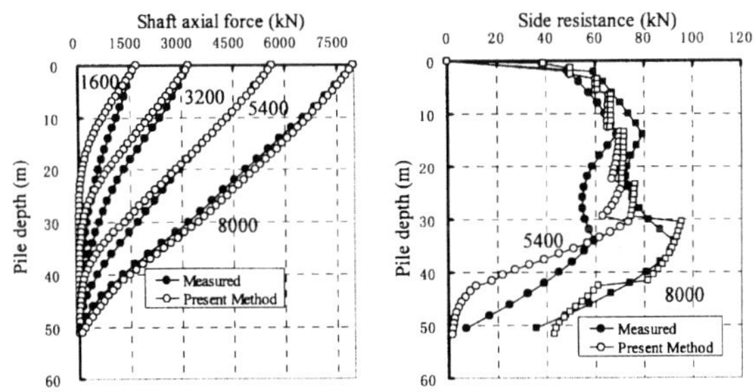

(b) Distribution of shaft axial forces **(c) Distribution of side frictions**

FIG.11. Comparisons of calculated and measured results after side-grouting

It can be seen from Fig. 10 and Fig. 11, the calculated results are close to the measured results. Hence, the analysis method for evaluating the load-displacement responses of the uplift piles with side-grouting is feasible.

UPLIFT BEARING CAPACITY OF PILES UNDER EXCAVATION CONDITION

Static Load Tests on the Uplift Piles of Shanghai World Expo 500kV Underground Transformer Substation Project

Before the construction of the Shanghai World Expo 500kV Underground Transformer Substation, in-situ uplift load tests on side-grouted piles and base-enlarged piles were carried out. These tests include three base-enlarged piles which were numbered T1, T2, T3 and three side-grouted uplift piles which were numbered T4, T5, T6. The dimensions of the tests piles are shown in Table 6. Considering the actual top elevation of the pile is -33.5 m, all six test piles were constructed by double-pipe technology, which aims to make the pile shaft from the ground to elevation -33.5m be set apart from the soil and keep no side resistance in the top section during the load test. Physical and mechanical parameters of the soils and piles are summarized in Table 7.

Table 6. Dimensions of the Test Piles

Type of pile	Test pile No.	L (m)	d (mm)	Effective length (m)	d_1 (mm)	H_0 (m)
Base -Enlarged	T1, T2, T3	82	800	48.5	1500	2
Side-grouted	T4, T5, T6	82	800	48.5	/	/

Table 7. Physical and Mechanical Parameters of the Soils and Piles

Soil Layer	Layer thickness (m)	γ (kN/m^3)	Φ (°)	c (kPa)	Es (MPa)	ν_s
①Fill	1.3	18	22	0	2.53	0.33
②Silty clay□	2	18.1	12.4	18.2	4.06	0.33
③Mucky-silt clay□	7.2	16.8	15.2	3.6	2.11	0.35
④Muddy clay□	6.1	16.2	14.1	5.6	2.13	0.35
⑤$_1$ Clay	4.9	18.2	11.1	9.4	2.90	0.33
⑤$_2$ Silty clay□	4.3	18.5	11.2	18.9	3.66	0.33
⑥Silty clay	4.8	19.6	9.7	29.9	7.44	0.32
⑦$_1$ Sandy clay□	8.2	19.1	20.0	14.4	16.63	0.3
⑦$_2$ Silt	6.2	19.3	22.3	10.9	20.76	0.3
⑧$_1$ Silty clay	14.9	18.4	16.6	14.9	9.03	0.33
⑧$_2$ Interbedded silty clay and silt	13	18.5	17.4	11.8	13.16	0.3
⑧$_3$ Interbedded silty clay and silt	5	19.4	14.5	15.6	22.30	0.3
⑨$_1$ Medium Sand	8.7	20.1	22	9.8	50.71	0.2
⑨$_2$ Grit	23.8	20.3	33	5.3	66.57	0.2
Pile	82	25.0	/	/	35000	0.167

The obtained load-displacement curves of uplift piles from in-situ test are shown in Fig.12. The maximum load of six tested piles was 8000kN. As shown in Fig.12, the maximum average uplift displacement of three base-enlarged uplift piles is about 58mm, and the maximum average uplift displacement of three side-grouted uplift piles is about 41 mm. All the load-displacement curves do not show apparent turning points, which indicate all uplift piles were still in active state and their ultimate bearing capacities are larger than 8000kN. At the same test load level, the uplift displacements of the side-grouted piles was less than the base-enlarged uplift piles, which indicates the interaction of pile-soil is well improved by side-grouting.

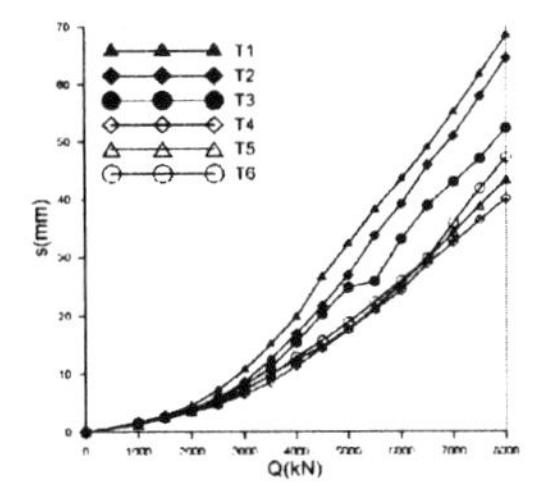

FIG.12. Load-displacement curves of the test piles

Elasto-plastic Finite Element Simulations of In-situ Uplift Pile Tests

In this section, the behavior of the uplift test pile of Shanghai Expo 500kV Transformer Underground Substation is simulated by FEM. The size of axisymmetric model used in the simulation is 160m in vertical and 120m in horizontal direction. Four-node parametric element is chosen to model the pile and soil. The pile is assumed to be elastic and the soil is assumed to be elasto-plastic and the plastic state is described by the Mohr-Coulomb model. As for the modeling of the interface of pile and soil, the Coulomb friction model is used to calculate the side resistance. As shown in Fig.13, the model is discretized by a finite element mesh composed of 7812 elements and 8084 nodes. The lateral displacements of the left and right sides are constrained, the bottom boundary is fixed in vertical direction and the top surface is set to be free

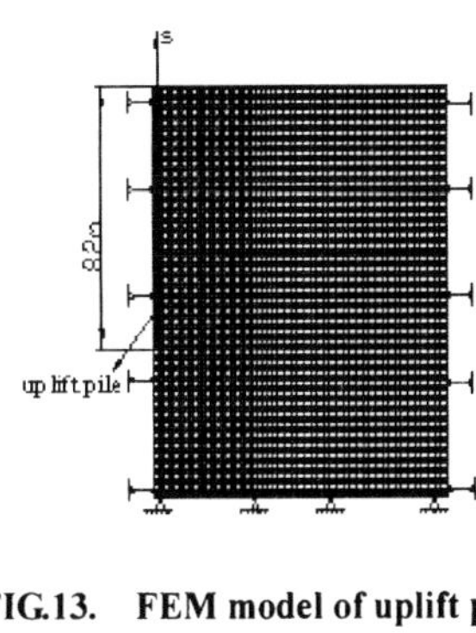

FIG.13. FEM model of uplift pile

The FEM model is validated and relevant parameters are obtained by comparing the simulation with the in-situ tests. The validated FEM model and parameters are then used to study the variation of uplift bearing capacity caused by excavation. Considering that the purpose of this numerical simulation aims to verify the limit equilibrium method for evaluating the bearing capacity loss due to excavation, while not to simulate the process of real construction. Hence, excavation could be assumed to be completed in one step; at last, the obtained displacement results are applied to the pile-soil model step by step, and then uplift process could be simulated.

The elasto-plastic FEM model previously defined is used to simulate the in-situ pile tests of Shanghai World Expo 500kV underground substation. The maximum uplifting displacement of uplift pile, which is 240 mm, is divided into 200 steps. The friction coefficients of the base-enlarged pile and the side-grouted pile are $\mu = 0.1$ and $\mu = 0.125$.

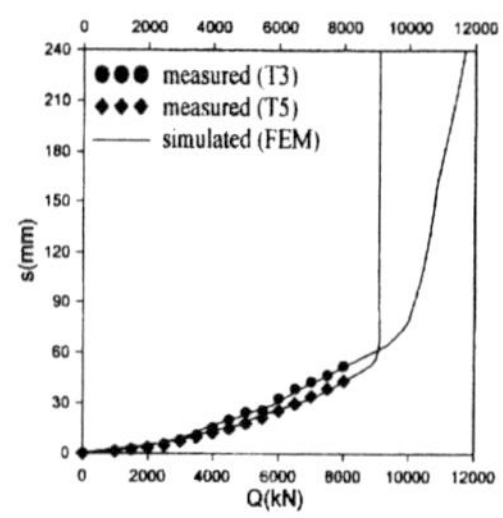

FIG. 14. Comparison between in-situ test data and FEM simulation

The comparison of numerical simulations and in-situ test data are shown in Fig. 14. When the load is less than 8000kN, the simulated results of the side-grouted pile (T5) and base-enlarged pile (T3) compare well to the experimental results. When the load is larger than 8000kN, the load of the side-grouted pile increases with the uplift displacement until it reaches 48mm, which corresponds to the turning point in the load-displacement curve. The ultimate bearing capacity of the grouted pile is chosen to be 9013kN. Similarly, the load of the base-enlarged pile does not increase when the pile top uplift displacement reaches 60mm, which again corresponds to the turning point in the load-displacement curve. The ultimate bearing capacity of the base-enlarged pile is chosen to be 9528kN.

Fig.15 and Fig.16 show the comparison of measured and computed distribution of side resistance and axial force along the depth of grouted pile and base-enlarged pile during uplift loading. As shown in Fig. 15, the simulations of the side-grouted pile when the uplift displacements are 3.6 mm, 12 mm, 25.2mm and 43.2mm are in good accordance with the experimental results. As shown in Fig. 16, the simulations of the base-enlarged pile when the uplift displacements are 4.8mm,

16.8mm, 33.6mm and 51.6 mm are also in good agreement with the experimental results. When the uplift displacement is 48mm for the side-grouted pile and 60mm for the base-enlarged pile, the side resistances do not increase any more, which indicate that the limit state is reached.

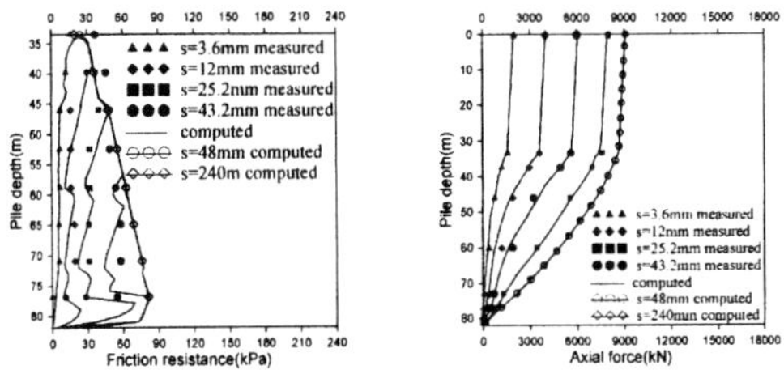

(a) Side resistance distribution **(b) Axial force distribution**

FIG. 15. Distribution of side resistance and axial force along the depth of side-grouted pile during uplift loading process

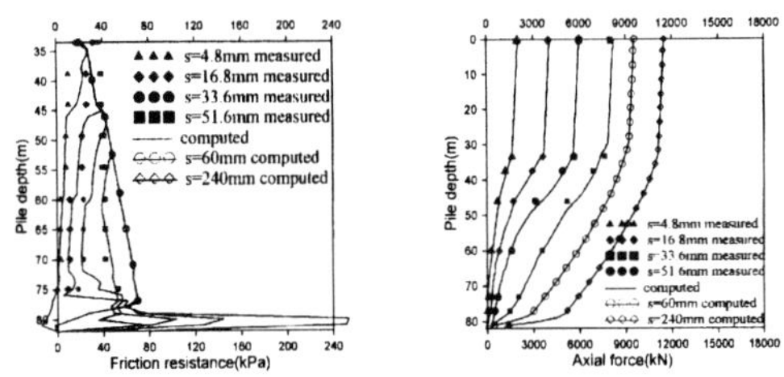

(a) Side resistance distribution **(b) Axial force distribution**

FIG. 16. Distribution of side resistance and axial force along the depth of base-enlarged pile during uplift loading process

From the above comparative study of experimental results and FEM simulated results of load-displacement curve, side resistance and axial force along the pile depth, we can conclude that the FEM is appropriate for modeling the loading process of uplift piles and the corresponding parameters chosen in this paper is reasonable.

Bearing Capacity Loss of Uplift Piles under Deep Excavation by Limit Equilibrium Method

Previous studies on the ultimate bearing capacity of the uplift piles were based on the assumption that the pile top is on ground surface, while this is not consistent with the actual engineering condition where the pile top is embedded with a certain depth. Based on the limit equilibrium method presented in Section 2, the impact of additional stress at any depth, which can be calculated by the Mindlin solution, is taken into account to calculate the ultimate bearing capacity of the uplift piles under deep excavation. The calculation model in the condition of excavation is shown in Fig. 17

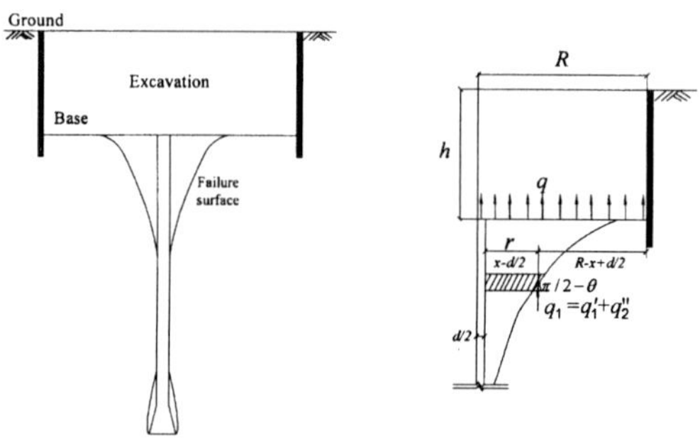

FIG. 17. Layout of analysis model (axisymmetric)

It is assumed that the vertical additional stress above the pile tip z, which is generated by the excavation is q_1. Employing the unified approach introduced in the second Section, the ultimate bearing capacity of uplift pile under the condition of excavation can be obtained. Hence, the key point is to calculate q_1 on the limit failure surface and it can indeed be easily estimated by using the Mindlin solution. As shown in Fig. 17, the vertical soil stress can be calculated by

$$\sigma_z = \frac{q}{8\pi(1-v_s)}\{\frac{(1-2v_s)(L-z-h)}{R_1^3} - \frac{(1-2v_s)(L-z-h)}{R_2^3}\theta + \frac{3(3-4v_s)(L-z)(L-z+h)-3h(L-z+h)[5(L-z)-h]}{R_2^5} + \frac{3(L-z-h)^3}{R_1^5} + \frac{30h(L-z)(L-z+h)^3}{R_2^7}\}; \tag{12}$$

where $q=\overline{\gamma}_s h$, v_s is the Poisson's ratio of soil, h is excavation depth, $R_1=\sqrt{(L-z-h)^2+r^2}$, $R_2=\sqrt{(L-z+h)^2+r^2}$, r is the radius of the calculation point. For the convenience of integration, the calculation for q_1 is divided into two parts, i.e. q_1' from 0 to $x-\frac{d_1}{2}$ and q_1'' from $x-\frac{d_1}{2}$ to $R+\frac{d_1}{2}-x$ (or from 0 to $-(R+\frac{d_1}{2}-x)$), as given in the following equations

$$q_1' = \int_0^{x-\frac{d_1}{2}} \sigma_z dr \tag{13}$$

$$q_1'' = \int_0^{R+\frac{d_1}{2}-x} \sigma_z dr \tag{14}$$

with R being the excavation radius. The vertical additional stress q_1 can then be obtained by adding two parts together

$$q_1 = q_1' + q_1'' \tag{15}$$

Taking into account Eq.(15), Eq.(7) which is used for the uplift bearing capacity calculation of the uniform-section without influence of the enlarged base can be modified into

$$\begin{aligned}\frac{\partial P_3}{\partial z} &= 2\left[\bar{\gamma}_s(L-z)-q_1\right]\pi x\cot\theta+\left(\gamma_s-\bar{\gamma}_s\right)\pi[x^2-(\frac{d}{2})^2] \\ &+\gamma_p\pi(\frac{d}{2})^2+2\pi x\left\{c+\left[\bar{\gamma}_s(L-z)-q_1\right](\cos\theta+K_0\sin\theta)\tan\phi\right\}\end{aligned} \tag{16}$$

Integrating Eq.(16) within the limit of H_1 to L, we can obtain the ultimate bearing capacity of the uniform-section without the influence of the enlarged base under excavation condition.

Likewise, the bearing capacity of the other two sections of the belled uplift pile under excavation can also be obtained easily.

Comparisons of FEM and limit equilibrium method

In the following part, the FEM is used to study the impact of deep excavation on the ultimate bearing capacity of the uplift piles. Table 8 lists the results obtained from the ultimate equilibrium method and FEM for different excavation radius R. The results of limit equilibrium method and FEM are quite close, although there are some deviations in some case but they are less than 10%.

Table 8. Comparisons of the Bearing Capacity of Uplift Piles under Excavation between Limit Equilibrium Method and FEM

Pile type	L(m)	Limit equilibrium method			FEM		
		R=0	R=65m	$R\to\infty$	R=0	R=65m	$R\to\infty$
Side-grouted pile	24.5	3749	2547	975	3742	2456	1080
	36.5	6152	4427	2161	6198	4565	2288
	48.5	9099	6880	3880	9013	7000	3750
Base-enlarged pile	24.5	4669	3344	1568	4450	3268	1511
	36.5	6914	5175	2856	6838	5091	2750
	48.5	9580	7482	4590	9528	7490	4634

Fig. 18 and Fig. 19 show the calculated side resistances obtained from the limit equilibrium method and FEM for different excavation radius and depth respectively. The results of the limit equilibrium method and FEM are basically consistent. With increasing excavation radius and decreasing pile length, the side resistance and bearing capacity decrease. Since some assumptions are applied to simplify the calculation, the results obtained from limit equilibrium method deviates slightly from the FEM, while the results are still acceptable.

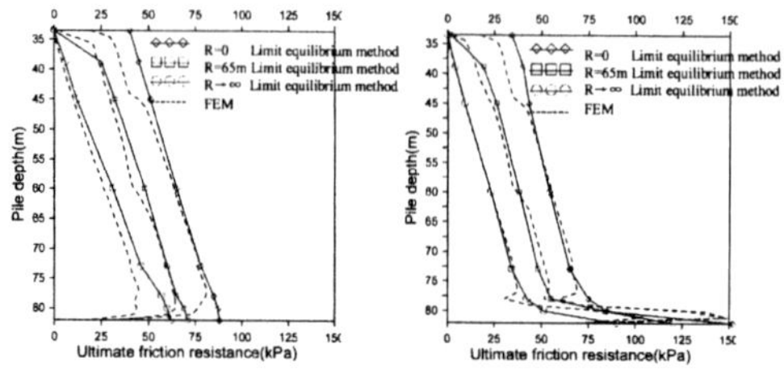

(a) Side-grouted pile (b) Base-enlarged pile

FIG. 18. Comparison of side resistance for two types of piles between limit equilibrium method and FEM for different excavation radius

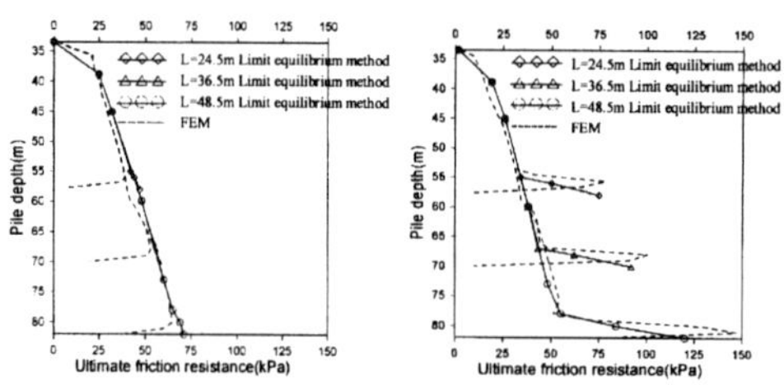

(a) Side-grouted pile (b) Base-enlarged pile

FIG. 19. Comparison of side resistance for two piles between limit equilibrium method and FEM for different pile length under excavation radius R=65m

Bearing Capacity Loss Ratio Caused by Deep Excavation

It is shown from the above analysis that excavation leads to the bearing capacity loss of the uplift piles. For the engineers caring about the loss ratio of different single piles in a pile-raft foundation, in this paper, different excavation radius are set to approximately model different location of piles. Fig. 20 shows the relationship between the bearing capacity loss ratio for both types of piles and excavation radius. The bearing capacity loss ratio for both types of piles gradually increases with the increasing of excavation radius and approaches to a steady state when excavation radius comes to a certain value which is close to the embedded depth of pile. The average capacity loss ratio is a little more than 50% in the condition of fully excavation ($R\rightarrow\infty$). The central pile shows the greatest loss ratio of bearing capacity. This indicates that excavation has greatest influence on the central pile, while the corner pile is relatively less affected by the excavation. Comparing to the base-enlarged pile, the side-grouted pile has greater loss ratio of bearing capacity in the condition of same excavation radius.

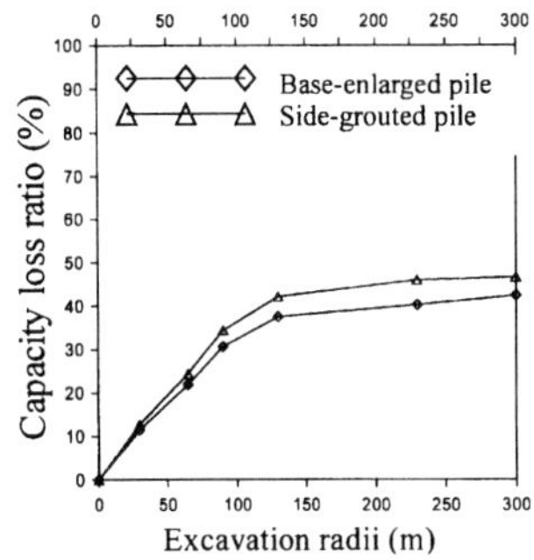

FIG. 20. Bearing capacity loss ratio for two types of pile vs. excavation radius

Fig.21 shows the bearing capacity loss ratios for both types of piles varying with pile length under the same excavation condition. The bearing capacity loss ratio for both types of piles decreases linearly with pile length, which implies that the increasing of pile length can effectively reduce the bearing capacity loss of piles in excavation condition. Comparing to the base-enlarged pile of the same pile length, the side-grouted pile shows greater bearing capacity loss.

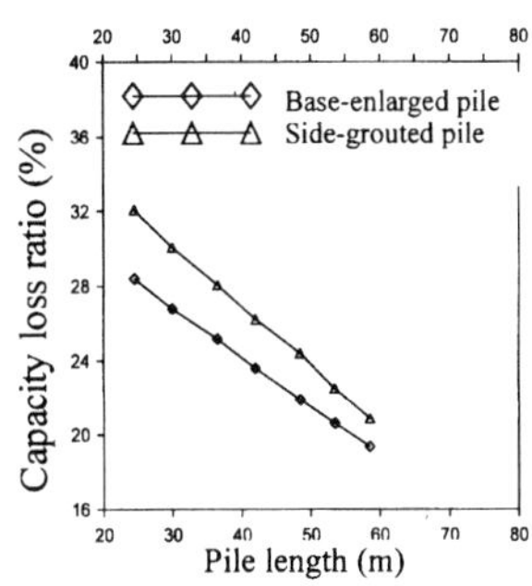

FIG. 21. Bearing capacity loss ratio of two types pile vs. pile length

CONCLUSIONS

The paper outlines the simplified methodology for analyzing the ultimate bearing capacity and deformation characteristics of the uplift piles. Both uniform-section uplift piles with side-grouting and enlarged-base uplift piles are examined. A unified approach to estimating the ultimate bearing capacity of uplift piles with enlarged-base is introduced. Under the ultimate state, the failure surface consists of an elliptic local failure surface on the enlarged base and an exponential function of failure surface for the section of straight-shaft. The proposed limit equilibrium method has been verified with the in-situ measured data of pile load tests. In order to evaluate the bearing capacity loss of uplift piles under deep excavation, the method was further extended to take into account of the change of ground stress field and the normal stress of pile-soil interface under deep excavation. Comparisons with finite element simulations have further demonstrated the feasibility of the proposed method. In order to overcome the shortcoming of the limit equilibrium method which is not capable of predicting the deformation properties of uplift piles, a load-transfer method based on normalized τ-z curves of side-grouted uplift piles was introduced. The normalized τ-z curves were obtained from the in-situ measurements of pile side resistances. Comparisons with field measurements have revealed that the proposed load-transfer method is reliable and effective. Research works conducted in this paper is providing a useful tool for the design of various uplift piles even under the condition of deep excavation.

ACKNOWLEDGMENTS

The first author acknowledges the financial support provided by the National Natural Science Fund for Distinguished Young Scholars (Grant No. 50825803).

REFERENCES

Balla,A.(1961). "The resistance to breaking out of mushroom foundation for pylons." *Proceedings of the 4th International Conference on Soil Mechanics and Foundation Engineering*, Paris, France, 1961: 69-76.

Chattopadhyay,B.C. and Pise,P.J.(1986). "Uplift Capacity of Piles in Sand." *Journal of Geotechnical Engineering*, Vol. 112(9): 888-904.

Goel,S. and Patra,N.R. (2007). "Prediction of load displacement response of single piles under uplift load." *Geotech. Geol. Eng.*, Vol. 25: 57-64.

He,S.M.(2001). "Study on bearing capacity and failure of uplift pile." *Rock and soil mechanics*, Vol. 22(3): 308-310.

Li,J.J., Huang,M.S., Mu,L.L., Wang,W.D. and Chen,Z.(2008). "Research on computation methods of uplift capacity of enlarged base pile in layered soils." *Rock and Soil Mechanics*, Vol. 29(7): 1997-2003.

Li,J.J., Huang,M.S., Wang,W.D. and Chen,Z.(2009). "Analysis on uplift capacity of long enlarged-base pile in soft ground." *Rock and Soil Mechanics*, 30(9):2643-2650 .

Meyerhof,G.G. and Adams,J.I.(1968). "The ultimate uplift capacity of foundations." *Canadian Geotechnical Journal*, Vol. 5(4): 225-244.

Sun,X.L., Yang,M. and Mo,H.H.(2008). "Displacement of base-enlarged tension piles calculated by load transfer method." *Chinese Journal of Geotechnical Engineering*, Vol. 30(12): 1815-1820.

Shanghai Foundation Design Code, DGJ08-111-1999.

Wang,S.J., Zhang,M. and Zhang,J.Z.(2001). "On Mindlin stress formulas." *Engineering Mechanics*, Vol. 18(6): 141-148.

Wang,W.D.,Wu,J.B. and Wang,M.(2007a). "Design and analysis of the uplift pile with enlarged base pile in the ultra-large underground engineering." *Proceedings of the 10th Soil Mechanics and Geotechnical Engineering colloquium of China Civil Engineering Society*, Chongqing, China, 2007: 801-807.

Wang,W.D., Wu,J.B.,Xu,L. and Huang,S.M.(2007b). "Full-scale field tests on uplift behavior of piles with enlarged base." *Chinese Journal of Geotechnical Engineering*, Vol. 29(9): 1418-1422.

Wu,J.B., Wang,W.D. and Huang,S.M.(2008). "Uplift mechanism of enlarged base of pedestal piles by numerical analysis." *Rock and Soil Mechanics*, Vol. 29(8): 2115-2120.

Settlement and Load Sharing of Piled Raft of a 162 m High Residential Tower

Kiyoshi Yamashita[1], Junji Hamada[2] and Yutaka Soga[3]

[1]Deputy General Manager, R&D Institute, Takenaka Corporation, Chiba, Japan; yamashita.kiyoshi@takenaka.co.jp
[2]Chief Researcher, R&D Institute, Takenaka Corporation, Chiba, Japan; hamada.junji@takenaka.co.jp
[3]Senior Chief Engineer, Building Design Department, Nagoya Office, Takenaka Corporation, Nagoya, Japan; soga.yutaka@takenaka.co.jp

ABSTRACT: This paper presents a case history of design and performance of a piled raft supporting a 162 m high residential tower. The 47-story building is a reinforced concrete structure and the average contact pressure over the raft is 600 kPa. The raft is founded on diluvial sand-and-gravel at a depth of 4.3 m below ground surface. Because the building has a base isolation system, differential settlement of the foundation is rigorously restricted. To reduce the differential settlement, a piled raft foundation consisting of thirty-six 50-m long cast-in-place concrete piles was adopted. To confirm the validity of the foundation design, field measurements were performed on the settlement of the foundation, axial loads of the piles, contact pressures between raft and soil and pore-water pressure beneath the raft from the beginning of construction to eight months after the end of construction. Based on the field measurement results, the foundation design was found to be appropriate.

INTRODUCTION

In recent years, there has been an increasing recognition that the use of piles to reduce raft settlements can lead to considerable economy without compromising the safety and performance of the foundation (Poulos, 2001). In Japan, piled raft foundations have been used for many buildings since a piled raft was first used for the four-story office building in Urawa in 1987 (Yamashita & Kakurai, 1991). However, there exist not so many case histories on the monitoring of load sharing between raft and piles as well as settlement (Katzenbach et al., 2000). As for piled rafts with end-bearing piles there exist a few case histories. Thus accumulation of field evidences, especially on soil-structure interaction behaviour by monitoring full-scale structures is required to develop more reliable design method (Mandolini et al., 2005)

This paper presents a case history of design and performance of a piled raft with end-bearing piles. The piled raft is used to reduce differential settlement of the foundation supporting a 47-story building and consists of 50-m long cast-in-place concrete piles embedded to very dense sand-and-gravel. To confirm the validity of the

Photo 1. 47-story isolated residential tower in Nagoya

foundation design, field measurements were performed on the settlement of the foundation, axial loads of the piles, contact pressures between raft and soil and pore-water pressure beneath the raft from the beginning of construction to eight months after the end of construction.

BUILDING AND SOIL CONDITIONS

The 47-story residential tower of 162 m in height above the ground surface and measures 50 m by 34 m in plan is located in Nagoya (Photo 1). FIG. 1 shows a schematic view of the building and foundation with soil profile. The building is a reinforced concrete structure with base isolation system of laminated rubber bearings.

The soil profile down to a depth of 4 m from the ground surface consists of fill and alluvial silt. Below the depth of 4 m, there lie diluvial medium-to-dense sand-and-gravel and sand with SPT N-values of 20-30 with some thin layers of silt to a depth of 17 m, underlain by medium silt. Between depths of 27 m to 43 m, there lie dense sand, very dense sand-and-gravel and medium-to-dense sand with SPT N-values 20 to 60 or higher. Below a depth of 43 m, there lies medium silt. Below a depth of 53 m, there lies very dense sand-and-gravel. The groundwater table appears approximate 2.5 m below the ground surface. The shear wave velocity measured by a P-S logging varies from 180 m/s at the foundation level to 510 m/s in the sand-and-gravel below a depth of 53 m.

FOUNDATION DESIGN

Because the building has a base isolation system, differential settlement of the foundation is rigorously restricted, i.e. maximum settlement of the foundation is less than 30 mm and maximum angular rotation of the raft is less than 1/1000 radian under working load condition. The average contact pressure over the raft is 600 kPa. The reinforced concrete raft located 4.3 m below ground surface is founded on the medium-to-dense sand-and-gravel. If a raft foundation was adopted, the calculated differential settlement would be greater than the allowable value due to the two silt layers, each with a thickness of approximate 10 m.

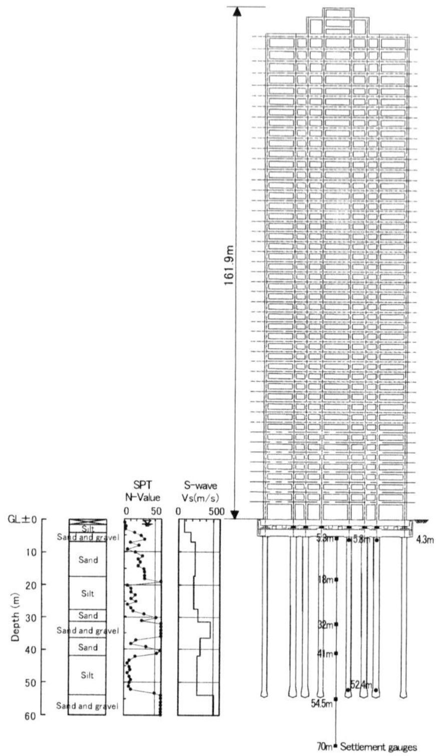

FIG.1. Schematic view of the building and foundation with soil profile

Therefore, to reduce the differential settlement, a piled raft foundation consisting of thirty-six 50-m long cast-in-place concrete piles was proposed, where the piles were embedded to the very dense sand-and-gravel below a depth of 53 m. In designing the piled raft, numerical analysis was carried out to obtain the settlement of the foundation and load sharing between raft and piles by means of the simplified method based on elasticity (Yamashita et al., 1998). In the simplified method piles and soil are represented by interacting springs and finite elements are used to analyze the raft members. To determine the design axial loads of the piles, the field evidences by monitoring full-scale structures were also considered for reference (Kakurai et al., 1987; Yamashita & Kakurai, 1991; Yamashita et al., 1994; Yamashita et al., 2008). As a result of the analysis, the maximum settlement of the foundation and maximum angular rotation of the raft were 26 mm and 1/1400 radian, respectively, and the ratios of the load carried by the piles to the effective load were fixed to be 0.8-0.9 in the design where the effective load is the total load minus buoyancy.

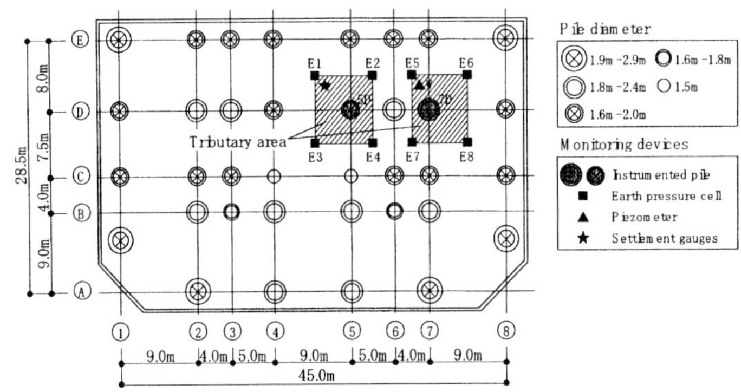

FIG.2. Layout of the piles with locations of monitoring devices

All piles were underreamed except two piles where the column loads are relatively small especially under seismic load condition. The nominal compressive strength of concrete is 48 N/mm^2. The piles have a diameter varying from 1.5 to 1.9 m and a 1.8 to 2.2 m enlarged bell at the pile bottom. The construction procedure of the piles was as follows: a shaft part was first excavated to the pile bottom by a conventional earth drilling method with polymer drilling fluid and then the pile bottom was enlarged by an underreaming tool. FIG. 2 shows the foundation plan with a layout of the piles.

INSTRUMENTATION

The locations of the monitoring devices are shown in FIG. 2. The two piles, 5D and 7D, were provided with a couple of LVDT-type strain gauges at depths of 5.8 m (at pile head) and 52.4 m (at pile toe) to investigate the load distribution along the piles as shown in FIG. 1. Surrounding the instrumented piles, eight earth pressure cells E1-E8 and one piezometer were installed beneath the raft. The vertical ground displacements below the raft were measured using differential settlement gauges. LVDT-type transducers were installed beneath the raft at depths of 5.3 m, 18.0 m, 32.0 m, 41.0 m and 54.5 m to measure the relative displacements to the reference point at a depth of 70 m from the ground surface as shown in FIG. 1.

The measurement of axial loads in the piles, the contact pressures between raft and soil and pore-water pressure beneath the raft started just after casting of the 0.5 m thick foundation slab. The measurement of vertical ground displacements below the raft began right before the beginning of the excavation for the foundation construction in December 2006.

OBSERVATIONS

FIG. 3 shows the measured vertical ground displacements at depths of 5.3 m, 18.0 m, 32.0 m, 41.0 m and 54.5 m relative to the reference point at a depth of 70 m where the initial values of the displacements were taken right before casting the foundation slab.

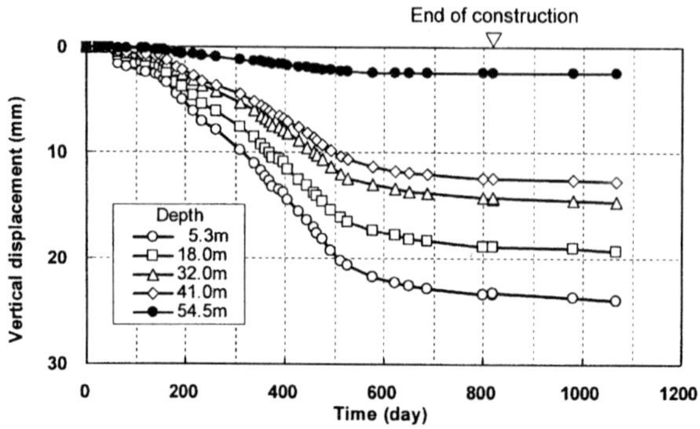

FIG.3. Measured vertical ground displacements

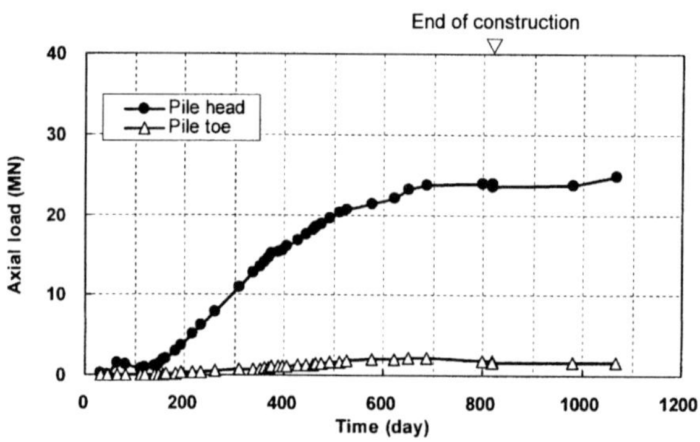

(a) Pile 5D

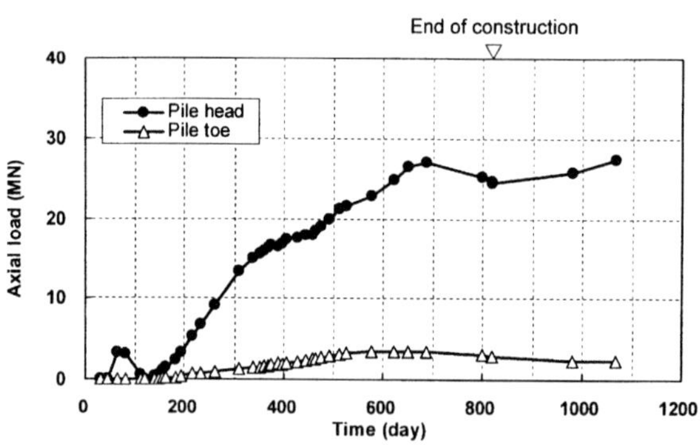

(b) Pile 7D

FIG.4. Measured axial load of the piles

The ground displacement measured at a depth of 5.3 m is approximately equal to the foundation settlement. The measured ground displacement increased with construction progress and reached 23.4 mm at the end of construction. Thereafter the displacement increased very slightly and reached 24.0 mm at eight months after the end of construction.

FIG. 4 shows the development of the axial loads measured on the piles 5D and 7D. At the end of construction, the pile-head loads were 23.6 MN and 24.4 MN on the piles 5D and 7D, respectively. Meanwhile the loads transferred to pile toes were relatively small and the ratios of the pile-toe load to the pile-head load were 0.07 and 0.12 on the piles 5D and 7D, respectively. Thereafter the measured pile-head loads were almost stable for both piles.

FIG. 5 shows the development of the measured contact pressures between raft and soil and pore-water pressure beneath the raft. The measured contact pressures differ considerably and the values were 28-98 kPa from the earth pressure cells E1-E4 near the

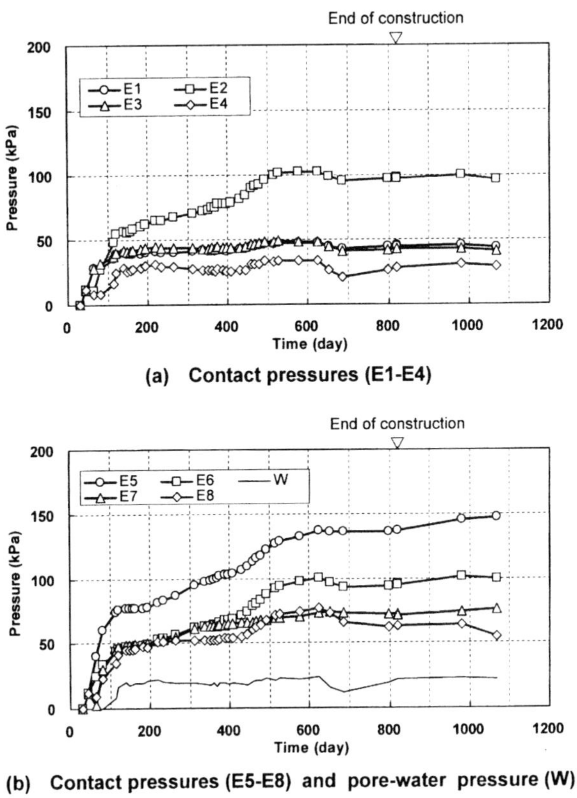

FIG.5. Measured contact pressures and pore-water pressure beneath the raft

pile 5D and 63-137 kPa from the earth pressure cells E5-E8 near the pile 7D at the end of construction. The measured pore-water pressure was approximately 20 kPa that is consistent with the groundwater table from the soil investigation. Thereafter the measured contact pressures were almost stable.

FIG. 6 shows the time-dependent load sharing among the piles, soil and the buoyancy on the tributary areas of the columns 5D and 7D shown in FIG. 2. At eight months after the end of construction, the sum of the measured pile-head load and raft load on the tributary area is 27.7 MN and 32.1 MN on the columns 5D and 7D, respectively. The raft load was obtained based on the mean value of the contact pressures (E1-E4) near the column 5D and that of the contact pressures (E5-E8) near the column 7D. The design vertical loads of the columns 5D and 7D, which correspond to the sum of dead load and

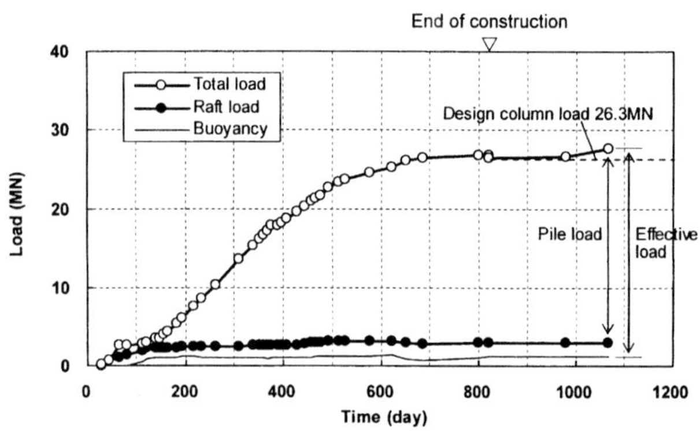

(a) Column 5D

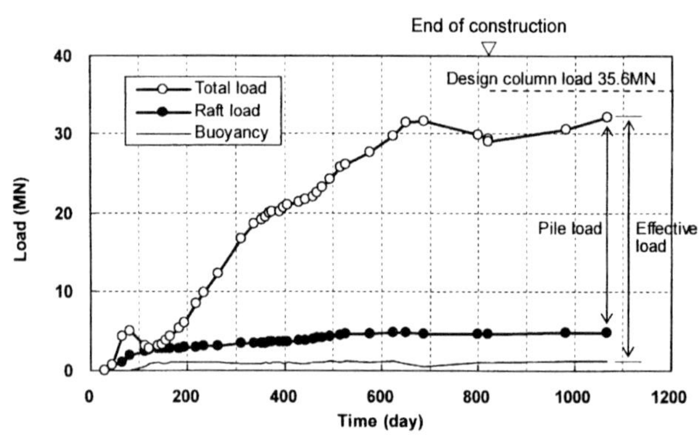

(b) Column 7D

FIG.6. Time-dependent load sharing between raft and piles on the tributary area

live load, are 26.3 MN and 35.6 MN, respectively. Therefore, the sum of the measured pile-head load and raft load on the tributary area is consistent with the design loads of both columns. The ratios of the load carried by the piles to the effective load on the tributary area of the columns 5D and 7D at the end of construction were estimated to be 0.93 and 0.87, respectively. Thereafter the ratios of the load carried by the piles to the effective load increased very slightly. At eight months after the end of construction, the ratios of the load carried by the piles to the effective load on the tributary area of the columns 5D and 7D were estimated to be 0.94 and 0.88, respectively.

CONCLUSIONS

Field measurements were performed on settlement and load sharing of a piled raft supporting a 162 m high isolated residential tower from the beginning of construction to eight months after the end of construction. At eight months after the end of construction, the measured settlement was 24 mm and the ratios of the load carried by the piles to the effective load were estimated to be 0.88-0.94, which were consistent with the design values of the foundation settlement and load sharing between raft and piles. Based on the measurement results, the foundation design was found to be appropriate.

ACKNOWLEDGEMENTS

The authors are grateful to Messrs. M. Yamada and T. Yokonami of Takenaka Corporation for their contribution to the structural design and field measurements of the piled raft foundation.

REFERENCES

Kakurai, M., Yamashita, K. and Tomono, M. (1987): Settlement behavior of piled raft foundation on soft ground, Proc. 8th ARCSMFE, pp.373-376.

Katzenbach, R., Arslan, U. and Moormann, C. (2000): Piled raft foundation projects in Germany, Design applications of raft foundations, Hemsley J.A. Editor, Thomas Telford, pp.323-392.

Mandolini, A., Russo, G. and Viggiani, C. (2005): Pile foundations: Experimental investigations, analysis and design, Proc. 16th ICSMGE, Vol.1, pp.177-213.

Poulos, H. G. (2001): Piled raft foundations: design and applications, Geotechnique 51, No.2, pp.95-113.

Yamashita, K. and Kakurai, M. (1991): Settlement behavior of the raft foundation with friction piles, Proc. 4th Int. Conf. on Piling and Deep Foundations, pp.461-466.

Yamashita, K., Kakurai, M. and Yamada, T. (1994): Investigation of a piled raft foundation on stiff clay, Proc. 13th ICSMFE, Vol.2, pp.543-546.

Yamashita, K., Yamada, T. and Kakurai, M. (1998): Simplified method for analyzing piled raft foundations, Proc. of the 3rd Int. Geotechnical Seminar on Deep Foundations on Bored and Auger Piles, pp.457-464.

Yamashita, K., Yamada, T. and Hamada, J. (2008): Recent case histories on monitoring settlement and load sharing of piled rafts in Japan, Proc. of the 5th Int. Symposium on Deep Foundations on Bored and Auger Piles, pp.181-193.

Effect of Pile Configuration and Load Type on Piled Raft Foundations Performance

Meisam Rabiei[1]

[1]Master Sience of Geotechnical Engineering , School of Civil Engineering , Amir Kabir University of Technology,Tehran, Iran.
Geotechnical Engineer in Bonyad Maskan Mazandaran, Sari, Iran.;meisam_rabiei@yahoo.com

ABSTRACT: In this paper effect of pile configuration and loading type on piled raft foundations performance was studied. The parametric study presented in this research work was carried out with a computer program ELPLA. Three basic pile configurations and three load distribution types were considered and the effect of loading type for each pile configuration on maximum moment in the raft, piles bearing factor (percentage of total load carried by piles) and piled raft settlement was studied. Pile configuration 1 has the pile uniformly distributed under whole raft area. Pile configuration 2 has piles under central area of the raft as well as under the edges of the raft. In pile configuration 3 the piles are placed only in the central area of the raft. It has been found that pile configuration and load distribution are very important and effective in piled raft settlement, maximum moment and piles bearing factor.

INTRODUCTION

Piled raft foundations provide an economical foundation option for circumstances where the performance of the raft alone does not satisfy the design requirements. The piled-raft foundation is a new design concept as one of the effective methods of foundation to reduce settlements of structures. The research works in either analytical models or laboratory tests for the piled-raft foundation were reported by several authors (Randolph 1994; Poulos 2001; Parakosa and Kulhawy 2001; Poulos et al 1997; Reul and Randolph 2003).

In recent years, there have been an increasing number of structures using piled rafts as the foundation to reduce the overall and differential settlements. For cases where a piled raft is subjected to non-uniform loading the use of piles with different sizes can improve the performance of the foundation. Extensive research work has been performed in the past to examine the behaviour of piled rafts. However, most of the research was focused on uniform piled raft configuration and load type, while the different pile configuration and load type has not received much attention.

The objective of this paper is to investigate the effect of pile configuration and load type on piled raft settlement, maximum moment and piles bearing factor (percentage of total load carried by piles).

MODEL VERIFICATION

Several methods of analyzing piled rafts have been developed, and some of these have been summarized by Poulos et al (1997). Three broad classes of analysis method have been identified:

- Simplified calculation methods
- Approximate computer-based methods
- More rigorous computer-based methods.

The parametric study presented in this research work was carried out with a computer program ELPLA. ELPLA is a program for analyzing raft foundations of arbitrary shape with the real subsoil model. The mathematical solution of the raft is based on the finite element method. This program can analyze different types of subsoil models, especially the three-dimensional Continuum model that considers any number of irregular layers. A good advantage of this program is the capability to handle the three analyses of flexible, elastic and rigid foundations. In addition, the mesh of the rigid and flexible foundations can be constructed to be analogous to the finite elements mesh of the elastic foundation.

Model verification was carried out for example represented in Figure 1. Figure 2 compares the computed load-settlement relationships (up to a total load of 18 MN) computed from ELPLA with various methods for the centre of the raft with 9 identical piles (Figure 1-b), one under each column.Variation of maximum positive bending moments, maximum settlement and differential settlement with raft thickness are respectively illustrated on Figure 2-b, Figure 2-c and Figure 2-d. There is reasonably good agreement between the computed results in this research with results obtained from other researchers (A Report Prepared on Behalf of Technical Committee TC18 on Piled Foundations).

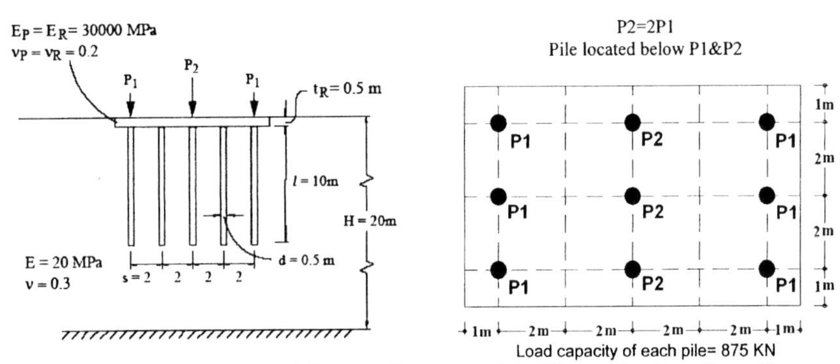

1-a: Model condition and material properties **1-b: Pile and load configuration**

Figure 1: Model configuration and properties for results verification

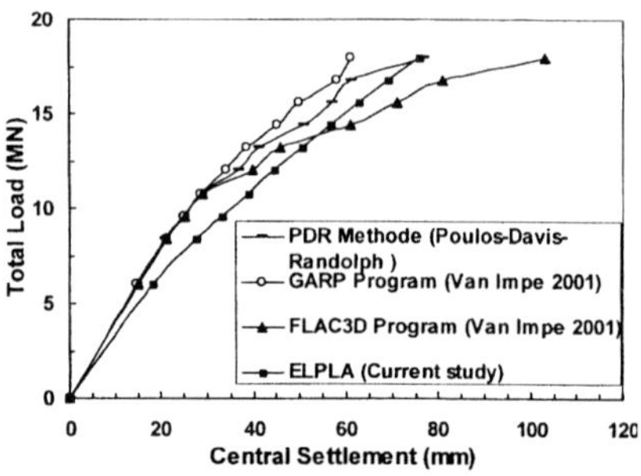

2-a:Comparison of various methods for load-settlement analysis

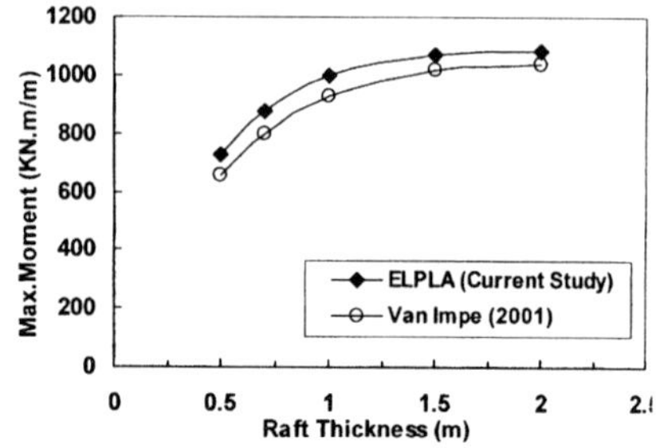

2-b:Effect of raft thickness on maximum bending momment

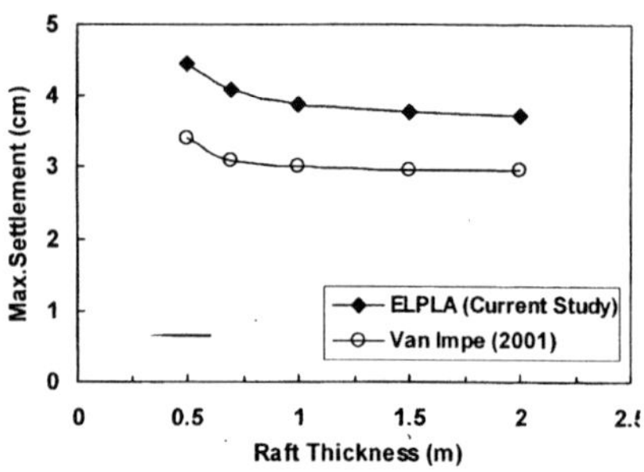

2-c: Effect of raft thickness on maximum settlement

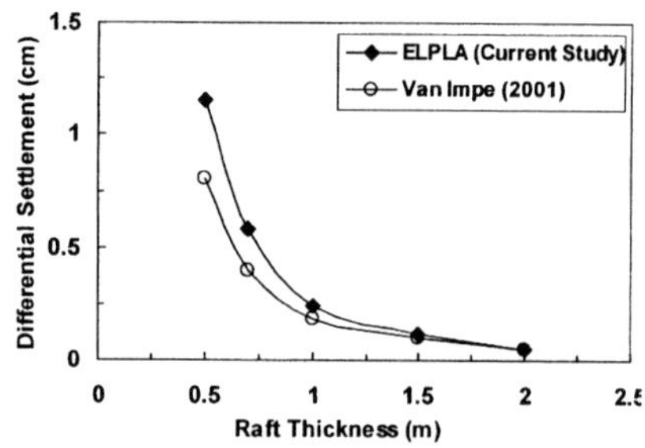

2-d: Effect of raft thickness on differential settlement

Figure 2: Model verification

SYSTEM CONFIGURATION

Material Parameters

The material parameters used in the analyses are summarized in Table 1.

Table 1. Material Parameters Used in Analysis

Parameter		Soil	Raft	Pile
E_s, E_r, E_p	MPa	20	22000	22000
ν_s, ν_r, ν_p	-----	0.15	0.2	0.2
γ	KN/m^3	19	25	25
Cohision	KN/m^2	30	-----	-----

Pile and Load Configuration

In the parametric study three basic pile configurations and three load configurations were investigated. Pile configuration, load type and model properties for parametric study are shown in Figure 3. Pile configuration 1 has the pile uniformly distributed under whole raft area. Pile configuration 2 has piles under central area of the raft as well as under the edges of the raft. In pile configuration 3 the piles are placed only in the central area of the raft.

As shown in Figure 3 three different load configurations have been studied. Load Type 1 describes a uniform load over the whole raft area, a load configuration frequently applied in parametric studies on the behavior of piled Rafts. In Load type 2 maximum of the load has been applied in the edge of the raft and the other load applied at the center of the raft. For Load Type 3 half of the load has been applied in a core area at the center of the raft and the other half at the edge of the raft. The core area is 25% of the total area of the raft. This is a typical configuration for tall buildings with stiff structural elements such as elevator shafts, stairways and shear wall at the center of the raft.

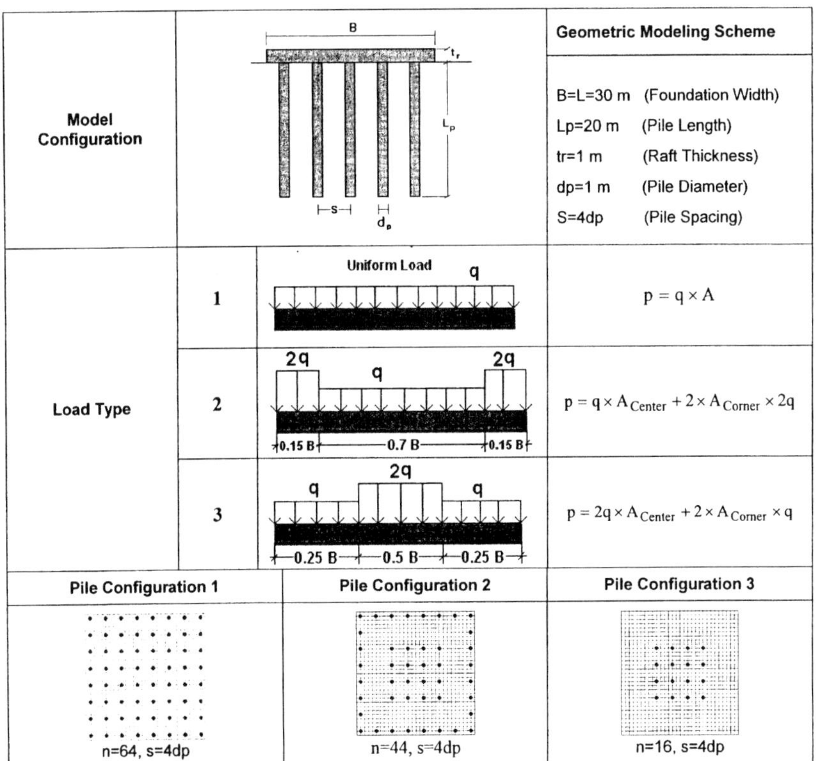

Figure 3. Model configuration and properties used in the analysis

DISCUSSION OF RESULTS

Figures 4, 5 and 6 show the effect of total load on piled raft behaviour for loading type 1, 2 and 3 respectively. In Figure 4 it can be seen for loading type 1 by increasing total load maximum settlement increased, pile bearing factor reduced, maximum positive and negative moment nearly constant in pile configuration 1, 2 and 3. In Figure 5 it can be seen for loading type 2 by increasing total load maximum settlement increased and maximum settlement in pile configuration 1 and 2 nearly equal, pile bearing factor reduced, maximum negative moment increaced in pile configuration 1, 2 and 3 and maximum positive moment reduced in pile configuration 2 but in pile configuration 1 and 3 by increasing total load positive moment nearly constant. In Figure 6 it can be seen for loading type 3 by increasing total load maximum settlement increased and maximum settlement in pile configuration 1, 2 and 3 nearly equal, pile bearing factor reduced, maximum negative moment nearly equal in pile configuration 1, 2 (in pile configuration 3 can not be seen regular variation for maximum negative moment) and maximum positive moment increaced in pile configuration 1,2 and 3.

Other case of parametric study was carried out for identical total pile length in each pile configuration 1,2 and 3. Figure 7 shows results for 320 m total pile length. Pile length is 5 m in pile configuration 1, 7.27 m in pile configuration 2 and 20 m in pile configuration 3.

In Figure 7 it can be seen for load type 1 and 2, pile configuration 1 and 2 are best choice for settlement reduction but for load type 3, pile configuration 3 has least settlement from other pile configurations.

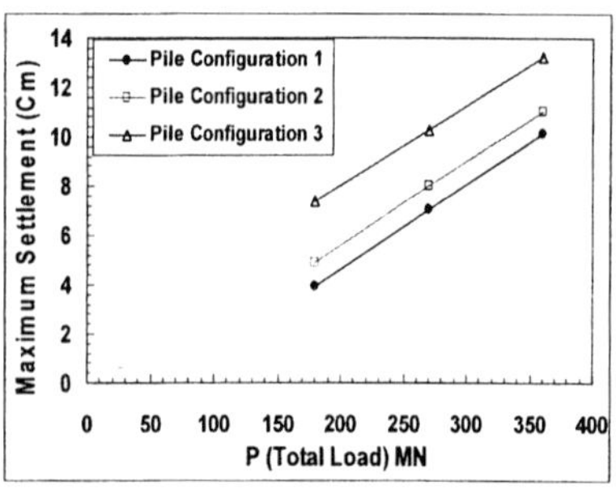

4-a. Effect of total load on piled raft maximum settlement

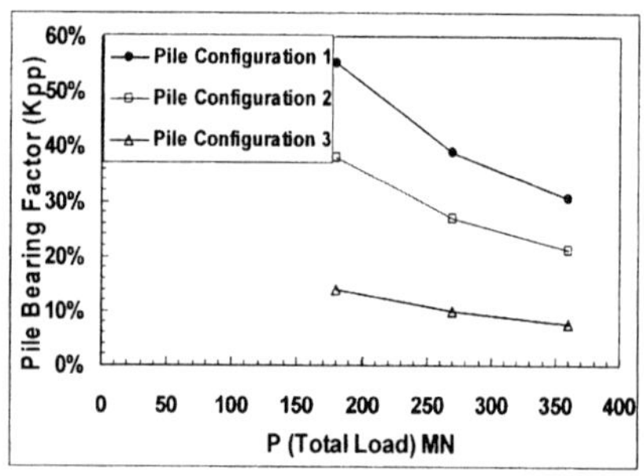

4-b. Effect of total load on pile bearing factor in piled raft

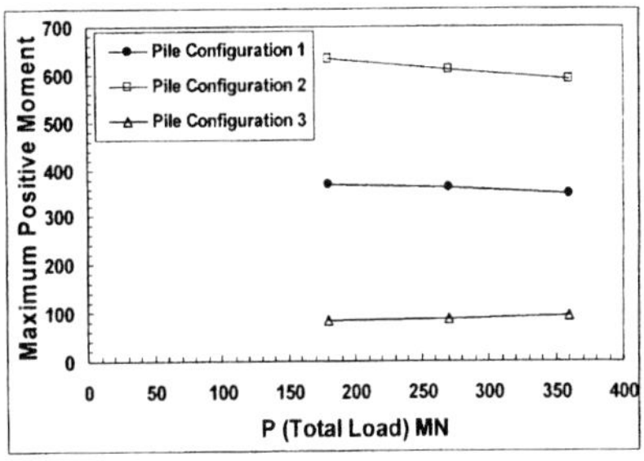

4-c. Effect of total load on piled raft maximum positive moment

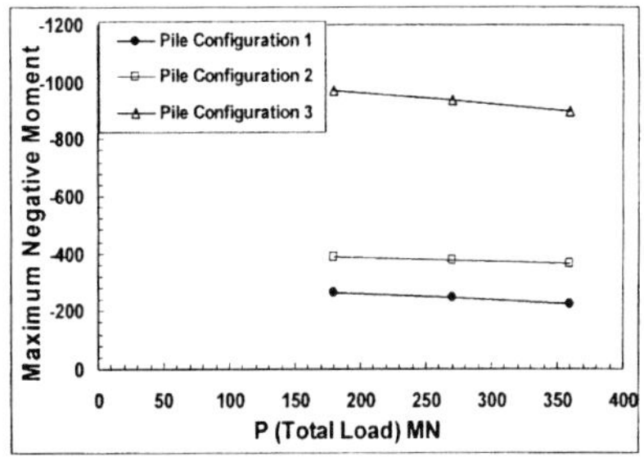

4-d. Effect of total load on piled raft maximum negative moment

Figure 4. Effect of total load on piled raft behaviour for load type 1

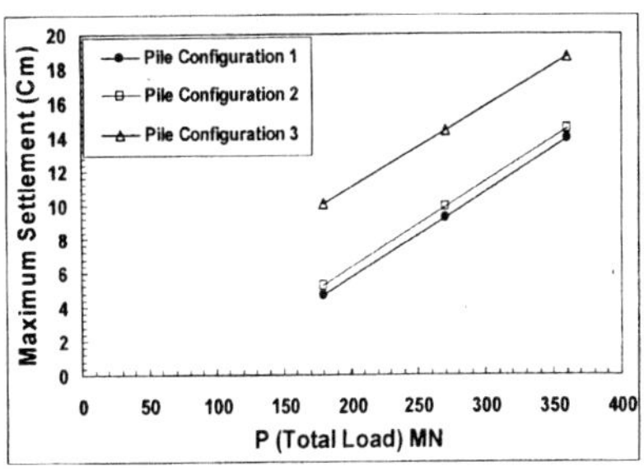

5-a. Effect of total load on piled raft maximum settlement

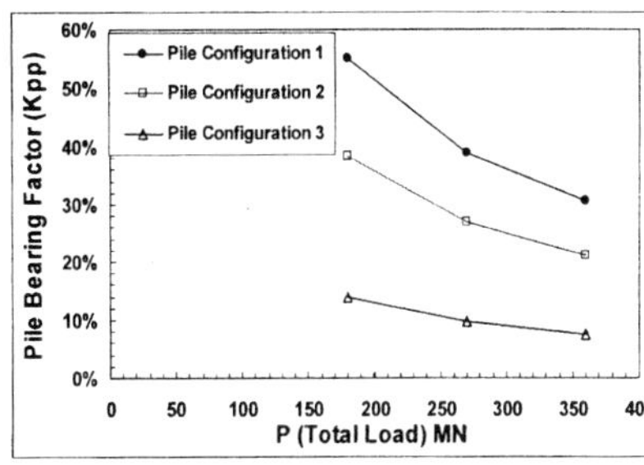

5-b. Effect of total load on pile bearing factor in piled raft

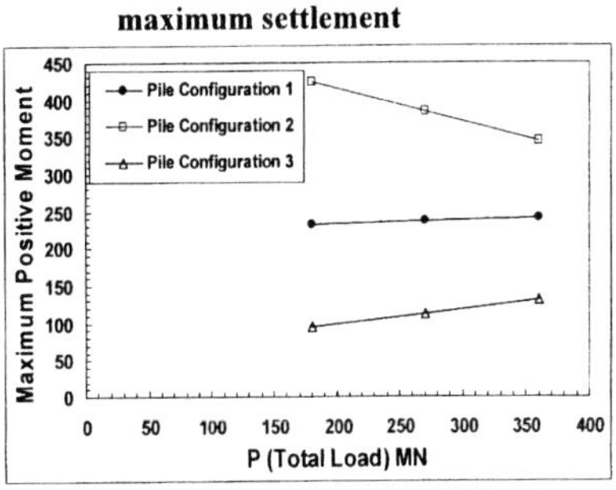

5-c. Effect of total load on piled raft maximum positive moment

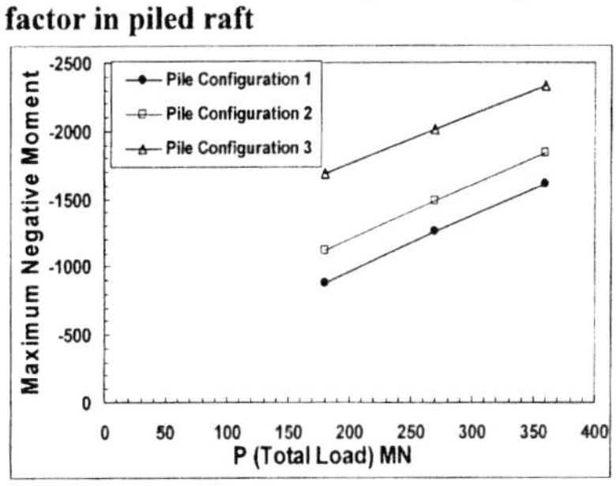

5-d. Effect of total load on piled raft maximum negative moment

Figure 5. Effect of total load on piled raft behaviour for load type 2

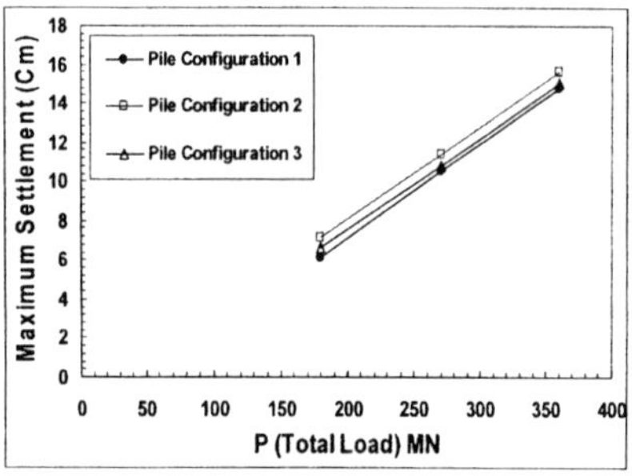

6-a. Effect of total load on piled raft maximum settlement

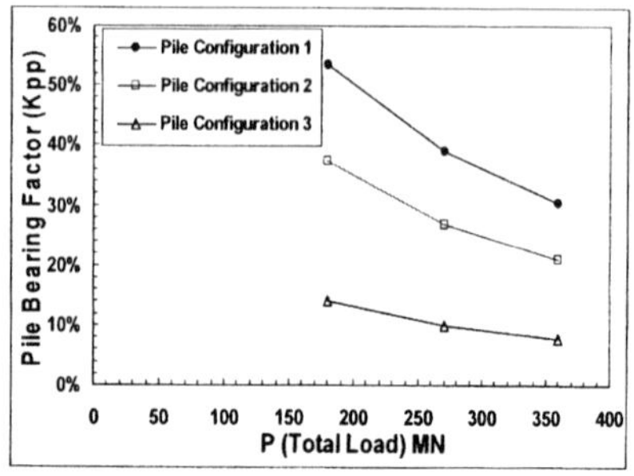

6-b. Effect of total load on pile bearing factor in piled raft

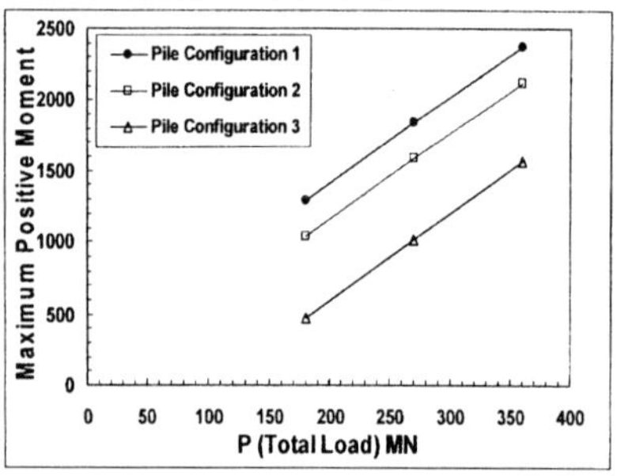

6-c. Effect of total load on piled raft maximum positive moment

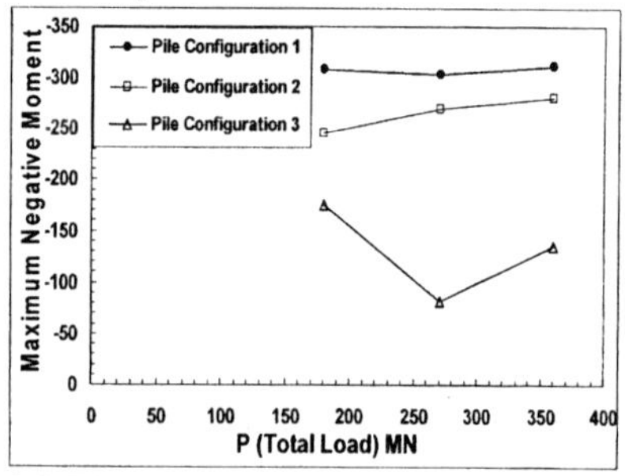

6-d. Effect of total load on piled raft maximum negative moment

Figure 6. Effect of total load on piled raft behaviour for load type 3

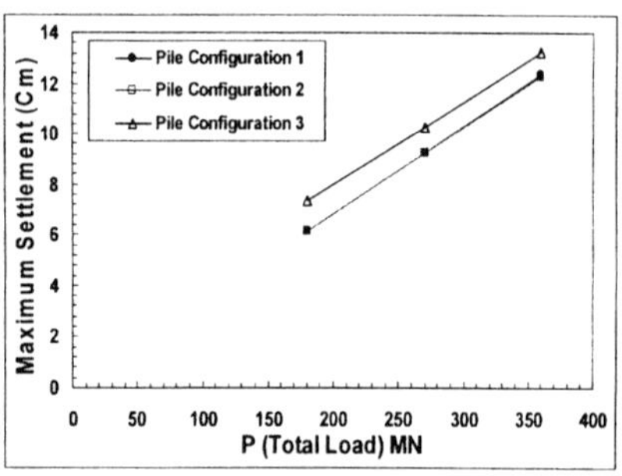

7-a. Effect of total load on piled raft maximum settlement (load type 1)

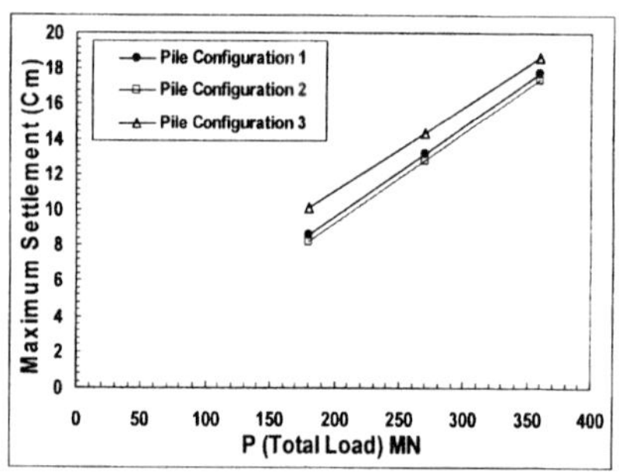

7-b. Effect of total load on piled raft maximum settlement (load type 2)

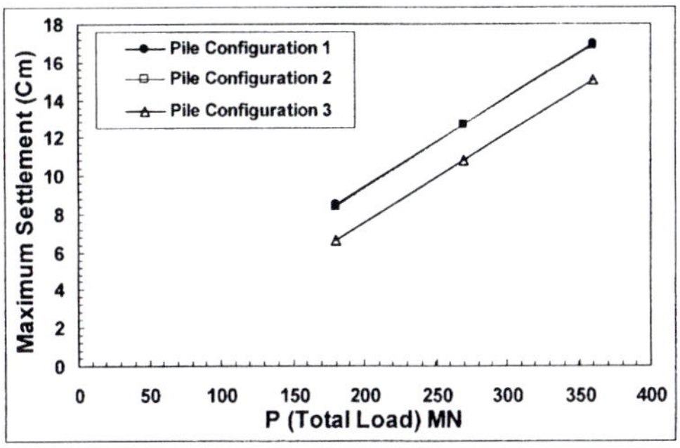

7-c. Effect of total load on piled raft maximum settlement for identical total pile length (load type 3)

Figure 7. Effect of total load on piled raft maximum settlement for identical total pile length

CONCLUSIONS

Piled raft foundations have the potential to provide economical foundation systems, under the appropriate geotechnical conditions. The design philosophy should be based on both ultimate load capacity and settlement criteria, with the key question to be answered being: "what is the minimum number of piles required to be added to the raft such that the ultimate load, settlement criteria are satisfied?" Use of some of the results outlined in this paper can be used to assist the foundation designer to provide a rational answer to this question.

REFERENCES

Kany, M, EL Gendy, M, EL Gendy, A. (2007): "Analysis and design of slab foundation-Program ELPLA."*GEOTEC Software, Zirndorf, Germany.*

Poulos H.G. (2001). "Piled raft foundations: design and applications." *Geotechnique*, 51(2), 95-113.

Poulos, H.G., Small, J.C., Ta, L.D., Sinha, J. and Chen, L. (1997). "Comparison of Some Methods for Analysis of Piled Rafts." *Proc. 14 ICSMFE, Hamburg*, 2:1119-1124

Prakoso, W.A. and Kulhawy, F.H. (2000). "Contribution to Piled Raft Foundation Design". *Jnl. Geot. and Geoenv. Eng.*, ASCE, 127(1): 17-24.

Rabiei, M. (2009). "Parametric Study For Piled Raft Foundations." *Electronical Journal of Geotechnical Engineering*, volume 14, Bundle A.

Randolph , M.F. (1994). "Design Methods for Pile Groups and Piled Rafts". *S.O.A. Report, 13 ICSMFE, New Delhi*, 5: 61-82.

Reul, O. & Randolph, M. F. (2003). "Piled rafts in overconsolidated clay: comparison of in situ measurements and numerical analyses." *Geotechnique 53, No. 3, 301–315*

Van Impe, W.F. and Lungu, I. (1996). "Technical Report on Settlement Prediction Methods for Piled Raft Foundations." *Ghent Univ., Belgium.*

A Study on Time-effects of Piled Raft System by Using Computational Methods

Chun-yi Cui[1], Mao-tian Luan[2], Mu-guo Li[3]

[1]Postdoctor, Institute of Geotechnical Engineering, School of Civil and Hydraulic Engineering, State Key Laboratory of Coastal and Offshore Engineering, Dalian University of Technology, Dalian 116024, China; Associate Professor, Department of Civil Engineering, Dalian Maritime University, Dalian 116026, China.
[2] Professor, Director of Institute of Geotechnical Engineering, School of Civil and Hydraulic Engineering, Dalian 116024, China
[3] Professor, School of Civil and Hydraulic Engineering, State Key Laboratory of Coastal and Offshore Engineering, Dalian University of Technology, Dalian 116024, China

ABSTRACT: On the soft grounds, effects of consolidation on performance of the system of piled rafts as well as superstructures cannot be overlooked due to nonlinear interaction characteristics. Conventional methods of engineering design and numerical procedures of two-dimensional interaction analysis cannot realistically simulate loading and deformation mechanism of the interaction system and cannot completely predict variations of deformations and internal forces of both structure and piled raft foundation. Therefore in order to well understand long-term behavior of the interaction system, time-dependent effect of soil deformations on performance of the interaction system of piled rafts and foundations as well as superstructure is evaluated in this paper. The nonlinear deformation and strength behavior of foundation soils are taken into account by using the elasto-plastic constitutive model based on Mohr-Coulomb's yield criterion while the consolidation effect of subsoil under loading is incorporated by numerically solving the coupling Biot's equations of consolidation. Numerical analyses are conducted for the structure-raft-foundation interaction system by using the finite element methods. Based on numerical analyses for a given interaction system, both the features of variation in time of internal forces, settlements and the characteristics of compatibility mechanism of interaction system are examined. It is shown that both nonlinear behavior of soft soils and consolidation characteristics of foundation remarkably affect the time-dependent performance of the interaction system. Such dependency is intimately related to variation of excess pore water pressure of subsoil.

INTRODUCTION

In soft soil sites, piled rafts are extensively and widely employed in engineering constructions such as high-rise buildings for their advantages of higher bearing capacity, lower differential settlement over other types of foundations. Loading applied to saturated soft soil layers by structures such as buildings causes the development of excess pore water pressure. Initially, the structure will undergo an immediate settlement as the excess pore pressure develops. However, with time, the excess pore pressure will dissipate as water flows from regions of high excess pore water pressure to regions of lower water pressure. As the excess pore pressure decreases, the effective stresses in the soil increase progressively, and this leads to further settlement of the foundation with time. The importance to consider the time-dependent behavior in analysis of foundations in soft soils due to consolidation was summarized by Sekhar et al. (2002). The time effects in soil-structure interaction analyses was first discussed by Wood et al. (1975) on the basis of one-dimensional Terzaghi's model of consolidation by virtue of finite difference method. Subsequently, the time variation characteristics and mechanism of differential settlement of raft on linear elastic foundation were presented by Booker et al. (1984) through study of a circular raft on linear elastic foundation on the basis of one-dimensional equation of Terzaghi's consolidation by virtue of analytic method. Two-dimensional Biot's consolidation theory was introduced by Lewis et al. (1989) to analyze a raft on linear elastic foundation by finite element formulation. Use of semi-numerical method to study the time effects of a piled raft foundation system was also reported by Zai et al. (1999). Recently, with the advancement in computational capability, it is possible to study much more complicated interaction problems considering time effects caused by consolidation. Finite layer method was introduced for interaction analysis of piled raft on nonlinear elastic foundation considering consolidation by Zai et al. (2005). The time-dependent response of the pile-raft-soil interaction system under vertical loading was analyzed by Cheng et al. (2004) using two-dimensional FEM based on Biot's theory of consolidation. This review of earlier work suggests that many attempts have been made to study the interaction problem regarding the time effects of soil consolidation, but very little work is available to analyze the time-dependent behavior of three-dimensional superstructure-piled raft-soil system in the three-dimensional space. Although, the influence of superstructure is considered in interaction by Wang et al. (2001) in which the supporting soil was represented by a simplified element model taking linear viscosity of material and process of consolidation into account by semi-numerical method; however, it still has some limitations when boundary conditions are complex and it can not authentically reflect the variations of excess pore pressure in subsoil. On the other hand, the stress-strain behavior of soil is highly nonlinear, but it is often idealized as a linear material. However, it was emphasized by Noorzaei et al. (2001) that for high loading level and for structures whose reliability is of prime importance, it would be better to resort to a more realistic constitutive model.

In the present work, nonlinear consolidation theory of three dimensions based on FE formulation proposed by Schiffman et al. (1969) is employed. Time-dependent behavior of a three dimensional interaction system of superstructure-piled raft-soil is considered. The nonlinear deformations and strength behavior of foundation soils are taken into account by using the elasto-plastic constitutive model based on Mohr-Coulomb's yield criterion, while the consolidation effect of subsoil under loading is incorporated by numerically solving the coupled Biot's equations of consolidation The numerical analysis is achieved by using finite element method with Newton-Raphson iterative scheme.

COMPUTATIONAL MODEL

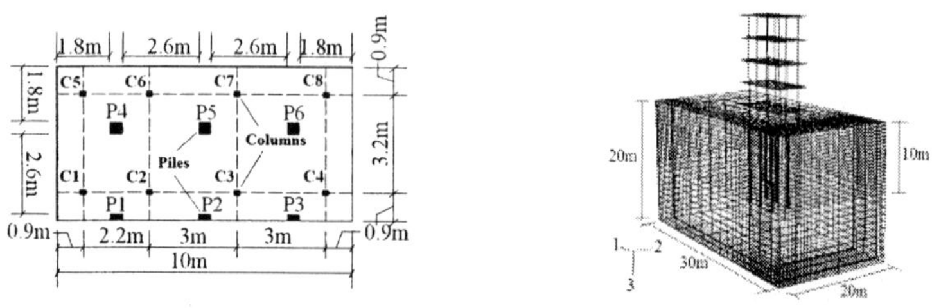

(a) Layout of Piles and Columns (b) FEM Model

FIG.1. The computational model used in FEM

Table 1 The soil parameters used in FEM analyses

Item	Soil Layer 1	Soil Layer 2	Soil Layer 1
Thickness/m	3	7	10
Coefficient of Permeability $k/(10^{-5}\text{m}\cdot\text{d}^{-1})$	10	9.6	7.5
Initial Modulus E_i/MPa	7.5	6.5	5
Cohension c'/kPa	10	13.5	18
Friction Angle ϕ'/°	21	19	18
Poisson's Ratio	0.3	0.3	0.35
Unit Weight of Soil/kN.m^{-3}	14	14	16
Porosity	0.5	0.5	0.5

Table 2 The other parameters used in FEM analyses

Item	Cross section (m^2)	Thickness (m)	Modulus of Elasticity (MPa)	Poisson's ratio
Columns	0.4×0.4	-	E_f=2×10^4	0.15
Beams	0.3×0.5	-	E_f=2×10^4	0.15
Floor slab	-	0.2	E_f=2×10^4	0.15
Raft	-	0.4	E_r=2×10^4	0.15
Piles	0.4×0.4	-	E_P=2×10^4	0.15

The model studied is a four-story, seven by three spans, three-dimensional space superstructure-piled raft-soil system. Figure 1 (a) shows the layout of piles and columns. Figure 1 (b) gives the perspective view of the system, including superstructure, piles, raft, and soil medium. Due to symmetry, only a quarter of the structure is shown and analyzed. The material and geometrical parameters used in numerical analysis are presented in Table 1 and Table 2. The boundary conditions are as follows. On the two outside vertical faces and the bottom plane, the normal components of displacement are constrained, and on the two sides of symmetry, the corresponding translational and rotational degrees of freedom are fixed. No flow of pore fluid through all the vertical and bottom sides is permitted. Drainage is possible on the surface of soil, but the raft and piles are impervious. The loads are applied suddenly, including all parts of gravities.

The pressure on slab is 10kPa. The pressure on the raft is 30kPa, and the unit weight of piled raft and superstructure is 25kN.m^{-3}. The excess pore water pressures generated by the constructions were ignored and they are induced only by the surcharge on the subsoil.

The numerical analyses are conducted on the platform of computer program ABAQUS. In the FEM model, all columns and beams are modeled as three-node quadratic beam elements B32; the raft and slabs are modeled as eight-node, reduced integration, quadratic shell elements S8R; three-dimensional eight-node, piles are modeled by reduced integration solid elements C3D8R; and the subsoil is modeled as three-dimensional eight-node, reduced integration porous solid elements C3D8RP.

VARIATION PATTERNS OF PORE PRESSURE AND DEFORMATION OF SOIL WITH TIME

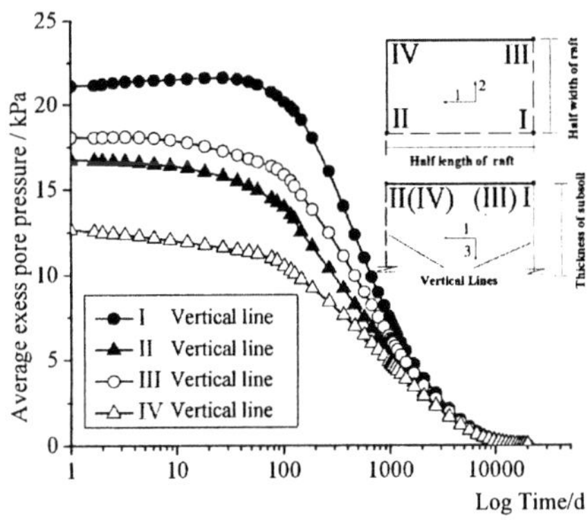

FIG.2. Variation of average excess pore water pressure along vertical typical lines with time

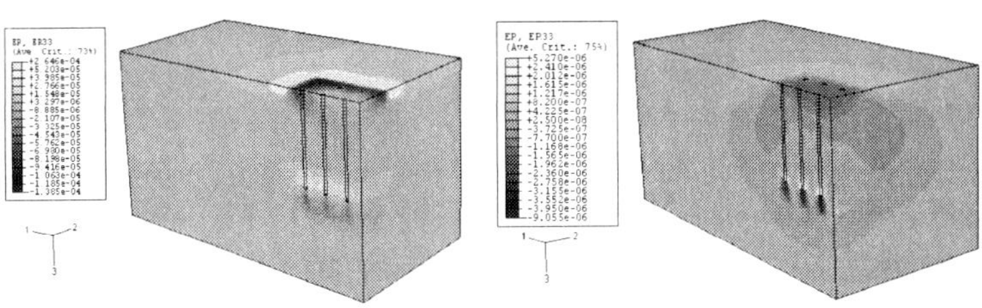

(a) t=50d (b) t=800d

FIG.3. Variation of the distribution of vertical strain rate of the foundation soils beneath piled raft with time

In Figure 2, variation of average value of excess pore water pressure along four specific vertical lines versus time is shown. The four vertical lines pass the points I, II,

III, VI respectively on the raft through the subsoil. Opposite to the other vertical lines, the average excess pore pressure along vertical line I increases slightly within a short period of time after loading, and then since about t=100d, the average excess pore pressure of vertical line I becomes to decrease more rapidly than those of other vertical lines. And the farther distance the vertical lines are away from the raft center, the more rapidly the average excess pore pressures of the line decreases. The phenomenon above of increasing excess pore pressure in local area of subsoil is so-called Mandel-Cryer effects, which is discussed by Sun (1999) and Zhu et al. (2001). In fact, after loading there are higher total stresses in subsoil at the edge of raft, which is reported by Balakrishna et al. (1992) and John (2001). In the earlier time of consolidation, due to the vicinity to the drainage surface, excess pore pressures of subsoil at the edge of raft dissipate more quickly in contrast to very small change in total stress distribution; thus there is a tendency for the variation of effective stress of subsoil at the edge of the raft to be greater than the variation of effective stress of subsoil at the center. With the time progressing, due to dissipation of excess pore pressures at the edge of raft and redistribution of total stress, the excess pore pressures at the center begin to dissipate in greater speeds than those at the edge of the raft. Thus there is a tendency for the variation of effective stress of subsoil at the center to be greater than the variation of effective stress of subsoil at the edge of the raft. It can also be seen from Figure 3 that when t=50d or in the earlier time of consolidation, there is a dominance of the vertical strains rates at the edge of the raft to those at the center; when t=800d or in the later time of consolidation, the vertical strains rates near the center of the raft and bottom of the piles have the dominance. It agrees with the variation pattern of dissipation of excess pore pressure and redistribution of effective stress discussed above.

TIME-DEPENDENT BEHAVIOR OF THE RAFT FOUNDATION

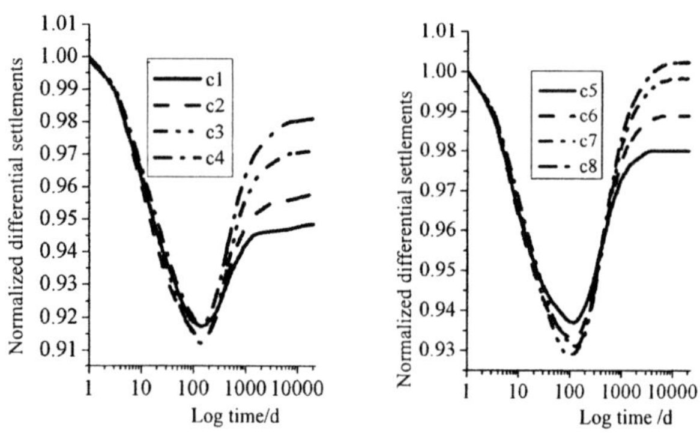

FIG.4. Variation of normalized differential settlements along the length of raft at the points of columns with time

Normalizing the differential settlement at different time by the initial values, it gives the time variation of differential settlement along the length of the raft at the points of columns as shown in Figure 4. Differential settlements along the length of the raft at the points of columns decrease in the beginning stage after loading and increase afterwards. This pattern of time variation of differential settlement is the same as that presented by Booker et al. (1984). And it agrees well with the features of time variation of the vertical strain rates and the corresponding dissipation of pore pressures. In other words, in the earlier time of consolidation after loading, there is a tendency for increment of settlements at the edge of the raft to be greater than the increment of settlements at the center, due to the dominance of the vertical strain rate at the edge of the raft; it is opposite in the later time of consolidation. Thus, differential settlement displays the time variation pattern as depicted in Figure 4.

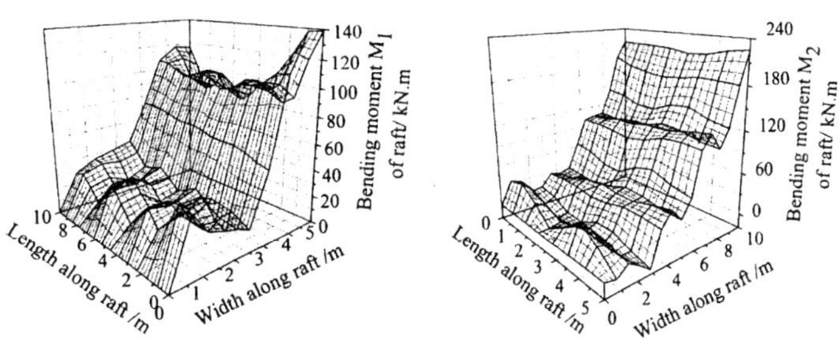

(a) Bending moments M_1 (b) Bending moments M_2

FIG.5. Distribution of bending moments of the raft when t=1d

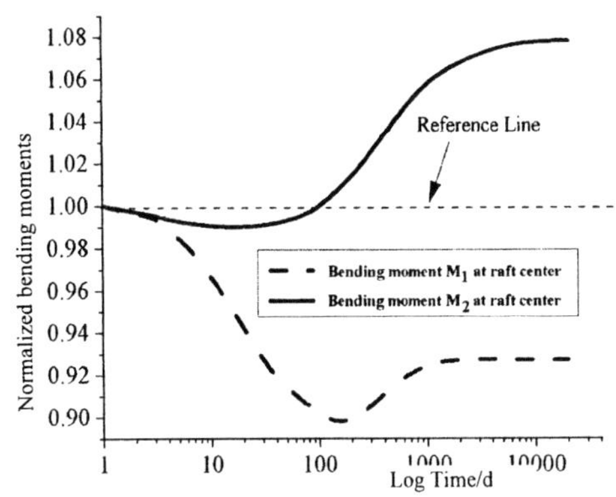

FIG.6. Variation of bending moments M_1 and M_2 at centroid of the raft with time

Shown in Figure 5 is the distribution of bending moments of the raft when t=1d. Obviously, bending moments in the central area of the raft is greater than those in the edge of raft. It is the typical feature of distribution of bending moment of the raft in engineering design and from actual measurements. Normalizing the bending moments at different time by the initial values, then it gives the time variation of bending moments in both direction 1 and 2 at centroid of the raft as shown in Figure 6. The bending moments at the centroid of the raft decrease in the beginning and increase afterwards. The time variation pattern of bending moments is the same as that of differential settlements. Furthermore, due to the difference of bending stiffness of the piled raft in direction 1 to direction 2, M1 decreases much more pronounced than M2 in the earlier time; inversely, M2 increases much more apparent than M1 later, where M1 and M2 represent bending moments in direction 1 and direction 2 respectively.

TIME-DEPENDENT BEHAVIOR OF THE PILES

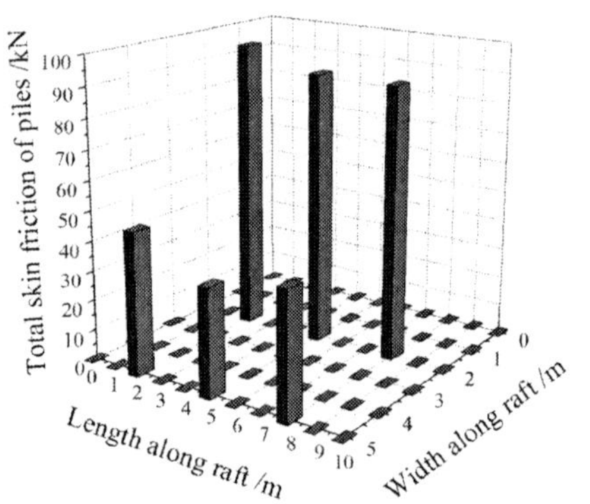

FIG.7 Distribution of total skin friction of piles when t=1d

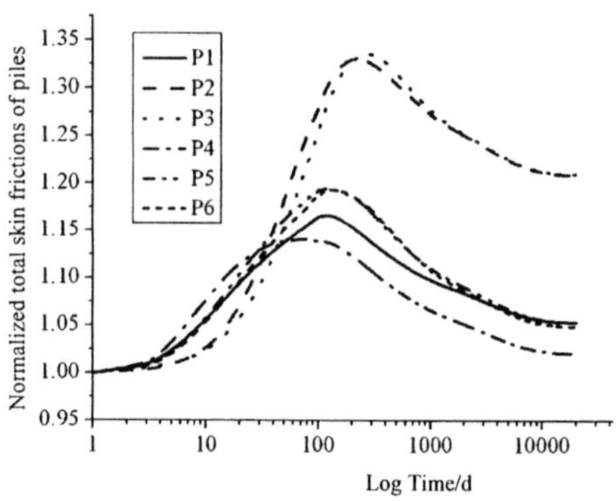

FIG.8 Variation of normalized total skin frictions of piles with time

Distribution of the total skin frictions of the piles at t=1d is shown in Figure 7. It can be seen that the total skin frictions of the piles near the edge, especially the corner pile P4, is higher than those near the center, which also is in accordance with the edge effects of raft. Normalizing the total skin frictions of the piles at different time by the initial values, the time variation curves of total skin friction of the piles are presented in Figure 8. Clearly, the total skin frictions of the piles increase at the beginning and then decrease till to the steady state. It behaves in an opposite manner to that of the bending moment and differential settlement. It obviously indicates the role of piles to adjust and compensate the raft in the pile - raft interaction. Since the level of total skin frictions of P2 and P3 are relatively lower than the others and the load imposed on the system is constant, the adjustment on the distribution of load sharing causes rather more loading change and more obvious time variation feature of the total skin frictions in P2 and P3 near the center.

TIME-DEPENDENT BEHAVIOR OF THE STRUCTURAL PERFORMANCE

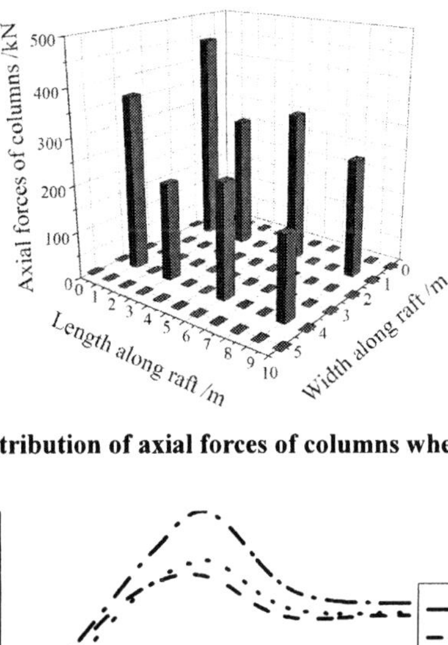

Fig.9 Distribution of axial forces of columns when t=1

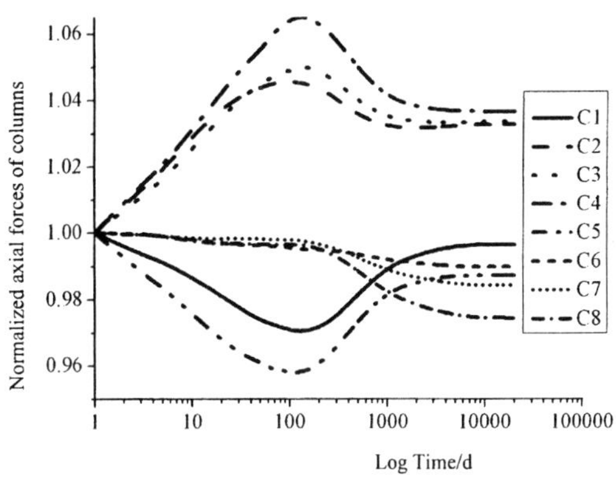

FIG.10 Variation of axial forces of columns with time

Presented in Figure 9 is the distribution of the axial forces of all columns on the raft when t=1d. It is obvious that the axial forces of the columns near the edge, especially the corner column C5, are higher than those near the center. And it also agrees with the edge effects of raft. Normalizing the axial forces of columns at different time by the initial values, the time variation of axial forces of the columns is illustrated in Figure 10. It can be seen that axial forces of the columns C2, C3 and C4 increase in the beginning and then decrease till reaching the steady state. They exhibit an opposite feature to that of the columns C1 and C5. On the contrary, columns C6, C7 and C8 display an invert feature. It is closely associated with the distribution of differential settlement on the raft. It was shown by Wang et al. (2001) that differential settlement was equivalent to apply a "tension" to the columns near the center. Consequently, the axial forces of columns near the center show the feature of time variation opposite to that of differential settlement. Furthermore, since the sum of axial forces of all columns is constant and differential settlement have little effect on columns near the edge, the axial forces of the other columns display invert or opposite time variation features.

CONCLUSIONS

An attempt has been made to take into account the time-dependent behavior of the superstructure-piled raft-subsoil interaction system on a deep consolidation layer in three-dimensional space by the proposed numerical method based on FEM. For the cases studied, several conclusions can be drawn from the results obtained.

(1) The time-dependent behavior of subsoil caused by consolidation has obvious effects on the entire interaction system of superstructure-piled raft-subsoil. In practical engineering, the time effects of subsoil on the interaction system would not be negligible in the long run and the time-independent methods may be not sufficient. Especially when drainage conditions of subsoil are poor, these time effects become rather sustained.

(2) The pattern of time variation of excess pore water pressure in subsoil is intimately associated with features of the internal forces and deformations, such as bending moment and differential settlement of the raft, and total skin frictions of piles.

(3) It is suggested that for a better estimation of the internal forces and deformations in the superstructure-raft-pile system, when the structures are important, both effects of consolidation in subsoil and the superstructure stiffness on the interactive behavior should be taken into account in analysis. Future studies may be able to analyze the coupled time effects of consolidation and rheology on the system, when creep characteristics of subsoil are significant.

ACKNOWLEDGMENTS

The author would like to express theirs thanks to the financial supports for this study from National Natural Science Foundation of China through the grants 50809009 and 50639010.

REFERENCES

Sekhar, C. D. and Rana, R. (2002): A critical review on idealization and modeling for interaction among soil-foundation-structure system, *Computers and Structures*, 80, 1579-1594.

Wood, L. A. and Larnach, W. J. (1975): The interactive behavior of a soil-structure system and its effect on settlements, *Symposium. on Recent Development in the Analysis of Soil Behavior and their Application to Geotechnical Structures*, University of New South Wales, Kensington, N. S. W., Australia, 12: 75-87
Booker, T. R. and Small, T. R. (1984): The time-deflection behavior of a circular raft of finite flexibility on a deep clay layer, *International Journal for Numerical and Analytical Methods in Geomechanics*, 8, 343-357.
Lewis, R. W. and Tran, D. V. (1989): Application of soil-structure interaction to off-shore foundations with specific reference to consolidation analysis, *International Journal for Numerical Methods in Engineering*, 27, 195-213.
Zai, J. M. and Zai, J. Z. (1999): Analysis and design of foundation in high-rise building, Beijing: China Construction Industry Press.
Cheng, Z. H., Ling, D. S. and Chen Y. M. (2004): Time effects on piled raft foundation under vertical loading. *China Civil Engineering Journal*, 37(2), 73-77.
Wang, J. H., Chen, J. J. and Pei, J. (2001): Interaction between superstructure and layered visco-elastic foundation considering consolidation and rheology of soil, *Journal of Building Structures*, 35(4), 489-492.
Noorzaei, J., Viladkar, M. N. and Godbole, P. N. (1995): Elasto-plastic analysis for soil-structure interaction in framed structure, *Computers & structures*, 55(5), 789-795.
Schiffman, R L., Chen, T. A. and Jordan, J. C. (1969): An analysis of consolidation theory, *Journal of the Soil Mechanics and Foundation Division, ASCE*, 95(1), 285-309.
Booker, J. R. (1974): The consolidation of a finite layer subject to surface loading. *International Journal of Solids and structures*, 10, 1053-1065.
Van, L. H. and Vermeer, P. A. (1990): Automatic step size correction for non-associated plasticity problem. *International Journal for Numerical Methods in Engineering*, 29, 578-579.
Thomas, H. R. and Zhou, Z. (1998): An analysis of factors that govern the minimum time step size to be used in the finite element analysis of diffusion problems, *Communications in Numerical Methods in Engineering*, 14(9), 809-819.
Hibbitt, Karlsson and Sorensen, Inc. (2000): ABAQUS Theory Manual . USA: *Hibbitt, Karlsson & Sorensen, Inc.*
Sun, J. (1999): Rheology of geotechnical materials and its application, Beijing: *China Construction Industry Press*, 369-372.
Zhu, J. G. and Yin, J. H. (2001): Deformation and pore- water pressure responses of elastic viscoplastic soil, *Journal of Engineering Mechanics, ASCE*, 127(9), 899-907.
Balakrishna, C. K., Srinivasa, B. R., Murthy, T. S. (1992): Stress distribution beneath rigid circular foundations on sands. *International Journal for Numerical and Analytical Methods in Geomechanics*, 16(1), 65-72.
John, C. S. (2001): Practical solutions to soil-structure interaction problems. *Progress in Structural Engineering and Materials*, 3(3), 305-314.
Godbole, P. N., Viladkar, M. N., Noorzaei, J. (1990): Non-linear soil-structure interaction analysis using coupled finite-infinite elements. *Computers & structures*, 36(6), 1089-1096.
Viladkar, M. N., Godbole, P. N. and Noorzaei, J. (1991): Soil-structure interaction in plane frames using coupled finite-infinite elements. *Computers & structures*, 39(5), 535-546.

Structural Characteristics and Design Technical Keys of Wind Turbine Foundation in Shanghai Donghai-bridge Offshore Wind Farm

Yifeng LIN[1,2], Xuan ZHOU[2]

[1] Department of Geotechnical Engineering Tongji University , Shanghai 200092, China; [2]Shanghai Investigation Design & Research Institute, Shanghai 200434 China; 2. hhhlyf@126.com

ABSTRACT: As the first large-scale wind farm of Asia, the wind turbine foundation of Donghai-bridge Offshore wind farm is in possession of the following four engineering characteristics such as marine structure, tower structure foundation, dynamic equipment foundation and complex soft subgrade foundation, which makes the foundation design a key technical issue. The common issues of offshore wind turbine foundation design are discussed in this paper. , while the characteristics of wind turbine foundation, the nonlinear dynamic interaction of foundation and subgrade, dynamic simulation of the system comprised of wind turbine-tower-subgrade-foundation and the fatigue analysis of foundation structure are emphatically analyzed.

INTRODUCTION

As the most mature technique in new energy field, the wind power generation is developing fast with the most feasible condition to explore in large-scale and the brightest future to commercialization. With the extension of wind power technology from onshore to offshore, the offshore wind power generation is becoming a focus of new energy field. Compared with the onshore wind farm, the advantages of offshore wind farm are the more abundant wind energy resources, the higher developing efficiency, the lower environment pollution and no need to take use of farmland. There is a wide coastline in China, and the offshore wind energy resources are abundant, while the electric load centers mostly centralize in the eastern coastal area, the developing future of offshore wind farm is wide. As the first large-scale wind farm of Asia, Donghai-bridge Offshore wind farm with a total installed capacity of 100MW is located in the eastern area of Donghai Bridge. Twenty wind turbines with a capacity of 5MW each unit will be installed.

The marine hydrographic condition, meteorological condition and geological condition are complicated for building offshore wind farm, which make the design and construction of wind turbine foundation more difficult. To design and construct the offshore wind turbine foundation in an economy way is essential to the wide application of offshore wind farms. The design institute of Donghai-bridge offshore wind farm take a monographic study of foundations, making a profound research on foundation structure type selection and structure calculation analysis, and predominating the characteristics and key design technique of wind turbine foundation.

WIND TURBINE FOUNDATION DESIGN

Geological Condition

Donghai-bridge offshore wind farm is located in the eastern area of Donghai Bridge, the north of wind farm is 8km away from the coastline and the south of wind farm is 13km away from the coastline. The layout of the 20 wind turbines is in three lines(see Fig 1), the distance between turbines is 750m (along the bridge) in north-south direction and 1.2km in east-west direction. The water depth is 9.9m~11.9m, and the elevation of the seabed mud line is -10.00m~-10.67m. The subgrade is mainly silt on surface. Fig 2 shows the soil cross section profile.

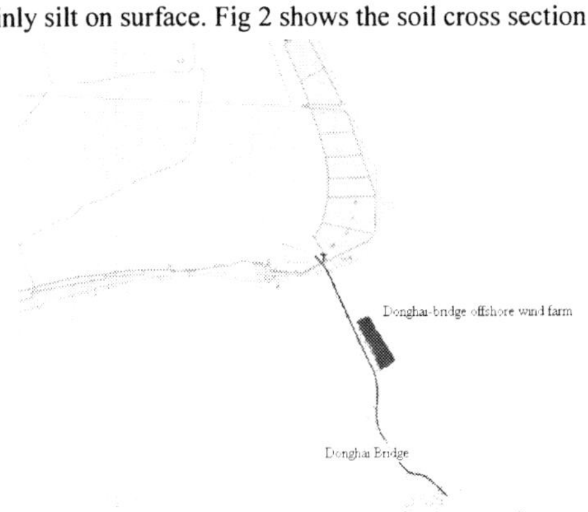

Figure 1 Location of Donghai-bridge offshore wind farm

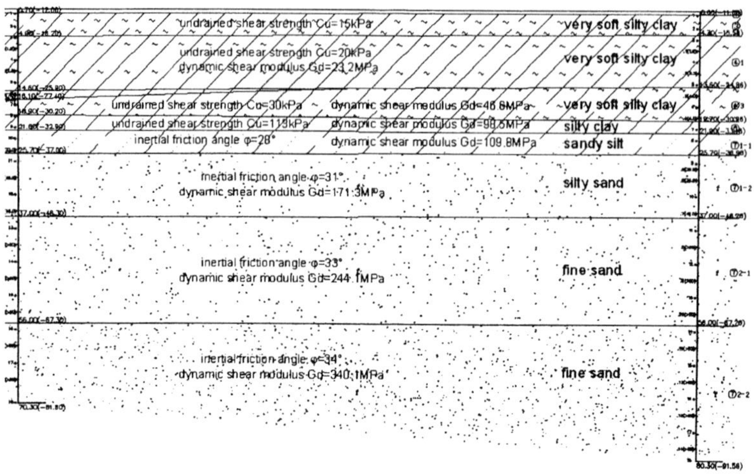

Figure 2 Soil cross section profile

Structure Scheme Design of Wind Turbine Foundation

A quadrupod combined foundation scheme is recommended in this project. The detail descriptions of this foundation type are as follows: four steel piles are driven and fixed in the seafloor and steel sleeves connecting the pile top are then used to support the quadrupod structure. The quadrupod bears the tower loads, and transfers the loads to the four steel pipe piles. During construction, 2m thickness high-strength geotechnical macadam was placed to the base of the four steel sleeves to enhance the bearing capacity of ground soil. The precast steel member of the quadrupod is then settled to the seafloor. Finally, the four piles are driven to the seabed through the steel sleeves. The diameter of each pile is 2.2m. The length of the pile is 55m. The pile top elevation is -3m, and the elevation of pile tip which is driven into the fine sand layer is -58m, and the layout of the piles is square with a spacing of 16m. The ectotheca of the steel sleeves is allocated with specialized pile-fixed member to connect the piles, after leveling the superstructure, adopting high-strength grout to complete the fixation between the steel piles and sleeves. The upper quadrupod is precast member including twelve horizontal and inclined steel connecting tubes with a diameter of 2m, which are used to connect four steel sleeves and a vertical steel pipe with a diameter of 6.5m in center. The sketch of quadrupod combined foundation is shown in Figure 3.

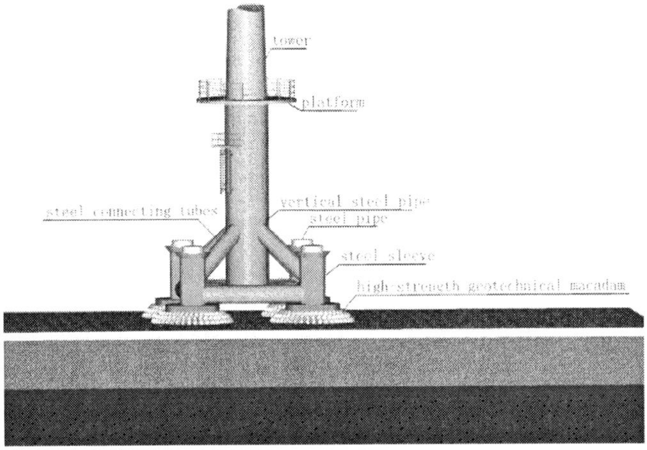

Figure 3 The sketch of quadrupod combined foundation

FOUNDATION CHARACTERISTICS AND DESIGN TECHNICAL KEYS

Engineering Characteristics of Foundation

(1) The offshore wind turbine is subjected to wave and current, which indicates a characteristic of marine engineering;

(2) As the tower which supports wind turbine is a towering structure, the wind turbine foundation is with a structure characteristic of towering structure;

(3) As wind turbine is a dynamic equipment, the wind turbine foundation is with a characteristic of dynamic equipment foundation.

The characteristics above are the significant distinctions. Besides, the seabed of Donghai-bridge offshore wind farm is thick soft soil, while wind turbine foundation is kind of a complicated soft subgrade foundation. That's why the foundation of Donghai-bridge offshore wind turbine is with the characteristics combined of marine engineering, towering structure foundation, dynamic foundation and complicated soft subgrade foundation, which bring a series of key technical problems to wind turbine foundation design. Fig 4 represents the engineering characteristics of the foundation.

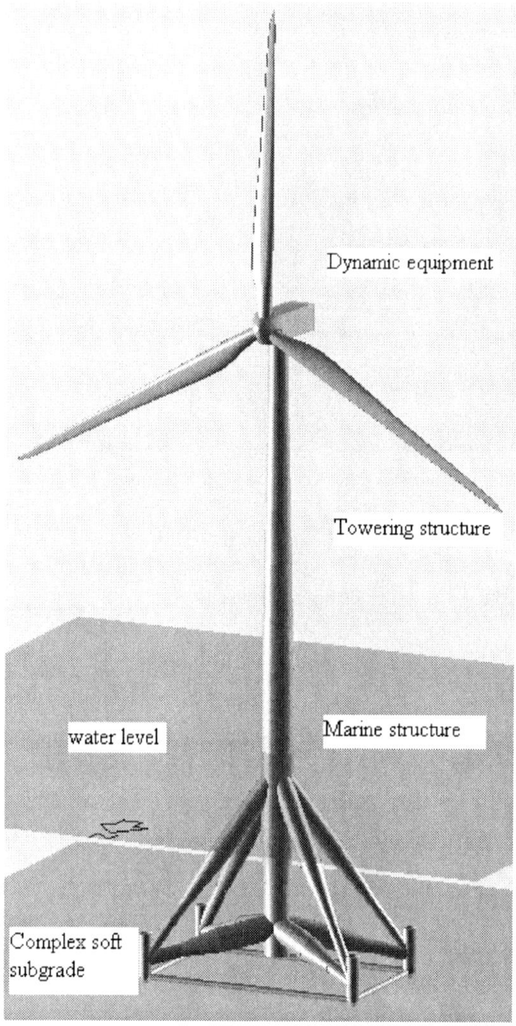

Figure 4 Engineering characteristics of foundation

As a marine steel structure, the damage form and failure location of Donghai-bridge offshore wind turbine foundation are complicated, the probably damage form are cumulative fatigue damage, the failure may be the foundation or subgrade. Either the maximum loads exceeding the ground ultimate bearing capacity or the strength reduction led by the long-term stress cycle induced by wave loads and wind loads can cause the damage to subgrade. The supporting structure of wind turbine adopts steel-frame structure welded with a series of circular steel tubes, while the damage can be led by the member yielding under ultimate loads, or led by cumulative fatigue damage of the steel joint caused by the long-term dynamic magnification induced by wave and current. On the other hand, the water depth is approximately 10m, and the cantilever length above the seabed is long, which leads a comparatively larger horizontal deformation when the foundation is subjected to coupled wave, current and wind loads. It is indeed a big problem to control the horizontal deformation of the wind turbine foundation as the soft soil in this site is 20m in depth. Therefore, choosing a suitable model to do static and dynamic calculation and predominating stress and deformation status are the key link of foundation design.

As a towering foundation, the offshore wind turbine foundation is bearing an overturning moment of 1.5×10^5 kN • m, while the vertical loads acting upon the foundation is small compared to typical towering structures. These lead to an enormous vertical compression force on the compressing side and an enormous vertical tension force on the tension side. Building a steel pipe piles foundation bearing such stress on the soft ground is a quite difficult task, which becomes a key technique of this project.

Wind turbine foundation is kind of a dynamic foundation to support the wind turbine, other than satisfying the strength and stability requirements of the foundation itself, it also needs to satisfy the frequency response requirements of the total system comprised of wind turbine-tower-foundation-subgrade to avoid over shock and to ensure the safe operation. Because of the special requirements of wind turbine equipments, the permitting frequency range of the system is narrow and is 0.27Hz~0.31Hz with a strict variation of 0.04Hz. Therefore, a careful design should be taken in choosing foundation type and structure layout, and the calculation parameters and model should be rated strictly. The frequency and mode should be overally mastered to ensure the foundation design satisfying the requirements of dynamic characteristic.

Different from the common dynamic equipment foundation and onshore wind turbine foundation, there is a cantilever member above seabed, and the length of which depends on the water depth. Thus the natural frequency is reduced and the foundation dynamic magnification effect induced by wave and wind loads is amplified. Therefore, it is difficult to ensure the structure safety by quasi-static analysis, and the dynamic analysis method is required. On the other hand, the pile foundation represents obvious nonlinear characteristic. These bring forward a strict requirement of dynamic analysis, which becomes another key technique of this project.

The wind turbine foundation of Donghai-bridge offshore wind farm is with the characteristics including marine engineering, towering structure foundation, dynamic equipment foundation and soft soil foundation. This type of foundation is lacking of adequate theoretic research and practical experience, and the domestic research of offshore wind turbine foundation is on an early stage. So, it is an enormous challenge to design the wind turbine foundation of Donghai-bridge offshore wind farm. On the basis of understanding and mastering the characteristics of wind turbine foundation, using both the basic theory of marine engineering, foundation engineering, structure dynamics and soil mechanics and the technology of numerical analysis to study the key techniques such as bearing and deformation of pile foundation, dynamic response, fatigue damage and etc.

Bearing and Deformation of Pile Foundation

The bearing loads of wind turbine foundation are dynamic and circular, and the strength of soft subgrade is reduced under cyclic loads, while the plastic deformation of subgrade soil occurs because of the large horizontal deformation. These characteristics can be represented by the structure model, the subgrade model and analysis method. The p-y curves are used to calculate the horizontal force on pile foundation, refer to Design of Offshore Wind Turbine Structures (DNV-OS-J101) for the equation to determine the p-y curves. Using finite element analysis software ANSYS to model the wind turbine foundation, with the nonlinear spring element and APDL (ANSYS Parameter Design Language) secondary development to simulate the nonlinear interaction between pile and soil. The horizontal deformations of tower base and pile on mud line are respectively 82mm and 18mm under ultimate limit states, and the equivalent stiffness of foundation can satisfy the requirements of wind turbine equipments. The maximum pile axial compression force and pile axial tension force of which are 14700kN and 7200kN, respectively, while the pile axial compression-resisting force and pile axial tension-resisting force are 15000kN and 7300kN(calculated in accordance with pile foundation code), respectively, which can satisfy the load bearing requirements.

Dynamic Modal Analysis of System

The wind turbine-tower-foundation-subgrade is an interactional dynamic system, and an uniform mechanical model should be used in dynamic analysis to identify the dynamic characteristics and response of the high coupled dynamic system. On the primary design stage, the modal analysis of the system consisting of wind turbine-tower-foundation-subgrade is conducted. The analysis results were used to judge whether the structural design satisfies the requirements of dynamic characteristic or not. The following two situations should be considered in modal analysis:

(1) On the preliminary operational stage, while the structure with no corrosion, the subgrade strength with no reduction and no marine organism attached to the

foundation, the frequency of the system corresponds to the upper limit of structural natural frequency;

(2) After a long-term operation, the structure is subjected to corrosion, the subgrade strength is reduced under dynamic cyclic loads and the marine organism is attached to the structure, the frequency of the system corresponds to the lower limit of structural natural frequency.

Simplify the wind turbine structure to a multiple-degree-of-freedom system, modeling the structure and tower with thin-walled pipe element, and modeling the subgrade with nonlinear spring element. Alpha and beta damping is used for structure and subgrade in the dynamic analysis. According to the wind turbine mass, stiffness and location supplied by manufacturer, the wind turbine is simplified to a concentrated mass point. The modal analysis shows that the primary natural frequency range of the structural system is 0.280Hz~0.294Hz, which can satisfy the requirements of natural frequency.

Fatigue Analysis of Foundation Structure

The wind turbine foundation scheme of offshore wind farm is spatial steel-frame consisted of steel pipes, and the stress concentration may occur in the pipe joints. With the wind and wave dynamic cyclic loads, the stress concentration of pipe joints has an obvious impact on the fatigue strength of joints, and the fatigue analysis of pipe joints is a key link of wind turbine foundation design. The S-N curve used in fatigue analysis is chosen in accordance with the regulations in Design of Offshore Wind Turbine Structures (DNV-OS-J101). Fig 5 shows the S-N curves for welded joints.

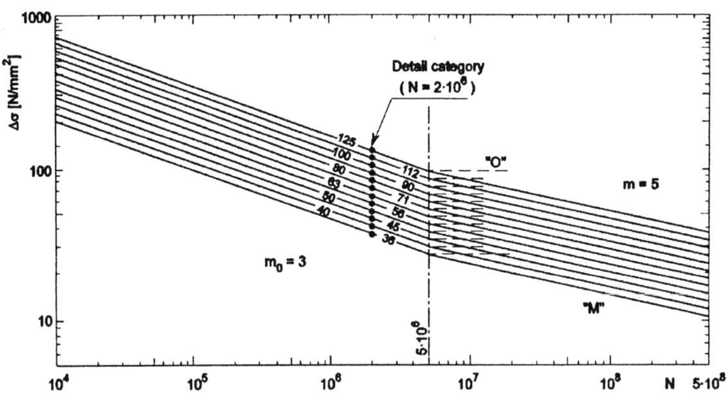

Figure 5 S-N curves, welded joints

According to the equivalent fatigue loads supplied by manufacturer, with a reference of the related wave information, using the thin-walled shell element to calculate the stress of wind turbine foundation. The fatigue stress of structure is $\triangle_s$=75.5MPa, the corresponding fatigue life N =3.47×10^8> fatigue life cycle [N]= 2×10^8, indicating that the wind turbine can satisfy the design requirements under fatigue loads.

CONCLUSIONS

As the first large-scale wind farm of Asia, the wind turbine foundation of Donghai-bridge Offshore wind farm is a milestone in the wind power generation history.

As for the special soft subgrade conditions and the characteristics of marine engineering, towering structure foundation, dynamic equipment foundation, there are significant characteristics and prominent technical problems in designing wind turbine foundation.

The technical key of offshore wind turbine foundation design lies in the analysis on nonlinear dynamic interaction of wind turbine foundation and subgrade, as well as time domain and frequency analysis of wind turbine-tower-subgrade-foundation and the fatigue analysis of foundation structure. The design institute has mastered the related key solution through monographic study combined with design, which formulates the technical foundation for the design of offshore wind turbine foundation.

REFERENCES

Shanghai Investigation, Design and Research Institute (2006). "Feasibility study report of Donghai-bridge offshore wind farm."

Wang Yuechong, Xiao Linhao (1983). "Design and examples of marine structures." *Marine Publication.*

Qian Jiahuan, Yin Zongze (1996). "Geotechnical principle and calculation." *Water Conservancy and Electric Power Publication.*

ANSYS Corporation (1997). "Advanced Analysis Technologies Guide."

DNV-OS-J101, Design of Offshore Wind Turbine Structures. June 2004.

Coupled Foundation-Superstructure Analysis and Influence of Building Stiffness on Foundation Response

Yat Fai Leung[1], Neil A. Hoult[2], Assaf Klar[3] and Kenichi Soga[4]

[1]Ph.D. Student, University of Cambridge, Department of Engineering, Trumpington Street, Cambridge, CB2 1PZ, United Kingdom; YFL27@cam.ac.uk
[2]Assistant Professor, Queen's University, Department of Civil Engineering, Ellis Hall, Kingston, K7L 3N6, Canada
[3]Senior Lecturer, Faculty of Civil and Environmental Engineering, Technion – Israel Institute of Technology, Haifa 32000, Israel
[4]Professor of Civil Engineering, University of Cambridge, Department of Engineering, Trumpington Street, Cambridge, CB2 1PZ, United Kingdom

ABSTRACT: This paper proposes a simple method to include superstructure stiffness in foundation analyses. The method involves extracting a small "condensed structural matrix" from finite element models of the superstructure, which can then be incorporated into pile group or piled raft analyses using common approaches such as elastic continuum or load transfer methods. The matrix condensation method directly couples structural and geotechnical analyses, and eliminates the need for iterative analyses between structural and geotechnical engineers. Effectiveness of the approach is illustrated through analyses of several buildings designed with a typical floor plan but with varying heights. The parametric study illustrates that superstructure stiffness can have a significant influence on foundation settlement estimates, and the stiffening effects are dominated by the lower stories of the superstructure. The proposed method aims to bridge the gap between structural and geotechnical analyses. Also, being a computationally simple and accurate approach, it is applicable to parametric or optimization studies that would otherwise involve large amounts of analyses.

INTRODUCTION

Geotechnical engineers often make assumptions about 'perfectly rigid' or 'flexible' pile caps connecting the piles, while real pile groups often behave in a manner somewhere in between these two extremes, depending on the stiffness of the superstructure. In some cases, in order to evaluate the holistic building response, structural and geotechnical engineers engage in a tedious iterative process (Chamecki, 1956; Lee & Harrison, 1970), where they continuously update the input information in their geotechnical and structural analyses based on results from their counterparts, until these results converge to within a reasonable tolerance. Other approaches to consider superstructure effects include approximating the superstructure as a beam or set of beams with equivalent stiffness to the superstructure (Meyerhof, 1953), or approximating the foundations as Winkler springs in the structural analyses (Miyahara & Ergatoudis, 1976). Most of these methods either involve iterations of settlement evaluations, or are approximations of the structure and/or foundation behavior.

This paper introduces the matrix condensation method, which allows direct coupling of superstructure stiffness into foundation analyses, in order to model the combined foundation-superstructure response without such iterations or approximations. The condensed matrix of a superstructure is obtained through a simple procedure that can be performed using common structural finite element programs. In this paper, a typical building configuration is studied using this technique to illustrate the potential impacts of superstructure stiffening effects on foundation settlement and bending moment estimates. Moreover, the influence of the building height on the stiffening effects is studied by analyzing superstructures with various numbers of stories.

CONDENSED SUPERSTRUCTURE STIFFNESS MATRIX

In many building projects, structural engineers construct building models using finite element packages. The complete structure stiffness matrix will consist of all the members in the building structure. Using these models, a 'condensed' structure matrix ($\mathbf{K}^s$) can be extracted by applying a unit displacement at the base of each column that connects to the foundation in sequence, thus extracting the reaction forces at all of the supports due to the unit displacement. For example, the component $\mathbf{K}^s_{i,j}$ in the condensed matrix represents the reaction force at support i due to a unit displacement applied at support j (Fig. 1). Unlike the complete structural stiffness matrix, the condensed structure matrix is fully populated, with a size of $n \times n$, where n is the number of columns or supports connecting the superstructure and the foundation.

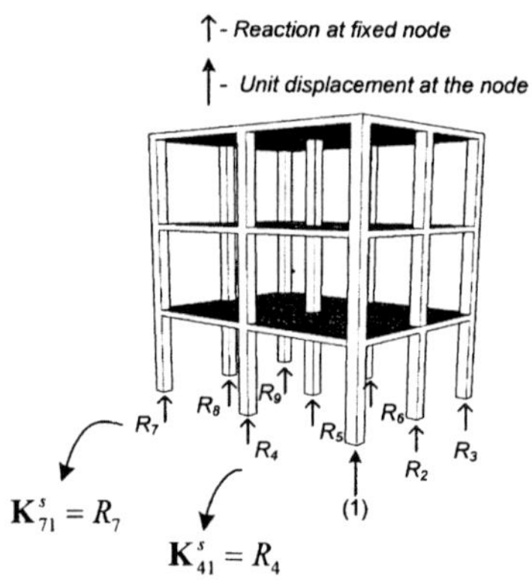

FIG. 1. Structure condensation process from finite element simulation.

With the condensed matrix obtained through the above procedure, it can then be coupled with the foundation stiffness in a holistic analysis of foundation and superstructure response:

$$(\mathbf{K}^p + \mathbf{K}^r + \mathbf{K}^s)\boldsymbol{u} = \boldsymbol{p}^w + \boldsymbol{p}^g \tag{1}$$

where $\mathbf{K}^p$ is the pile stiffness matrix and $\mathbf{K}^r$ is the raft stiffness matrix in cases when the piles are connected to a foundation raft; $\boldsymbol{u}$ is the vector of displacements at the pile/raft nodes, while $\boldsymbol{p}^w$ is the loading due to the self weight and live loads acting on the superstructure (obtained from the structural finite element program) and $\boldsymbol{p}^g$ is the ground reaction force acting on the pile and raft elements.

In the current work, $\mathbf{K}^r$ is evaluated using the finite element method (Zienkiewicz & Taylor, 2005). The foundation is analyzed using the continuum approach. Soil-pile-raft interaction ($\boldsymbol{p}^g$) is evaluated using Mindlin's solution (Mindlin, 1936). Pile-soil/raft-soil contact forces are limited by the pile and raft capacities, respectively. The method is formulated into a computer code with details described in Leung et al. (2009a, b), where the effectiveness of the approach is validated against a case study in London.

ANALYSIS OF HYPOTHETICAL HIGH-RISE CONCRETE BUILDING

A typical building system of a high-rise structure is investigated using the proposed method. The structural floor plans and section sizes of the building are specified based on standard structural design procedures (gravity and lateral load analysis with appropriate Eurocodes), to ensure that they represent realistic structural systems. The hypothetical case involves a 30-story concrete structure, representative of a typical high-rise condominium building. Fig. 2(a) shows the superstructure floor plan of the building, consisting of mainly column-slab components. The column dimensions were set to be 350 mm×350 mm, and the slab thickness of 200 mm was assigned to all 30 stories. The central core walls, which form the two elevator shafts, comprise 200 mm concrete shear walls. Additional 200 mm shear walls are designed to increase the lateral stiffness, in order to limit the deflection of the structure under typical wind loads prescribed by the European design standards. The long-term concrete Young's modulus is assumed to be 14 GPa in the structural finite element analyses. This takes into consideration the creep behavior of concrete. The complete superstructure is modeled using a finite element program, where $\boldsymbol{p}^w$ and $\mathbf{K}^s$ are extracted according to the abovementioned procedures (Fig. 1).

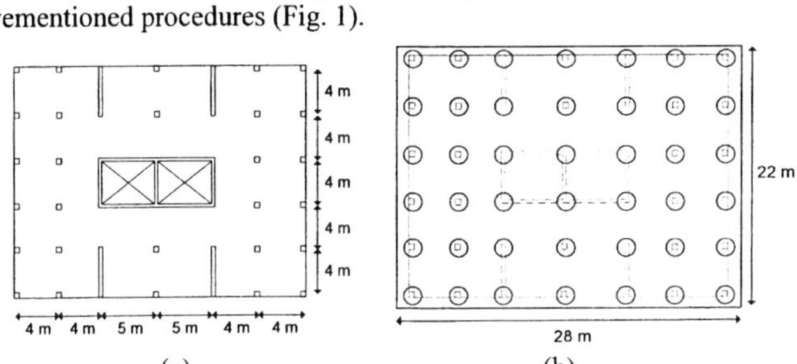

FIG. 2. (a) Superstructure plan, and (b) foundation layout of 30-story building.

The ground conditions assumed for the scenario consist of a homogeneous soil with Young's modulus of 50 MPa, Poisson's ratio of 0.3 and a pile shaft resistance of

50 kPa. These parameters are typical for 'stiff' overconsolidated clays. The building is designed to be founded on a piled raft system (Fig. 2(b)) supported by 42 piles directly underneath the columns/corners of shear walls. Each pile is 30 m long and has a diameter of 1 m, where the raft thickness (t) is 0.75 m. These raft design parameters correspond to a raft-soil stiffness ratio, K_{rs} (Horikoshi & Randolph, 1997), of 0.05, making it slightly flexible compared to most existing piled raft foundations. Based on the foundation and soil information, the foundation stiffness matrices ($\mathbf{K}^p$ and $\mathbf{K}^r$) and pile-soil interaction forces ($\boldsymbol{p}^g$) can be evaluated (Equation 1).

Fig. 3 compares the settlements at the raft center, the differential settlements (center-corner) and maximum bending moments estimated by two piled raft analyses, one with consideration of $\mathbf{K}^s$ and the other without. Coupling $\mathbf{K}^s$ in the foundation analyses represents a more realistic model, and the estimated differential settlements are roughly 30% lower (t = 0.75 m). The bending moment estimates are also significantly lower. In this case, neglecting the stiffness effects of the superstructure can lead to significant overestimations of the differential settlements and bending moments experienced by the foundation, as well as the potential distortion of the superstructure.

The importance of superstructure stiffness on the foundation behavior depends on the flexural stiffness of both the superstructure and the foundation system. The presence of shear walls in the building substantially increases the superstructure stiffness, and hence its contribution to the overall behavior. On the other hand, a thicker (stiffer) raft will reduce the impact of the superstructure by increasing the relative stiffness of the foundation, whereas a thinner (more flexible) raft has the opposite effect. For example, with a raft thickness of 1 m in the above analyses, the difference in differential settlement estimates becomes 22% with and without $\mathbf{K}^s$. A 0.5 m thick raft results in a difference of 40% with and without $\mathbf{K}^s$ (Fig. 3).

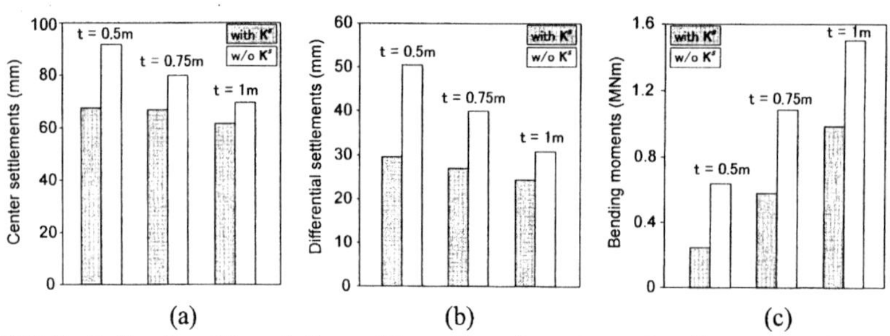

FIG. 3. Estimates of foundation settlements and maximum bending moments (M_{xx}) with and without including $\mathbf{K}^s$.

SUPERSTRUCTURE INFLUENCE FOR DIFFERENT BUILDING HEIGHTS

With the same superstructure layout shown in Fig. 2(a), buildings of different heights were analyzed to reveal the respective impacts of superstructure stiffness on the foundation response. Apart from the 30-story building described earlier, buildings with 10 and 20 stories are analyzed (t = 0.75m). The pile lengths were changed to 10 m and 20 m accordingly, to reflect realistic piled raft designs which aim to control the maximum differential settlements to approximately 15-25 mm.

Fig. 4 compares the analysis results for 10-, 20- and 30-story buildings. A larger contribution of superstructure stiffening can be expected from a taller building, but Fig. 4 shows that the stiffening effects are not directly proportional to the building height. The relationship appears to be nonlinear, with increasing building height only adding minimal stiffening effects beyond 20 stories.

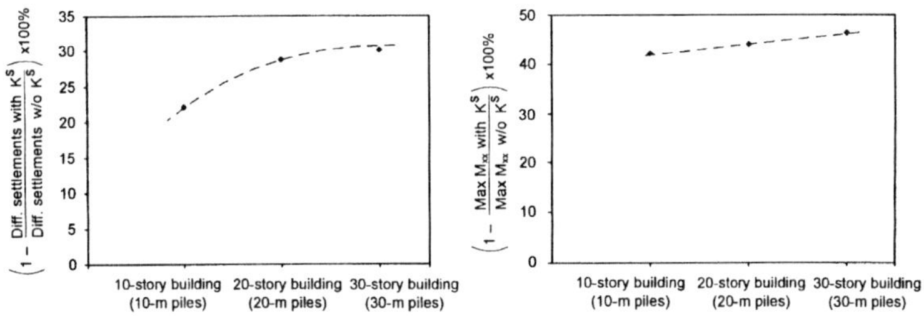

FIG. 4. Superstructure stiffening effects on buildings of different heights.

STIFFNESS CONTRIBUTION BY DIFFERENT STORIES

While Figs. 3 and 4 illustrate the significance of superstructure stiffening effects on buildings of different heights, Fig. 5 shows a breakdown of the influence provided by different stories of the 30-story building. The investigation was performed by analyzing the 30-story building at various construction stages (1, 5, 10, 20, and 30 stories), and comparing the influence of superstructure stiffness at each of the stages. Reductions in differential settlements (center-corner) and bending moments (maximum M_{xx}) due to superstructure stiffening are shown in Fig. 5, which shows a nonlinear relationship: the stiffening effects are mainly controlled by the lower stories, and additional stiffening is less significant at upper levels. Specific results of such analyses will vary with different buildings and foundation conditions, but the general trends (Figs. 4 and 5) should prevail in the majority of high-rise buildings.

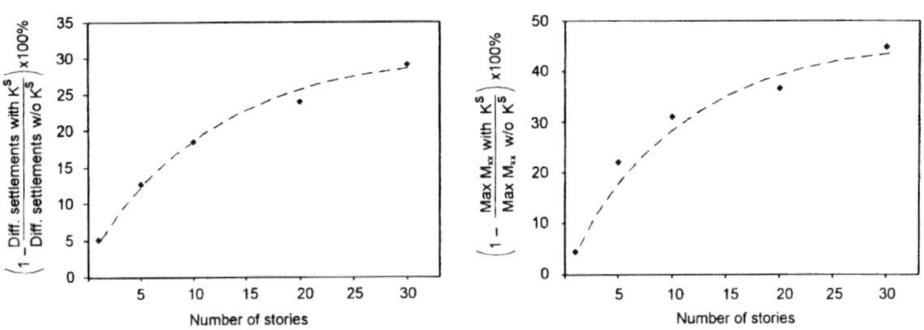

FIG. 5. Superstructure stiffening contributed by different stories.

CONCLUSIONS

This paper introduces the matrix condensation method, which couples superstructure stiffness with the foundation stiffness for a holistic assessment of the building response. The condensed matrix can be easily obtained using common structural finite element packages. It requires a minimal amount of information and can be conveniently stored and transmitted between structural and geotechnical engineers. The method eliminates the need for crude assumptions of structural/geotechnical parameters or iterations between structural and geotechnical analyses, and is particularly useful for sensitivity and optimization analyses of building foundations.

The preceding analyses have demonstrated the effectiveness of the approach. They also show that consideration of superstructure stiffening effects are important for a typical high-rise building constructed in concrete, with core walls near the center of the building. Obviously, the relative importance of superstructure effects depends on many factors, such as the superstructure layout, foundation configurations and soil conditions. The example scenario illustrated in this paper may provide some 'benchmark values' upon which practitioners can base their engineering judgements.

ACKNOWLEDGMENTS

The work presented in this paper is part of the Smart Foundations project (EP/D040000/1) funded by the Engineering and Physical Sciences Research Council (EPSRC). Their financial support is gratefully acknowledged. The first author would also like to express his gratitude to the Cambridge Overseas Trust (COT) for additional funding supporting his studies at the University of Cambridge.

REFERENCES

Chamecki, S. (1956). "Structural rigidity in calculating settlements." *J. Soil Mech. & Found. Div.*, Vol. 82 (SM1): 1-19.

Horikoshi, K. & Randolph, M. F. (1997). "On the definition of raft-soil stiffness ratio for rectangular rafts." *Géotechnique* Vol. 47(5): 1055-1061.

Lee, I. K. & Harrison, H. B. (1970). "Structure and foundation interaction theory." *J. Struct. Div.*, Vol. 96 (ST2): 177-197.

Leung, Y. F., Hoult, N. A., Klar, A. & Soga, K. (2009a). "Matrix condensation method for coupled foundation-superstructure analysis of buildings" Submitted to *Géotechnique*.

Leung, Y. F., Klar, A. & Soga, K. (2009b). "A theoretical study on pile length optimization of pile groups and piled rafts." Accepted for publication in *J. Geotechnical & Geoenv. Engrg.*

Leung, Y. F., Soga, K and Klar, A. (2008). "Optimisation analysis on pile length distribution of squared pile groups." *Proc. 2nd BGA Int. Conf. on Found.*, ICOF2008, Dundee: 505-516.

Meyerhof, G. G. (1953). "Some recent foundation research and its application to design." *The Structural Engineer*, Vol. 31: 151-167.

Mindlin, R. D. (1936). "Force at a point in the interior of a semi-infinite solid." *Physics*, Vol. 7: 195-202

Miyahara, F. & Ergatoudis, J. G. (1976). "Matrix analysis of structure-foundation." *J. Struct. Div.*, Vol. 102 (ST1): 251-265.

Zienkiewicz, O. C. & Taylor, R. L. (2005). *The finite element method for solid and structural mechanics.* 6th Ed., Butterworth-Heinemann.

Observation of Pile-Soil-Interaction during Cyclic Axial Loading using Particle Image Velocimetry

Hans-Georg Kempfert[1], Sebastian Thomas[2], Berhane Gebreselassie[3]

Institute of Geotechnics and Geohydraulics, University of Kassel, Germany
[1]geotech@uni-kassel.de; [2]sebastian.thomas@uni-kassel.de; [3]berhaneg@uni-kassel.de

ABSTRACT: Two phenomena characterize cyclic axial loaded piles: accumulation of plastic pile displacements and degradation of pile capacity with increasing load cycles. The reason for this complex pile-soil-interaction is still not well understood. Hence, load and displacement controlled small-scale model pile tests were performed to investigate pile and soil behaviour from a qualitative point of view. Particle Image Velocimetry (PIV) was used to visualize the grains movement near the pile. A volume change was generally observed in all the tests depending on the mode of cyclic load. Shear localization appears in a small shear zone parallel to the pile shaft and remains approximately constant with increasing load cycles. Next to the shear zone a shear band can be observed which expands outwards with increasing load cycles. Moreover, a different pile and soil behaviour was observed under cyclic load compared to the static load test.

INTRODUCTION

The phenomena of accumulation of plastic displacements and degradation of pile capacity with increasing load cycles is observed in many field and model tests, e.g. Kempfert/Lauffer (1991), Lehane et al. (2003) and LeKouby et al. (2004). It is known that the reduction in pile skin friction is a reason for the degradation of pile capacity. In this research work, small-scale model pile tests on bored piles were performed to identify the mechanisms of pile-soil interaction during cyclic axial loading based on Particle Image Velocimetry (PIV). The program VidPIV (ILA 2004) was employed to visualize grain movement near the pile.

PIV was developed originally for flow field measurement of gaseous and fluid media in the field of hydromechanics. In geotechnical engineering it has been successfully employed to assess deformation fields in granular soils (White/Bolton 2004, Hauser 2005).

The quality of the test results obtained by PIV analysis depends on the accuracy of the measurement. However, it must be noticed that exact measurements are not possible due to systematic and random errors as defined in DIN 1319-1:1995-01. These errors were studied systematically with a different testing program and the error is estimated to be about 0.03 mm. This seems sufficient enough for a qualitative analysis of the small-scale model test results.

EXPERIMENTAL PROGRAM

Testing setup and procedure

Small-scale model pile tests were performed using a 60 cm high, 40 cm long and 20 cm deep box. The front side of the box is made of 15 mm thick Plexiglas wall whereas all the other sides are made of steel sheets. A groove of 3 mm was cut into the Plexiglas to avoid intrusion of sand grains between the wall and the pile. A quadratic model steel pile (7.5 x 7.5 cm) with rough surface (surface injected with model sand) was used. A void was left below the pile tip in order to avoid the influence of base resistance. After inserting the pile, a dry silica model sand with grain size of 0.064 to 4.00 mm ($d_{50} = 0.5$ mm, $e_{min} = 0.48$, $e_{max} = 0.84$) was placed and compacted by tamping to reach a medium density ($I_D \approx 0.5$). A surcharge of 20 kPa was applied on the surface using a rigid wooden plate. The cyclic axial load had been applied using a hydraulic press at a loading frequency of 1 Hz. After every 10 load cycles, the cyclic loading was stopped and a picture had been shot using a digital camera with a resolution of 3456x2304 pixels. It is assumed that the observed grain movement around the pile (spatial) are the same as that on the front wall (plane condition).

Testing program

One static displacement-controlled tension test (PIV-01) was performed to evaluate the ultimate static pile resistance R_{ult} and the corresponding pile displacement in order to allow a comparison of grain movement with cyclic tests. Furthermore, one displacement-controlled test (PIV-02) was carried out with cyclic amplitude of s' = 0.25 mm and up to 100,000 load cycles. Finally, five load-controlled tests with cyclic load level of $F_{cycl} = 0.4$ kN (double amplitude) and different average load levels F_{avg} from -0.2 kN to 0.2 kN were conducted to identify pile and soil behaviour under different cyclic loading modes (Table 1).

Table 1. Testing program of the load-controlled tests

Test no.	Load mode	F_{avg} (kN)	F_{cycl} (kN)
PIV-03	symmetric two-way	0	0.40
PIV-04	unsymmetric two-way tension	0.1	0.40
PIV-05	unsymmetric two-way compression	-0.1	0.40
PIV-06	one-way tension	0.2	0.40
PIV-07	one-way compression	-0.2	0.40

TEST RESULTS

Soil behaviour in displacement-controlled test

Fig. 1a shows velocity vectors after 100 load cycles evaluated using PIV analysis for the test PIV-02. Different grain movement along the pile causes strain localization near the pile. These shear strains (Fig. 1b) are derived from the velocity field shown in Fig. 1a. Hence, they can be positive or negative depending on the direction of the velocity field in global coordinate system. It can be seen from Fig. 1b that there is a development of a thin shear zone near the contact area between pile and soil. Next to it, a shear band can be observed. There is a high concentration of shear strains in the shear zone and relative a small concentration in the shear band. Outside these two regions no shear strains are observed. The width of the shear zone remained constant at $d_{SZ} = 10\ d_{50}$ with increasing load cycles, whereas the width of the shear band increased with the number of cycles and reaches 100 d_{50}.

Segregation of grains was observed after 100,000 load cycles in the shear zone in the upper third part of the pile as indicated in Fig. 2a. This might be the result of particle breaking and migration of fine grains downwards. Such segregation of particles may be the reason for the different stress-strain relationship during cyclic loading. Even at low stress level in the small-scale tests, the breaking of particle seems possible, in particular by large number of load cycles. A similar phenomenon was observed from results of cyclic triaxial tests by Donohue et al. (2009). The influence of abrasion seems to be not relevant, since a washed rounded model sand was employed.

The net grain movements after every ten load cycles were evaluated and shown in Fig. 2b. The accumulation of the grain movement results in decreasing void ratio which in turn causes compaction of the soil.

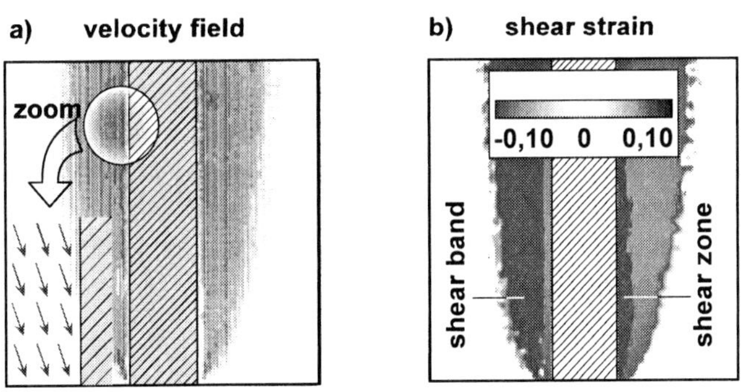

FIG. 1. a) Velocity vectors after 100 load cycles for test PIV-02, b) shear strains derived from velocity vectors

Soil behaviour in load-controlled tests

Similar soil behaviour was also observed in the load-controlled tests. The cyclic loading results in soil compaction near the pile in the tests PIV-03 to PIV-05 and PIV-07. A shear zone was developed during the first 10 load cycles and remained constant with further load cycles. The width of the shear zone was approximately 4 to 12 mm (Fig. 3b), and depends on the mode of cyclic loading. The width of the shear zone is smaller in one-way tests than in two-way tests. The shear band extends further with the number of load cycles but it also remains smaller in one-way test.

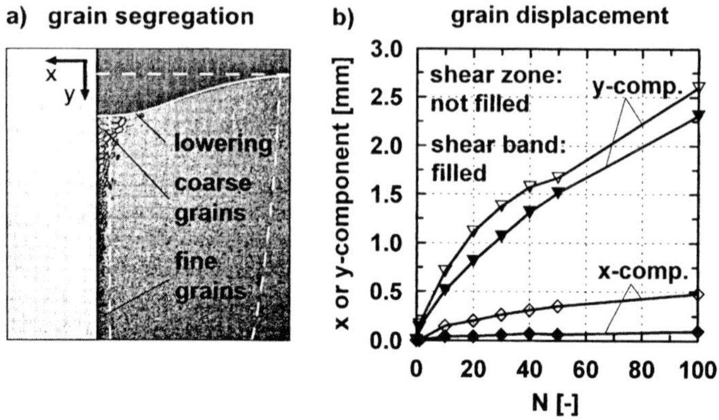

FIG. 2. a) Grain segregation in the shear zone, b) grain movement parallel or perpendicular to the pile shaft

Different soil behaviour was observed in the one-way tension test PIV-06. Like in the static tension test a dilatant soil behaviour was observed. No shear zone was developed near the pile. However, the width of the shear band is slightly larger than from the static test at pile failure. The shear strains registered are also larger as compared to the shear strains in static tests as well as two-way tests. Similarly, the width of the shear band is smaller than in two-way tests (Fig. 3b).

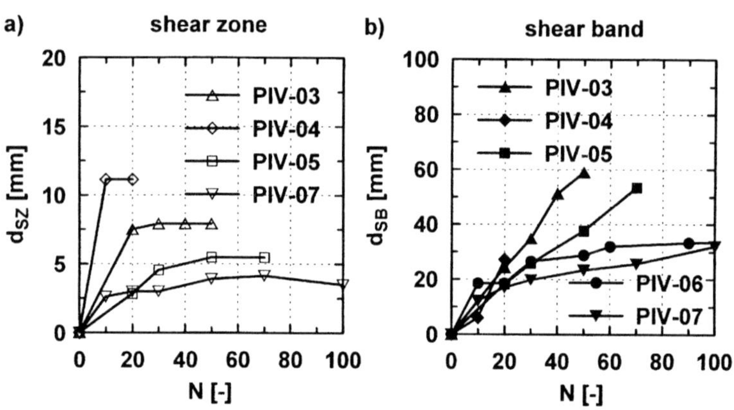

FIG. 3. Development of a) shear zone and b) shear band

Pile behaviour under cyclic loading

The ultimate pile resistance (in tension) in the static test PIV-01 was determined to be R_{ult} = 0.53 kN and the corresponding pile displacement is s = 1.5 mm. It is assumed that the pile failed under the cyclic loading if the pile displacement exceeds s = 1.5 mm. The number of load cycles N_f at failure depends on the mode of cyclic load as indicated in Table 2. All tests failed by pull out except test PIV-07 which is not failed after 1,000 load cycles. As it can be seen in Table 2, the two-way tests failed generally earlier than one-way tests.

Table 2. Pile failure at load cycle N_f in load-controlled tests

Test No.	Load mode	N_f
PIV-03	symmetric two-way	50
PIV-04	unsymmetric two-way tension	20
PIV-05	unsymmetric two-way compression	80
PIV-06	one-way tension	90
PIV-07	one-way compression	not failed after 1,000

There is an accumulation of plastic pile displacements in one-way tests (Fig. 4a). The test PIV-07 shows a cyclic stability and it was not failed after 1,000 load cycles, while test PIV-06 shows a cyclic failure after approximately 90 load cycles. The elastic pile displacements s_{el} are constant with increasing load cycles. Therefore, the cyclic pile stiffness ($k = F_{cycl}/s_{el}$) remains constant (Fig. 4b).

The rate of increase of plastic pile displacements s_{pl} is slower at the beginning in two-way tests than one-way tests. However, after only a few load cycles the plastic displacements are increased rapidly until pile failure. The elastic pile displacements were getting larger in these tests and hence a decrease in cyclic pile stiffness with increase of load cycles (Fig. 4b). Such pile behaviour was also observed from field tests by Karlsrud et al. (1986).

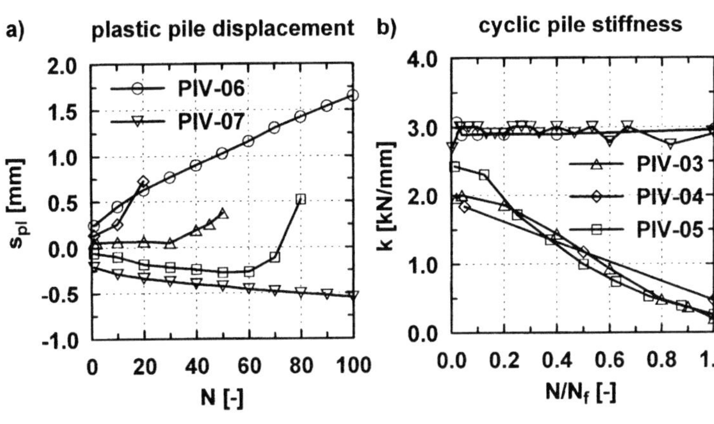

FIG. 4. Pile behaviour in one-way and two-way tests: a) plastic pile displacement and b) cyclic pile stiffness (Note: the legends are for both diagrams)

CONCLUSIONS

The paper presents some results of displacement and load controlled cyclic small-scale model pile tests in medium dense dry sand. The grain movement was visualized using PIV. It was observed that the pile and soil behaviour depends mainly on the mode of the cyclic load. Moreover, a compaction of soil near the pile mostly occurs during cyclic axial loading. Only during one-way tension test a dilatant soil behaviour was observed. Shear localization occurs in a small shear zone near the pile shaft, whose width remains constant with the number of load cycles. Next to the shear zone a wider shear band was observed, whose width grows further with increasing cyclic loading.

The investigation in this research work helps to better understand the different hypothesis in the literature regarding the pile-soil interaction behaviour under cyclic axial loading. Furthermore, the paper shows that there is no a single conceptual model for the pile behaviour under cyclic loading. The research project is not yet concluded and new findings will be published in the future.

ACKNOWLEDGMENTS

The authors would like to thank the German Research Society (DFG) for the financial support of this research.

REFERENCES

DIN 1319-1:1995-01. "Grundlagen der Messtechnik - Teil 1: Grundbegriffe." *Deutsches Institut für Normung.*

Donohue, S., O'Sullivan, C. and Long, M. (2009). "Particle breakage during cyclic triaxial loading of a carbonate sand." *Geotechnique*, Vol. 59 (5): 477-482.

Hauser, C. (2005). "Boden-Bauwerk-Interaktion bei parallel-wandigen Verbund-systemen." *Berichte des Lehr- und Forschungsgebietes Geotechnik, Bergische Universität Wuppertal*, Nr. 29, Dissertation.

ILA (2004). "Introduction to VidPIV - User Manual." *ILA - Intelligent Laser Applications GmbH*, Juelich.

Karlsrud, K., Nadim, F. and Haugen, T. (1986). "Piles in clay under cyclic axial loading - Field tests and computational modelling." *Proceedings of the 3rd International Conference on Numerical Methods in Offshore Piling*: 165-190.

Kempfert, H.-G. and Lauffer, J. (1991). "Pfahlprobebelastung in weniger tragfähigen Böden unter statischer und wechselnder Belastung." *Geotechnik*, Nr. 14: 105-112.

Lehane, B.M., Jardine, R.J. and McCabe, B.A. (2003). "Pile group tension cyclic loading: Field test programme." *Imperial College Consultants (ICON)*, Research Report Nr. 101.

LeKouby, A., Canou, J. and Dupla, J.C. (2004). "Behaviour of model piles subjected to cyclic axial loading." *Proceedings of the International Conference of Cyclic Behaviour of Soils and Liquefaction Phenomena*, Bochum: 159-166.

White, D.J. and Bolton, M.D. (2004). "Displacement and strain paths during plane-strain model pile installation in sand." *Geotechnique*, Vol. 54 (6): 375-397.

Analysis of Bridge Response Considering Water-soil-pile-structure Interaction under a Scoured Condition

Cheng Lin [1], Caroline R. Bennett [2], A.M.ASCE, Robert L. Parsons[3], M.ASCE, and Jie Han[4], M.ASCE

[1]Graduate Research Assistant, University of Kansas, Civil, Environmental, and Architectural Engineering (CEAE) Dept., Lawrence, KS 66045; cheng@ku.edu
[2]Assistant Professor, University of Kansas, CEAE Dept., Lawrence, KS 66045; crb@ku.edu
[3]Associate Professor, University of Kansas, CEAE Dept., Lawrence, KS 66045; rparsons@ku.edu
[4]Associate Professor, University of Kansas, CEAE Dept., Lawrence, KS 66045; jiehan@ku.edu

ABSTRACT: Scour removes soil and reduces the capacity of a pile foundation, and thus may cause instability of a bridge. An analysis of bridge response under scoured conditions requires consideration of water-soil-pile-structure interaction. However, limited research has been done to integrate the analysis of water-soil-pile-structure interaction for a bridge due to scour. In this study, an attempt was made to consider such integration. A pile group was treated as an equivalent single pile, and the nonlinear response between soil and the equivalent single pile was modeled using a *p-y* curve. The influence of water flow was considered by applying stream pressure to the bridge piers and the pile foundation at different scour depths. Interaction between the piles and superstructure was analyzed by performing iterative calculations of their respective stiffness matrices until convergence was obtained. Kansas Bridge 45 was selected as a case study for evaluation of a typical scour-susceptible bridge under scoured conditions. The results show that scour significantly increased lateral deflection of the pile head. Considering pile cap had positive effects on the decrease of the lateral deflection of the pile groups. The investigation was also made into the influence of boundary conditions of bridge abutment on the lateral bridge responses.

INTRODUCTION

Scour is removal or erosion of streambed materials around bridge foundations due to flowing water, usually categorized into long-term bed degradation, contraction, and local scour. Contraction scour occurs in a contracted part of the river caused by the presence of bridges or revetments. Local scour is attributed to the acceleration of water around the pier or abutment. Rate of scour varies in different soils due to different soil resistance to the erosion, exhibiting the highest in sand, the second in clay, and the lowest in rock. Much research has been done on the prediction of scour depths in different soils (for example, Rachardson and Davis, 1995; Briaud et al., 1999). When the maximum scour depth is predicted, the scour effects can be considered in the bridge design. Many countermeasures were developed to protect bridges from the damage of scour, *i.e.* riprap, partially grouted riprap, articulating concrete block systems, gabion mattresses, grout-filled mattresses, and geotextile sand containers (Lagasse et al. 2007).

Based on technical advisories and guidance from the Federal Highway Administration (FHWA), new bridges should be designed to resist potential damage from scour events, while regular inspection and evaluation of existing bridges is required to determine whether an extreme event such as flood can result in damage (Lagasse et al. 2007). However, evaluation of bridge performance under scour is a complicated process because it depends on interaction between water, soil, piles, and the bridge superstructure which currently was not fully understood. Scour erodes away soils in the vicinity of pile foundations, thus changing the soil conditions and increasing the unsupported pile length. Such changes in pile foundations affect behavior of the bridge superstructure. Therefore, scour may significantly alter the consequences of force effects on structures (AASHTO 2007).

During a flood, lateral response of a bridge is one of the most significant issues for consideration. This study attempted to investigate the lateral response of a bridge considering water-soil-pile-superstructure interaction under scour. Water pressure due to a 100-year flood was estimated and analytically applied to a bridge. Soil-pile foundation interaction was described using the *p-y* curve method. The superstructure of the bridge was analyzed under lateral transient loads. Integration of the soil-pile foundation analysis and the superstructure analysis was then carried out at various scoured depths. Bridge 45 in Kansas in the United States was studied as an example to investigate the effects of scour on the bridge response.

SOIL AND PILE INTERACTION

The pile foundation in a bridge is often composed of a group of piles. A group effect exists in pile groups having close spacing (less than 6 times of pile diameter) between individual piles due to the overlap of soil resistance. Using a *p*-multiplier to reduce the lateral soil resistance accounting for this group effect, Mokwa and Duncan (2000) proposed the Group Equivalent Pile (GEP) method to analyze the behavior of a laterally loaded pile group.

In the Group Equivalent Pile (GEP) approach (Mokwa and Duncan 2000), a pile group may be modeled as an equivalent single pile having the same properties as each individual pile except for the moment of inertia. The moment of inertia is taken as being equal to the sum of moments of inertia of all the individual piles. This equivalent single pile is then loaded with the total actual load. The group effect is considered by modifying the *p-y* curve of each single pile by a *p*-multiplier to obtain the GEP *p-y* curve as expressed in Eq. (1):

$$p = \sum_{i=1}^{N} p_i f_{mi} \qquad (1)$$

where p is the soil resistance to piles; p_i is the p-value for the ith single pile; f_{mi} is the p-

multiplier for the ith single pile, depending on the position of the pile, the pile spacing, the soil type, and the installation method, as shown in Figure 1; and N is the number of piles in the group.

When an embedded pile cap is included, the cap can be modeled by enlarging the top portion of the GEP based on the actual dimensions of the cap. The *p-y* curve of the embedded cap can be estimated using the hyperbolic load-deflection relationship. The details of the *p-y* curve for the pile cap were reported by Mokwa and Duncan (2000) and Lin et al. (2009).

The LPILE 5.0 Plus software was used with the GEP method in this study to analyze the pile group although two other software packages, GROUP and FB-PIER, are also available for analysis of pile groups. As Mokwa and Duncan (2000) pointed out, GROUP and FB-PIER have limitations when an externally-generated pile cap *p-y* curve is used.

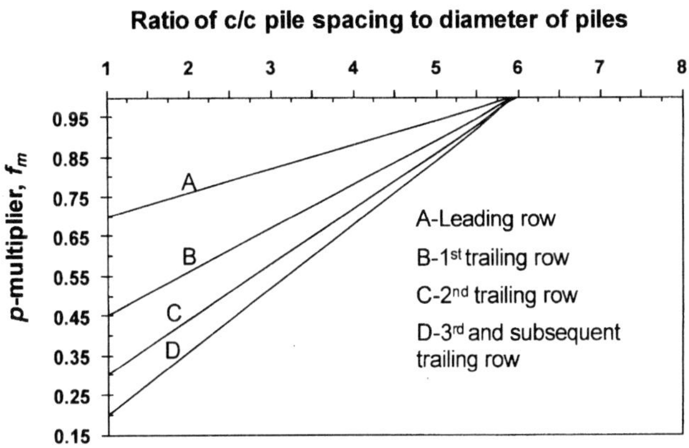

FIG. 1. Design chart for the p-multiplier (Mokwa and Duncan 2000)

INTEGRATION OF SUBSTRUCTURE AND SUPERSTRUCTURE

Superstructure in this study refers to the structure above the pile foundations, including bridge decks, girders, and piers. Substructure is limited to pile foundations. The analysis of the bridge superstructure was first conducted at a fixed boundary condition between the bridge pier and the pile foundation using software STAAD Pro. 2007. The boundary condition was set with fixity because the stiffness provided by pile foundation is relatively big; in addition, this initial support condition could be set easily and consistently for different integrations at various scour depths instead of assuming specific stiffness values. A study was conducted based on these two initial support conditions for the bridge pier (i.e., fixity and assumed values of stiffness) and resulted in similar final results when convergence was achieved. Reactions at the pier bases were obtained and then used as loads applied to the pile foundation for analysis. The responses of the pile head under the applied loads were calculated. Stiffness of the pile head was computed by dividing the loads by the corresponding displacements (i.e., spring supports to the superstructure were assumed). When the base supports of the superstructure (the stiffness from the pile analysis) were updated, new reactions from the superstructure were calculated, which were subsequently assigned to the pile foundation as new loads. Based on the above procedure, iterations were performed between the superstructure and the substructure until the difference between displacements at the base of the superstructure (i.e., the bottom of the pier) and at the top of the substructure (i.e., the pile head) was insignificant (i.e., less than 5% in this study), or the change of two successive stiffness values during the iteration was considerably small (<5%).

CASE STUDY

Bridge 45 is situated in Jewell County, Kansas in the United States and carries State Highway K14 over a local creek. The five-span bridge was constructed in 1956 and has a total length of 113 m. Four W33x141 girders are used to support the bridge deck as shown in Figure 2. Bridge 45 has eight circular tapered piers with diameters of 0.93 m at the top and 1.39 m at the bottom, each supported by a group of eight HP10x42 piles with spaced at distances three times the width of a single pile as shown in Figure 3.

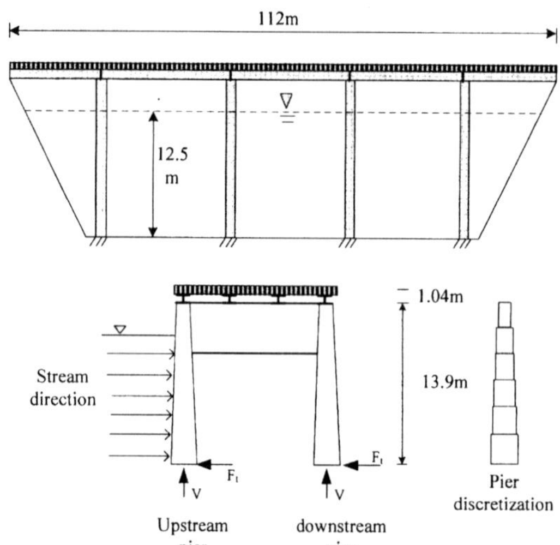

FIG. 2. Bridge K45 superstructure

This study considered two lateral loads applied on the bridge: wind loads and stream loads. The self-weight of the superstructure was assigned as a vertical load. All load factors within the load combinations used in this study were taken as 1.0. The stream loads with debris at 100-year flood design were calculated and then applied to the superstructure. The elevation of the 100-year flood design based on the bridge plan was 12.5 m above the base of piers, as depicted in Figure 2. The debris load and the wind load were calculated according to AASHTO (2007), and only applied to piers in the upstream row due to a relative small distance of two piers in the transverse direction. Therefore the responses of the piers in the upstream and downstream rows were different. The tapered piers were divided into multiple sections as shown in Figure 2 to simplify the structural analysis.

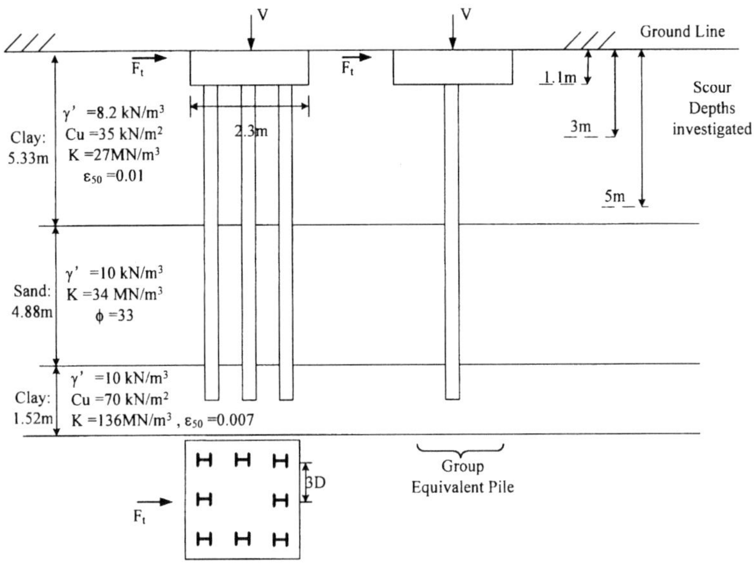

FIG. 3. Cross section of the pile foundation and the scour depths investigated

The basic information of the pile foundation and the scour depths analyzed are presented in Figure 3. The pile length is 9.8m and the square pile cap is 2.3m wide and 1.1m thick. Each pile is embedded 0.3m into the pile cap. The top of the pile cap is at the same elevation as the ground line for all the piers for simplicity. Three scour depths were assumed for the purpose of investigating the lateral responses of the bridge structure, while the vertical responses were ignored because axial soil resistance is not able to be considered in LPILE 5.0 Plus. Three soil layers were encountered in this site. Properties of the soils are summarized in Figure 3. Based on the GEP method, the modified p-value was calculated to be 5.5 times the p-value of a single pile using the plot shown in Figure 1 and Eq. (1). The stream loads were also taken into account for the pile foundation when scour proceeded below the elevation of the top of the pile cap.

The pile foundation and the superstructure were assumed to be rigidly connected during the iterative analyses. Consequently, the rotation was ignored and the rotational stiffness was not examined during the iterations. Vertical displacement was assumed as zero, and only lateral stiffness was computed during the iterations. However, the vertical load was considered during the calculations because the vertical load was expected to influence the lateral response of the piles.

RESULTS AND DISCUSSION

The response of the bridge under scour was evaluated after the integration. However, only the responses of the pile head and the base of the bridge pier are presented and discussed here.

Mokwa and Duncan (2000) pointed out that the pile cap resistance to a lateral load influenced the behavior of the pile foundation; however, very few researchers have considered this effect. This study investigated the response of the bridge with and without a pile cap.

Figure 4 presents the deflection of the pile head under the upstream piers at various scour depths when the boundary condition of the bridge end abutments is fixed. The results show that deflection of the pile head increased with the scour depth. Numerical results that considered the pile cap effect obtained less deflection of the pile head as compared with those that ignored the pile cap effect. This comparison shows that the pile cap could provide significant lateral resistance, increasing the lateral-load carrying capacity of the pile foundation.

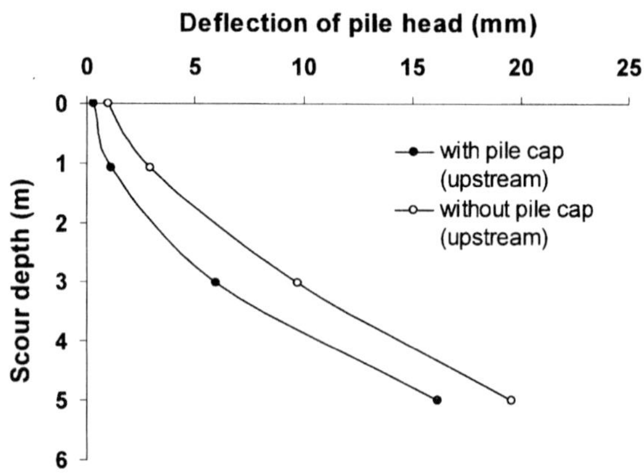

FIG. 4. Influence of pile cap on deflection of the pile foundation under scour

The influence of the bridge end abutment boundary conditions was also investigated in this study. Two types of abutments with extreme conditions (denoted as rollers and fixity) were considered. The roller model had a single-point constraint and was free in translation and rotation. The fixity model contained rigid supports with constraints in every direction. Figure 5 shows that the deflection of the pile cap increased exponentially with increased scour depth. If the end abutments were fixed, the pile cap lateral deflection was much smaller than that if the abutments were supported by rollers. The actual boundary conditions of the bridge should lie between these two bounding scenarios. Figure 5 also shows that the upstream piers had a slightly higher pile cap deflection than the downstream piers. This difference was caused by the fact that more lateral loads were applied onto the upstream piers than the downstream piers (i.e., the wind loads and the debris loads were only applied onto the upstream piers).

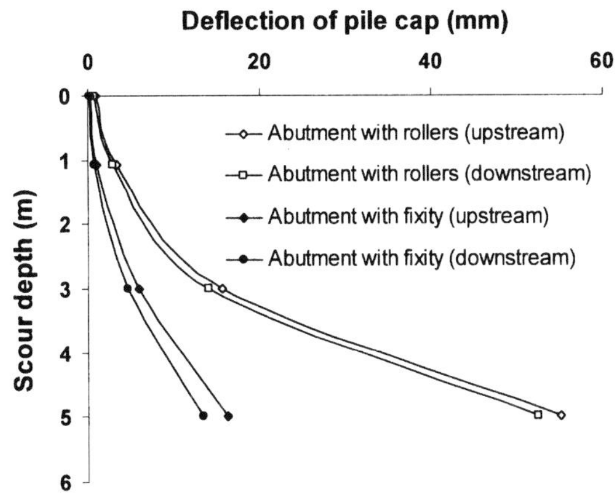

FIG. 5. Deflections of the pile cap at different abutment boundary conditions

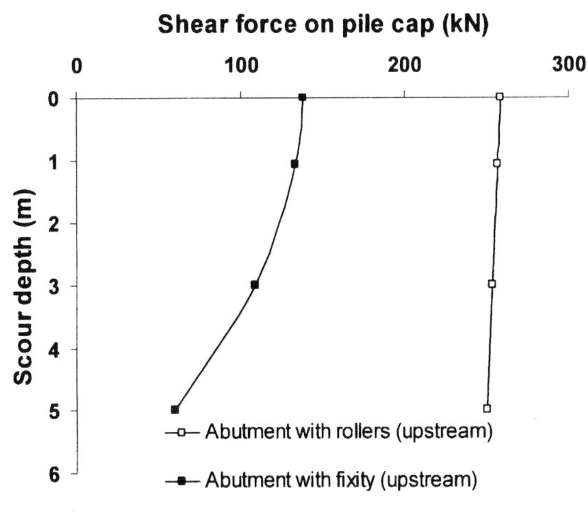

FIG. 6. Shear forces on the pile cap at different abutment boundary conditions

Figure 6 shows that for the abutments with roller boundaries, the pile cap carried higher shear forces than that for the abutments with fixed boundaries. This result is because at the same external forces on the bridge, the roller boundaries of the abutment provided less support to the bridge thus requiring the pile foundations of the interior piers to provide more shear force to balance the external loads as compared with the fixity boundaries of the abutments.

CONCLUSIONS

Analysis of an entire bridge structure under scour conditions was performed in this study. Integration of the substructure and the superstructure was implemented. Rotation of the pile head was neglected by implementing a rigid connection between the pile foundation and the bridge pier. Lateral stiffness was calculated to integrate the substructure and superstructure lateral behavior. This study showed that scour significantly increased lateral deflection of the pile head. It was found that the presence of a pile cap provided significant lateral resistance for the pile foundation; hence lateral deflections were reduced as compared with the case without a pile cap. The case in which end abutments were assigned fixed boundaries resulted in less deflection of the pile cap than the case in which end abutments were assigned roller boundaries. The response of an actual bridge should be between these two conditions.

ACKNOWLEDGMENTS

This research project is funded by the Kansas Department of Transportation (KDOT) through the KTRANS program. Mr. John Jones at KDOT is the project monitor. The authors would like to express their appreciation to KDOT for this support. The results and opinions presented in this paper are the authors' and do not reflect the policy and recommendation of KDOT.

REFERENCES

AASHTO (2007). *AASHTO LRFD Bridge Design Specifications*. American Association of State Highway and Transportation Officials, Washington, DC.

Briaud, J.L., Ting, F. C. K., Chen, H. C., Gudavalli, R., Perugu, S., and Wei, G. (1999). "SRICOS: Prediction of scour rate in cohesive soils at bridge piers." *Journal of Geotechnical and Geoenvironmental Engineering*, 125(4), 237–246.

Lagasse, P. F., Clopper, P. F., Zevenbergen, L. W., and Girard, L. G. (2007). "Countermeasures to Protect Bridge Piers from Scour." *NCHRP Report No. 593,* Transportation Research Board, National Research Council, Washington, D.C.

Lin, C., Bennett, C.R., Parsons, R.L., and Han, J. (2009). "Evaluation of The Group Equivalent Pile Approach for laterally loaded Shafts." *Proc. of the 34th Annual Conference on Deep Foundations*, Kansas City, MO.

Mokwa, R. L. and Duncan, J. M. (2000). "Investigation of the resistance of pile caps and integral abutments to lateral loading." *Research Report Submitted to the Virginia Transportation Research Council, Report No. VTRC 00-CR*, Charlottesville, VA.

Richardson, E. V., and Davis, S. R. (1995). ''Evaluating scour at bridges.'' *Report No. FHWA-IP-90-017 (HEC 18)*, Federal Highway Ad-ministration, Washington, D.C.

Nonlinear Dynamic Soil–Micropile-Structure Interactions: Centrifuge Tests and FEM Analyses

Ahmad Mahboubi[1] and Ali Nazari-Mehr[2]

[1] Associate professor, Power and Water University of Technology, P.O. Box 16765-1719, Tehran, Iran; mahboubi@pwut.ac:ir
[2] M.Sc., Power and Water University of Technology, P.O. Box 16765-1719, Tehran, Iran; nazari_mehr@yahoo.com

ABSTRACT: Micropiles are grouted and small diameter piles that are often used in foundation retrofit. Experimental evidences have indicated that micropiles behave well under seismic loading due to their high flexibility. Moreover, observations in the 1995 Kobe Earthquake indicate a good performance of friction piles under seismic loading. However, the seismic behavior of micropiles is not fully understood due to the limited number of full and model-scale tests, as well as the limited amount of numerical modeling studies for micropiles. This paper focuses on Finite Element modeling of single and micropile groups under both static and dynamic loading. The micropile/soil interface was modeled either with perfect bonding or with frictional interface elements. The lateral vertical boundaries of the soil were set up in such a way that the reflection of the arrival waves at the boundaries was avoided. The problem includes soil non-linearity, gapping and slippage between the micropile and the soil, complex boundary conditions and possible pile non-linearity. The structure is described by a single degree of freedom system composed of a concentrated mass and a column. The results of this study show the influence of micropiles inclination on dynamic amplification and on the seismic-induced internal forces in micropiles.

INTRODUCTION

Micropiles are drilled and grouted small diameter replacement piles that are commonly used in foundation retrofit. Micropiles are reinforced and typically have diameters less than 300 mm. In most applications, micropiles behave as friction (e.g. floating) flexible piles. The advantages of using micropile systems include (a) their high flexibility during seismic conditions, (b) micropiles can be installed in low overhead clearance (less than 3.5 m), in all types of soils and ground conditions, (c) minimal disturbance is caused during construction, (d) inclined micropiles can be easily constructed, (e) they are able to resist axial and lateral loads, (f) only small volumes of earth to be excavated due to small diameter, (g) little disturbance is caused during drilling through an existing structure due to their small diameters, and (h) they can be drilled with boring machines that do not cause much noise. Despite the increased use of micropiles, the seismic behavior of a single micropile and a micropile group is not fully understood due to the limited number of full- and model-scale tests, as well as the limited amount of numerical modeling studies of micropiles.

The Finite Element (FE) method provides a tool to understand the seismic behavior of micropiles. FE analyses can be used to systematically alter the parameters that affect the seismic response of micropiles. However, the dynamic analysis of soil micropile-structure interaction is a very complex problem. The problem includes soil non-linearity (e.g. variation of soil shear modulus and damping with strains), gapping and slippage between the micropile and the soil, complex boundary conditions (especially at the vertical lateral boundaries), and possible pile non-linearity.

The response of a structure subjected to seismic or dynamic loading primarily depends on the characteristic of the site response, the external loading, the mechanical properties of the surrounding soils, and the structure itself. An extensive review of the current literature on site response and soil-structure interaction (SSI) was done by Wong (2004).

CENTRIFUGE MODEL

Benslimane (2000) performed a series of centrifuge tests on single micropile, micropile groups, and micropile network. Various micropile configurations, inclinations, number of micropiles, and loading levels were conducted.

INSTRUMENTATION AND DATA ACQUISITION

Figure 1 shows the instrumentation used. Linear variable differential transformers (LVDTs) were placed at different locations: lv1 and lv2 were placed to record any lateral soil profile displacement (not expected), lv3 was placed for the pile cap settlements, lv4 was placed for the lateral pile cap displacements, and lv5 was placed to record surface settlements. Accelerometers were used at different locations for acceleration–time history measurements. The labeled input accelerometer is used to monitor the input base motion: acc1 to acc5 record the wave propagation along the soil profile and the free field accelerations, and acc6 measures the pile head accelerations to characterize the structural response of the soil-micropile system. Pairs of half bridge circuited strain gauges were installed on the surface along the model piles to monitor bending (sg2, sg4, and sg6) and axial (sg1, sg3, and sg5) strains. (Juran *et al.*, 2001)

SOIL PROPERTIES

Dry Nevada Sand 120 was used at a relative density of 57 percent. Tests conducted on the sand yielded maximum and minimum void ratios of 0.51 and 0.88, specific gravity of 2.67, an average particle size (D_{50}) of 0.13 mm, and a coefficient of uniformity (D_{60}/D_{10}) of 1.6. The grain size distribution curve is shown on figure 2. Maximum value of internal friction angle (ϕ) for this soil was founded 35° for relative density of 57% and 33° and 36° for 40% and 60% of relative density, respectively. (Juran *et al.*, 2001)

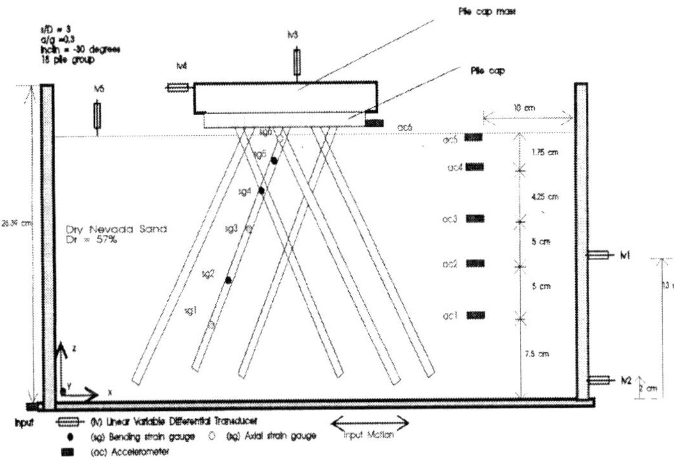

FIG. 1. RPI laminar test box and model setup (Juran *et al.*, 2001)

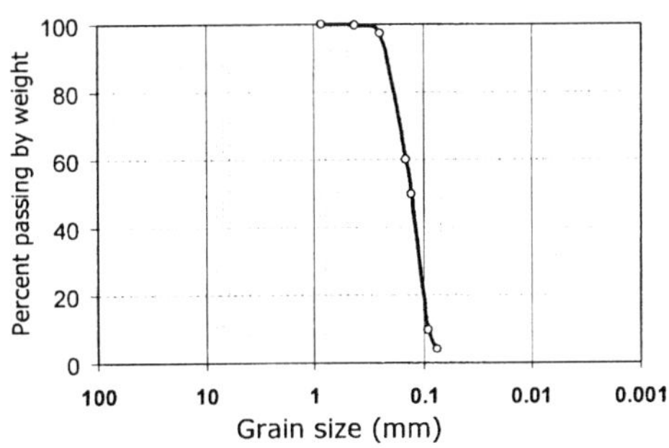

FIG. 2. Grain size distribution of Nevada sand used in centrifuge tests

MODEL PREPARATION

The procedure for model preparation was as follows. First, a latex membrane 0.02 cm thick was used to line the inside of the laminar box, prevent leakage of the contents, and reduce the side friction. During model construction, the external side of the container was sealed and connected to a vacuum pump to remove air from the outer chamber. Dry sand with relative density of 57 percent was poured in layers of 50-mm thickness or less. The amount of sand needed for each layer to achieve required density was weighed. At appropriate stages, transducers (accelerometers and LVDTs) were placed in the soil. (Juran *et al.*, 2001)

Characteristics of the different models in centrifuge tests are shown in table 1. Centrifuge tests are done with a scale of 20, which means that an acceleration of 20g is applied in vertical direction. For investigation about the effect of superstructure on the behavior of the micropile, three different models were adopted. The first one is a micropile with pile cap and without surcharge, the second model is micropile + pile cap + 50% of failure load and the third one is similar to second one with this difference that 90% of the failure load is applied. Horizontal vibration of 100 cycle of uniform sinusoidal acceleration with a frequency of 40 Hz which is equivalent of 2 Hz in real scale is applied to the model. The box walls composed of the laminar flexible plates to absorb the waves and to stimulate the semi-infinite condition of the model.

Table 1. Models characteristics

Model No.	Micropile config.	Micropile Diameter (cm)	L/D	Inclination	S/D	Weight of pile cap (kN)	Failure load (kN)	Dimensions of pile cap and the structure over it (cm)					
								Without structure		Cap + 50% failure load		Cap + 90% failure load	
								cg	L×W×H	cg	L×W×H	cg	L×W×H
1	Single	13	32	Vertical	--	4	32	38	64×64×36	--	H=30, D=100	65	64×64×36
2	Group (2×1)	13	32	Vertical	3	10	56	39	96×96×38	62.5	96×96×26	78	100×96×56
					5	16	200	39	128×128×38	72.5	100×200×40	95	100×200×84
3	Group (2×1)	19	22	Vertical	3	16	64	39	96×96×26	60.3	100×96×56	65	100×96×56
				30°		8	152	39	100×100×38	--	103×103×58	--	--

(cg = distance between load point to soil surface)

FINITE ELEMENT MODELING

In this section we are trying to simulate the centrifugal test by the FE Method. ABAQUS is used for the simulation of the centrifuge tests. The soil layer thickness considered as 5 meters and the micropiles had a length of 4 meters. The piles were considered to be of concrete material with elastic behavior. The soil is considered elasto-plastic. Elasticity is considered as linear elasticity and the plastic behavior was assumed to be of Drucker-Prager. Rayleigh damping which includes mass and stiffness proportional damping was used to account for the wave energy attenuation through the soil medium. Table 2 summarizes the parameters used for different parts of the models.

Table 2. Material parameters used for model parts in all analyses.

Model part	Material type	Material property
Soil	Sand	E_s=150 MPa, ρ_s=1700 kg/m^3, ϕ=35°, ν_s=0.45, k_0=0.42
Micropiles	Concrete	D=13 and 19 cm, ρ_p=2500, Ep= 210 MPa, ν_p=0.20, Lp= 4 m

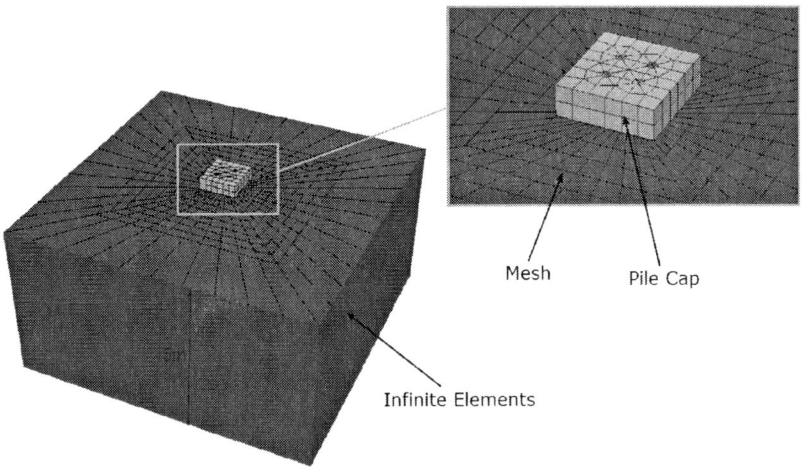

FIG. 3. Example of mesh generation of a FE model, micropile group 2×2

BOUNDARY CONDITIONS

The procedure of introducing the boundary conditions for every model required two types of different boundaries which were silent boundaries for the sides of the models and quiet boundaries applicable for the bedrock. Since the boundaries at sides of every model had to be transmitting, infinite elements were used at those boundaries to transfer the seismic wave energy completely and to avoid "Box effect" which could lead to erroneous results. On the other hand, for the bedrock boundary simulation, Kelvin elements were used as a portion of the seismic wave energy had to be reflected. The dashpots and springs were arranged in three perpendicular directions. The Kelvin elements constants were considered to be independent of the frequency of excitation since this dependence within the range of the frequency of interest was low. Dynamic sinusoidal load with uniform acceleration and frequency of 2 Hz is applied at the bottom of the model.

COMPARISON OF NUMERICAL SIMULATION AND CENTRIFUGE TESTS

Figures 4 and 5 show the comparison between micropile displacements and flexural moments found by numerical simulations and the results of centrifuge test. The figures show a good concordance of the simulated results and the experimental results.

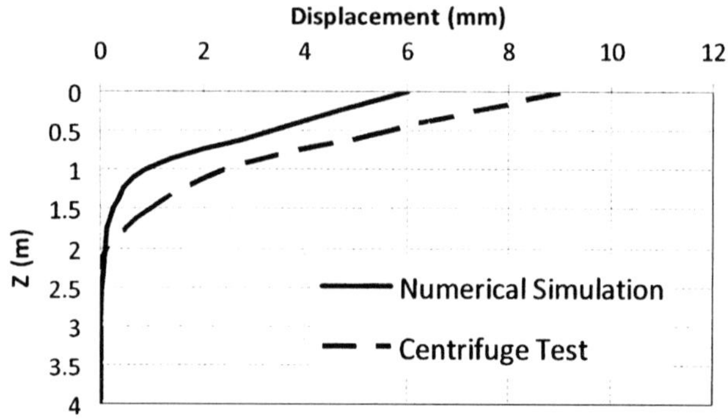

FIG. 4. Comparison of the displacements results from the centrifuge test and numerical simulation (model 1, a/g=0.03, 90% failure load)

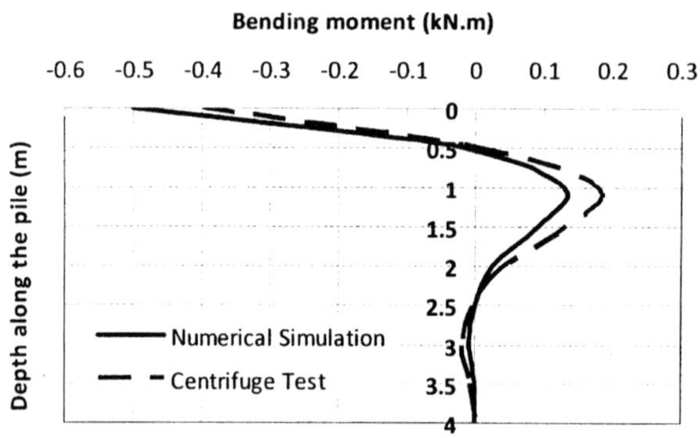

FIG. 5. Comparison of the bending moments results from the centrifuge test and numerical simulation (model 1, a/g=0.03, 90% failure load)

PARAMETRIC STUDY

In this section we are going to investigate the effect of different parameters on the dynamic behavior of the micropiles. The parameters whose effects will be investigated are: slenderness ratio (ratio of length to diameter, L/D), ratio of micropile spacing to micropile diameter (s/D) in a pile group and micropile inclination.

EFFECT OF THE RATIO OF LENGTH TO DIAMETER (L/D)

The effect of slenderness ratio (L/D) is shown on figure 6, it shows the evolution of flexural moments and pile displacements (model 1 and 2) and pile group of 2×1 for two values of 22 and 32 of slenderness ratio (L/D). The ratio of spacing to diameter (s/D) of the micropile group is 3 and it is subjected to a load on the superstructure equal to 90% of failure load with a vibration with amplitude of 0.3g.

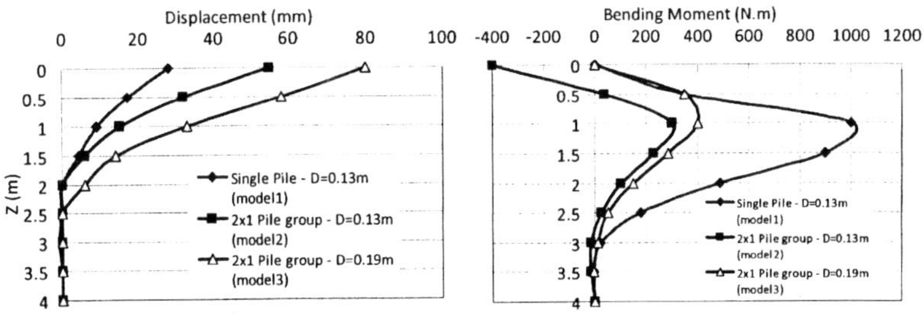

FIG. 6. Effect of (L/D) on displacement and bending moment distribution along the micropile (a/g=0.03, s/D=3, with 90% of failure load)

EFFECT OF MICROPILE SPACING TO DIAMETER IN A PILE GROUP

As the previous investigation, the effect of the ratio of micropile spacing to its diameter (s/D) on dynamic behavior is studied. The results are shown on figure 7. It shows the evolution of flexural moments and pile displacement. A group of 2×1 with ratios of pile spacing to diameter (s/D) of 3 and 5 are shown. All simulations are done with the vibration of 0.3g and under the external load of 90% of failure load. The obtained results show:

- The evolution of flexural moments and micropile displacement for the single pile and group of 2×1 with s/D=5, under the same loadings, are similar. It means that due to large spacing of the micropiles the interaction between piles is omitted.
- Flexural moments and micropile displacements are smaller for the group of 2×1 with s/D=3 than the single pile and the group of 2×1 with s/D=5.

EFFECT OF PILE INCLINATION

Using the model no. 3 a parametric study for investigating the effect of pile is done. The different inclinations 0°, 10° and 30° are examined. Figure 8 shows the effect of pile inclination on displacements and the bending moment distribution. It can be concluded that in the case of the inclined micropiles the applied load is rather transmitted by the axial loads. Figure 9 shows the comparison of the experimental and simulation results for the model no. 3 (group 2×1, α=30°) with 50% of failure load and an acceleration of 0.3g. The comparison shows a good concordance between the numerical results and experimental results.

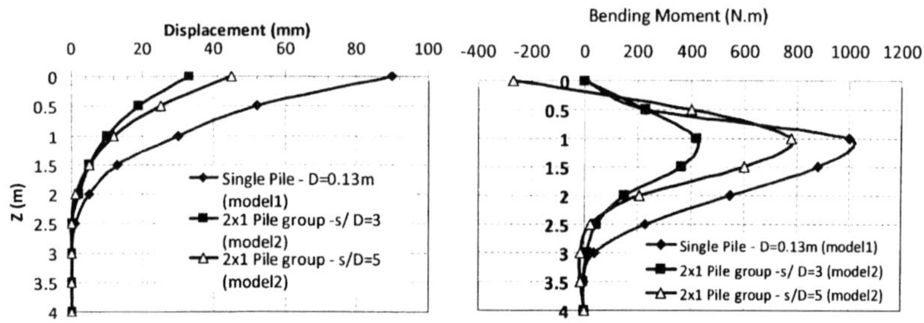

FIG. 7. Effect of (s/D) on displacement and bending moment distribution along the micropile (a/g=0.03, D=0.13 m, with 90% of failure load)

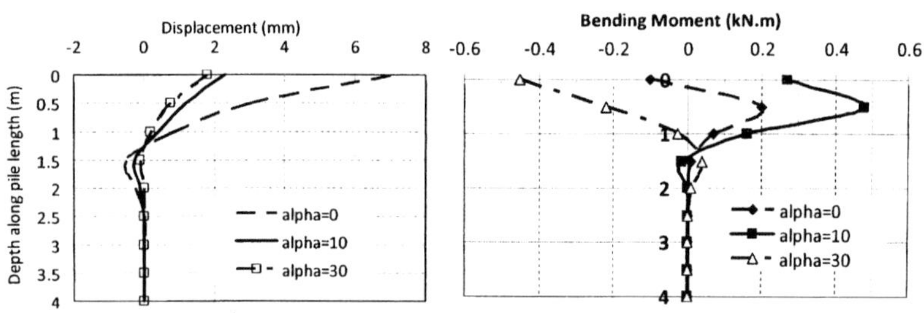

FIG. 8. Effect of pile inclination on displacement and bending moment distribution along the micropile (micropile group 2×1, a/g=0.3, a/g=0.03, with different values of inclination)

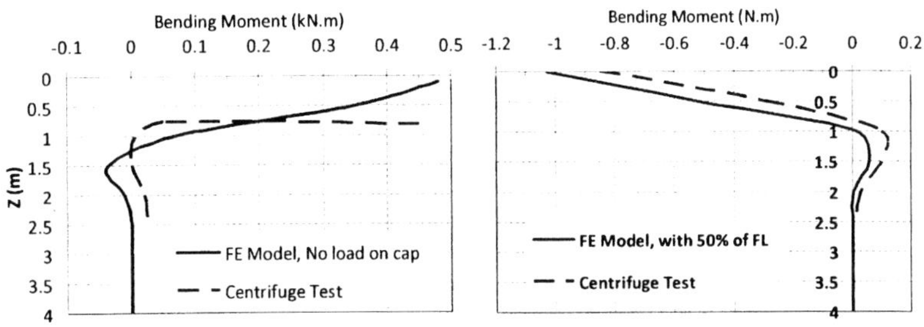

FIG. 9. Effect of micropile inclination on bending moment distribution along the micropile (group of 2×1, a/g=0.3, s/D=3, α=30°, with 0 and 50% of failure load)

CONCLUSIONS

From the numerical simulations we can deduct the following points:

• Soil-micropile systems show a flexible behavior. For the applied sinusoidal excitation, because of the low flexural stiffness of micropile, the displacements of the micropile, unless for the higher depths, are similar to displacements of the free boundaries. This causes that in the case of an earthquake, micropile exerts lower stress to soil.

• Experimental and numerical results show that the inclination of the micropile can cause:

a) Reducing the pile cap displacement
b) Increasing the axial forces
c) Increasing the flexural moment in the point of junction to pile cap. For the same reason it is recommended to use the flexible connections between the micropile and the pile cap.

REFERENCES

ABAQUS, Dassault Systèmes, 2006

Benslimane, A. (2000), "Dynamic behavior of micropile systems Subjected to Sinusoidal Ground Motions, Centrifugal Model Studies", Thesis submitted in partial fulfillment of requirements for the degree of Doctor of philosophy at The Polytechnic University.

Juran, I., Benslimane, A., and Hanna, S. (2001). "Engineering analysis of the dynamic behavior of micropile systems" Transportation Research Record, 1772, Paper No. 01-2936, pp. 91-106.

Wong, J. C. (2004). "Seismic behavior of micropiles", M.Sc. dissertation, Washington State University, Pullman, WA.

The Effect of Soil-Pile Separation on the Transfer and Impedance Functions

Kamran Panaghi[1] and Ahmad Mahboubi[2]

[1]MSc student, Power&Water University of Technology, P.O. Box 16765-1719, Tehran, Iran; kmn.pnghi@gmail.com

[2]Associate professor, Power&Water University of Technology, P.O. Box 16765-1719, Tehran, Iran; mahboubi@pwut.ac.ir

ABSTRACT: This study presents the results of modeling the soil-pile interaction phenomenon under acceleration-time history seismic loading. The studies are carried out for three sets of single piles with different diameters which consider the effect of gapping between soil and pile in more details. The Drucker-Prager soil constitutive law is used to model soil behavior while the behavior of the concrete piles is assumed to be elastic. Analyses were conducted by the commercial program ABAQUS 6.8 and the results were converted to the corresponding frequency magnitudes that are of more interest for the frequency dependent transfer and impedance functions regarding the kinematic and inertial interactions, respectively. The results showed that the effect of gapping on the kinematic interaction is to increase the displacement of the pile head compared to free-field. In addition, gapping causes reduction in the stiffness and damping of the system in general.

INTRODUCTION

Recent developments in the analysis of soil-structure interaction phenomenon make it necessary to study this issue with more rigors. In the past, the input ground motion was directly imposed to the base of the structure and it was assumed to be in favor of safety demands in building codes. This was due to the assumption that the interaction between soil and the deep foundation causes the seismic waves to attenuate as there is damping between the two body surfaces and therefore compared to a corresponding fixed base structure this idea seemed to be conservative. However some researchers believe that depending on the site and seismic excitation, the kinematic soil-pile interaction may amplify the lateral loading that applies to the structure (Gazetas and Kavvadas 1993). Maheshwari *et al.* (2004) studied the effects of nonlinearity of soil and separation at the soil - pile interface on the dynamic behavior of a single pile and pile groups. For the loading on a pile cap, they concluded that the nonlinearity increases the response for most of the frequencies of excitations while decreasing the dynamic stiffness of the soil-pile system. Maheshwari *et al.* (2004) mentioned that the effects of nonlinearity on the responses are dependent on the frequency of excitation with nonlinearity causing an increase in response at low frequencies of excitation. Studies on how soil nonlinearity, geometry of piles, gapping and loading intensity affect the magnitude of transfer and impedance functions have resulted in vast data resources available for researchers. Since gapping is one of the major parameters in analyses and its impact on the interaction of soil-structure whilst the soil behavior is plastic is still not much known, this work is an attempt to deal with the effect of soil-pile separation on the kinematic and inertial interaction more elaborately.

FINITE ELEMENT MODELING

Three different free-head single piles with diameters of 0.4, 0.5 and 0.64 m were used in analyses. The piles had a length of 8m and therefore the slenderness ratios of the models were 20, 16 and 12.5, respectively. The piles were considered to be of concrete material with elastic behavior. These free head piles were first used for the computation of transfer functions in the kinematic interaction study and then used for the inertial interaction analyses by introducing a structural mass of 50 tons for every pile. The soil mass model was considered to have side dimensions of 20 m and formed a cubical shape. The soil behavioral law was assumed to be of Drucker-Prager with the triaxial extension strength to the triaxial compression strength ratio of 0.8. The soil parameters simulated dense sand which due to its nature, the effects of pore pressure initiation and dissipation could be disregarded. Rayleigh damping which includes mass and stiffness proportional damping was used to account for the wave energy attenuation through the soil medium. Table 1 summarizes the parameters used for different parts of the models.

Table 1. Material parameters used for model parts in all analyses.

Model part	Material type	Material property
Soil	Sand	$E = 80 \times 10^6\ N/m^3, \nu = 0.35, \rho = 1900\ kg/m^3, \varphi = 35^\circ, \psi = 5^\circ.$
Piles	Concrete	$E = 2.1 \times 10^{10}\ N/m^3, \nu = 0.2, \rho = 2400\ kg/m^3$
Structure	Fictional	$\rho = 50000\ kg/m^3$

BOUNDARY CONDITIONS

The procedure of introducing the boundary conditions for every model required two types of different boundaries which were silent (completely transmitting) boundaries for the sides of the models and quiet (partially transmitting) boundaries applicable to the bedrock. Infinite elements were used at the sides of every model to transfer the seismic wave energy completely and to avoid "Box effect" which could lead to erroneous results. On the other hand, for the bedrock boundary simulation, Kelvin elements were adopted as a portion of the seismic wave energy had to be reflected. The dashpots and springs were arranged in three perpendicular directions. The Kelvin elements constants were considered to be independent from the frequency of excitation since this dependence within the range of the frequency of interest was low. Figure 1 shows the half model of soil-pile system developed for the inertial interaction analysis. It can be seen that sides of the model look neutral in shape due to use of infinite elements.

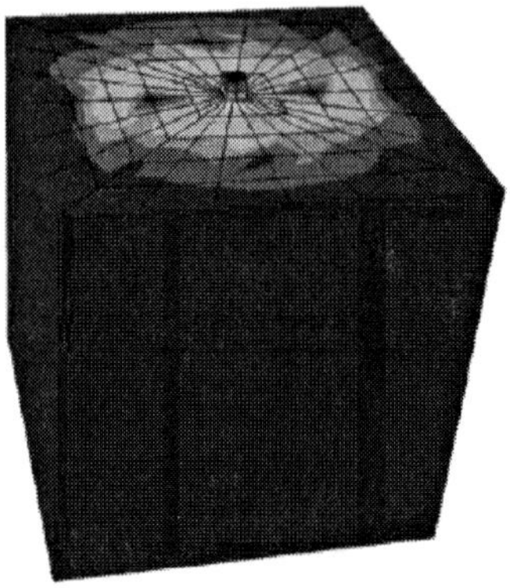

FIG. 1. Single pile embedded in soil with infinite extent.

LOADING

An acceleration time history related to one of the seismic events records in Iran (Zarrat) was used as input bedrock motion. Since the duration in which the highest acceleration amplitudes occurred was important and due to computational efficiency reasons, only 10 seconds of the 43.525 second seismic event was chosen. The maximum acceleration amplitude in the record was approximately 0.3g.

KINEMATIC INTERACTION STUDY

The input earthquake motion forces the structure and the soil in free-field to oscillate. Kinematic interaction in soil-structure systems is studied by means of frequency dependent transfer functions. The magnitude of transfer function is defined as the ratio of the maximum pile head displacement to the corresponding value of free-field. The following formula is the mathematical description of the function:

$$TF(f) = \frac{U_p(f)}{U_{ff}(f)} \tag{1}$$

Where $U_p(f)$ and $U_{ff}(f)$ are the maximum lateral response at the pile top and the free field for a given excitation frequency, respectively. Transfer functions for three kinds of single piles and in two approaches regarding soil-pile separation were calculated and the results are shown in figures 2 and 3. It can be seen that gapping slightly increases the magnitude of transfer functions. In addition, since the stiffness of piles due to the increase in their diameter increases, the pile head displacements and therefore the transfer function magnitudes decrease.

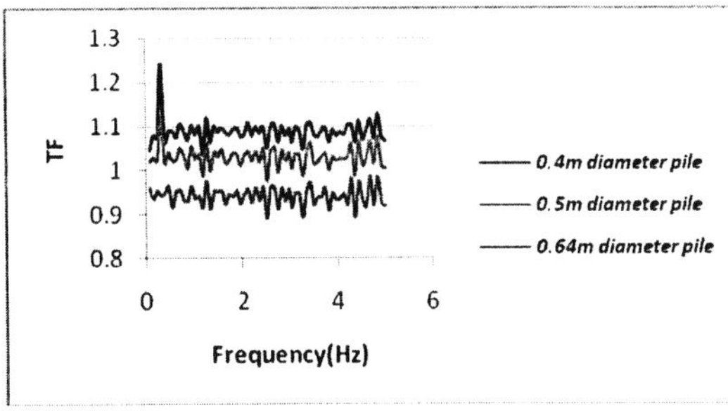

FIG 2. Transfer function versus frequency considering separation between soil and pile.

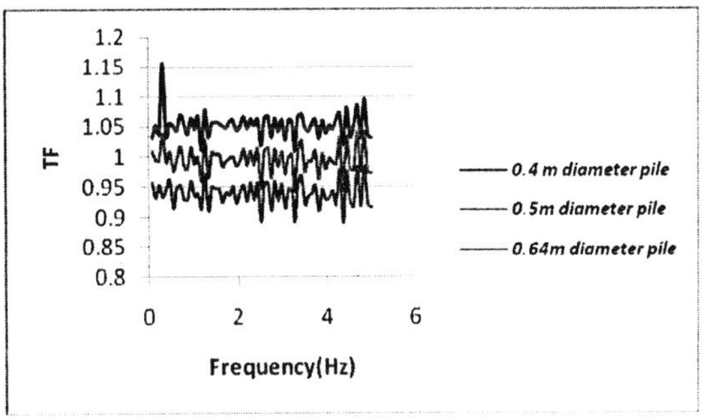

FIG. 3. Transfer function versus frequency considering perfect bond between soil and pile.

INERTIAL INTERACTION

Inertial interaction was studied by means of impedance functions. The following formula was used to calculate the functions:

$$K(\omega) = \frac{FT(P(t))}{FT(U(t))} = \frac{P(\omega)}{U(\omega)} \qquad (2)$$

In the above equation, $P(\omega)$ and $U(\omega)$ are Fourier transforms of the excitation forces and displacements, respectively. The impedance functions in the frequency domain were extracted as complex numbers, in which the real and imaginary parts were corresponding stiffness and damping of the system, respectively. The magnitudes were normalized by static state stiffness related to every single pile. It is noteworthy that the damping calculated herein includes internal and radiation damping. The results are shown in Figures 4 through 7. It can be seen that separation causes both real and imaginary parts of the impedance functions to decrease. In addition, as the pile diameter increases, both real and imaginary parts increase which indicates increase in dynamic stiffness of the system.

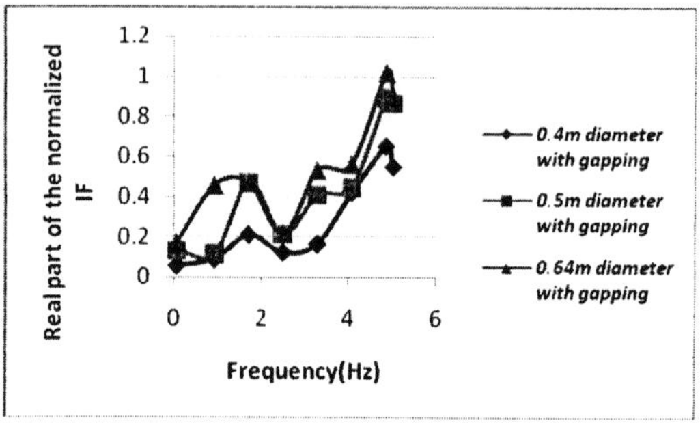

FIG.4 . Real part of the normalized impedance functions versus frequency considering soil-pile separation.

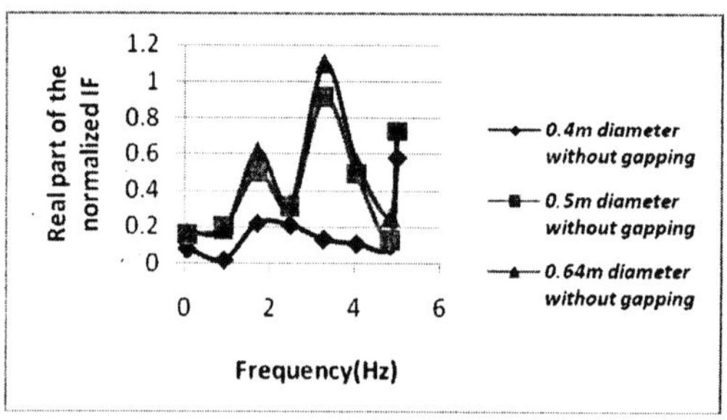

FIG.5 . Real part of the normalized impedance functions versus frequency considering no separation between soil and pile.

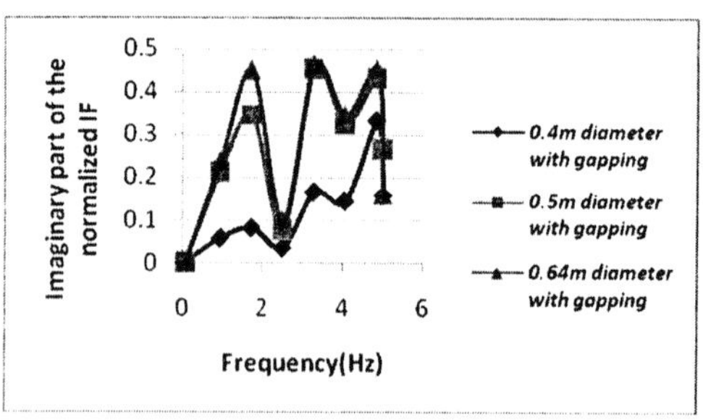

FIG.6 . Imaginary part of the normalized impedance functions versus frequency considering soil-pile separation.

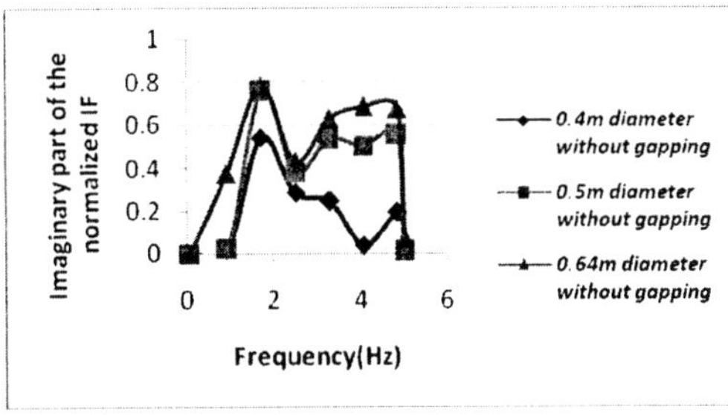

FIG.7 . Imaginary part of the normalized impedance functions versus frequency considering no separation between soil and pile.

CONCLUSIONS

The effects of soil-pile separation on the kinematic and inertial interaction of soil-single pile systems were investigated. The following conclusions can be drawn:

• Soil-pile separation slightly increased the ratio of pile head displacement to the free-field's corresponding magnitude. In addition, the effect of pile diameter on transfer function is to reduce its amount as it increases.

• Soil-pile separation reduces the real and imaginary parts of the normalized impedance functions. These reductions can be attributed to less support of the surrounding soil for the reduction in the system stiffness and the fact that damping lessens due to loss of contact in the upper part of the soil-pile system.

• As pile diameter increases, the real part of the impedance function corresponding system stiffness increases. This effect is also observed regarding the imaginary part of the function.

REFERENCES

Balendra, S. (2005). Numerical modeling of dynamic soil-pile-structure interaction. MS thesis, Department of Civil and Environmental Engineering, Washington State University.

Kavvadas, M. and Gazetas, G. (1993). "Kinematic seismic response and bending of free-head piles in layered soil." *Geotechnique*, 43 (2): 207-222.

Maheshwari, B. L., Truman, K. Z., El Naggar, M. H., and Gould, P. L. (2004). "Three Dimensional Nonlinear Analysis for Seismic Soil-Pile-Structure interaction." *Soil Dynamics and Earthquake Engineering*, 24: 343-356.

Maheshwari, B. L., Truman, K. Z., El Naggar, M. H., and Gould, P. L. (2004). "Three-Dimensional Finite Element Nonlinear Dynamic Analysis of Pile Groups for Lateral Transient and Seismic excitations." *Canadian Geotechnical Journal*, 41: 118-133.

Kramer, S. L., 1996, Geotechnical Earthquake Engineering, Prentice Hall, Englewood Cliffs N.J

Numerical Simulation of Pile-Soil-Structure Interaction under Dynamic Loading

S. N. Moghaddas Tafreshi

Associate Professor, K.N. Toosi University of Technology, Tehran, Iran; nas_moghaddas@kntu.ac.ir

ABSTRACT: In this paper, nonlinear pile-soil-structure interaction under dynamic loads is formulated by using free-field soil analysis and beam on Winkler foundation. The free-field motions are calculated separately through a site response analysis using DYFRA program which was developed by the author to prepare the data in the form of displacement time history. Linear beam column finite elements are used to model the piles and structural elements. Nonlinear modeling of soil media is done by introducing a rational approximation to continuum with nonlinear interface springs along the piles. In order to calibrate the parameters used in the model and also verify the proposed formulation, the result of a shaking table tests are used. The numerical analyses show good consistency with the test results, in terms of recorded spectral acceleration of soil layers and acceleration time history of superstructures above the piles.

INTRODUCTION

Various approaches have been used for the dynamic response analysis of pile-supported structures. A 2-D or 3-D finite element analysis is definitely a powerful method, but modeling a soil–pile system and setting numerical parameters for the entire model is indeed laborious and the computational effort can be very time consuming (Kimura and Zhang 2000; Yang and Jeremic 2003). Moreover, direct methods require both soils and structures to be treated with equal rigor and complex variations of soil profile in a 2-D or 3-D space should be provided for the analysis. Hence, there still remains an important place for simple approaches even in these days that highly complex numerical solutions are available for difficult problems. It has been customary in engineering practice to assume that a pile is supported by distinct side-soil springs (Winkler hypothesis) (Matlock 1970; Tahghighi and Konagai 2007).

In the present study, a seismic design method for soil-pile-superstructure interaction is established based on large shaking table tests on pile-structure model (Maymand et al. 2000). Initially, the free field motions are calculated separately through a site response analysis using DYFRA program (developed by the author). Secondly, the motions in the form of displacement time history are used as input boundary conditions for a beam (pile) on nonlinear Winkler model to evaluate the response of the pile. This model is composed of a linear elastic beam-column representing the pile, non-linear p-y springs and linear dashpots representing the surrounding soil. The results are also compared with those of the physical model to confirm that our simulations can predict the behavior of pile with acceptable accuracy.

SHAKING TABLE TEST

As mentioned before, the study presented here focuses on simulation of the seismic behavior of soil-pile superstructure interaction which is performed based on test results on a large shaking table (Maymand et al. 2000). In this experimental program, a series of scaled physical model tests have been performed at U.C. Berkeley on the shaking table to examine the seismic response of soil-pile-structure interaction.

A model soil with appropriately scaled stiffness and strength properties was developed for the experiments, consisting of 72% Kaolinite, 24% Bentonite, and 4% type C fly ash (by weight). The model soil has a unit weight of 14.8 KN/m^3, a plasticity index of 75% and undrained shear strength of 4.8 kPa. The shear strength profile corresponded to a lightly over-consolidated soft to medium stiff clay (layers 1-8), overlying a hard clay bearing layer (layer 9). The soil characteristics and parameters used in the analysis are represented in Table 1.

Table 1. Soil Properties of Layers (Maymand et al. 2000)

Layer Number	Soil Material Type	Thickness of Layers (m)	Maximum Shear Modulus (Mpa)	Total Unit Weight (kN/m^3)	Shear Wave Velocity (m/sec.)	Vertical Effective Stress (kpa)
1	1	0.2	0.49	14.80	18.09	0.51
2	1	0.3	0.82	14.80	23.25	1.62
3	1	0.3	1.22	14.80	28.43	3.04
4	1	0.5	1.55	14.80	32.04	4.94
5	1	0.5	1.97	14.80	36.18	7.33
6	1	0.5	3.64	14.80	49.10	9.72
7	1	0.3	5.06	14.80	57.89	11.61
8	1	0.2	9.48	14.80	79.25	12.74
9	2	0.2	46.09	18.00	158.49	13.74
OUTCROP			89.70	22.00	200.00	14.37

The flexural rigidity of the prototype pile was computed as 79.120$kN.m^2$. Accordingly, the model piles were fabricated using 51mm diameter aluminum tubing with a wall thickness of 0.71 mm, which provided the correctly scaled flexural rigidity (EI). The fixity conditions of the pile are known to be significant in lateral response, so the piles are fixed against rotation at the head, and fixed against translation at the tip. This corresponds to a pile driven into a firm stratum at the base and cast into the pile cap. It consisted of four single piles (S1, S2, S3 and S4) with head masses ranging from 4.5 to 72.7 kg, embedded in 2.0 m of model soil. The vertical arrays of accelerometers were located within 0.5 m of the piles, but were out of the line of shaking.

The model was subjected to a series of seismic events including sine sweeps and earthquake records. A scaled version (amax=0.20g) of the motion recorded at Yerba Buena Island during the Loma Prieta earthquake was used as the shaking table command signal for the test analyzed herein (Fig. 1).

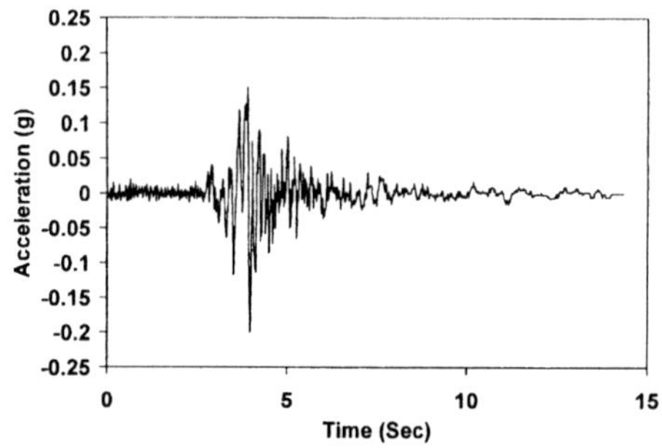

FIG. 1. Scaled acceleration of the motion recorded during the Loma Prieta earthquake (Maymand et al. 2000)

SIMULATION OF THE SHAKING TABLE RESULTS

For the uncoupled analysis, the solution proceeds in two steps, namely, the computation of free field motions and the analysis of the pile-structure response. The free field motions are calculated separately through a site response analysis and the motions in the form of displacement or acceleration time history are then used in the second stage as input boundary conditions in nonlinear Winkler model.

FREE-FIELD SOIL ANALYSIS

At a large distance from pile foundation (the so called free-field), the motions of these piles have a smaller effect on soils and the one-dimensional wave propagations are adequately assumed for the behavior of layered soil deposits. As a result, the equivalent linear method of analysis in the frequency domain was used for describing the site response. A computer program was written for the free field procedure formulations in frequency domain dependent approach in DYFRA program which can be used to solve the ground response problem. In simple terms, the input motion is represented as the sum of a series of sine waves. A relatively simple solution for the response of the soil profile to sine waves of different frequencies (in term of transfer function) is used to obtain the response of the soil deposit to each of the input sine waves. The overall response is obtained by summing the individual response to each of the input sine waves.

The shear modulus and the damping factor of the soil layers are the two main parameters which should be determined by calibration. The calibration is carried out by DYFRA program. In the first step the value of shear modulus and damping factor are selected to correspond with the small value of shear strain (=0.001%). In the next steps the value of maximum shear strain, effective shear strain and its shear modulus and damping factor are obtained. This procedure continues until the difference of the two obtained values in two consecutive steps becomes negligible and consequently, the value of shear modulus and damping factor obtain for free-field soil analysis.

Fig. 2 shows the comparison of the recorded spectral acceleration from the experimental data of large-scale seismic soil-pile-structure tests and the result of free-field analysis (obtained of DYFRA program) for second and sixth layers. The results show good consistency and satisfactory accuracy. For all of the layers, the best coincidence of spectral acceleration is obtained for high periods. Overall, the analysis can predict the values of spectral acceleration for magnitudes of period greater than 0.2 second with good accuracy. In other words, the maximum difference takes place for high frequencies (f >5Hz.). The peak acceleration happened in a period less than T=1 sec. This fact confirms the presented typical spectral acceleration by API code and also notability of local site effect in soft clay.

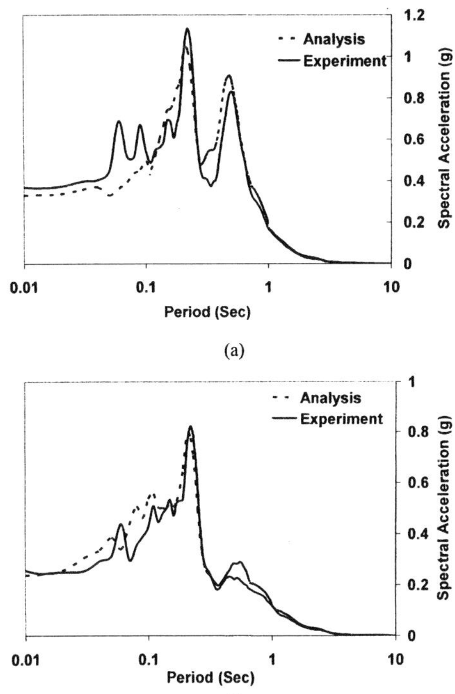

FIG. 2. Comparison of the recorded spectral acceleration from the test and DYFRA program (a) second layer; and (b) sixth layer

Fig. 3 shows the recorded displacement for fourth and eighth layers obtained of DYFRA program. The results state that the magnitude of displacement (and maximum displacement) reduces from second layer to eighth layer, significantly. Also the maximum displacement for all the layers happened in time of 4.2 sec.

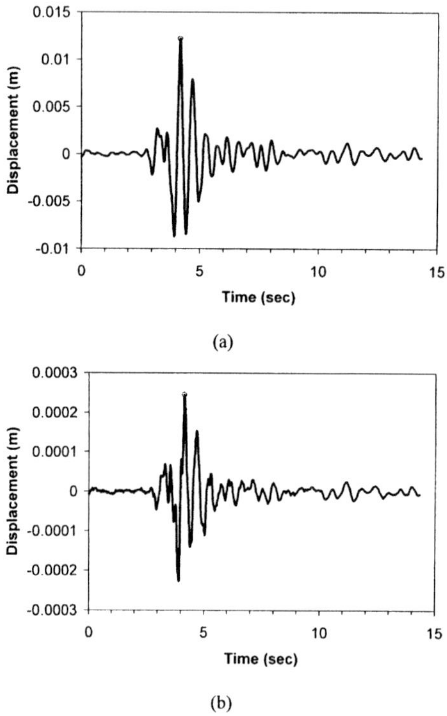

FIG. 3. Recorded displacement from DYFRA program (a) fourth layer; and (b) eighth layer

WINKLER SIDE-SOIL SPRING

As described before, the problem herein is that of a single pile embedded in layered soft clay and subjected to lateral motion induced by a shaking table. The surrounding soil is considered to be a homogeneous deposit with shear wave velocity Vs, density ρs, shears strength S=Su(z), which varies with depth. The soil–pile interface is modeled as a Winkler foundation interacting with the pile through nonlinear springs at near surface and with linear springs at deep depth; along with these elements linear dashpots are used. Fig. 4 shows the model of pile-superstructure, non-linear p-y springs and linear dashpots representing the surrounding soil layers. The outputs of the free field analysis for each layer in the form of displacement time history are applied as input boundary conditions in the free end of the nonlinear Winkler side-soil springs in ANSYS 5.4 program (Kimura and Zhang 2000). The length of the pile is divided to 9 segments (8 segments embedded in the soil and one segment is out of the soil surface). The masses in the head of the piles S1, S2, S3, and S4 to model the effect of superstructure are respectively, 4.5, 11.25, 45, 72.7 kg. The schematic layout model of the piles and superstructure and the uncoupled seismic soil-pile-superstructure diagram are shown in Fig. 4.

The Winkler model requires a definition of parameters for the nonlinear springs as well as the viscous dashpots. The formulation used here is based on the works of (Gazetas and Dobry 1984; Makris and Gazetas 1992; Matlok 1970; Maymand et al. 2000) which is not illustrated here to limit the length of the paper. According to the formulation mentioned, the nonlinear spring equation and the dashpot equation of the constitutive model are consistent with the visco-plastic analysis at the limit of small deflections and the plasticity analysis at the zero-frequency limit. The coefficients J, n and also the near field parameters (shear wave velocity; Vs and Poisson's ratio; υ) should be determined by calibrating the model against results of the shaking table experimental data for the piles S1, S2, S3, and S4. Table 2 shows the values of these parameters which are obtained by calibration.

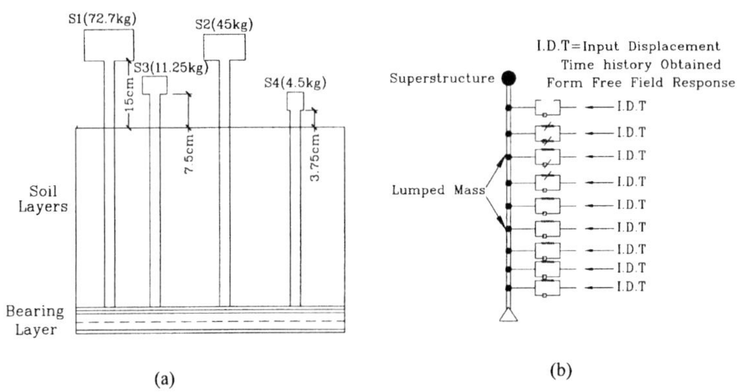

FIG. 4. (a) Schematic model of piles and superstructure; (b) Uncoupled seismic soil- pile-superstructure diagram

Table 2. The values of calibrated parameters

Pile Number	J	n	Vs (m/sec.)	ν
S1	0.0047	0.25	23.25	0.4
S2	0.0027	0.55	28.43	0.4
S3	0.0021	0.85	32.04	0.4
S4	0.0019	1	36.18	0.4

Then the values of calibrated parameters in Table 2 are introduced to ANSYS 5.4 (Kohnke 1997) program as model parameters to investigate the behavior of the piles. The results of the analysis are included a time history of displacement and acceleration of superstructure, and also acceleration of lumped mass at the nodes of pile.

Fig. 5 compares the experimental data with the computed acceleration of superstructure for the piles S3 and S4. The output values show that, the maximum acceleration occurs, respectively, in the pile S1, S2, S3 and S4. The reason is that the concentrative mass on the head of the piles and the free length of the piles above the level of soil increases from the pile S4 to S1 (see Fig. 4-a). The computed results are in a reasonable agreement with the experimental values, indicating that the proposed p–y model is promising in estimating acceleration of superstructure, distributions of the bending moment and displacement of piles and thus may be convenient to be used as a tool for designing pile foundations in engineering design of piles.

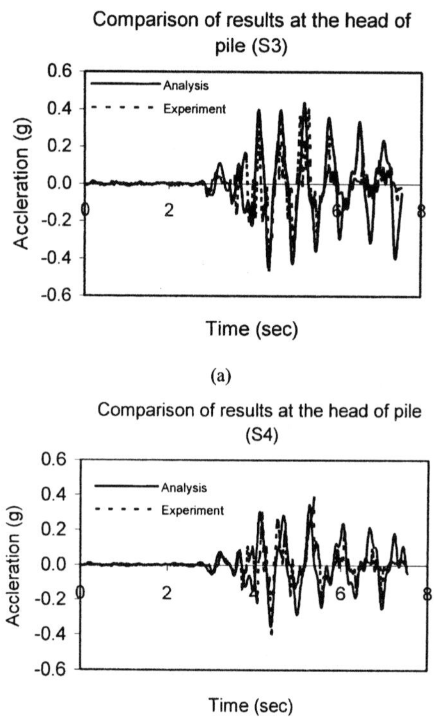

(b)

FIG. 5. Comparison of the recorded acceleration of superstructure from the test and Winkler side-soil springs model (a) pile S3; and (b) pile S4.

CONCLUSION

The uncoupled nonlinear Beam-on-Winkler foundation model and site response analysis used in the method proposed in this research provide a convenient method for an extensive study of soil-pile-structure interactions. The responses of free-field and single piles analysis were calculated and compared with the results observed in the large shaking table. The results show that the estimated spectral accelerations of the free-field analysis for various soil layers are in good agreement with those obtained from the experiments. The peak acceleration for all layers of soil happened in a period less than 1 second; this fact confirms the notability of local site effect in soft clay. Overall, these results suggest that the result of DYFRA program is promising for estimating the spectral acceleration of soil layers with a reasonable degree of accuracy. Comparisons between the obtained acceleration for the superstructure of the Beam-on-Winkler foundation model and those of the experiments reveal that there is a good agreement between the two approaches. These agreements clarify that the proposed method is a reliable method for predicting the soil-pile-superstructure behavior subjected to seismic loading.

Although the results obtained in the present paper are encouraging for considering the uncoupled Beam-on-Winkler foundation model in investigating the soil-pile-structure interaction, but it should be noted that the numerical results are obtained for only one type of soil (soft clay). Hence the values of calibrated parameters in Table 2 are limited to soil characteristics and the tests conditions used and other studies with similar grading, plasticity index and characteristics.

REFERENCES

Badoni, D.; Makris, N. (1996) “Nonlinear response of single piles under lateral inertial and seismic loads” *Soil Dynamic and Earthquake Eng.*, 15 (1): 29-43.

Gazetas, G.; Dobry, R. (1984) “Horizontal Response of Piles in Layered Soils” *J.Geo. Eng., ASCE,* 110 (1): 20-40.

Kimura, M.; Zhang, F. (2000) "Seismic evaluations of pile foundations with three different methods based on three-dimensional elasto-plastic finite element analysis." *Soils Found,* 40(5), 113–32.

Kohnke, P. (1997) "ANSYS Theory Reference Release 5.4, 8th Edition, ANSYS, Inc"

Makris, N.; Gazetas, G. (1992) "Dynamic Pile-Soil-Pile Interaction – Part II: Lateral and Seismic Response." *Earthquake Eng.Struct.Dyn,* 21(2), 145-162.

Matlock, H. (1970) "Correlations of design of laterally loaded piles in soft clay." Proceedings of the offshore technology conference, Houston, TX, Vol. 1 (1204): 577–94.

Maymand, P.; Reimer, M.; Seed, R. (2000) "Large Scale Shaking Tables Tests of Seismic Soil-Pile Interaction in Soft Clay." *Proc.12th World Conf.Earthquake Eng. New Zealand,* Vol.5, No.0915.

Tahghighi, H.; Konagai, K. (2007) "Numerical analysis of nonlinear soil–pile group interaction under lateral loads." *Soil Dynamic and Earthquake Eng.*, 27 (5): 463–474.

Yang, Z.; Jeremic, B. (2003) "Numerical study of group effects for pile groups in sands." Int J Numer Anal Methods Geomech, 27 (5): 1255–76.

Influence of Pile Geometry on Internal Sand Displacement around a Laterally Loaded Pile using Transparent Soil

Jinyuan Liu[1], M.ASCE, P.E., P.Eng., Mingliang Liu[2], and Hongmei Gao[3]

[1]Assistant Professor, Ryerson University, Toronto, Canada M5B 2K3; jinyuan.liu@ryerson.ca
[2]Ph.D. Candidate, Geo-Institute, Hohai University, Nanjing, China 210098; mliu@hhu.edu.cn
[3]Ph.D. Candidate, Geo-Institute, Hohai University, Nanjing, China 210098;hongmei54@163.com

ABSTRACT: The influence of pile geometry on internal sand movement around a laterally loaded pile was visualized in this paper by utilizing transparent soil and digital image correlation (DIC) methods. Transparent sand was made of amorphous silica gel and a pore fluid with a matching refractive index. An optical system consisting of a laser light, a camera, a loading frame, and a computer was developed to illuminate a targeted section inside a transparent soil model. A distinctive laser speckle pattern was generated by the interaction between the laser light and transparent soil. Two laser speckle images before and after deformation were used to calculate the displacement field using DIC. Two different cross-sectioned piles were used in the tests: a square-sectioned pile and a circular pile. Both piles were made of acrylic and had a diameter or width of 12.7 mm. Two different failure planes were observed in the tests, where a larger zone was mobilized by the circular pile compared to the one in the square-sectioned pile.

INTRODUCTION

The pile foundations have been used widely to resist horizontal loadings in many structures. The loading capacity of a laterally loaded pile has attracted the interests of engineers for more than five decades. Many methods have been developed to analyzed the laterally loaded pile, including Broms' Method (Broms 1964), the Elastic Method (Poulos 1971), and the P-y Curve Approach (Matlock 1970; Reese 1977), and the Strain Wedge Method (Ashour et al. 2002). However, all these methods mentioned above focus only on the loading capacity or deformation of the laterally loaded pile itself, there is limited information available on soil deformation around the pile due to the non-linear behavior of soil and the complex soil-pile interaction. Finite element method (FEM) can be used to investigate soil movement around a pile under lateral loading. However, the FEM results have not been verified due to the limited field data available. This paper is to visualize the influence of pile geometry on internal sand movement around a laterally loaded pile using transparent soil.

TRANSPARENT SOIL

Transparent Soil

Visualization of soil movement can significantly improve the understanding of geotechnical problems. Many visualization techniques have been used in geotechnical engineering research. However, most of these investigations so far are intrusive measurements. Though non-intrusive measurement techniques including computerized axial tomography and magnetic resonance imaging were attempted in geotechnical experiments, nevertheless, routine applications of these techniques are limited by their high cost and their technical limitations.

Transparent materials, including photoelastic material or glass, have been used to model natural soil. However, their studies are limited by the fact that these materials cannot accurately model soil strength and deformation behavior and also have a poor quality of transparency.

Mannheimer and Oswald (1993) demonstrated that transparent substances made of amorphous silica powder and a pore fluid with a matching refractive index exhibit macroscopic geotechnical properties similar to natural clay. Later on, a different kind of transparent soils made of silica gel was developed to model sand. Transparent soil used in this paper was a commercial product of *Multisorb Technologies Inc.* without further processing. The silica gel is an angular fine silica gel with an aggregate diameter of 0.5-1.5 mm. Fig. 1 shows typical shear stress vs. horizontal displacement curves from direct shear tests. The angle of friction was measured at approximately 33° from direct shear test. More details about their properties can be found in Iskander et al (2002b). Transparent soil was used to model loose saturated sand in this study.

The liquid used in this research was a 50:50 blend by weight of colorless Drakeo® mineral oil and a Magiesol® 47 normal-paraffinic solvent supplied by *Penreco*. The viscosity and density of the oil blend at room temperature (23°C) were 5.0 cP and 804 kg/m3, respectively. Transparent soil has been used to model natural soil in scaled experiments by many researchers (Welker et al. 1999; Gill and Lehane 2001; Toiya et al. 2007).

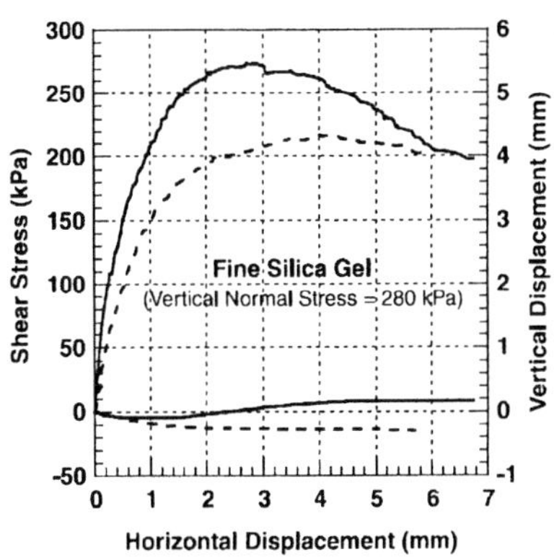

FIG. 1. Shear stress vs. displacement curves of silica gel (Sadek et al. 2002)
(Dashed lines are for loose samples and solid lines for dense samples)

Digital image correlation

The image processing technique used in this research is called digital image correlation (DIC), also known as particle image velocimetry (PIV). It is a classic pattern recognition technique where two images are compared to obtain the relative displacement between them. DIC is widely used in many engineering fields to obtain spatial deformation patterns, including geotechnical engineering (Gudehus and Nubel 2004). DIC has been applied and validated in measuring internal soil displacement within a transparent soil model (Sadek et al. 2003).

In this paper, the PIVview2c software programmed by PivTec GmbH (PIVTEC 2006) was used to calculate the internal soil displacement field. This software has features to allow users to select the window size, cross-correlation algorithm, peak function, etc. Unless noted, the parameters used in this research were final window size 32 x 32 pixels, the multiple-correlation algorithm, and the multi-grid interrogation method.

TEST PROGRAM

Optical Set-up for Internal Soil Deformation Measurement

The optical set-up consisted of a mono complementary metal–oxide–semiconductor (CMOS) camera, a laser light source, a line generator lens, a loading frame, and a PC, as shown in Fig. 2. The camera was PixeLink PL-B741E model camera with a resolution of 1280 x 1024 pixels. It was controlled by the PC through a in-house developed software driver using Matlab. The laser system had a wavelength of 632.8 nm and an output power of 35 mW. The line generator lens was utilized to create a laser light sheet from the laser beam to slice the sample. The lateral loading arrangement consisted of a load cell with a capacity of 100 N attached to a linear variable differential transducer (LVDT) with a linear strike of ±25 mm. A data acquisition system was developed to acquire the loads and displacements during pile loading.

A Plexiglas mould with dimensions of 150 mm (wide) × 300 mm (long) × 200 mm (height) was used in the investigation. Two different shaped piles were used in the tests: a square-sectioned pile and a circular pile. Both model piles were made of acrylic and had a diameter or width of 12.7 mm. These piles were embedded approximately 160 mm in transparent soil. The pile was connected to the loading frame through a string at a loading point, which was about 30 mm above the sample height. The camera was set 15 cm above the model with its optical axis perpendicular to the horizontal plane. Load was applied through a screw mechanism by manually rotating a handle.

Sample Preparation

First, silica gel was immersed in the pore fluid. Second, a vacuum was applied to de-air the mix until it turned transparent. Third, silica gel was packed into the Plexiglas mould to form a transparent soil sample. In this test, loose condition was modeled where no compaction was involved during sample preparation. Fourth, the pile was manually driven into the sample to the desired depth and the string was attached to the pile and adjusted to provide a horizontal loading to the pile. The circular pile was tested first and then replaced by the square-sectioned pile at the same location. Soil disturbance was noticed and should be avoided in future research.

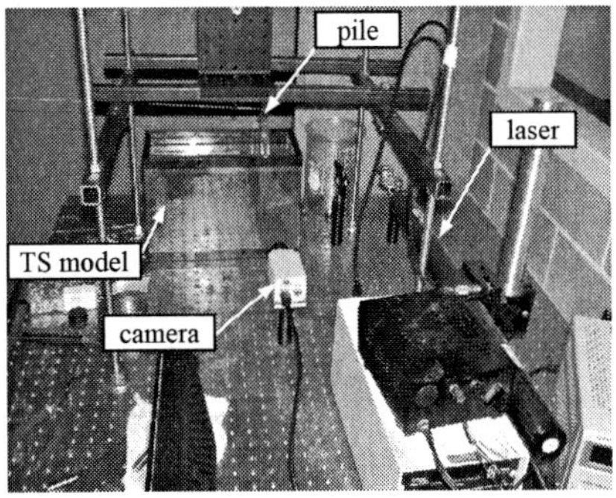

FIG. 2. Test set-up for measuring internal sand displacement

RESULT ANALYSES

Load vs. Displacement Curve

The load vs. displacement curves of the model piles measured at the loading point are shown in Fig. 3. The curves resembles typical load vs. displacement curve for a laterally loaded pile in loose sand. The load increased linearly at the initial stage to represent the elastic behavior of soil and then the rate of load increase gradually reduced with increasing displacement. There was a fluctuation in the load during pile movement. It was believed the high compressibility of silica gel and silica particle movement around the pile contributed to this phenomenon. A higher load capacity was expected in the square-sectioned pile compared to the circular pile. However, the similar ultimate pile capacities were due to soil disturbance during sample preparation.

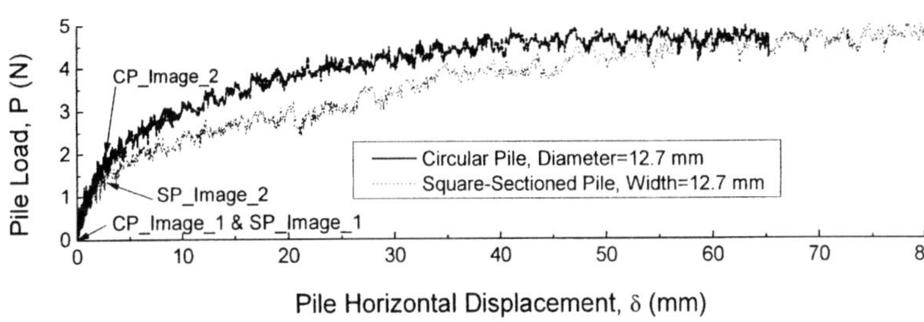

FIG. 3. Load vs. displacement curves of the model piles

2-D Internal Soil Displacement Field

The laser light sheet was used to target an interested cross-section inside a transparent soil model. The interaction between the transparent soil and the laser light sheet produced a distinctive laser speckle pattern. Two pairs of images (CP_Image_1 and CP_image_2 for the circular pile and SP_Image_1 and SP_image_2 for the square-sectioned pile) were taken to calculate soil movement in the initial stage.

The soil displacement field from this relative displacement was calculated using PIVview software and shown in Fig. 4, where the background image was the laser speckle image after soil movement.

In order to show clearly soil movement, the displacement fields were re-plotted in Fig. 5, where the displacements were limited to only the displacement at every four grids in the horizontal direction. As expected, the soil moved away from the pile during pile movement. Two different soil displacement fields were observed. A bell-shaped with a larger region soil was mobilized by the circular pile compared to the less soil movement in the square-sectioned pile. The failure plane was defined by linking the points which had horizontal displacements half the maximum horizontal displacements at each vertical section and the influence zone was defined by linking the points which started to have soil movement triggered by the pile movement. A similar method has been used by other researchers in identifying the rupture plane in soil around an uplifting earth anchor (Ilamparuthia and Muthukrishnaiah 1999). The failure plane is normally related to the shear strain fields. However, it was assumed in this study that the failure plane from soil displacement is similar to the one from the shear strain field. More research is needed to validate this assumption in the future.

The inclined angle with the horizontal varied between 24° and 28° for the circular pile and between 20° and 24° for the square-sectioned pile. These values were close to $2/3\ \phi$, where ϕ was measured at approximately 33°. Because the piles blocked the view, the movement at the peak load was not detected under the current test configuration. The pile movement itself was not accurate due to the fact that the pile was out of focus in the current setting. The extent of the influence zone from the anchor could not also be identified, as shown in two dashed lines in Fig. 5, which can be solved by studying a vertical cross-section along the pile centerline in the future.

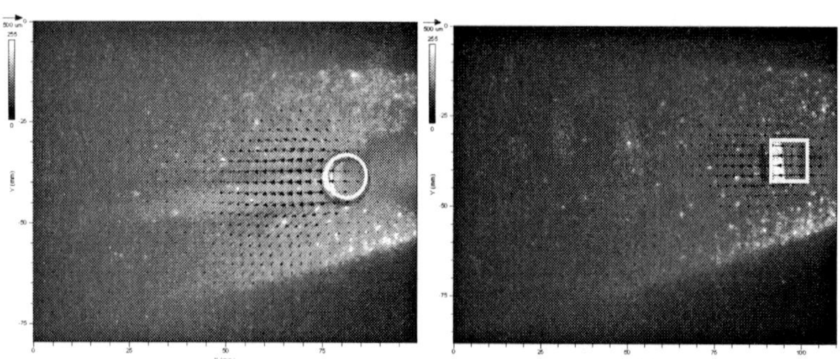

(a) (b)

FIG. 4. Internal sand displacement field around a laterally loaded pile

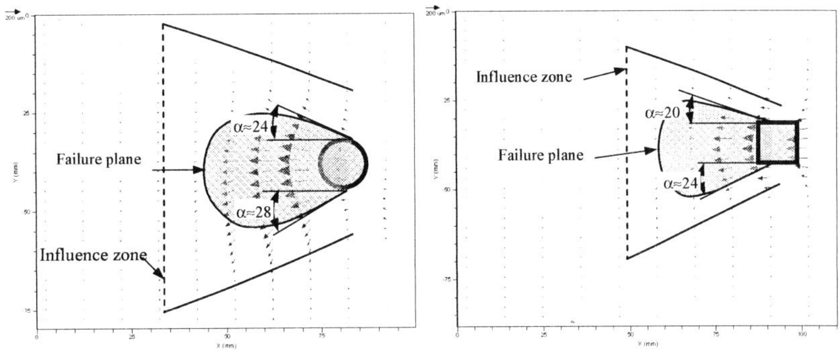

(a) (b)

FIG. 5. Internal sand displacement field around a laterally loaded pile

CONCLUSIONS

The influence of pile geometry on internal soil displacement field around a laterally loaded pile was modeled in this research using transparent soil and digital image correlation method. Transparent sand made of amorphous silica gel and a pore fluid with a matching refractive index was used in the study. An optical system consisting of a laser light, a camera, a loading frame, and a computer was developed to optically slice a transparent soil model. The displacement from a pair of images was calculated using digital image correlation (DIC). Two different failure planes were observed in the tests, where a larger zone was mobilized by the circular pile compared to the one in the square-sectioned pile.

This paper is limited to a qualitative analysis of soil movement around a laterally loaded pile because of the lack of soil and pile deformation at the peak load stage due to the limitations of the current test setting. A detailed quantitative analysis and comparison with the results from natural sand is being investigated with an improved test configuration. In summary, transparent soil and the developed optical system can be used to explore opportunities for more advanced non-intrusive deformation measurements for various soil-structural interaction problems.

ACKNOWLEDGEMENTS

The authors want to acknowledge the financial support of start-up fund awarded to the first author from Ryerson University and the one-year scholar awarded to the second and third authors from China Scholarship Council for their study at Ryerson University.

REFERENCES

Ashour, M., Norris, G., and Pilling, P. (2002). "Strain wedge model capability of analyzing behavior of laterally loaded isolated piles, drilled shafts, and pile groups." *ASCE, J. of Bridge Eng.*, 7(4), 245-254.

Broms, B. B. (1964). "Lateral resistance of piles in cohesionless soils." *ASCE Proc., J. of Soil Mech. & Found. Div.*, 90(SM3, Part 1), 123-156.

Gill, D. R., and Lehane, B. M. (2001). "An optical technique for investigating soil displacement patterns." *GTJ*, 24(3), 324-329.

Gudehus, G., and Nubel, K. (2004). "Evolution of shear bands in sand." *Geotechnique*, 54(3), 187-201.

Iskander, M., Lai, J., Oswald, C., and Mannheimer, R. (1994). "Development of a transparent material to model the geotechnical properties of soils." GTJ, 17(4), 425-433.

Iskander, M., Sadek, S., and Liu, J. (2002). "Optical measurement of deformation using transparent silica gel to model sand." *IJPMG*, 2(4), 13-26.

Ilamparuthia, K. and Muthukrishnaiah K. (1999). "Anchors in sand bed: delineation of rupture surface." *Ocean Engineering,* 26, 1249-1273.

Mannheimer, R. J., and Oswald, C. J. (1993). "Development of transparent porous media with permeabilities and porosities comparable to soils, aquifers, and petroleum reservoirs." *Ground Water*, 31(5), 781-788.

Matlock, H. (1970). "Correlations for design of laterally loaded piles in soft clay." *2nd Annual Offshore Tech. Conf.*, Houston, 1, 577-594.

Murchison, J. M., and O'Neill, M. W. (1984). "Evaluation of p-y relationships in cohesionless soils." *Analysis and Design of Pile Found.*, 174-191.

Poulos, H. G. (1971). "Behavior of laterally loaded piles: I-single piles." *JSMFD*, 97(5), 711-731.

PIVTEC (2006). P*IVVIEW2C/3C User Manual*, Ver. 2.4

Reese, L. C. (1977). "Laterally loaded piles: Program documentation. " *JGGE*, 103(4), 287-305.

Sadek, S., Iskander, M., Liu, J. (2002). "Geotechnical properties of transparent silica." *Can. Geotech. J.* 39, 111–124.

Sadek, S., Iskander, M., Liu, J. (2003). "Accuracy of Digital Image Correlation for Measuring Deformations in Transparent Media." *J. of Computing in Civil Engineering*, 17(2), 88–96.

Toiya, M., Hettinga, J., and Losert, W. (2007). "3D imaging of particle motion during penetrometer testing : fffrom microscopic to macroscopic soil mechanics." *Granular Matter*, 9(5), 323-329.

Welker, A. L., Bowders, J. J., and Gilbert, R. B. (1999). "Applied research using a transparent material with hydraulic properties similar to soil." *GTJ*, 22(3), 266-270.

Soil-water Coupling Analysis of Real-scale Field Test for 9-pile Foundation Subjected to Cyclic Horizontal Loading

Y. Jin [1], X. Bao[1], Y. Kondo[1] and F. Zhang[2]

[1]Postgraduate School of Nagoya Institute Technology, Nagoya, Japan
[2]Nagoya Institute of Technology, Nagoya, Japan (E-mail: cho.ho@nitech.ac.jp)

ABSTRACT: Group-pile foundations are used extensively for supporting buildings, bridges, and other structures to transfer structural loads safely to the ground and to avoid excess settlement or lateral movement. Therefore, it is necessary to understand the behaviors of the group-pile foundations and superstructures during a major earthquake. In this paper, numerical simulation of real-scale group-pile foundation subjected to horizontal cyclic loading is conducted using a program named as DBLEAVES. In the analysis, nonlinear behaviors of ground and piles are described by *subloading* t_{ij} model (Nakai and Hinokio, 2004) and axial force dependent model (AFD model, Zhang and Kimura, 2002) which considered the axial-force dependency in the nonlinear moment-curvature relations. The purpose of this paper is to verify the applicability of the proposed numerical method by comparing the numerical results with the test results.

INTRODUCTION

Horizontal forces may form a major part of the loading system for structures supported on pile groups and may often be repetitive or cyclic in nature, and include forces due to earthquake shocks. It is known that during a strong earthquake, the dynamic behavior of a group-pile foundation is related not only to the inertial force coming from the superstructures but also to the deformation of the surrounding ground. Therefore, it is necessary to understand the behaviors of the group-pile foundations and superstructures during a major earthquake.

Numerous tests either in model scale or in real-scale on group-pile foundation subjected to lateral loading can be found in literatures in order to elucidate the ultimate state of the foundations during strong earthquakes, e.g., Tokimatsu et al. (2007) conducted shaking table test using E-defense to estimate the effects of dynamic soil-pile-structure interaction on pile. Needless to say, a full-scale loading test is the most accurate way to determine the mechanical behaviors of deep foundations though it might be extremely expensive and time consuming. On the other hand, numerical simulation also plays a very important role in determining the behaviors and a large number of numerical studies have been done in this field. Kimura and Zhang (1999) developed a three-dimensional static and dynamic finite element analysis code (DGPILE-3D) in order to investigate the static and dynamic interaction between soil and pile foundation.

The purpose of this paper is to provide an applicable numerical way of evaluating the mechanical behavior of a pile foundation subjected to cyclic lateral loading up to an ultimate state. Three-dimensional (3D) finite element analyses of a real-scale group-pile foundation (Kosa et al. 1998) subjected to horizontal cyclic loading is conducted using a program named as 「DBLEAVES」 (Ye, 2007). In this research, nonlinear behaviors of ground and piles are described by *subloading* t_{ij} model proposed by Nakai and Hinokio (2004) and AFD model proposed by Zhang et al. (2002), respectively. The numerical analyses are conducted by the total stress and effective stress methods. By comparing the numerical results with the test results, the applicability of the proposed numerical method is verified.

RESULTS AND DISCUSSION

Brief Description of Real-scale Test for 9-pile Foundation Subjected to Horizontal Cyclic Loading

The static horizontal cyclic loading test of 9-pile foundation in real-scale was conducted in Kishiwada, Osaka Prefecture. An elevated highway bridge is supported by a group-pile foundation made of cast-in-place reinforced concrete piles. The plan view of the test site is presented in FIG. 1. The test piles with 1.2m in diameter and 30.4m in length are installed in a 3×3 pattern with a 3m pile spacing center to center. Reaction piles have pairs of group piles 3×2 and 3×3. The surface layer of the ground at the test site is a very young reclaimed layer with a thickness of 13m, constructed only three years before the loading test. The cyclic horizontal loading up to a maximum value of 20.5MN was applied with oil jacks installed between the test pile-group and the reaction pile-groups as shown in Fig.1.

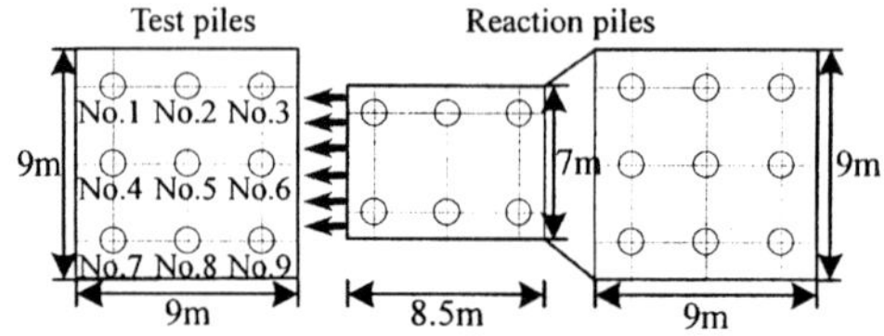

FIG. 1. Schematic layout of real scale group-pile in plan view

Determination of the Material Parameters

Theoretical simulations of drained triaxial compression test were conducted to determine the material parameters of the reclaimed layer. The particle size of the test specimens was adjusted by the removal of coarse grain with diameter greater larger than 50mm. This remold specimens were collected from 11~12m below the ground. FIG. 2 shows the comparison of experimental and theoretical results of triaxial compression test under different four confining stresses. The test results at low confining stress, as shown in FIG. 2(a) and (b), are reproduced considerably well in stress-strain relations while in the volumetric strain, a small discrepancy between the test and the theory results exists. The results at high confining stress, as shown in FIG. 2(c) and (d), are reproduced quite well quantitatively in both results. On the whole, the constitutive model is extremely precise in describing the behaviors of the soil. The material parameters of the ground are listed in Table 1. Only the bottom layer of the ground was supposed to an elastic material in the numerical analysis.

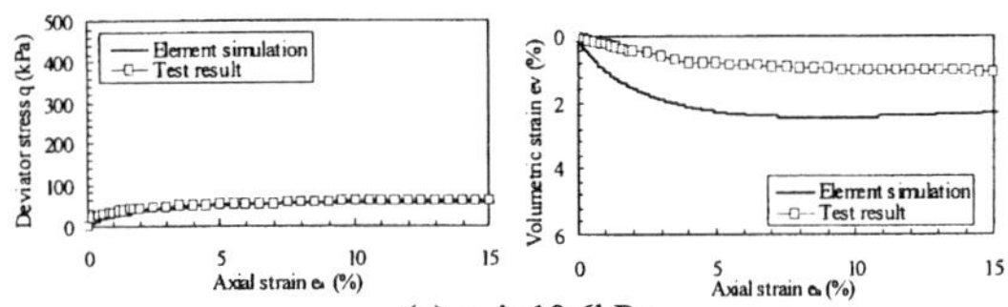

(a) σ_3'=19.6kPa

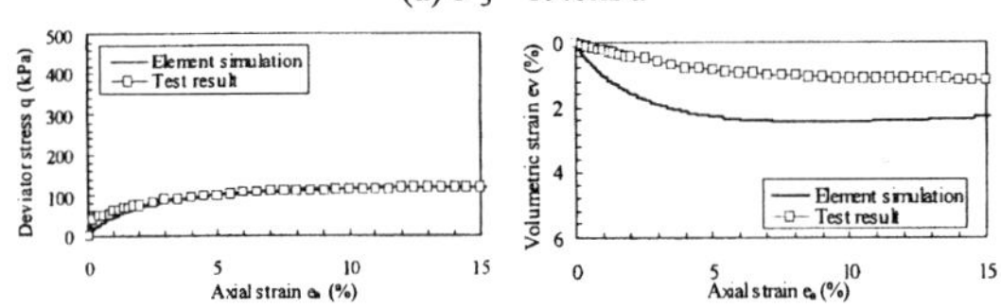

(b) σ_3'=39.2kPa

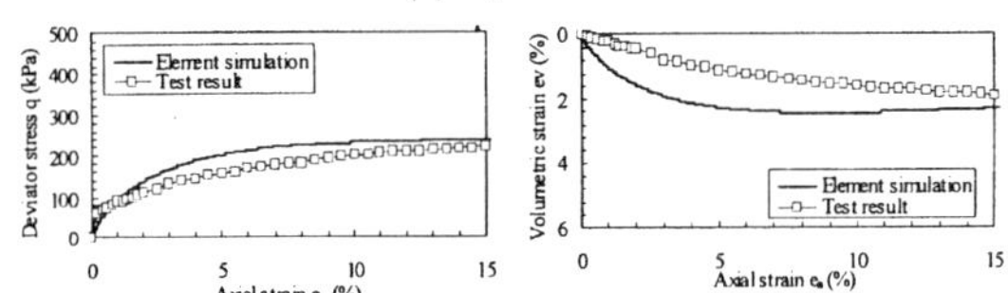

(c) σ_3'=78.4kPa

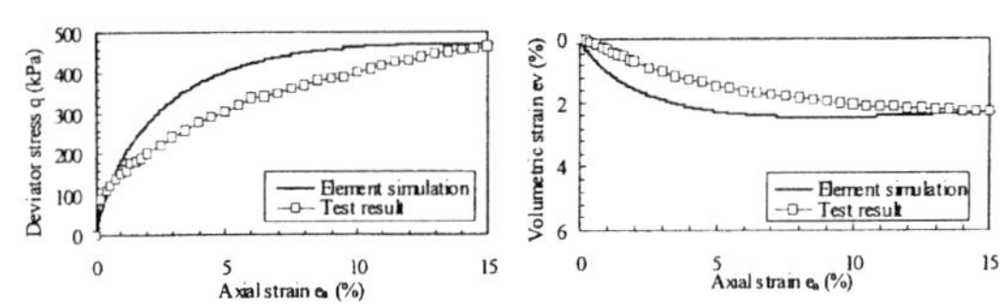

(d) σ_3'=156.8kPa

FIG.2. Theoretical simulation and test results

Table1. Parameters of ground

Soil	E (kPa)	ν	λ	κ	N	R_f	a	β	OCR	ρ (tf/m^3)	*k (cm/s)*
B1	-	0.3	0.089	0.002	0.9	3.9	500	1.5	3	1.97	6.56E-01
B2	-	0.3	0.089	0.002	0.9	3.9	500	1.5	3	1.97	3.61E-02
B3	-	0.3	0.089	0.002	0.9	3.9	500	1.5	3	1.97	1.09E-01
As1	-	0.3	0.23	0.046	0.8	3.69	500	1.5	10	1.80	1.00E-02
Ac1	-	0.38	0.23	0.046	0.7	4	500	1.5	2	1.70	1.00E-06
Dc	1.00E+05	0.3	-	-	-	-	-	-	-	1.80	1.00E-07

The pile is modeled by hybrid element proposed by Zhang et al. (2000), composed of beam element and solid element, to take into consideration the volume of pile. The beam element bears most of the loading acting on the pile, while the neighboring solid elements bear less loading but occupy the volume of the pile. The sharing ratio of bending stiffness between the beam element and the solid elements is 9 to 1. In order to consider the influence of axial force on bending stiffness of pile, AFD model is employed in the analysis. The material parameters of the pile are listed in Table 2. The concrete footing above pile heads is modeled with elastic elements.

Table2. Parameters of piles

1. Physical properties of RC
Compressive strength of concrete : σ_C=3.8×10^{4}kPa
Young's modulus of concrete : E_C=2.5×10^{7}kPa
Young's modulus of steel : E_S=2.1×10^{8}kPa
Yield stress of steel :σ_y=3.8×10^{5}kPa
2. Arrangement of the reinforcement :
D29-24 (upper part : 14.5m from the surface of the ground)
D22-12 (lower part:15.9m)
Overburden of the reinforcement : 15cm

FEM Mesh and Boundary Conditions

FIG. 3 shows the geologic profile of ground and FEM mesh. The ground is composed of 6 layers based on the soil property chart. The water table is located 1m below the ground surface. The initial stress of the ground is regarded as stratification bedding without considering the effect of pile installation.

Because of the symmetric condition of the geometry and the load, only half of the domain is taken under consideration. The boundary conditions of the ground are fixed at the bottom, sliding at the two sides whose normal direction is parallel to y-axis and x-axis. The boundary condition of the pile in the calculation is that the head of the pile is rotation fixed with the footing and the tip of the pile is free.

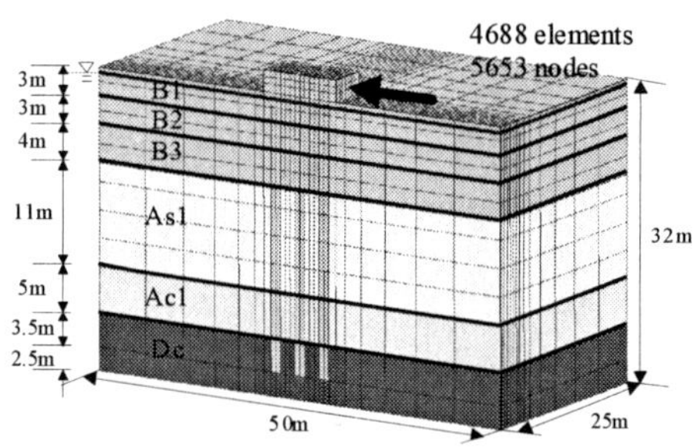

FIG. 3. FEM mesh and ground profile

Numerical Results

The cyclic horizontal load with a maximum value of 20.5MN adopted in the numerical analysis is the same as the test load shown in FIG. 4. In the simulation of one-side cyclic lateral loading test, a concentrate lateral load is applied at the center of the side surface same as the field test, where all the nodes on the side surface placed 0.9m from the ground surface are kept to be equal to each other in the x, y, z directions as a rigid, as shown in FIG. 5.

FIG. 6 shows the lateral load and displacement relations at node No.1 shown in FIG. 5. The result of total stress analysis underestimates the test result at the maximum and residual displacements while the result of effective stress analysis coincides well with the test result until sixth cycle and the maximum and residual displacements are close to the test results. In the field test, many cracks occurred on the ground surface around sixth loading cycle. In calculation, however, it is impossible to describe the cracks, which might be the reason why a discrepancy between the test and the calculated results occurred after sixth cycle in effective analysis, as shown in FIG.6.

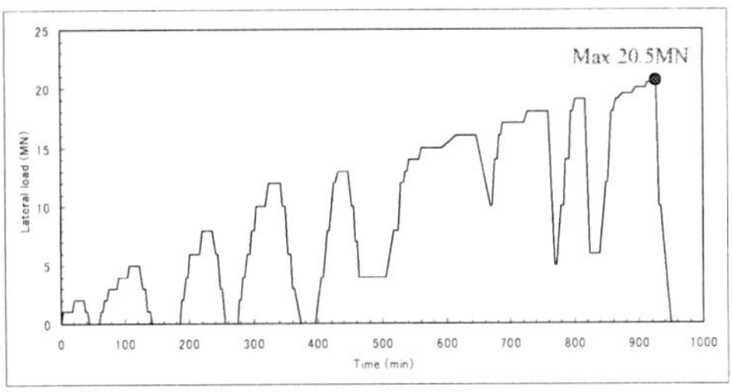

FIG. 4. Time history of lateral load in field test

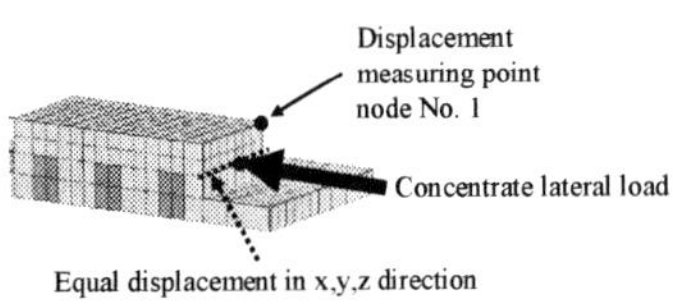

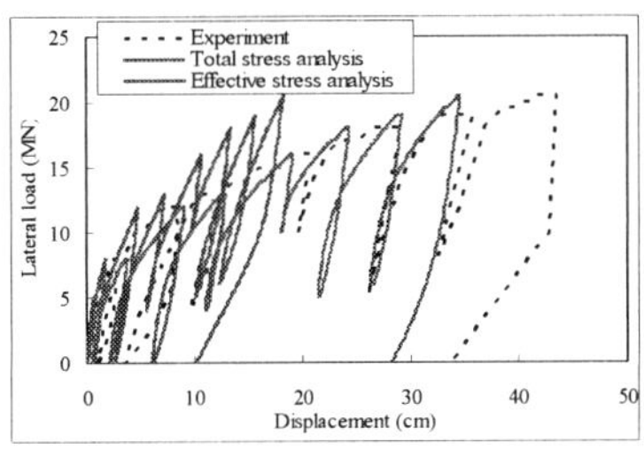

FIG. 5. Loading method **FIG. 6. Lateral load-displacement relations**

FIG. 7(a) shows the position of the piles in the pile group, among which No.4 pile located in front row is discussed in detail. It is known from FIG. 7(b) and (c) that test and calculated results agree well with each other and that the effective stress analysis performs better than total stress analysis when the maximum value and its occurring depth are concerned. FIG. 8 shows comparisons of test and calculated results of sectional forces at the loading of 8MN. The maximum bending moment are in the order of front pile, middle pile and rear pile. It is seen from the figure that the front pile bears larger bending moment than the rear pile because of the difference of axial forces, that is, compressive axial force in the front pile and tensile axial force in rear pile. It is seen that both the calculated and test results clearly reflect the phenomenon and agree well with each other, showing that the mechanical behavior of pile group can be described properly by using the axial force dependent (AFD) model in the numerical simulations.

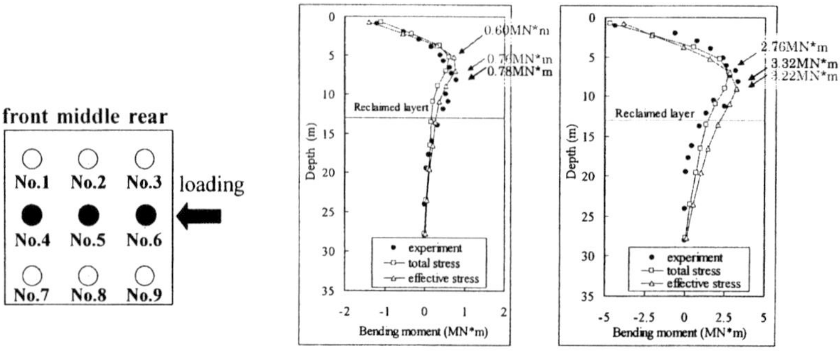

(a) Arrangement of test piles (b) Load=8MN (c) Load=20.5MN

FIG. 7. Comparisons of test and calculated results of bending moment in pile No.4

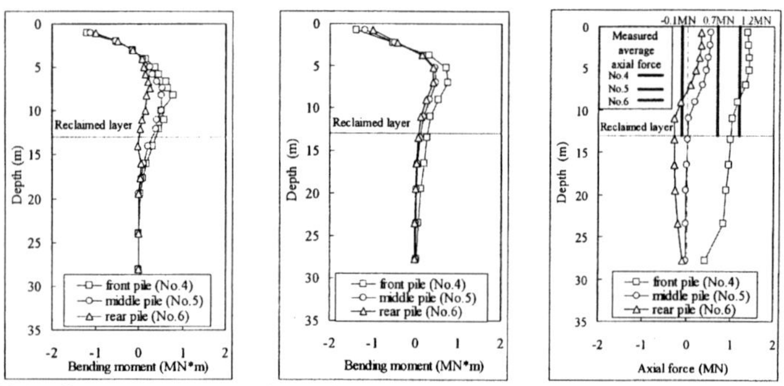

(a) Bending moment (Test) (b) Bending moment (Calculation) (c) Axial force

FIG. 8. Comparisons of test and calculated results of sectional forces at the loading of 8MN

Residual lateral displacements at the measuring points, obtained from field measurement and numerical analyses, are listed in Table 3. The calculated lateral displacements at the ground surface show reasonably good agreement with the field measurements.

FIG. 10 denotes the distributions of volumetric strain ε_v at the maximum loading of 20.5MN. In both analyses, the ground in the area near the front of the footing is compressive in a wide area; while the ground in the area behind the footing is expanded in a narrow area. In the field test, many cracks were observed in the ground near the footing, while in the calculation, large volumetric strain happened at the same place where cracks occurred in the field test.

FIG. 11 shows stress paths of ground at different positions during cyclic loading in which J_2 is the second invariant of the deviatoric stress tensor. The stress path of the element 1 located in front of the footing increases and decreases repeatedly with a relatively large stress range during the cyclic loading. The degree of change in the stress, however, is different in which, effective stress analysis gives a much larger change in the stress path, showing that dilatancy affects directly the stresses due to the consideration of soil-water coupling effect. As to the behaviors of elements 2 & 3, effective analysis gives a quite different description, compared to the total stress analysis, about the stress paths in the way that after the stress decreased to some extent, it will turn to increase again due to the dilatant behavior of sand after undergone some shear strain. In the total stress analysis, however, it is impossible to describe this behavior because dilatancy dose not affect the stress directly.

Table 3. Residual lateral displacement at the measuring points

Measuring point	Test resuslt (m)	Total stress result (m)	Effective stress result (m)
K-1	0.13	0.12	0.15
K-2	0.07	0.03	0.05
P-1	0.11	0.12	0.15
P-2	0.07	0.07	0.11
P-3	0.12	0.10	0.10
P-4	0.08	0.05	0.09

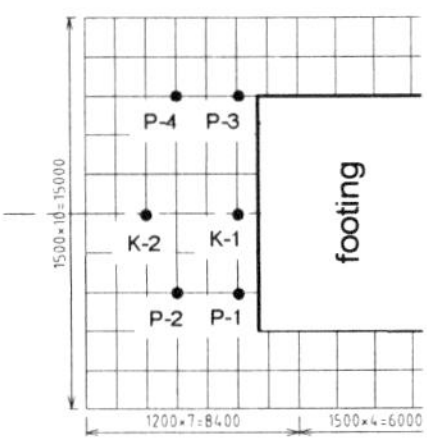

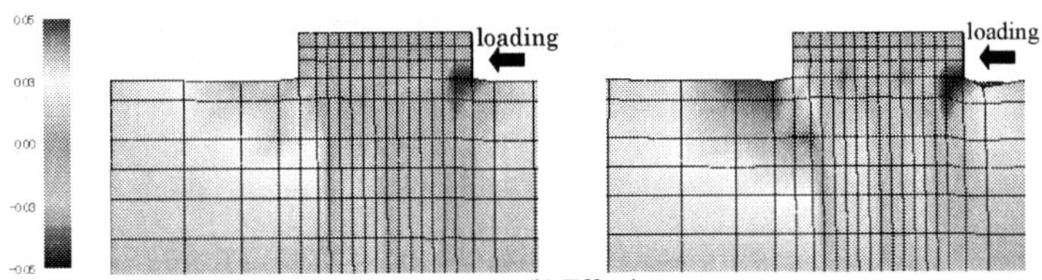

(a) Total stress (b) Effective stress

FIG. 10. Distributions of volumetric strain at the loading of 20.5MN

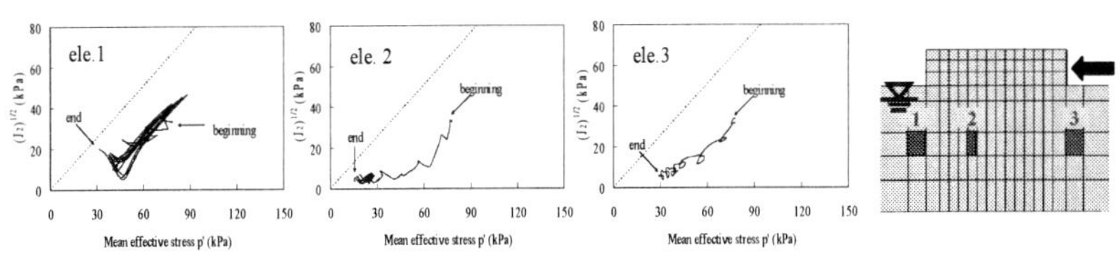

(a) Total stress analysis

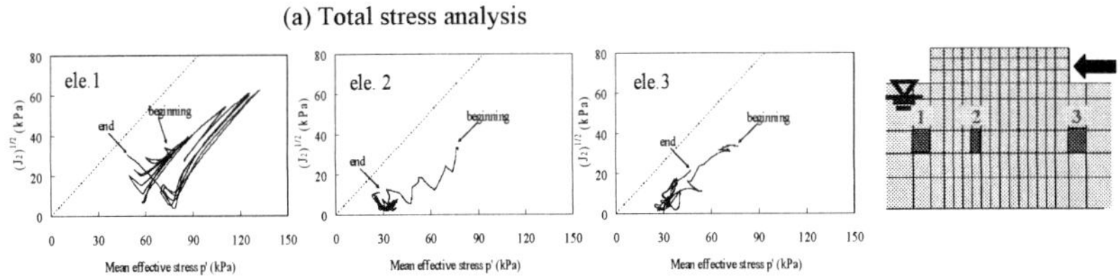

(b) Effective stress analysis

FIG. 11. Stress paths of the surrounding ground during cyclic loading

CONCLUSIONS

In this study, 3D finite element-finite difference soil-water coupling analyses on a real-scale group-pile foundation subjected to horizontal cyclic loading are conducted to investigate the mechanical behaviors of group-pile foundation. By comparing the test and the numerical results, the applicability of the proposed numerical method is verified and the following conclusions can be given.

1. Simulations of drained triaxial compression test are conducted to determine the material parameters of the reclaimed layer. The element simulation is extremely precise in describing the behaviors of the soil.
2. In the lateral load and displacement relations, the total stress analysis underestimates the test result, while the effective stress analysis considering soil-water coupling well predicts the test result at the maximum and residual displacements.
3. The calculated results of bending moment and axial force also agree well with the field observation. The difference of the bending moment due to axial force can be properly simulated by the calculation based on AFD model. Although the shapes of the distributions of the bending moment and the axial force are similar to the test results in both analyses, the results of the effective stress are much closer to the test results.
4. The lateral displacement of the ground surface at the end of the test shows reasonably good agreement with the field measurements.

From the results mentioned above, the applicability of the proposed numerical method is encouraging and therefore it is quite confident to say that the method can be used to predict the mechanical behaviors of group-pile foundation to a satisfactory accuracy, particularly with the effective stress analysis.

REFERENCES

Kimura, M. and Zhang, F. (1999). "Seismic evaluation of pile foundation by static and dynamic 3-D finite element analyses" *Proc. of the 11th Asian Regional Conf. of Int. Society for Soil Mechanics and Geothch. Engrg.*, Vol. 1, 507-510

Kosa, K. Suzuki, N., Kimura, R., Kimura, Y. and Morita, Y. (1998). "Lateral loading test of full-scaled pile foundation focused on ultimate behavior", *JSCE*, No.596/III-43, pp. 249-260, 1998. (in Japanese)

Nakai, T. and Hinokio, M. (2004). "A simple elastoplastic model for normally and over consolidated soils with unified material parameters", *Soils and Foundations*, Vol.44, No.2, 53-70

Tokimatsu, K., Suzuki, H., Tabata, K. and Sato, M. (2007). "Three dimensional shaking table tests on soil-pile-structure models using E-defense facility", *4th International Conference on Earthquake Engineering*, June 25-28, Thissaloniki, Greece

Ye, B. (2007). "Experiment and Numerical Simulation of Repeated Liquefaction -Consolidation of Sand", *Doctoral Dissertation, Gifu University*

Zhang, F., Kimura, M., Nakai, T. and Hoshikawa., T. (2000). "Mechanical behavior of pile foundations subjected to cyclic lateral loading up to the ultimate state", *Soils and Foundations*, Vol.40, No.5, 1-17.

Zhang, F. and Kimura, M. (2002). "Numerical prediction of the dynamic behaviors of an RC group-pile foundation", *Soils and Foundations*, Vol.42, No.3, 77-92.

Pile Diameter Effect on Bending Behaviour of Piles in Clay under Earthquake Loads

R. Ayothiraman[1], A.M. ASCE and G. Chandra Prakash[2].

[1]Assistant Professor, Department of Civil Engineering, Indian Institute of Technology Delhi, New Delhi–16. araman@civil.iitd.ac.in
[2]Formerly PG Student, Department of Civil Engineering, Indian Institute of Technology Guwahati, Guwahati–781039. prakash267@yahoo.co.in

ABSTRACT: Piles installed in seismically active regions are subjected to predominantly lateral loads due to earthquakes. Design of laterally loaded pile is mainly governed by its bending behaviour, which may be amplified under earthquake load compared to static loads. The influence of pile diameter on seismic response of single pile embedded in clay is investigated with a help of a three-dimensional finite element model. The model developed was validated through experimental/numerical results reported in literature. Subsequently a comprehensive parametric study was carried out by varying soil and pile parameters. The results of the numerical study are interpreted to estimate bending behaviour of piles under earthquake loads. It is found that the soil/pile characteristics, particularly, pile diameter have pronounced effect on seismic response of piles in clay.

INTRODUCTION

Pile foundations have been frequently used to support various structures and buildings including bridges since prehistoric time, mainly to by-pass the soft and weak shallow soils and to transfer loads from superstructure to appropriate depths. The coincidence of major structures sited on soft soils in earthquake prone areas also results in significant demand on these deep foundations. Investigation into the analysis of the damages to the civil engineering structures experienced during past and recent earthquakes such as San Francisco (1906), El Centro (1940), Alaskan (1964), Niigata (1964), Koyna (1967), Loma Prieta (1989), Latur (1993), Kobe (1995), Turkey (1999), Taiwan (1999) and Bhuj (2001) clearly demonstrate that the extent of damage to the structures is predominantly dependent on the type of foundation and soil conditions, in particular, liquefaction and/or strain-softening behaviour of soft soils existing in shallow depths (Bhattacharya & Bolton, 2004). This is observed even in the most recent earthquakes; Kashmir (2005) & Peru (2007). Earthquake load is acting predominantly in the lateral direction, but the stiffness in the lateral direction is very low in comparison to the vertical stiffness and hence the lateral capacity/stiffness of pile governs the design in most cases, in which the lateral load such as earthquake load is dominant. The lateral capacity, stiffness and its bending behaviour are mainly dependent upon the characteristic of top soil layers within a few meters of depth, which mainly consists of weak deposits such as soft clay or loose sand. Studies reported in the literature investigated the effect of pile length (*L*) and diameter (*d*), i.e. in the form of *L/d* ratio; however, in most cases, pile length was varied for a constant diameter to simulate different *L/d* ratios. But, the zone of stressed soil mass within the top soil layers is a function of pile diameter, from which the pile derives its passive resistance and hence it is very essential to investigate the influence of pile diameter rigorously on bending behaviour of pile under earthquake loads. This paper presents result of three-dimensional finite element analysis carried out to study the influence of pile diameter on seismic bending behaviour of piles in clay.

SEISMIC SOIL-PILE INTERACTION

Finite Element Modelling

Finite element method is well suited for analyzing problems, such as a pile/pile groups in layered soil, which is not easily handled with analytical or semi-analytical formulations. Also, hysteretic damping, Rayleigh damping and wave absorbing lateral boundaries can be introduced to account for damping characteristics and appropriate boundary conditions. In the present study, 3D finite element formulation is considered to model the seismic soil-pile interaction using ANSYS. The soil has been discretized as eight-noded brick elements and pile as a beam element. Each of the 8 nodes has three translational degrees of freedom in the nodal x, y, and z-directions. The pile is completely embedded in the soil. Bentley and El Naggar (2000) and Maheshwari et al. (2005) had used these elements successfully for modeling the dynamic response of piles; hence the similar model elements are used in the present study. Soil is modeled as elastic-perfectly plastic material with Mohr-Column failure criterion. Pile is modeled as linear elastic.

To simulate an infinite soil medium, springs and dashpots are attached on the side walls of the foundation which provide the proper boundary conditions. It is noted that these are used in all three directions along the boundary. The coefficients of the springs and dashpots are derived separately for the horizontal and vertical directions based on the predominant frequency of loading. The constants of the springs and dashpots in the two horizontal directions were calculated using the solution developed by Novak & Mitwally (1988). This also simulates the radiation damping/wave absorbing conditions at the boundary. Material damping is considered in the analysis. Additional hysteretic damping may develop due to the nonlinearity, but dissipation of seismic energy through inelastic deformation tends to overshadow the dissipation of the energy through hysteretic damping and is therefore neglected. The developed 3-D FE model is shown in Figure 1.

Validation of Model

The efficacy of the proposed model is demonstrated by comparing the predicted response of pile with the experimentally measured single pile response. Three well instrumented full-scale experiments are selected where the pile response was recorded beyond the elastic limit. Kramer et al. (1990) conducted lateral monotonic and cyclic tests on a cylindrical steel pile installed in the Mercer Slough peat near the eastern shore of Lake Washington in Bellevue, Washington. Jennings et al (1984) and Brown et al. (1988) conducted dynamic experiments on solid and hollow piles embedded in medium dense and saturated silty sand deposits respectively. The soil properties and pile characteristics are summarized in Table 1, and these properties were used in the analysis to predict the pile response.

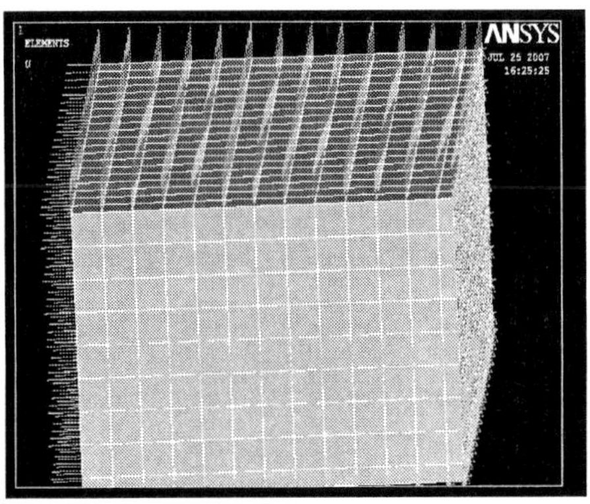

FIG. 1. 3D–finite element model.

Table 1. Properties of Soil and Pile

Test/Year	Soil Properties	Pile Properties
Kramer et al. (1990)	ν_s= 0.49; $V_{s(tip)}$ = 30 m/s ρ_s =1120 kg/m3; c_u=14.4 kN/m2	L = 14.9 m; d = 0.20 m E_pI_p = 3.81 × 10^3 kN-m^2
Jennings et al. (1984)	ν_s = 0.49; $V_{s(tip)}$ = 125 m/s ρ_s = 1600 kg/m^3; ϕ'=38.5^0	L = 6.75 m; d = 0.45 m E_pI_p = 0.8 × 10^5 kN-m^2
Brown et al. (1988)	ν_s = 0.48; $V_{s(tip)}$ = 160 m/s ρ_s = 1600 kg/m^3	L = 13.1 m;d = 0.273 m t = 0.0093 m; E_pI_p = 7.3 × 10^3 kN-m^2

The pile responses, such as pile deflection behaviour at 120kN load and p-y curves at different depths are obtained from the analysis. The results are compared with the experimental data of the respective literature and the results of numerical analysis reported by Badoni and Makris (1996). Figure 2 & 3 show the comparison of predicted and measured pile response and it is found from the figures that the agreement of the prediction with the measured values is very good.

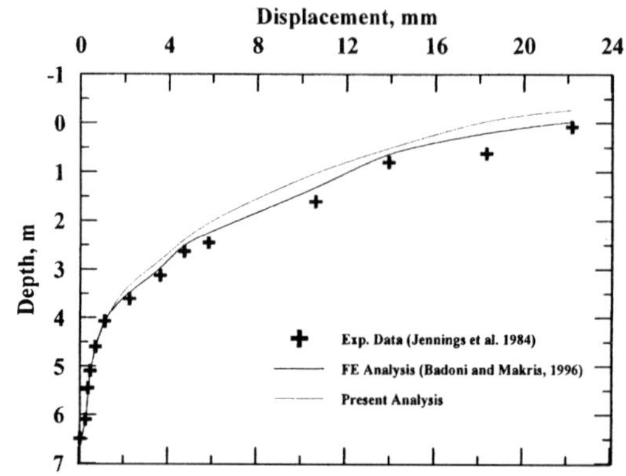

FIG. 2. Comparison of predicted and measured pile deflection and for lateral force of 120kN.

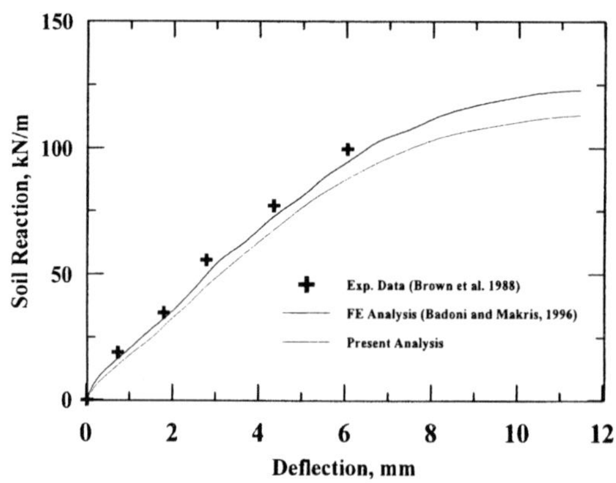

FIG. 3. Comparison of predicted and measured p-y curve at a depth, 1.22m.

Parametric Studies

Parametric analysis for single pile under seismic loads was performed by varying the different parameters like pile diameter (d = 0.3, 0.5, 0.7 & 1.0 m), pile slenderness ratio (L/d = 10, 20, 30 & 40) and soil modulus (E_s = 20, 40, 60 & 80 MPa) etc. using the proposed FE model. A constant material damping of 20% is used in the analysis for all cases, whereas the radiation damping at the boundary of the model is modeled with springs and dashpots; their constants are calculated using

Novak & Mitwally (1988). The seismic loading is applied at the base of the 3-D model, i.e. at the bottom face of the soil block/model as transient motion. For the transient motion, an acceleration time history for the El Centro Earthquake 1940 (N-S Component), with peak ground acceleration equal to 0.32g was used.

RESULTS AND DISCUSSION

Pile Deflection Profile

Pile deflection profiles are plotted against the normalized pile depth (*z/d*), i.e. the ratio of desired depth along the pile (*z*) to pile diameter (*d*). Typical pile deflection profile for *L/d*=10; E_s = 20 MPa is shown in Figure 4. It is observed from Fig. 4 that the pile deflection is reducing with an increase of pile diameter. This is due to fact that the passive resistance zone increases with an increase of pile diameter and pile length.

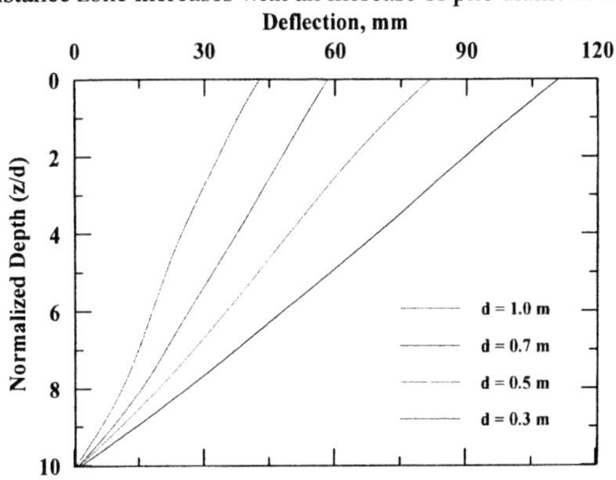

FIG. 4. Typical pile deflection profile for *L/d*=10 at E_s = 20 MPa.

Maximum Pile Deflection

The maximum deflection of the pile was measured from the pile deflection profiles of different cases. The variation of maximum deflection with L/d ratio of pile for different diameters at Es = 20 MPa is shown in Figure 5. It is inferred from Fig. 5 that the maximum deflection decreases substantially with an increase of pile diameter and length. This may be due to increase of zone of passive resistance under lateral seismic shaking. This observation is consistent with pile embedded in other soil modulus conditions (soft to medium stiff clay). However it is noted that at higher soil modulus values, the rate of reduction of maximum deflection is relatively less when compared to the one in very low soil modulus (E_s = 20 MPa). This is probably because of higher hysteretic damping in addition to radiation damping that prevails in very soft clay (E_s = 20 MPa), thus leading to high rate of reduction of maximum deflection.

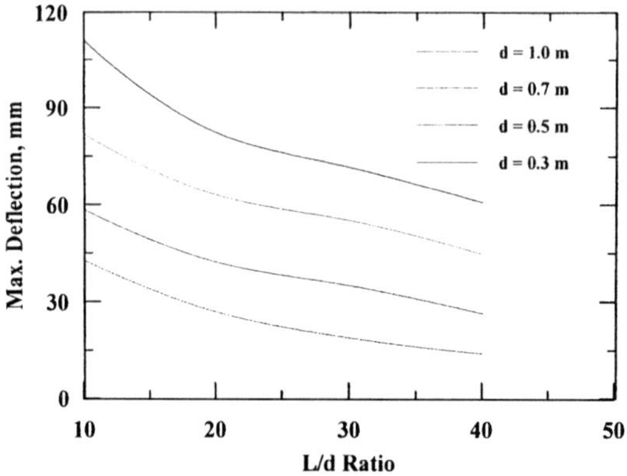

FIG. 5. Max. Deflection Vs *L/d* ratio for different pile diameters at E_s = 20MPa.

Pile Bending Moment

The predicted pile bending moment profiles are plotted against the normalized depth of pile (*z/d*) and the typical bending moment profile for Es = 40 MPa is shown in Figure 6. It is observed that bending moment profiles are almost following the same trend, but it is only the magnitude of bending moment that increases with increase in the pile diameter. This may be due to the increase of the flexural rigidity with pile diameter which indirectly causes the increase in bending moment. Also it is found that the bending moment increases with pile length, since the long piles undergo larger bending because of its flexible nature.

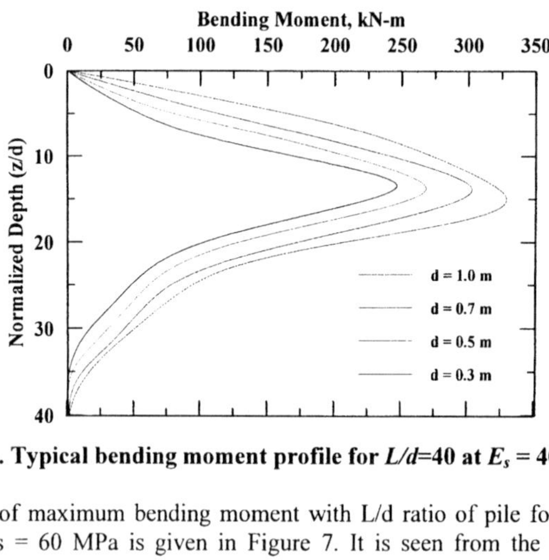

FIG. 6. Typical bending moment profile for *L/d*=40 at E_s = 40 MPa.

The variation of maximum bending moment with L/d ratio of pile for different pile diameter at Es = 60 MPa is given in Figure 7. It is seen from the figure that the maximum bending moment substantially increases with pile length. Also the figure indicates that the pile diameter has pronounced effect on the magnitude of maximum bending moment of piles under seismic loads, particularly for long flexible piles. Similar observation is noticed for piles embedded in other consistencies of clay.

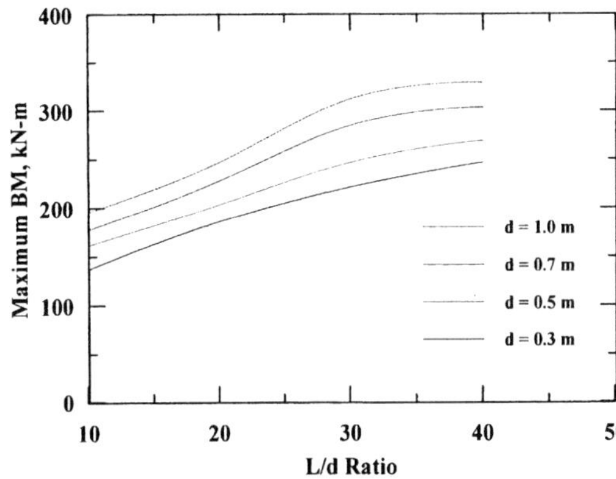

FIG. 7. Maximum BM Vs *L/d* ratio for different pile diameters at E_s=60 MPa.

Effective Pile Length

The effective pile length is defined as a pile length from the ground surface at which the pile deflection is zero. The effective pile length values, i.e. the normalized depth of zero deflection were obtained from the pile deflection profiles and are referred as normalized effective pile length. The variation of effective pile length with *L/d* ratio of pile for different diameters at E_s = 20 MPa is shown in Figure 8.

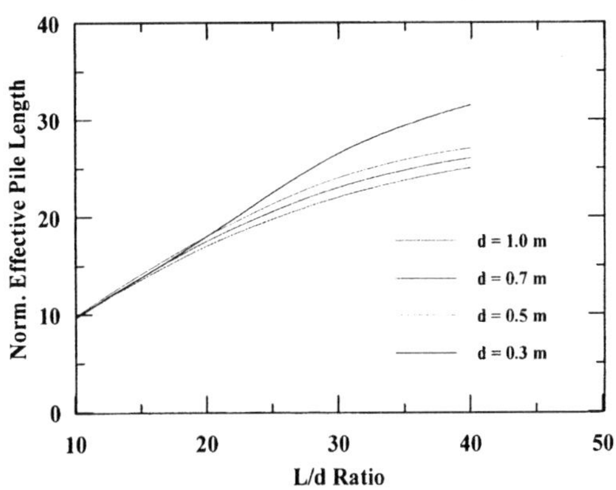

FIG. 8. Normalized effective pile length Vs *L/d* ratio at Es = 20 MPa.

It is inferred from Fig. 8 that the effective pile length increases with pile length for all soil modulus conditions, but the rate of increase is high for low soil modulus. This is because that the almost full length of pile undergoes vibration during seismic shaking for piles embedded in very soft clay having low soil modulus. It is also inferred from Figure 8 that the pile diameter does not have significant effect on the effective pile length of short piles (L/d<20), but has a significant effect for long piles (L/d>30) embedded in all soil modulus conditions considered in the analysis. Moreover, the effective pile length is an important parameter for long piles only, i.e. for L/d>30, which is significantly affected by the pile diameter.

The variation of normalized effective pile length with modulus ratio (E_p/E_s), for different L/d ratio of pile and constant diameter of pile of 0.5 m is given Figure 9. It is noticed from the figure that the effective length significantly reduces with decrease of modulus ratio, i.e. with an increase of soil modulus for all L/d ratios. The effective length estimated using existing analytical and semi-analytical expressions suggested by various authors (Krishnan et al. 1983; Gazetas, 1984; Poulos & Hull, 1989; Gazetas, 1991; Ayothiraman & Boominathan, 2008) for any kind of dynamic loads are also presented in Figure 9.

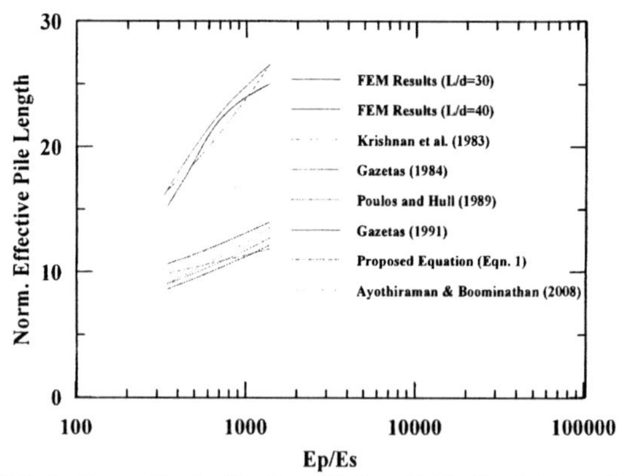

FIG. 9. Normalized effective pile length Vs Modulus ratio.

It is found from Figure 9 that these equations highly underestimate the effecting pile length of long flexible piles under earthquake loads. In other words, one can say that the effective pile length under earthquake load is significantly amplified even compared to other dynamic loads. Hence an equation is developed by performing multiple regression analysis to estimate the effective pile length of long piles under earthquake load as given below:

$$\frac{L_e}{d} = 2.20\left(\frac{E_p}{E_s}\right)^{0.345} \qquad (1)$$

In which, L_e = Effective pile length under earthquake load, d is pile diameter, E_p is Young's modulus of pile and E_s is Young's modulus of soil. The regression analysis gives a correlation coefficient of 0.9465. The proposed equation can be used to find out the effective length of single piles subjected to earthquake load. It is to be noted here that the effect of different ground motion characteristics and magnitude of earthquake are not studied in Fig. 9 and Eqn. 1, hence the proposed equation needs to be modified by taking it into account of these effects in further study.

SUMMARY AND CONCLUSIONS

The seismic response of single pile predicted by a 3D finite element model is presented in this paper and the various response quantities such as pile deflection and bending behaviour are discussed. An equation is developed to estimate the effective pile length under earthquake load, which can be used to find out the minimum pile length requirement under earthquake load. Some of the major conclusions arrived based on this comprehensive study are given below:

- Pile length/diameter, *L/d* ratio and soil modulus significantly influences the behavior of piles under earthquake loads.
- Maximum pile deflection is consistently reducing with increase in the pile length, pile diameter and soil modulus. However, it is found that the rate of reduction of maximum deflection with increase of pile length at higher soil modulus, i.e. stiff clay, is not significant.
- It is also found that the bending moment under earthquake load is substantially increasing with increase of pile length and diameter.
- The effective pile length is significantly influenced by the soil modulus and pile diameter, especially for long piles.
- An equation is developed to estimate the effective pile length under earthquake loads for single piles. It is to be noted that the effect of magnitude of seismic excitation on the proposed equation is not investigated by varying different earthquake ground motion data in the present study and hence the equation may need to be modified, if necessary based on further investigation.

REFERENCES

Ayothiraman, R. and Boominathan, A. (2008). Dynamic bending behaviour of piles in clay. *Journal of Geomechanics and Geoengineering*, Under Review.

Badoni, D., and Makris, N. (1996). Nonlinear response of single piles under lateral inertial and seismic loads. *Soil Dynamics and Earthquake Engineering*, 15, 29–43.

Bhattacharya, S. and M. D. Bolton. (2004). A fundamental omission in seismic pile design leading to collapse, in *Proceedings of the 11th International Conference on Soil Dynamics and Earthquake Engineering*, Berkeley, USA, 1, 820 – 827.

Bentley, K. J. and El Naggar, M. H. (2000). Simplified BNWF model for nonlinear seismic response analysis of offshore piles with nonlinear input ground motion analysis. *Canadian Geotechnical Journal*, 32, 183–196.

Brown, D. A., Morrison, C. & Reese, L. C. (1988). Lateral load behavior of pile group in sand. *Jl. Geotechnical Engineering*, ASCE, 144, 30-45.

Gazetas, G. (1984). Seismic response of end-bearing piles. *Soil Dynamics and Earthquake Engineering*, 3, 82 – 93.

Gazetas G. (1991). Foundation vibrations. *Foundation Engineering Handbook*, 2nd Edition, Van Nostrand Reinholds, 553-593.

Jennings D. N., Thurston, S. J. and Edmonds, F. D. (1984). Static and dynamic lateral loading of two piles. *Proc 8th World Conference on Earthquake Engineering*, San Francisco, CA. (3), 561-68.

Kramer, S. L., Satari, R. and Kilian, A. P. (1992). Evaluation of in situ strength of a peat deposit from laterally loaded pile test results. *Transportation Research Record No. 1278*, Transp. Res. Board, Washington, D.C., 103-109.

Krishnan, R., G. Gazetas, and A. Velez. (1983). Static and dynamic lateral deflection of piles in non-homogeneous soil stratum, *Geotechnique*, 33, 307 – 325.

Maheshwari, B. K., K. Z. Truman, P. L. Gould, and M. H. El Naggar. (2005). Three-dimensional nonlinear seismic analysis of single piles using finite element model: Effects of plasticity of soil. *Journal of Geomechanics*, ASCE, 5, 35-44.

Novak, M., and Mitwally, H. (1988). Transmitting boundary for axisymmetrical dilation problems. *Journal of Engineering Mechanics*, ASCE, 114(1): 181–187.

Poulos, H. and Hull, T. (1989). The role of analytical geomechanics in foundation engineering. *Foundation Engg: Current Principles and Practices*, ASCE, 2, 1578-1606.

Predictions of Pile Setup and its Resistance Factors for South Louisiana

X. Wang[1], Eric Steward[2]

[1] Associate Professor, Department of Civil Engineering and Construction Engineering Technology, Louisiana Tech University, Ruston, LA 71272; xwang@latech.edu
[2] PhD Candidate, Trenchless Technology Center, Louisiana Tech University, Ruston, LA 71272; ejs018@latech.edu

ABSTRACT

A growth-rate based model has been briefly introduced to predict pile setup for the south Louisiana clayey soils based on the PDA data collected by the Louisiana Department of Transportation and Development (LADOTD) from the construction of the LA-1 relocation project. This paper presents the development procedures of the rate-based model. Predictability of this model was evaluated by using the dynamic, static, and statnamic testing results, and by comparing it with the conventional Skov-Denver model. In the paper, preliminary work was done to incorporate the pile setup into the regular LRFD method by providing independent resistance factors for the rate-based and the Skov-Dever prediction models, respectively.

INTRODUCTION

Geotechnical engineers and researchers (Long et al. 1999, Bullock et al. 2005) have reported for many years that the axial capacity of a driven pile may increase over time, which is usually referred to as pile setup or freeze. For instance, shaft resistances of some driven piles at the LA-1 relocation project gained as much as 30 to 100% of the initial resistance during the first week of driving. Pile setup has been studied for several decades, and empirical, semi-empirical, analytical and numerical techniques have been proposed and developed.

Although great progress has been made in understanding the short term and long term pile resistance increases, geotechnical engineers currently still rely on empirical correlations and personal judgment and experience developed in local areas to estimate pile setup. Therefore, there is a need for developing a reliable design methodology that will account for the benefit of pile setup so that a more cost-effective pile design may be used in the future. An accurate prediction model for pile setup is expected to be developed for use. For this purpose, a growth rate-based model was developed in the research.

The data used for this study include 115 restrike records from 95 production piles and the records of restrikes, static, and stanamic load tests of nine fully instrumented test piles. All these piles were driven in the typical soft clays in southern Louisiana. The upper 25 feet of the soils include some peats and organic rich clays. The mudline is around 2~4 feet below the water surface (Soil Testing Engineers, Inc. 2004).

A BRIEF INTRODUCTION TO THE PILE TESTING DATA

All the production pile restrike data came from the Phase 1B of the LA-1 relocation project, which consists of the construction of a 4-mile long high-level bridge with connecting ramps and interchanges. The 16", 24", & 30" prestressed concrete (PSC) piles were used extensively in the project. A total of 115 restrike records on 95 piles have been gathered from the four construction segments: the North Connector, the South Connector, the Mainline-span over Bayou Lafourche, and the Ramp N1. All the data from the nine load-tested piles were collected at the four locations along the new LA-1 alignment for the Phase-1 project.

Skin friction, tip, and total capacities are available for each restrike event. Those pile capacities were obtained from the signal matching (CAPWAP) analyses. Observations indicate that the end bearing appears to be constant or has insignificant setup as compared with the shaft capacity. Therefore, the prediction models in this paper are only for shaft resistances. The production piles are not at the same diameter and were driven to different depths and embedded into various soil strata. As such, instead of the total or friction resistance, the average unit skin friction was used as the parameter of interest in which the 24-hour restrike data was used for the reference parameters (S_o and t_o) for the production piles, in most cases. The normalized unit skin friction (S/S_0) was obtained by taking the measured average unit skin friction at the restrike time divided by the average unit skin friction at the reference time. The normalized time (t/t_0 or T/T_0) was defined as the ratio of the restrike time to the reference time.

RESULTS OF THE SKOV-DENVER MODEL DEVELOPMENT

The empirical model of semi-logarithmic relationship proposed by Skov and Denver (1988) is commonly employed by many researchers and engineers. It reads:

$$\frac{Q}{Q_0} = A \log_{10}\left(\frac{t}{t_0}\right) + 1 \qquad (1)$$

where, A is the dimensionless setup parameter, t_0 is the reference time, and t is the time elapsed since the end of initial driving. A and t_0 are the parameters used to characterize piles and soils. Q and Q_0 are either the total or shaft pile capacity at time t and t_0, respectively. For the purpose of this study, Q and Q_0 can be replaced by S and S_0, the unit skin friction at any time and the unit skin friction at the reference time, respectively.

The least-square method was conducted using all the pile testing data (restrikes of the production piles, restrikes, static and statnamic tests of the test piles) at the LA-1 relocation project. The Skov-Denver model is developed as follows:

$$S(t) = S(t_0)(0.570\log\left(\frac{t}{t_0}\right) + 1) \tag{2}$$

where, t_0 is taken as 24 hours. Similar models were also developed on the different construction segments of the LA-1 project site, and the nine test piles. Overall, the pile setup parameters A range mainly from 0.5 to 0.7, as shown in Table 1.

Table 1: Skov-Denver Models from the Restrike Data of the Production Piles and the Nine Test Piles.

Data Source	Mainline -S	North Connector	South Connector	Ramp-N1	The combined data	The nine test piles
Parameter A	0.362	0.717	0.547	0.521	0.547	0.670
Reference time t_0 = 24 hours for the production piles, and 20.6~24.7 for the test piles.						

DEVELOPMENT OF THE RATE-BASED MODEL

An ideal pile setup prediction model should be able to predict the ultimate capacity – the maximum capacity that can be reached after pile installation, and the time it takes to achieve the pile resistance designers intend to use. The Skov-Denver model cannot predict ultimate pile resistances. Motivated by the expectation, the pile setup data were re-examined and a rate-based model was proposed and developed.

There are a total of six restrike records from the piles at Bent NC29, and nine pile capacity records from the cylinder test pile at test site No. 3. Restrike time, skin friction resistance, the calculated average unit skin friction, and the unit growth rate of the skin friction are presented in Tables 2 and 3, respectively. The unit growth rate is defined as:

$$q(t) = \frac{1}{S(t)}\frac{dS(t)}{dt} \cong \frac{1}{S(t)}\frac{(S(t+\Delta t)-S(t-\Delta t))}{2\Delta t} \tag{3a}$$

$$\cong \frac{1}{S(t)}\frac{(S(t+1)-S(t-1))}{T/T_0(t+1)-T/T_0(t-1)} \tag{3b}$$

The growth rates are calculated and presented in the last column of each table, with Equation 3b for Table 2, and Equation 3a for Table 3, respectively. It can be observed that the unit skin friction growth rate is the largest immediately after the pile installation. It reduces with the increase in the skin friction resistance. The resistance must eventually stop growing after a certain period of time, which indicates that a theoretical ultimate shaft capacity may be reached. Based on the observation, it is reasonable to assume that the unit skin friction growth rate is a function of the initial growth rate R and the magnitude of the unit skin friction and it decreases with time until it reaches zero. It is written as (Mesterton-Gibbons 1989):

$$q(t) = \frac{1}{S(t)}\frac{dS(t)}{dt} = R\left(1 - \frac{S(t)}{S(\infty)}\right) \tag{4}$$

where, $S(\infty)$ is the ultimate unit skin friction based on the data displayed in Table 2. A similar equation can be established for the shaft resistances presented in Table 3, and then $S(\infty)$ is the ultimate skin friction. In equation (4), S(t) might also represent the unit skin friction, or shaft resistance, and t can be replaced by a dimensionless time factor t/t_0 (or T/T_0). Solving differential equation (4), the closed-form solution is:

$$S(t) = \frac{S(\infty)S(0)}{S(0)+\left(S(\infty)-S(0)\right)e^{-R(t-t_0)}} \tag{5}$$

where, S(0) is the unit skin friction (or skin friction) at the reference time of $t_0 = 24$ hours. Using all the collected data for the established Skov-Denver model, the rate-based model was developed based on the normalized unit skin friction by following the least squared method. Its equation is written as:

$$S(t) = \frac{1.846S(t_0)}{1+0.846e^{-0.261(\frac{t}{t_0}-1)}} \tag{6}$$

In addition, the rate-based model was also developed for the four segments of the LA-1 site, and the combined data of the nine test piles, respectively. All the model parameters are given in Table 4.

Table 2: Restrike Records and Skin Friction Growth Rates for Bent NC29.

Pile	Time	T/T_0	R_{skin}	Unit skin S(t)	$\Delta S/\Delta(T/T_0)/S$
	hrs		kN / kips	kN/m² / kips/ft²	
NC29-03	24	1	947.4 / 213	11.13114 / 0.23248	
NC29-03	144	6	1,205.4 / 271	14.19402 / 0.29645	0.01915379
NC29-03	672	28	1,925.9 / 433	18.47163 / 0.38579	0.018356009
NC29-02	744	31	1,570.1 / 353	22.67070 / 0.47349	0.005170061
NC29-03	1728	72	2,006.0 / 451	23.62878 / 0.49350	0.000924159
NC29-02	1728	72	2,001.6 / 450	23.56606 / 0.49219	

Table 3: The Unit Skin Friction Growth Rates for the 54-in Cylinder Pile at Testing Site No. 3.

Event	t (hours)	Ru (kips)	Rs (kips)	$\Delta S/\Delta(t)/S$
End of Driving	0.0	378	287	
Restrike 2 hrs	2.0	696	596	0.1733781
Restrike 4 hrs	3.9	798	690	0.0185149
Restrike 24 hrs	24.7	1027	886	0.0078698
Restrike 48 hrs	44.2	1112	971	0.0030227
Restrike 72 hrs	72.4	1169	1026	0.0017709
Restrike5 days	117.4	1247	1104	0.0007025
Restrike 12 days	287.7	1337	1193	0.0006005
Load Test	384.0	1395	1295	

As shown in Equation 6 and Table 4, the ultimate normalized unit skin frictions vary slightly. These values are stable and close to 2.0, with the smallest number of 1.4 at the South Connector, and the largest number of 2.2 at the North Connector. It implies that the ultimate skin friction was around twice as much as the skin friction measured at the 24-hour restrike. Based on the model presented as Equation 6, at the site of LA-1B relocation project, the shaft capacity generally gained 90~95% of the ultimate shaft capacity two weeks after the pile installations.

Table 4: The Two-parameter Rate-based Models for the Different Data Sources.

Data source	North Connector	South Connector	Main line-S	Ramp N1	All the production piles	The nine test piles
Initial growth rate R	0.238	4.131	0.069	0.221	0.213	0.308
Ultimate normalized unit skin friction S_∞/S_0	2.161	1.414	1.857	1.872	1.865	1.840

RESISTANCE FACTORS ACCOUNTING FOR PILE SETUP EFFECT

Due to the limited availability of pile setup data, the reliability analysis was only conducted on one category: the concrete square piles at the LA-1 relocation project. The resistance factors were calculated using the Mean-Value-First-Order-Second-Moment method (MVFOSM). The research has employed the load statistics and the load factors from the latest AASHTO LRFD Specifications (AASHTO, 2007) to make the pile foundation design consistent with the bridge superstructure design. The load combination of dead load (QD) and live load (QL) for the Strength I Case was chosen for the calibration of the resistance factors. Two random variables, the load (Q) and the resistance I, are assumed to be lognormally distributed. As specified in the AASHTO LRFD Specifications, the resistance factor for MVFOSM is given as follows:

$$\emptyset = \frac{\lambda_R\left(\gamma_D\frac{QD}{QL}+\gamma_L\right)\sqrt{\frac{1+COV_{QD}^2+COV_{QL}^2}{1+COV_R^2}}}{\left(\lambda_{QD}\frac{QD}{QL}+\lambda_{QL}\right)EXP\left\{\beta_T\sqrt{\ln\left[\left(1+COV_R^2\right)\left(1+COV_{QD}^2+COV_{QL}^2\right)\right]}\right\}} \quad (7)$$

where, the factors of loads are: $\gamma_{QL} = 1.75$, $\gamma_{QD} = 1.25$, $\lambda_{QL} = 1.15$, $\lambda_{QD} = 1.05$, $COV_{QL} = 0.2$, and $COV_{QD} = 0.1$. In the pile foundation design, the level of safety should be consistent in the LRFD-based pile foundation design, and a constant target reliability index should be used. Four target reliability indices of 2.0, 2.33, 2.5 and 3.0 are selected, and the ratio QD/QL is taken as 1.0, 2.0, 2.5, 3.0, and 4.0, respectively.

After pile installation, pile capacities corresponding to different restrike times are different. Thus, different resistance factors should be used for different restrike times. However, because there is limited amount of pile testing data available, only one resistance factor was computed for all the measured pile capacities at different setup times. It is assumed that the bias factors of pile resistances are independent of the setup time, and thus all the bias factors at different setup times are combined to calculate the average bias factor λ_R, which is based on the predicted and measured shaft capacities. Values of the bias factors for the Skov-Denver and the rate-based models are 0.997 and 0.991, respectively. The coefficient of variation COV_R was calculated using the measured 24-hour restrike capacities, which is assumed to apply to the measured capacities at different setup time. It turned out to be 0.371. The calibrated resistance factors are as follows:

Table 5: Resistance Factors ϕ for the Two Prediction Models.

	Q_{DL} =1.0 β_T=2.0	Q_{DL} =2.0 β_T=2.5	Q_{DL} =3.0 β_T=3.0	Q_{DL} =4.0 β_T=2.33	Q_{DL} =2.5 β_T=2.33
Skov-Denver	0.63	0.50	0.41	0.51	0.53
Rate-based	0.62	0.50	0.40	0.51	0.52

CONCLUSIONS

A new growth-rate based pile capacity prediction model has been proposed. The model indicates that the pile capacity growth rate will approach zero eventually, suggesting the leveling off of the capacity increase with time. The model has provided the ultimate pile capacity as expected by foundation engineers. In general, the piles at the LA-1 relocation project reaches about 90~95% of the ultimate shaft capacities within two weeks after installation. The ultimate shaft capacities of the piles were about twice the measured shaft capacities at the 24-hour restrike. Predictions from the rate-based model are very consistent with those from the Skov-Denver models. Resistance factors for the Skov-Denver and the rate-based prediction models are very close, and they are between 0.5 and 0.6 for the regular target reliability indices and the dead to live load ratios.

ACKNOWLEDGMENTS

The research work presented herein was performed as part of Louisiana Transportation Research Center (LTRC) contract No. 04-1GT. The support and assistance of the LTRC office and LA DOTD personnel are gratefully acknowledged. The first author would like to thank Dr. Ching Tsai and Dr. Doc Zhang for their comments, suggestions and the enormous information they have provided.

REFERENCES

Bullock, P.J. and Schmertmann, J.H., McVay, M.C., and Townsend, F.C. (2005). "Side shear setup. I: Test piles driven in Florida." *Journal Geotechnical and Geoenvironmental Engineering*, Vol. 131, No. 3, pp. 292-300.

Mesterton-Gibbons, M. (1989). *A Concrete Approach to Mathematical Modelling*, Addison-Wesley Publishing Company.

Long, J.H., Kerrigan, J.A., and Wysockey, M.H. (1999). "Measured Time Effects for Axial Capacity of Driven Piling." *Transportation Research Record 1663,* Paper No. 99-1183, pp. 8-15.

Skov, R. and Denver, H. (1988). "Time-dependence of bearing capacity of piles." *Proc. 3rd Int. Conf. on the Application of Stress-Wave Theory to Piles*, B.G. Fellenius, ed., BiTech Publishers, Vancouver, BC, 879-888.

Soil Testing Engineers, Inc. (2004). *Geotechnical Interpretation Report, Phase 1B, LA-1 Relocation*. LaFourche Parish, Louisiana. Baton Rouge, Louisiana.

Experimental Study of Model Piles Socketed into Rock with a Hanging Base

Ningzhong Xi[1], Jinli Liu[1] and Hong Yang[2]

[1]Institute of Foundation Engineering, China Academy of Building Research, Beijing, 100013, China.
[2]Parsons Brinckerhoff, 303 2nd St, 700N, San Francisco, CA 94103, USA; yangh@pbworld.com

ABSTRACT: To investigate the effect of pile-rock adhesion condition and rock stiffness on both socket friction in rock and side friction in the overlying soil of piles socketed into rock, pile load tests were performed on four instrumented model piles with hanging bases (i.e. the pile ends were unsupported). Cement grout and lean concrete were used to simulate strata of weak and hard rock, respectively; and slurries of different concentrations were applied to the pile-rock interfaces to simulate different pile-rock adhesion conditions. The overlying soil layer was comprised of compacted silt. The load test results indicate that the side resistance of the piles in the soil layer could fully develop even when the pile tip displacement was relatively small; the side resistance distribution was of a saddle shape in general with higher values at the top and bottom and lower values in middle of the soil layer.

INTRODUCTION

Side resistance behavior of piles socketed in rock have been studied and reported by many authors, e.g. Williams and Pells (1981), Johnston and Haberfield (1992), and Nam and Vipulanandan (2008). According to a recent comprehensive review performed by Zertsalov and Konyukhov (2007), the bearing capacity of a pile in rock is determined by many factors such as the unconfined compressive strength of the rock, the ratio of the diameter of the pile to the depth of pile embedment and depth of rock socket, the ratio of the compression modulus of the rock mass to the elastic modulus of the concrete pile, the conditions at the contact between the perimeter surface of the pile and rock, as well as procedural characteristics of the work performed in installing the piles.

During the pile installation, slurry is often used to stabilize the shaft wall. The uncleaned debris or slurry will affect the adhesion at pile-rock interface and load-carrying characteristics of the piles. Further, the stiffness (or strength) of the rock also affects the load-carrying characteristics of the piles. To investigate these two factors (pile-rock adhesion condition and rock stiffness) on side resistance behavior of the piles, four model piles with a hanging base (i.e. the pile tip was separate from the rock and unsupported) were constructed, instrumented and load-tested, and the results are presented and discussed in the paper. The hanging base helped to eliminate the complication of the end bearing of the pile, thus the axial load was fully carried by the pile side (pile shaft).

TEST METHOD

The model piles were 3.0 m long, 200 mm diameter precast concrete piles. The piles were embedded in 2.80 m thick compacted silt overlying a 200 mm thick simulated rock stratum. The rock stratum was formed using cement grout (unconfined compressive strength of 5.0 MPa) or lean concrete (unconfined compressive strength of 30.0 MPa) to simulate soft rock or hard rock, respectively, in which the lower portion of the pile (the 200 mm pile segment) was embedded. The entire test set-up (i.e. the pile, silt and rock) was constructed in a pre-excavated hole in a firm ground, thus the silt and rock was constrained in the ground laterally, and the rock was also supported by the firm ground (Figure 1). The silt was compacted to dense conditions in lifts and had a total density of about 19 kN/m^3 and moisture content of about 10 to 15%. The pile-rock interface was coated with thick or thin mud slurry (i.e. large or small amount of mud) to simulate weak or strong adhesion conditions, respectively, prior to placing the 200 mm thick "rock" stratum and the 2.80 m soil layer. The model piles and their socket conditions are summarized in Table 1.

Each of the piles had a hanging base, i.e. the end of the pile was unsupported and below the pile tip was void (about 40 mm high). Therefore, all the applied axial compression loads were carried by the pile side. The instruments used in the tests include strain gauges, load cells and settlement gauges. Pile top settlement was directly measured and the pile length decrease due to elastic compression was obtained by the strain gauge measurement. Thus, the corresponding pile tip settlement was calculated and the effect of the pile compression was disregarded. From the strain gauge measurement, the skin friction along the pile side at the various applied axial loads was also computed. Further, the pile resistance (capacity) based on the skin friction was calculated as a check, which agrees well with the applied axial load (the difference is generally less than 1%).

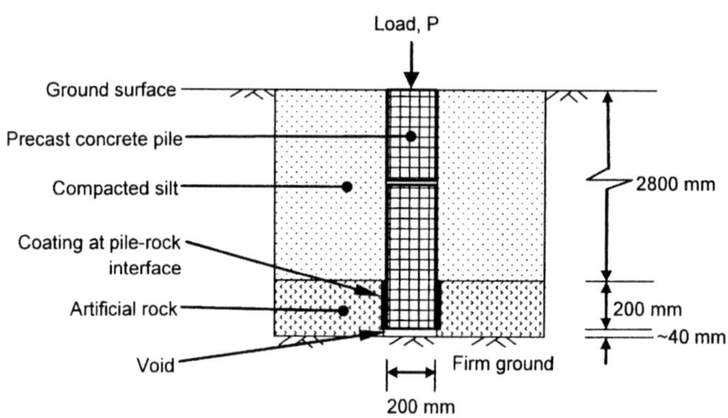

Figure 1. Schematic of set-up for the model pile

Table 1. Summary of the Model Piles and Socket Conditions

Pile No.	Pile length (m)	Pile dia. (mm)	Socket Condition			Pile-rock Interface	
			Depth (mm)	Material used	Rock simulated	Coating applied	Condition simulated
A	3.0	200	200	Concrete	Hard	Thick slurry	Weak adhesion
B	3.0	200	200	Cement Grout	Soft		
C	3.0	200	200	Concrete	Hard	Thin slurry	Strong adhesion
D	3.0	200	200	Cement Grout	Soft		

TEST RESULTS AND DISCUSSIONS

Settlement versus Load Curves and the Ultimate Loads. The pile tip settlements versus applied axial compression loads for the four piles are plotted on Figure 2. The results indicate that the failure modes of the piles are all brittle failures. The loads just before the steep drops on the load-settlement curves were taken as the ultimate loads, which are summarized in Table 2, together with the corresponding pile tip settlement at ultimate loads and the calculated average unit skin friction. It can be seen that the ultimate pile tip settlements are generally very small (less than 4 mm, or 2% of the pile diameter). The ultimate loads range from 324 to 513 kN. The piles and rock were exposed at the end of the tests, and no crush in the piles or cracks in rock was observed; only shear slip at the pile-rock interface occurred.

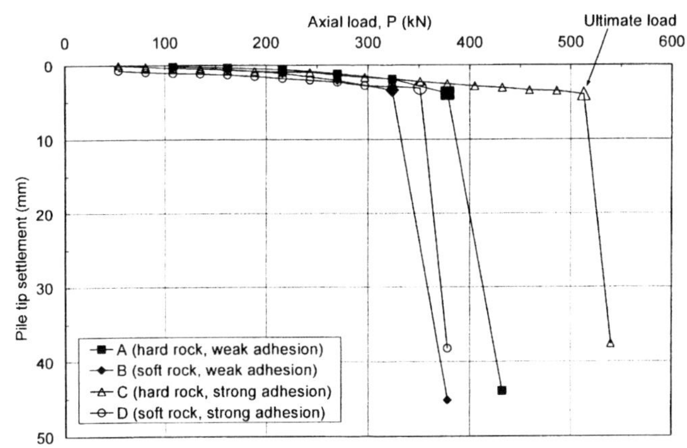

Figure 2. Load-settlement curves

Table 2. Summary of Load Test Results at Ultimate Load

Pile no.	Ultimate load, P_u (kN)	Ultimate total shaft resistance, P_{su} (kN)	Average ultimate unit skin friction, q_{su} (kPa)		Settlement of pile tip at ultimate load, S_{bu} (mm)
			Soil	Rock	
A	378	378	82.3	1,866	3.85
B	324	324	83.3	1,398	3.43
C	513	513	99.3	2,712	3.98
D	351	351	79.1	1,739	3.14

Relationship between of Skin Friction and Axial Load. The average unit skin friction developed along the pile-rock interfaces is plotted against the applied axial load as shown on Figure 3. It can be seen that the skin friction values increase with the increase in axial load regardless of the pile-rock interface adhesion conditions. The skin friction also increases as the stiffness and strength of rock increases, as would have been expected. Higher skin friction was developed when the pile-rock interface adhesion was stronger, as revealed by the curves for piles C and D versus the curves for piles A and B. However, for piles with weak pile-rock adhesion (piles A and B), the load-resistance characteristics were revealed from the beginning of loading, i.e. the skin friction values for the pile socketed in hard rock (pile A) were constantly larger than the pile socketed in soft rock (i.e. pile B). The skin friction at pile-rock interface increased with the increase in rock stiffness and the difference in skin friction was apparent. On the other hand, for piles with strong pile-rock adhesion (piles C and D), the difference in shaft-resistance characteristics due to the difference in rock stiffness was displayed progressively following the increase in axial loads, and the difference was not apparent at the early stage of loading. The shaft resistance of the pile socketed in stronger rock (i.e. pile C) continuously developed at the later stage until failure of the pile occurred.

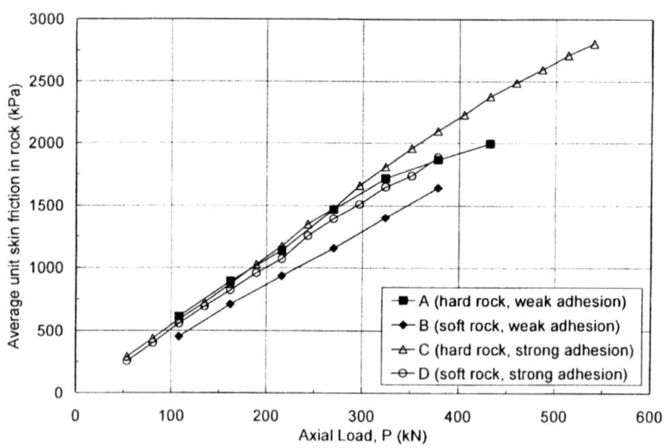

Figure 3. Load versus skin friction of the pile segment in rock

The average unit skin friction developed along the pile-soil interfaces are also plotted against the applied axial loads as shown on Figure 4. The figure indicates that considerable skin friction has developed in the soils, up to about 100 kPa. The figure also shows that the developed skin friction was influenced by both the rock stiffness and the pile-rock adhesion conditions. In general, with the same applied axial load, the skin friction values developed at pile-soil interface with stronger rock socket (i.e. piles A and C) are less than those with weaker rock socket (piles B and D). The skin friction values developed for piles with stronger pile-rock adhesion (piles C and D) were also smaller than those with weaker pile-rock adhesion (piles A and B). However, for piles with stronger rock socket (piles A and C), the skin frictions further increased with the increase in axial loads at the later stage until failure occurred.

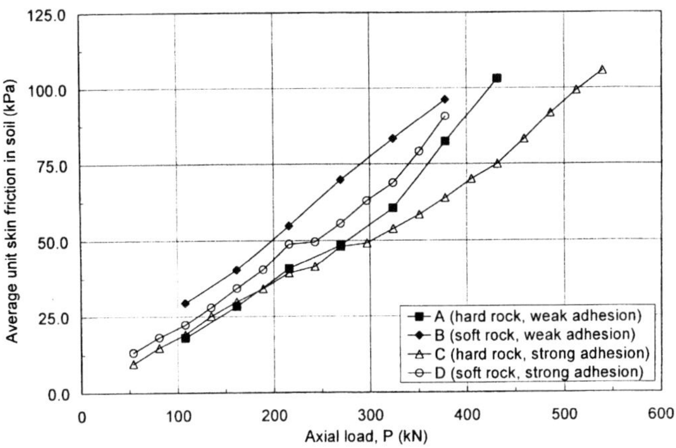

Figure 4. Load versus skin friction of the pile segment in soil

Relationship between Skin Friction and Pile Tip Displacement. The curves on Figure 5 illustrate the relationship between the skin friction developed at the pile-rock interface and the pile tip displacement. The figure shows that the curves are more gradual for piles with weaker pile-rock adhesion (piles A and B) than for piles with stronger pile-rock adhesion (piles C and D). However, brittle failures (abrupt slip between pile-rock interface) still occurred at the end for all the piles. For the same amount of pile tip displacement, the skin friction developed at pile-rock interface increased with the increase in rock stiffness, i.e. the skin frictions for piles A and C were larger than for piles B and D, respectively.

The plots of average unit skin friction developed in pile-soil interface against the pile tip settlement are illustrated on Figure 6. The figure indicates that brittle failures of the pile occurred in soil, which suggested that the failures of the piles were controlled by the rock sockets. A large portion of load was transferred to the soil after the limiting shear strength was reached along the pile-rock interface, which led to the abrupt increase in the relative displacement between the pile and soil. Thus, the shaft resistance of the pile segment in rock lost suddenly and pile slip occurred. At the early stage of loading, when the pile tip displacement was small, the shaft resistance of pile segment in soil developed concurrently with that in rock, regardless of the rock stiffness. At the later stage of loading, when the failure was impending, the shaft resistance of the pile segment in soil developed continuously with the increase in pile tip displacement until the abrupt failure induced by the loss of shaft resistance in rock.

Effect of Pile-Rock Interface Adhesion. The test results indicate that pile-rock adhesion greatly influences the shaft resistance in rock (Figure 5). For piles with stronger pile-rock adhesion (piles C and D), the shaft resistance was improved by 24% to 45% compared with piles with weaker pile-rock adhesion (piles A and B), as can be found based on the data in Tables 1 and 2. The pile-rock adhesion also has some influence on shaft resistance of the pile segment in soil as shown on Figure 6. With the same rock stiffness, the pile-rock adhesion condition has increasing effect on the shaft resistance in the soil, i.e. the stronger the pile-rock adhesion, the more apparent the brittle failure (e.g. pile C versus pile A).

Distribution of Skin Frictions in Soil. The skin frictions of the piles developed in the soil layer at the ultimate axial loads are plotted against pile elevation as shown on Figure 7. The test results indicate that the distribution of skin friction in the soil portion appears to be a saddle shape in general with high values at the top and bottom and low value in the middle.

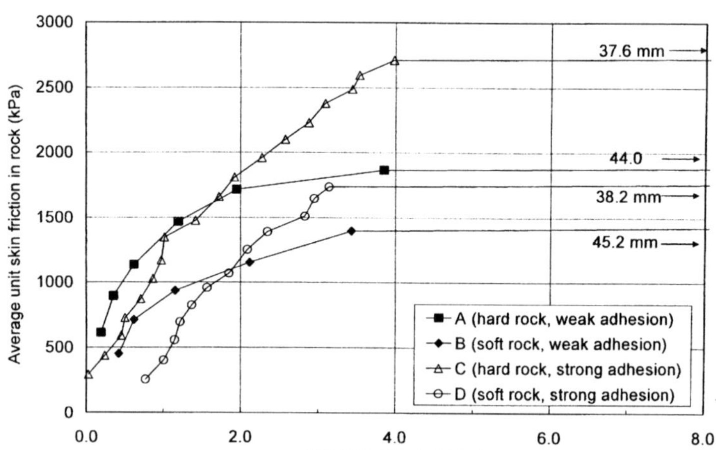

Figure 5. Pile tip displacement versus rock skin friction

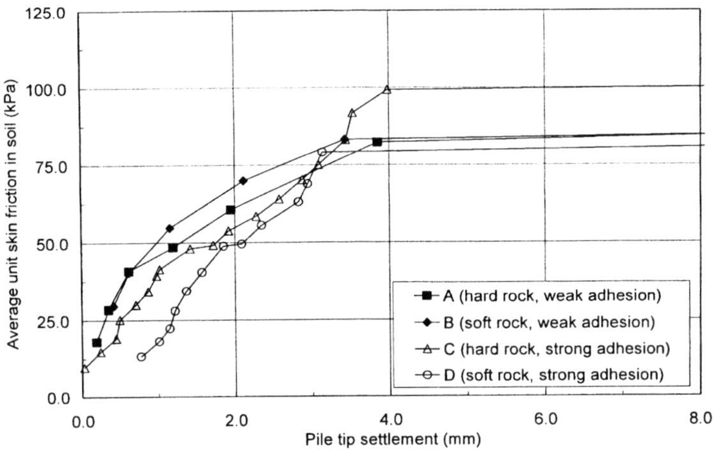

Figure 6. Pile tip displacement versus soil skin friction

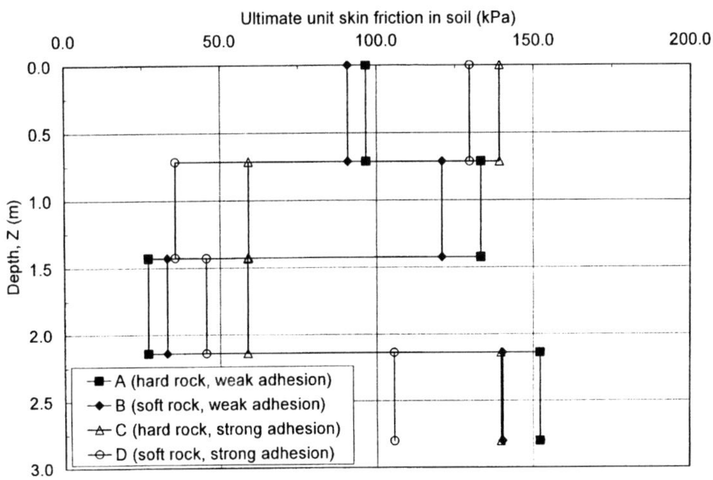

Figure 7. Distribution of skin frictions at ultimate loads

CONCLUSIONS

The following conclusions can be drawn from the study: (1) The stiffness of rock on which the pile tip is socketed influences the development of shaft resistance. The stiffer the rock, the higher the ultimate shaft resistance of the embedded portion of the pile, and the higher the ultimate shaft resistance of the pile in the overlying soil layer.

(2) Pile-rock adhesion condition has great influence on shaft resistance of the rock socket but has less influence on the shaft resistance of the pile in the soil. (3) Although the ultimate displacement of pile socketed into rock is small, shaft resistance of the soil layer can fully develop. The shaft resistance distribution in soil appears to be a saddle shape with high values occurring at the top and bottom and low value in the middle. (4) The shaft resistance in soil accounts for considerable amount of pile capacity.

ACKNOWLEDGMENTS

The insightful review comments provided by colleagues of the third author at Parsons Brinckerhoff Americas, Inc. are greatly appreciated.

REFERENCES

Johnston, I.W. and Haberfield, C.M. 1992. "Side resistance of piles in weak rock." Proc. of Conference on Piling - European Practice and Worldwide Trends, edited by M. J. Sands, Institution of Civil Engineers, Thomas Telford London, 52-58.

Nam, M.S. and Vipulanandan, C. 2008. "Roughness and unit side resistance of drilled shafts socketed in clay shale and limestone." J. of Geotech. & Geoenviron. Eng., 134(9), 1272-1279.

Williams, A.F. and Pells, P.J.N. 1981. "Side resistance of rock sockets in sandstone, mudstone and shale." Canadian Geotech. J., 18, 502-513.

Zertsalov, M.G. and Konyukhov, D.S. 2007. "Analysis of piles in rock." J. of Soil Mechanics and Foundation Eng., Springer New York, 44, 9-14.

The Behavior of a Single Pile under Cyclic Axial Loads

Z. Li[1], S. K. Haigh[2] & M. D. Bolton[3]

[1]Research student, Department of Engineering, University of Cambridge, UK. Email: zl247@cam.ac.uk
[2]Lecturer, Department of Engineering, University of Cambridge, UK. Email: skh20@cam.ac.uk
[3]Professor, Department of Engineering, University of Cambridge, UK. Email: mdb@eng.cam.ac.uk

ABSTRACT: Many piled foundations have been destroyed under significant cyclic loads in earthquakes. Centrifuge modelling of a single pile subjected to cyclic loads has been conducted to investigate the influence of cyclic loads on the axial performance of the single pile. Different pile installation procedures were applied to compare the axial behaviour of different piles under cyclic loads. Pile head permanent settlements accumulated due to cyclic axial loads, and these increased with the increasing load amplitude. Also the pile head axial secant stiffness decreased with the increasing number of axial load cycles, and with increasing amplitude. Furthermore, the axial pile performance is influenced significantly by different installation methods.

INTRODUCTION

Piled foundations are often subjected to significant cyclic loads arising from the swaying and rocking motions of superstructures during earthquakes. These cyclic loads caused the cyclic degradation of axial pile head stiffness and accumulated permanent settlements. Consequently, many piled foundations have failed catastrophically owii these cyclic loads, causing destructions of major pile-supported structures.

For monotonically loaded pile behaviour, dynamically-driven piles are stiffer bored piles caused by the influence of penetration on the stress state of soil surroun piles. Nowadays jacked piles become more popular to be used in urban constructior to the minimal noise and ground vibration. Compared to bored and driven piles, jacked piles show a stiffer base response due to stiffening effect of the final jacking stroke, and exhibit a stiffer shaft response resulting from enhanced shaft resistances without friction fatigue effect (White & Deeks, 2007).

Many researchers have paid much attention to the axial performance of bored or driven piles subjected to cyclic axial loads (e.g. Poulos, 1989). Nevertheless, the axial behaviour of jacked piles under cyclic loads is still not well understood at present.

In this research, a series of cyclically loaded pile tests was conducted in the centrifuge environment. Different pile installation methods were applied to capture the full range of construction-induced soil conditions available in the field. The pile head axial load-displacement response under cyclic loads is described; the pile head stiffness degradation and accumulated permanent settlements are also discussed.

METHODOLOGY

Test Apparatus and Model Pile

Single pile tests were conducted at 50g in the Turner beam centrifuge at the Schofield Centre, Cambridge University. The whole test package is presented in FIG 1(a). A new two-axis servo actuator was designed to apply vertical loads to the model pile. An introduction to the actuator is described in Haigh et al. (2010).

A stainless steel tubular model pile was used, with an outsider diameter of 12.7 mm and an internal diameter of 11.3 mm. The embedment depth and total length of the pile are 200mm and 250mm respectively. The pile tip is flat and close-ended. The shaft surface is smooth, and the pile head is fixed without any rotation. The pile was instrumented with a load cells at the pile head to measure axial forces, lateral forces and bending moments.

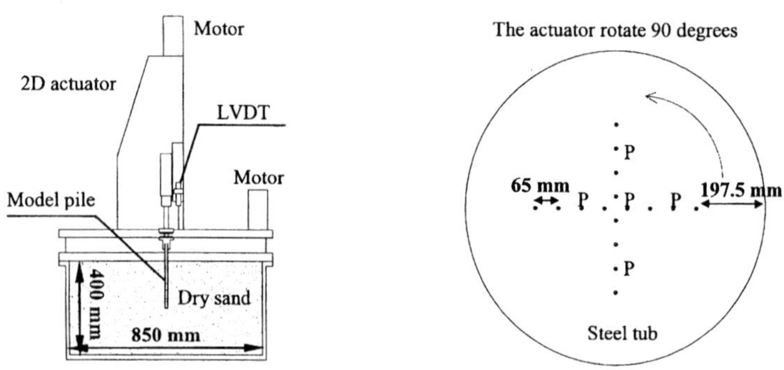

(a). Elevation view of the test package **(b). Test arrangement**

FIG. 1. Centrifuge test plan.

Instrumentation

Pile head axial displacements were measured using a Linearly Variable Differential Transformer (LVDT) mounted on the 2D actuator, with a maximum stroke length of 22 mm. Also the pile head vertical and horizontal displacements were recorded by two encoders fixed on the track of the 2D actuator.

Sand and Container

Dry Fraction E silica sand was used in this project, and it should have behaved like a continuum as would be the case in the prototype. The sand was pluviated into a cylindrical steel tub (850mm diameter and 400mm deep) by an automatic sand-pouring machine, and a dense homogeneous sand specimen with a relative density of 83% was achieved. The steel tub is large enough to ignore the boundary effect ($D_{container}/D_{pile}$=67).

CYCLIC AXIAL LOAD TESTS

Test Procedure

All test locations are presented in FIG 1(b). The interspaces between successive tests were about 15 times the pile diameter to nullify any effects due to effective stresses induced in the sand by preceding tests. The pile was pre-jacked into sand at 1g to simulate undisturbed soil conditions surrounding bored piles, or monotonically jacked at 50g, or cyclically jacked at 50g to model jacked piles in the field. After installation, ten displacement-controlled axial load cycles with an amplitude of 0.52mm or 1.3mm were conducted; and ten force-controlled load cycles were conducted subsequently, in which the pile head axial maximum and minimum force in each cycle was equal to those in the tenth displacement-controlled cycle.

Displacement-Controlled Load Cycles

Pile head axial force-settlement curves in the first ten displacement-controlled cycles for the jacked pile are shown in FIG 2(a). In these load cycles, pile head minimum force (P_{min}) had little change; while pile head maximum force (P_{max}) reduced with increasing number of cycles but at a reducing rate, as shown in FIG 3. The greatest reduction took place within the first ten cycles, which is consistent with Poulos' (1989) test results.

This reduction of P_{max} mainly derives from the loss of pile shaft friction. When the pile is subjected to cyclic axial loads, the pile shaft friction reduces due to a decrease in the normal stress caused by cumulative contraction of the sand within the shear zone close to the pile-soil interface (White & Lehane, 2004). In each load cycle, although dilation occurs in the shear zone firstly, the shear band contracts significantly after reversal of the loading direction, leading to some net contraction per cycle.

For the pre-jacked pile, P_{max} rose to around 900N at a settlement of 0.52mm in the first cycle, and decreased to 450N after ten cycles; while P_{max} of the monotonically or cyclically jacked pile was around 2000N in the first cycle and reduced to 1600N after ten cycles with an amplitude of 0.52mm. In the first cycle, P_{max} of either the monotonically or cyclically jacked pile is twice as large as that of pre-jacked pile due to the larger soil radial stress around the pile shaft and higher soil shear modulus underneath the pile base. After ten cycles the reduction of P_{max} is similar for the pre-jacked, monotonically and cyclically jacked pile, attributed to the reduction of pile shaft friction as discussed above. This indicates that the cumulative contraction of the sand within the shear zone surrounding the interface is not influenced by the original radial stress level of sand, and is thus not changed by different pile installation methods.

For the monotonically jacked pile, P_{max} rose to around 3000N in the first cycle and reduced to 1700N after ten cycles with an amplitude of 1.3mm. Although this P_{max} value is 1.5 times that found using an amplitude of 0.52mm in the first cycle, it then decreased much more significantly and became comparable to that found using an amplitude of 0.52mm after ten cycles. Thus the reduction in P_{max} increases with increasing amplitude of cyclic load, indicating that the reduction in radial stress in the sand increases rapidly with increasing amplitude of cyclic displacement. However, the ultimate mobilized radial stresses of sand surrounding the pile shaft are similar for different cyclic displacement amplitudes.

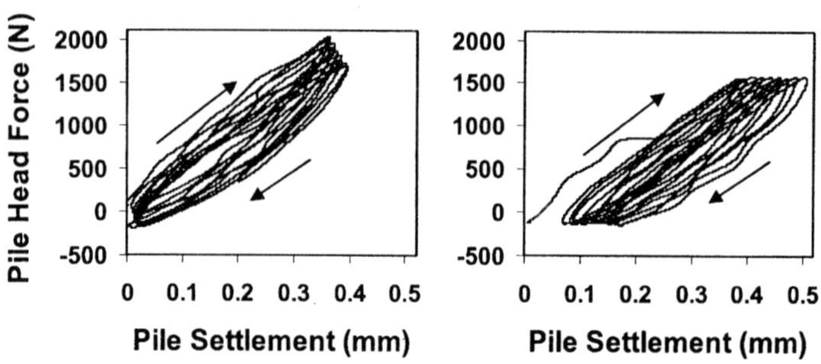

(a). Displacement-controlled cycles **(b). Force-controlled cycles**

FIG. 2. Pile cyclic axial force-displacement curves.

Force-Controlled Load Cycles

In force-controlled axial load cycles, the pile head maximum and minimum displacements in each cycle increased with the increasing number of cycles but at a reducing rate, as shown in FIG 2(b). In each cycle, in order to counteract the effect of the accumulated contraction of sand around the pile-soil interface, an extra increment of pile head axial displacement is required to mobilize a given maximum load, thus inducing an accumulated permanent displacement, as shown in FIG 4.

Accumulated permanent settlement (S_a) of the pre-jacked pile is the largest with a value of 0.23mm, followed by the monotonically jacked pile and cyclically jacked pile with a value of 0.124 mm and 0.08mm respectively. The monotonically or cyclically jacked pile had a stiffer base response than the pre-jacked pile due to the pre-loaded soil under the pile base. Also they have higher shaft resistance and therefore higher stiffness caused by the larger radial stress in the soil surrounding the pile shaft. Thus, the accumulated settlement of the monotonically or cyclically jacked pile is much smaller than that of the pre-jacked pile during force-controlled load cycles.

Additionally, for the monotonically jacked pile, the permanent settlement S_a is influenced by the amplitude of previous displacement-controlled load cycles. S_a of the monotonically jacked pile, previously subjected to ten load cycles with an amplitude of 1.3mm, is 0.24mm, around twice as large as that of the monotonically jacked pile subject to ten load cycles with the amplitude of 0.52mm. Although the load amplitudes of these two piles are similar in load-controlled cycles, it seems that the larger amplitude of cyclic axial loads to which a pile has previously been subjected, the larger is the displacement needed to achieve the same P_{max}, thus the larger is the permanent settlement induced during force-controlled cycles.

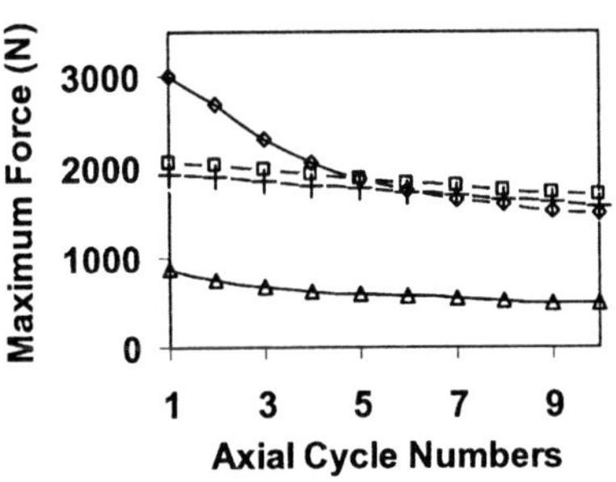

FIG. 3. The pile head maximum axial force.

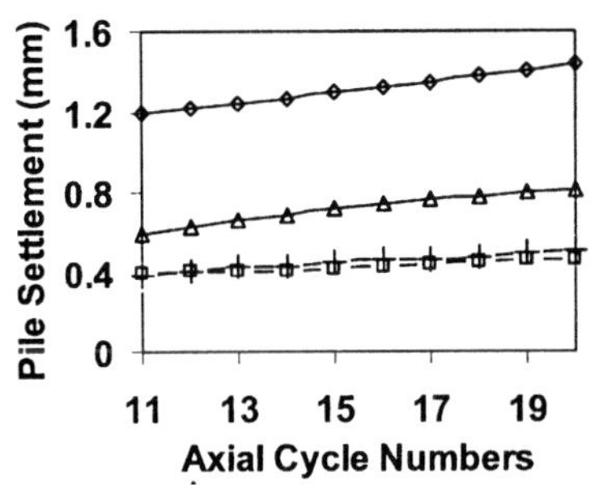

FIG. 4. The pile head settlement.

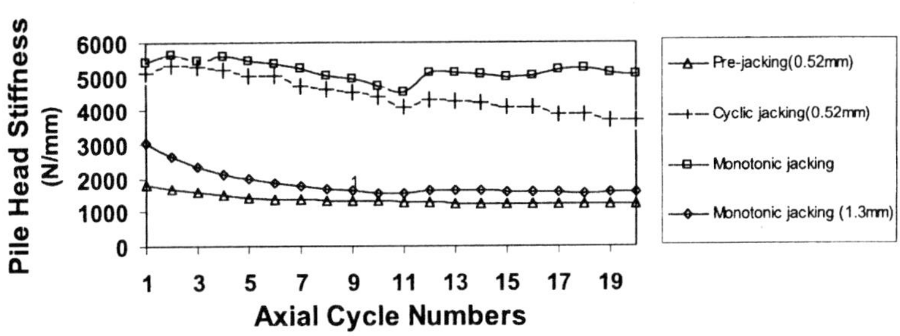

FIG. 5. Pile head axial secant stiffness.

Pile Head Axial Secant Stiffness

FIG 5 presents the pile head axial secant stiffnesses ($K_{s,a}$) obtained in cyclic axial load tests. In displacement-controlled axial load cycles, $K_{s,a}$ decreased with increasing number of axial load cycles; while in the succeeding force-controlled cycles, $K_{s,a}$ decreased very slowly with increasing number of cycles. Under cyclic axial loads with a given amplitude, the stiffness $K_{s,a}$ of the monotonically jacked pile is the largest, followed by the cyclically jacked pile and then the pre-jacked pile. This is because the jacking stroke creates the stiffer base response and higher radial stress along the shaft than the pre-jacked pile. Moreover, for the same type of pile, $K_{s,a}$ reduces with increasing amplitude of cyclic axial load.

CONCLUSION

A series of cyclically loaded axial pile tests was conducted in the beam centrifuge. Different pile installation methods were applied to investigate and compare the axial behaviour of different piles under cyclic loads.

Under displacement-controlled cyclic axial loads, the pile head maximum force (P_{max}) reduces with increasing number of axial load cycles, especially in the first ten cycles, due to the contraction and radial stress reduction of the sand within the shear zone close to the pile-soil interface. The decrement in P_{max} is not influenced by different pile installation methods, but it increases with increasing cyclic axial load amplitude, and the ultimate value of P_{max} is similar for different load amplitudes.

The accumulated displacement of the monotonically or cyclically jacked pile is much smaller than that of the pre-jacked pile during force-controlled load cycles due to the more highly stressed soil created by in-flight installation causing an increased stiffness in the soil surrounding the pile. Also the accumulated displacement increases with increasing load amplitude.

The pile head axial secant stiffness ($K_{s,a}$) of the monotonically jacked pile is the largest, followed by the cyclically jacked pile and then the pre-jacked pile. Also $K_{s,a}$ reduces with increasing cyclic axial load amplitudes.

These results provide a better understanding of the influence of cyclic loads on the performance of piled foundations, and offer an insight to further research to optimise construction methods and designs of piled foundations to resist live loads in service.

ACKNOWLEDGMENTS

This research was funded by Atkins Geotechnics and by Giken Seisakusho Ltd.

REFERENCES

Haigh S.K., Houghton N.E., Lam S.Y., Li Z. and Wallbridge P.J. (in printing). Development of a 2D servo-actuator for novel centrifuge modelling. ICPMG'10, Zurich.

Poulos, H.G. (1989). Cyclic axial loading analysis of piles in sand. Journal of Geotechnical Engineering, ASCE, Vol. 115, No. 6, pp. 836–852.

White, D.J. and Lehane B.M. (2004). Friction fatigue on displacement piles in sand. Géotechnique. 54, No.10, pp. 645-658.

White D. J. and Deeks A. D. (2007). Recent research into the behaviour of jacked foundation piles, International Workshop on Recent Advances in Deep Foundations, The Netherlands, n/a, pp. 3-26.

Axial Pile Resistance of Different Pile Types Based on Empirical Values

Hans-Georg Kempfert[1] and Patrick Becker[2]

[1]Professor, Institute of Geotechnics and Geohydraulics, University of Kassel, Germany; geotech@uni-kassel.de
[2]Research Assistant, Institute of Geotechnics and Geohydraulics, University of Kassel, Germany; p.becker@uni-kassel.de

ABSTRACT: This paper presented a comparison of pile capacity from axial load tests and that from empirical equations. Additionally, a correlation was developed based on the load test database to estimate skin friction and end bearing of piles based on cone penetration resistance or undrained shear strength. Up to 1000 pile load tests on displacement piles, such as precast concrete, cast in place, steel, and screwed cast in place piles, micro piles, etc. are evaluated. The result of this study has been already integrated in the national German recommendations for piles "EA-Pfähle".

INTRODUCTION

According to German national standard DIN 1054:2005, analytical calculation methods cannot generally be used for the determination of pile resistance, since no methods are available at present, which describe the mechanical model and the influences of the installation method of the different pile types. Therefore, the resistance-settlement behaviour has to be verified on the basis of pile load tests on the field or comparable pile load test results from the nearby area with similar underground conditions. If no pile load tests are carried out and empirical values from directly comparable load tests are not available, the characteristic axial pile resistance of a single pile can be determined from general empirical values of axial pile resistances according to DIN 1054:2005. Similar specifications can also be found in EN 1997-1:2008 (Euro code EC 7-1).

However, very limited empirical values for pile resistance are available for few pile types in the existing German pile standards DIN 4026, DIN 4014, DIN 4128 and DIN 1054:2005. This deficiency in empirical values has been taken as a motive to form a database of axial pile load tests on different pile types and analyze them statistically. The goal of the study was to derive a range of empirical value of base resistance and skin friction as much as possible for different pile types and hence to contribute to the economical evaluation of the axial resistances of piles.

STATISTICAL METHODS AND RANGE OF EMPIRICAL VALUES

The descriptive and statistical methods used for the derivation of axial pile capacity based on empirical values are widely known. For details, see Hartung et al. (2002) and Kempfert & Becker (2007).

In addition to the scattering of soil strength parameters due to the boundary conditions of the geological process, the pile load bearing behaviour is also influenced by the installation method. Since the empirical values for the pile resistance are available for few pile types in a very limited amount, the scatter of pile bearing capacity can be considered in the statistical analysis by using a range of quantile values as shown in Figure 1.

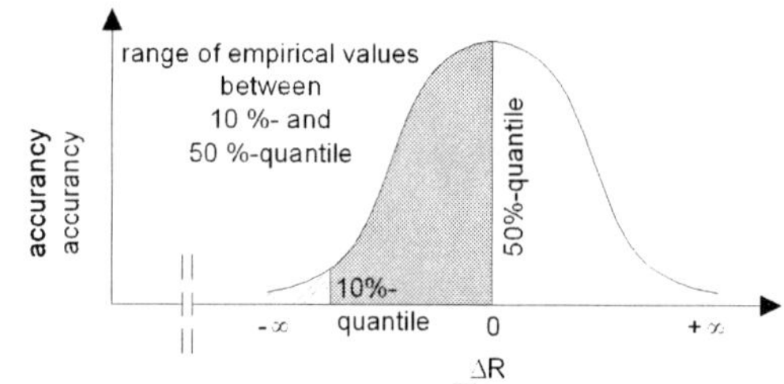

FIG. 1. Range of empirical values of pile resistances.

In the present analysis, empirical values for pile resistance have been derived at 10%, 20% and 50% quantiles. A 10% quantile means that 90% of the cases of the empirical determination of the axial pile resistance lies on the safe side and/or does not exceed the existing measured resistances. In contrast to this, a conservative average value at 50% quantile is usually taken for determination of the characteristic soil properties so far in practice.

The range of empirical values indicated in Figure 1 can vary depending on pile load tests and local boundary conditions and it only serves as first orientation.

PILES TYPES AND DATABASE

Data are collected mainly from static but also from dynamic load tests of different pile types (Table 2) and they are compiled in a database.

Table 2. Overview of pile load tests of the different pile types

Pile types	Construction type	no. of tests	
		static	dynamic
Driven precast piles	Reinforced and prestressed concrete piles	121	-
	Steel piles	98	80
Driven cast in place piles	Simplex pile	70	-
	Franki pile	300	-
Screwed cast in place piles	Atlas pile	124	-
	Fundex pile	52	-
Micro piles	Composite and cast in place piles	38	-

In the derivation of the axial pile resistance, only those pile load test results that have adequate information on the underground conditions, are exclusively used in order to attain a reliable correlation between the soil strength properties and the pile resistance. For details, see Kempfert & Becker (2007).

EMPIRICAL AXIAL PILE RESISTANCE

Determination of Pile Resistance

The characteristic axial bearing capacity of a single pile is given by

$$R = R_b + R_s = q_b \cdot A_b + \sum_{i=1}^{n} \left(q_{s,i} \cdot A_{s,i} \right) \tag{1}$$

where:

A_b area of pile base;
$A_{s,i}$ area of pile shaft in layer i;
q_b base resistance;
$q_{s,i}$ shaft resistance in layer i.

For the definitions of q_b and $q_{s,i}$ refer to Figure 4 and Figure 5 as well as the publication by Kempfert & Becker (2007). In the following the axial pile resistance R is referred to the ultimate limit state (ULS). Additional empirical values for the serviceability limit state (SLS) depending on different settlement criteria are given in EA-Pfähle (2007).

Depending on the qualitative relationships between soil strength and base resistance q_b and/or shaft resistance q_s of the correlation analysis, a regression model is developed to take into account the proportion of the two components of the axial pile bearing capacity depending on the pile type (see Kempfert & Becker (2007) for details). In the regression analysis, the functional relationships between axial pile load capacity and soil strength had been optimized iteratively until the difference between measured and calculated axial pile load capacity becomes or approaches to zero:

$$\Delta R = \frac{R_m - R_{cal}}{R_m} \equiv 0 \tag{2}$$

where:

ΔR Difference between measured and calculated axial pile resistance for the ultimate limit state (ULS),
R_m Measured value of the axial pile resistance from pile load tests,
R_{cal} Calculated value of the axial pile resistance according equation (1).

The results of the lower empirical pile resistances, which are already introduced in EA-Pfähle (2007), are shown in Figure 2 exemplary for driven precast piles. The difference between measured and calculated axial pile resistances ΔR for the ultimate limit state (ULS) falls between –2.38 % and 0.29 % for the lower and upper empirical values (not shown in the diagram) respectively.

Comparison of Empirical Axial Pile Resistance for Different Pile Types

Some issues of the empirical analysis of pile resistance are explained below with a comparative summary for the different pile types. In Figure 3a) a comparison of the total load capacity for the different pile systems is shown. Here, the variation between the results of the static pile load tests R_m and the calculated empirical pile resistances R_{cal} according to equation (1) is displayed with shaded areas for each pile system. The average values of the empirical analysis are indicated by symbols.

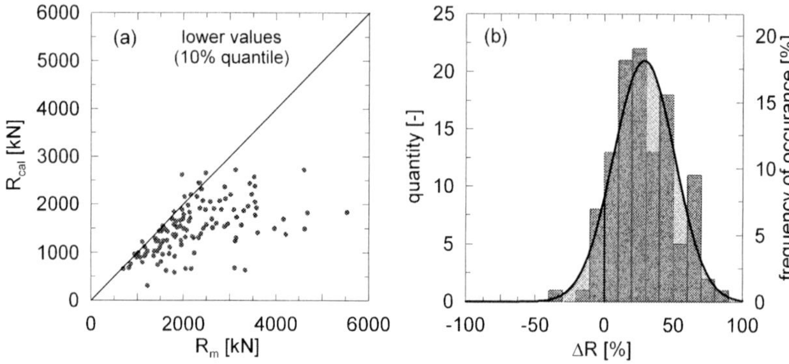

FIG. 2. a) Scatter plot and b) Histogramm of lower empirical values of driven precast piles in noncohesive soils at ULS.

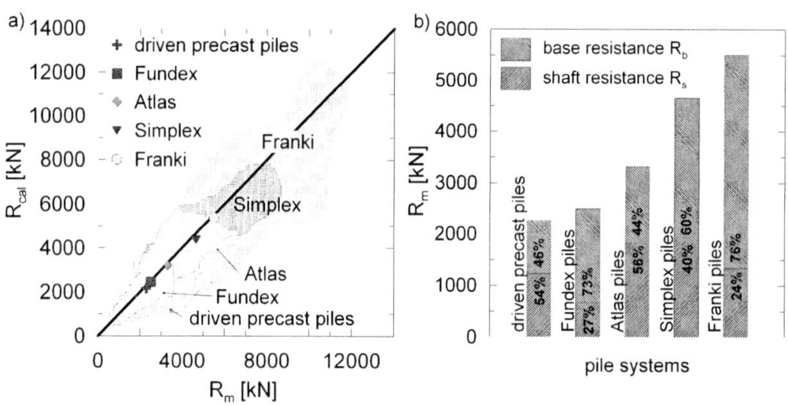

FIG. 3. a) Comparison of total load capacities (symbols = average values) and b) mean proportion of base and shaft resistances for different pile types.

The mean proportion of the base and shaft resistance of the different pile types is shown in Figure 3b). Depending on the installation method of the pile type the ratio of base resistance to the total load varies from 44 % to 76 %.

The comparison of the base resistance separately in Figure 4a) leads to an obvious relationship between the installation method, the pile system and the bearing capacity. Driven displacement piles generate a higher base resistance because of the displacement and compaction of the soil below the pile tip while driving the pile. The empirical values for the base resistance in noncohesive soils for driven precast piles and cast in place simplex piles are similar because the installation method is the same. On the other hand, the shaft resistance of the driven piles is lower than the other pile systems as it can be seen in Figure 4b) and Figure 5. Only the lower values are shown in Figure 5. The ratio between the upper and lower empirical values falls approximately between 1.3 and 1.4.

The empirical values for the base resistance of Fundex piles are smaller than the driven Simplex piles because of the effects of displacement and compaction of the installation method, with rotating and pushing movements. Both pile types have a pile tip with overlap, i.e. the diameter of the pile shaft is smaller than the diameter of the pile tip, which remains in the ground after the pile installation. This overlap leads to a loosening of the ground in the shaft area and a reduction of the empirical shaft resistances.

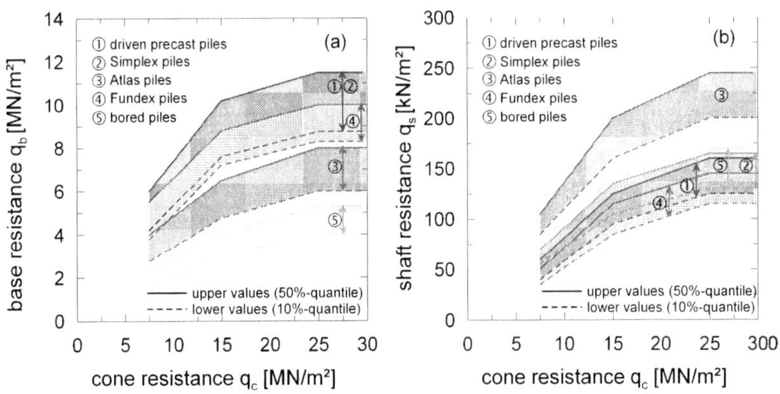

FIG. 4. Upper and lower empirical values of different piles in noncohesive soils for a) base resistance and b) skin friction.

The installation method of Atlas piles is characterised by a cutter head with a single-thread screw blade, which is twisted into the ground in the same way as the Fundex piles by a powerful rotary drive applying a large vertical pressure at the same time. The resulting screw-like shape of the Atlas pile shaft leads to larger values of shaft resistances.

The lower values of base resistances of bored piles and auger piles can be explained by the relaxation of the initial stress conditions in the area of pile tip due to the boring process.

All empirical values for the analysed pile system are given in EA-Pfähle (2007) for both ULS and SLS.

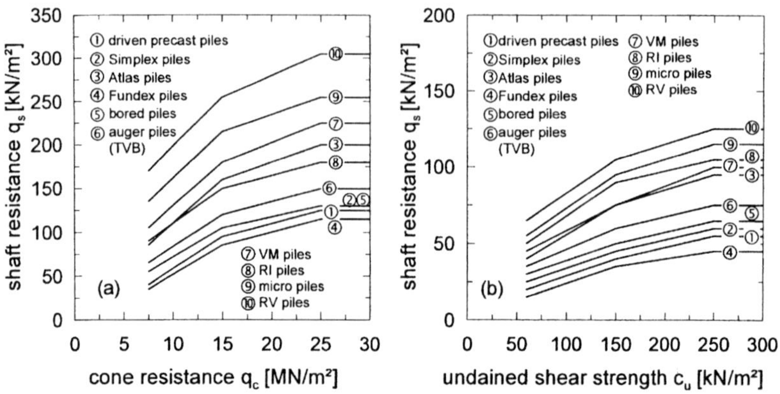

FIG. 5. Lower empirical values of skin friction for different piles in a) noncohesive soils and b) in cohesive soils.

CONCLUSIONS

The study of pile load tests provides to a large extent a secured range of empirical values for the base resistance and the skin friction for the different pile types as a function of the soil strength. Based on comparative statistical analysis of pile load tests on different pile systems, it becomes possible to derive a consistent analysis of bearing behaviour of pile types, which provides a safe and an economical pile bearing capacity depending on the extend of preliminary soil investigations.

The results of this study has been already integrated in the national German recommendations for piles "EA-Pfähle" of the German society of Geotechnics (DGGT).

REFERENCES

EA-Pfähle (2007). "Empfehlungen des Arbeitskreises Pfähle". *Deutsche Gesellschaft für Geotechnik (DGGT), Arbeitskreis 2.1. Ernst & Sohn, Berlin.*

Witzel, M. (2004). "Zur Tragfähigkeit und Gebrauchstauglichkeit von vorgefertigten Verdrängungspfählen in bindigen und nichtbindigen Böden." *Universität Kassel, Schriftenreihe Geotechnik, Heft 15.*

Hartung, J., Elpelt, B. and Klösener, K.-H. (2002). "Statistik, Lehr- und Handbuch der angewandten Statistik." *13. Auflage. München: Oldenburg.*

Kempfert, H.-G. and Becker, P. (2005). "Ableitung von Spitzenwiderstand und Mantelreibung vertikal belasteter Spundwände und Großrohrpfähle für die Anwendung im Hafenbereich." *Forschungsbericht – unpublished. Universität Kassel, Institut für Geotechnik und Geohydraulik.*

Kempfert, H.-G. and Becker, P. (2007). "Grundlagen und Ergebnisse der Ableitung von axialen Pfahlwiderständen aus Erfahrungswerten für die EA-Pfähle." *Bautechnik 84, Heft 7, S. 441-449.*

Optical Model Test on Work Mechanism of Single Pile with Cap

Yadong CHEN[1, 2], Xudong WANG[2], Yuexin SHE[1]

[1]Faculty of Architecture and Civil Engineering, Huaiyin Institute of Technology, Huaian, China; chenyundong2004@163.com

[2]College of Civil Engineering, Nanjing University of Technology, Nanjing, China; cewxd@njut.edu.cn

ABSTRACT: A deformation measurement system based on the digital image correlation method has been developed for studying the work mechanism of a single pile with a cap. A digital camera with high resolution is used to capture the gradual deformation failure process of the pile foundation under vertical load. Reliability and precision of this measurement system are validated by an example, which presents sub-pixel precision. The test results indicate that the settlement of the pile foundation varies gradually with total load. As the load increases, the ultimate state of the pile is earlier reached than that of the soil under the cap, and the additional increase in load is taken mainly by the soil. The displacement pattern of a pile with a cap is similar to that of a single pile without a cap. The plastic region of soil beneath the cap is developed after the pile reaches the ultimate state and the plastic region is transformed from cap edge to center as the load increases. A triangular compression zone beneath the cap occurs and is similar to those proposed by Terzaghi using the limit equilibrium method. The pile foundation is presented as the general shear failure. The test results indicate that this deformation measurement system is a useful tool for geotechnical physical model test to study global and localized deformation problems.

INTRODUCTION

The physical model test is an important method to study the nonlinear interaction behavior and work mechanism of pile foundations. Deformation measurement and analysis play an important role in these tests.

Due to space limitations and lack of measurement points, the traditional methods, such as setting displacement sensors or drawing mesh on the observed surface of the model, don't satisfy the requirements of fine mechanical properties analysis and measurement of the global deformation field.

Recently, with the development of high speed personal computer, image analysis techniques and digital cameras with high resolution, methods such as digital image correlation (DIC) have been used to measure planar deformation fields in pile model tests.

White (2003) used a deformation measurement system based on particle image velocimetry (PIV) and close-range photogrammetry to monitor the installation of a displacement pile. Li (2005) described the successful application of DIC and AutoCAD to study the pile barrier effect on lateral soil deformation. Based on model tests and image techniques, Wang (2006) studied the mobilization mechanism of toe resistance of piles in sand.

These experiments mainly focused on the effect of driving piles and the resulting displacement of surrounding piles. In this paper, the DIC method is discussed as a way to study the work mechanism of a single pile with cap in model tests. This study will concentrate on the gradual deformation process of soil around a pile. The failure pattern under ultimate load is also discussed to evaluate the nonlinear interaction behavior of pile-cap-soil.

THEORY OF DIGITAL IMAGE CORRELATION (DIC) METHOD

The optical method employs digital image correlation (DIC) technique, which is a velocity measurement technique originally developed in experimental mechanics. DIC removes the need for discrete target markers by tracking the inherent texture (i.e. spatial variation of brightness) of soil grains through a series of images. Images are recorded at discrete time intervals using a high resolution digital camera before each image is divided into a mesh of DIC patches. The displaced location of each patch in a subsequent image is obtained by determining the location of highest correlation between each patch and a larger search region from a following image. The correlation plane is evaluated at single pixel intervals by fitting a bicubic interpolation to the region close to the integer peak and the displacement vector is established to sub-pixel resolution. A validation experiments demonstrated that the system offers greater accuracy and precision when tracking the movement of sand.

TEST PROGRAM

The test container had a plan area of 100 cm×150 cm and was 100 cm deep as shown in FIG. 1. A model cap with a plan dimension of 200mm×65mm×20mm was used. The square pile has a width of 2 cm and a length of 40 cm. The plan view of the pile foundation is shown in FIG. 2.

The pile model test was performed in this container with a toughened glass plate at its front face to permit the visual observation of the displacement of soil beneath the cap under vertical load. Load was applied vertically through a jack mechanism. A load cell and a displacement sensor were connected to the reaction frame to measure load and deformation during the test. The digital camera used to capture the images for the DIC analyses was mounted securely on a tripod along a common vertical axis.

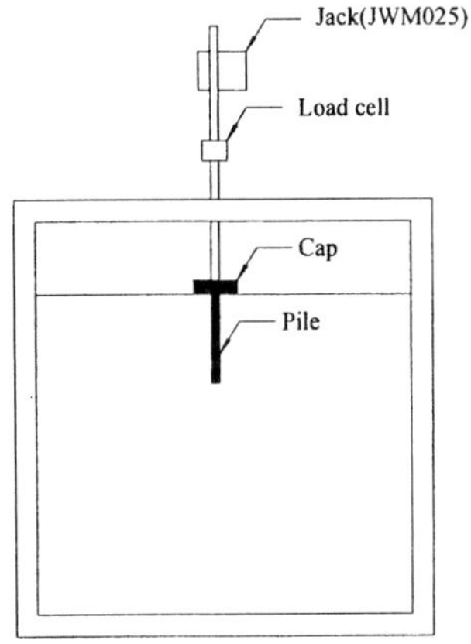

FIG. 1. Schematics of test setup

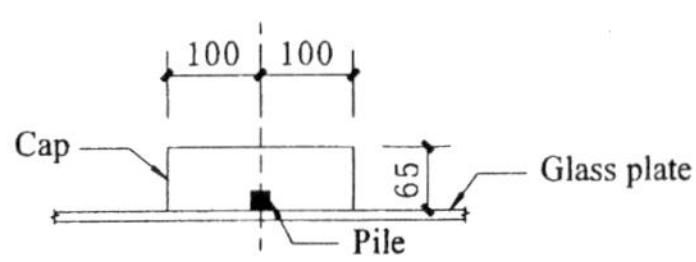

FIG. 2. Plan of a single pile with cap

Sand was used in this test and was rained into the testing container via a sieve. Some of the physical properties of the sand are shown in Table 1. Where ρ is the density, φ is the angle of internal friction, ω is the water content, D_r is the relative density, G_s is the specific gravity.

Table 1. Sand Properties

ρ (g/cm^3)	φ (^{0})	ω (%)	D_r (%)	G_s
1.54	34	0.27	0.67	2.65

The digital images were captured using a Samsung digital still camera, which has an image resolution of 2816×2112 pixels. The patch is 64×64 pixels in this test.

TEST RESULTS

FIG. 3 shows the load-settlement curves of the pile foundation under vertical loads. Where Q is the total load, Q_c is the load shared by soil immediately beneath the cap, Q_p is the load shared by the pile, S is the settlement of pile foundation.

As shown in FIG. 3, Q-S curve varies gradually as the load increases. Because of the pile, the settlement of cap is obviously reduced. The ultimate state of the pile is reached prior to that of the soil beneath the cap, and the additional increase in load is taken by the soil beneath cap.

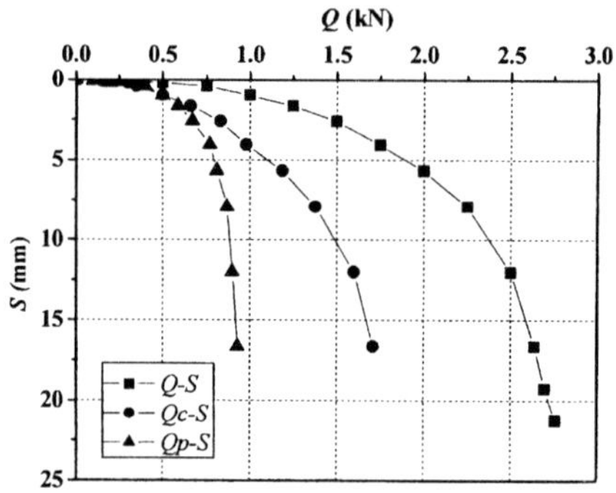

FIG.3. Load-settlement curves of a single pile with cap

FIG. 4 shows the displacement field of soil beneath cap. At the beginning of load, soil beneath cap is at the phase of linear deformation and displays vertical downward movement. When Q is equal to 1.75kN, the deformation region beneath the cap is being further extended, there are some incline downward movements, and the pile is approaching its ultimate state.

At Q=2.25kN, the settlement of the pile foundation is 7.925mm. Because of the pile skin friction, the displacement around the pile inclines downward, the soil under the pile tip is displaced out by the spherical cavity compaction pressure, which is similar to the deformation pattern of a single pile under axial load, and the pile reaches its ultimate limit state.

At Q=2.5kN, pile head displacement is increased significantly. The plastic zone is expanded gradually as the load increases, which transfers from the cap edge to center. The displacement surrounding the upper portion of the pile shaft continues vertical-down. A triangular zone immediately under the cap is gradually presented, which is similar to those proposed by Terzaghi(1943) using the limit equilibrium method. The total load shared by the pile is smaller than that by the cap. The bearing capacity of soil beneath the cap is fully mobilized after the loaded pile reached it ultimate limit state, the additional applied loads are resisted by the soil, and the failure pattern is mainly controlled by the soil under the cap. Some upward displacement occurs, which indicates that some uplift presents on ground surface.

At the last stage of loading, the displacement beneath the cap and surrounding pile is developed completely and the pile foundation subsides rapidly, which indicates the general shear failure.

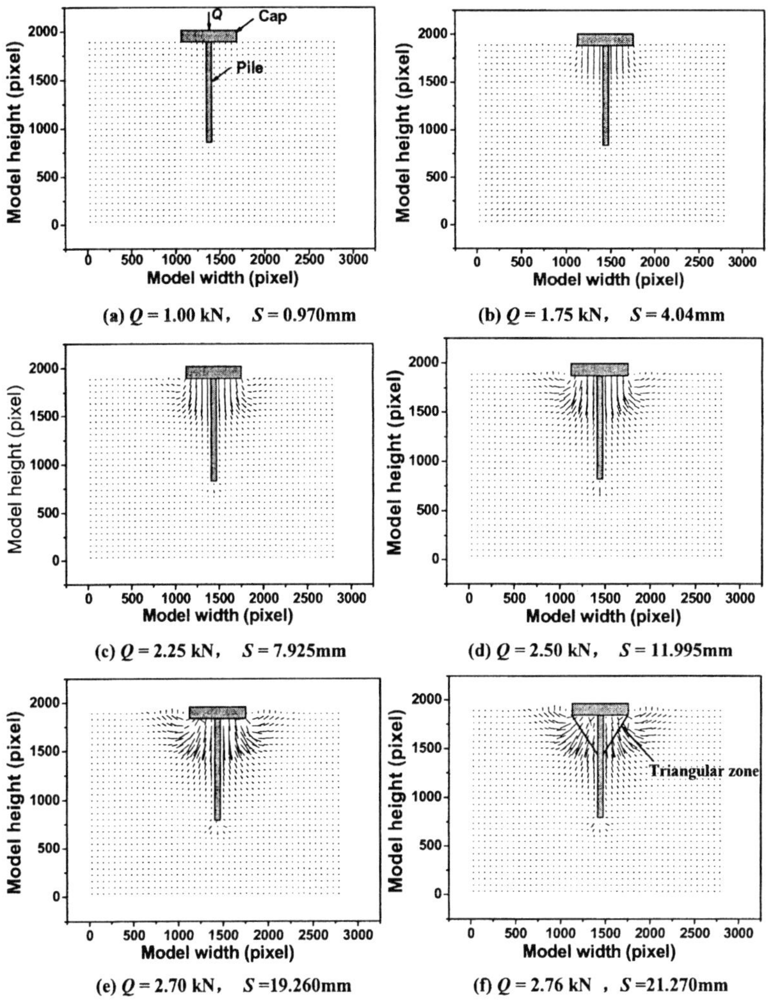

FIG.4. Vector graph of displacement field of soil beneath cap

CONCLUSIONS

The deformation and failure mechanisms to ultimate bearing capacity have been observed for a model pile with cap on sand using the DIC method. The test results indicate that the load transfer mechanism of a single pile with cap varies gradually as the load increases, the additional applied loads are taken by the soil beneath cap after the pile reaches its ultimate state, and the ultimate failure pattern of the pile foundation is controlled by the soil under the cap. The soil beneath cap is converted from the stage of linear deformation to elastoplastic deformation, and the plastic zone is transferred from cap edge to center. The triangular compression zone beneath the cap occurs and is similar to those deduced by the limit equilibrium method. The deformation pattern of soil under the pile tip is similar to that of the single pile under axial load, and the pile foundation is presented as the general shear failure.

This study focused on the measurement technique of DIC using in pile model tests. However, it should be noted that the study is valid only for pile in sand with flat tip. Piles with different shapes and soil types should be investigated further.

The test results show that the DIC method is a useful tool for geotechnical physical model test to study global and localized deformation problems and it may have a wide range of application. Such as in-situ deformation measurement and engineering forecasting in the future.

REFERENCES

Sydenstricker, R.M. and Landau, L. (2000). "A study of some triangular discrete Reissner-Mindin plate and shell elements." Computers & Structures, Vol. 78: 21-33.

White, D. J, Take, W. A., Bolton, M.D. (2003). "Soil deformation measurement using particle image velocimetry (PIV) and photogrammetry." Geotechnique, Vol. 53(7): 619-631.

Li, X. W. (2005). "Model test and analysis on nonlinear behavior mechanism of composite pile foundation." M.S.Thesis, Nanjing University of Technology, Nanjing.

Wang, H., Zhou, J., Deng, Z. H. (2006). "Model tests on pile-soil-cap interaction." Chinese Journal of Geotechnical Engineering, Vol. 28(10): 1253-1258.

Knappett, J.A., Haigh, S.K., Madabhushi, S.P.G. (2006). "Mechanisms of failure for shallow foundations under earthquake loading." Soil Dynamics and Earthquake Engineering, Vol. 26: 91-102.

Chen, Y. D. (2007). "Displacement field and work mechanism research of soil around pile based on digital image correlation method." M. S. Thesis, Nanjing University of Technology, Nanjing.

Terzaghi, K. (1943). Theoretical Soil Mechanics. Wiley, New York.

Analysis of Stability of Super-long Bored Pile Considering Resistance of Soil on the side of Pile

Jing-Pei Li[1], Ming Chen[2], Fa-Yun Liang[3]

[1]Professor, Department of Geotechnical Engineering, Tongji University, Shanghai 200092, lijp2773@tongji.edu.cn
[2]Ph.D Student, Department of Geotechnical Engineering, Tongji University, Shanghai 200092, tongjichenbo@163.com
[3]Associate Professor, Department of Geotechnical Engineering, Tongji University, Shanghai 200092; fyliang@tongji.edu.cn

ABSTRACT: The stability of super-long piles that are supported laterally along part of their length or their entire length by an elastic Winkler foundation is investigated for the case when the coefficient of subgrade reaction is the general power distribution and the skin friction develops along the embedded pile length. Based on the potential energy method, taking into account the characteristic of load-transfer of super-long pile, the deflection functions for six different boundary conditions developed using the Rayleigh-Ritz method are analyzed in this paper. The applicability of the proposed analytical solution is validated against results of pile load tests reported in the literate. Therefore, the proposed method is considered promising for practical use.

INTRODUCTION

With the development of pile foundation engineering in soft soil area in China, the use of slender super-long bored piles has increased considerablely over the past decade. For this kind of piles, the vertical bearing capacity is often determined by the allowable pile top settlement or the buckling stability of the pile shaft. Especially when the covering soil around the pile shaft is soft or easy to get liquefied after vibration, and free-stand length extending a considerable distance above the ground line, the check of buckling stability of piles should be carried out.

Several approaches were used for analyzing buckling of axially loaded piles. Early approaches by Brandtzaeg and Harboe (1957), Golder and Skipp (1957) used Euler stability theory with the analysis verified using a limited number of buckling tests. Davisson and Robinson (1965) used the governing differential equation for buckling deflection to estimate critical loads with the assumption of constant and linearly increasing subgrade moduli. Gabr and Wang (1997) developed a model for estimating the critical buckling loads of piles assuming uniform variation of skin friction as a function of depth and nine cases of boundary condition. Peng (1996) considered the elastic resistance of soil around piles and used the solution of differential equations based on Galerkin method, and gave some charts and formulas for practical use. Zou (2005) based on the simplified elastic beam model and assumed exponential distribution of subgrade reaction modulus, deduced the analytical solutions for the buckling length and buckling load of the pile shaft by applying the principle of minimum potential energy of the pile-soil system.

The researches mentioned above gave out some useful conclusions and were utilized in a few of practical projects. In order to take into account the diversity and complexity of the soil conditions around piles, this paper, in the basis of previous studies, considering the super-long piles load-transfer characteristics, a buckling model is developed assuming a general power distribution of the coefficient of horizontal subgrade reaction with depth, and the total potential energy equation of the pile-soil system is set up based on the simplified elastic beam model. Then, analytical solutions for the buckling load of piles shaft are deduced by applying the principle of minimum potential energy.

LATERAL SUBGRADE REACTION

Applying a slight adaptation to Terzaghi's (1955) work, a common equation to describe the horizontal subgrade distribution can be expressed as follows:

$$k_h = m_h z^{\omega} \zeta^{(1-\omega)} \tag{1}$$

Where k_h = coefficient of horizontal subgrade reaction kN/m^3; m_h = ratio between coefficient of horizontal subgrade reaction and depth below ground surface, (kN/m^4); z = depth below ground surface; ω = empirical power index equal to or greater than zero and describes distribution of k_h, and ζ = constant, Rowe (1956) suggested setting $\zeta = 1$ in order to facilitate the dimensionless of k_h.

Given the coefficient of horizontal subgrade reaction for a single soil layer in (1), then the lateral soil reaction can be expressed as (Gabr et al., 1997):

$$q(x) = k_h y(x) d_0 = m_h (h-x)^{\omega} \zeta^{1-\omega} y(x) d_0 \tag{2}$$

Where d_0 = the calculation width of pile, which can be calculated as the methods of Technical Code for Building Pile Foundation of China (JGJ 94-2008), circular piles: when $d \le 1m$, $d_0 = 0.9(1.5d + 0.5)$; when $d > 1m$, $d_0 = 0.9(d+1)$; square piles: when $b \le 1m, d_0 = 1.5b + 0.5$; when $b > 1m, d_0 = b + 1$; d = the pile diameter and b = the pile width; h = the embedded pile length; x = the distance from the pile tip; and $y(x)$ = lateral deflection due to pile buckling.

AXIAL LOAD TRANSFER ALONG THE PILE

The effect of axial load transfer on the critical buckling load has been investigated by Reddy and Valsangkar (1970), Gabr et al. (1997), and Heelis (2004). Based on their researches, considering the load-bearing characteristics of super-long piles, shaft friction, $f(x)$ is defined by the use of parameters λ_1 and λ_2, such that

$$f(x)=\frac{2p_0(1-\psi)}{L}\left[\lambda_1+(\lambda_2-\lambda_1)\frac{(h-x)}{h}\right] \tag{3}$$

In which it is assumed that

$$\lambda_1+\lambda_2=1 \tag{4}$$

Where λ_1 and λ_2 are defined at the top and bottom of the pile respectively. For example, $\lambda_1=\lambda_2=0.5$ suggests uniform friction along the pile shaft. In Eq. (3) ψ is defined as

$$\psi=\frac{p_0-p_f}{p_0} \tag{5}$$

Where p_0 is the vertical load at pile top, p_f is the shaft friction of piles under the vertical load. ψ, that is the proportion of the applied load in end-bearing at the base of the pile. The value of ψ between 0 and 1, a value of 0 implies purely pile skin frictional and 1 implies solely end-bearing resistance to the applied load.
Then the axial load distribution can be expressed as

$$p(x)=p_0-\int_{h-x}^{h}\frac{2p_0(1-\psi)}{L}\left[\lambda_1+(\lambda_2-\lambda_1)\frac{(h-x)}{h}\right]dx \quad (x\le h) \tag{6-a}$$

$$p(x)=p_0 \quad (x>h) \tag{6-b}$$

METHOD OF ANALYSIS

The coordinates system for length and deflection is shown in Fig.1 (a). Deflection functions for the six boundary conditions were chosen using the Rayleigh-Ritz method, as presented in Table 1. These deflection functions satisfy the various geometric boundary conditions of the analysis model.

Table 1 Deflection functions and boundary conditions (Gabr et al., 1997)

Boundary Condition		Deflection Function
Top	Tip	
Free	Fixed	$y(x)=\sum_{n=1}^{k}C_n\left[1-\cos\left(\frac{2n-1}{2L}\pi x\right)\right]$
Free	Pinned	$y(x)=C_0\frac{x}{L}+\sum_{n=1}^{k}C_n\left[\sin\left(\frac{n}{L}\pi x\right)\right]$
Fixed, Sway	Fixed	$y(x)=\sum_{n=1}^{k}C_n\left[1-\cos\left(\frac{n}{L}\pi x\right)\right]$
Fixed, Sway	Pinned	$y(x)=\sum_{n=1}^{k}C_n\left[\sin\left(\frac{2n-1}{2L}\pi x\right)\right]$
Pinned	Fixed	$y(x)=\sum_{n=1}^{k}C_n\left[\cos\left(\frac{2n+1}{2L}\pi x\right)-\cos\left(\frac{2n-1}{2L}\pi x\right)\right]$
Pinned	Pinned	$y(x)=\sum_{n=1}^{k}C_n\left[\sin\left(\frac{n}{L}\pi x\right)\right]$

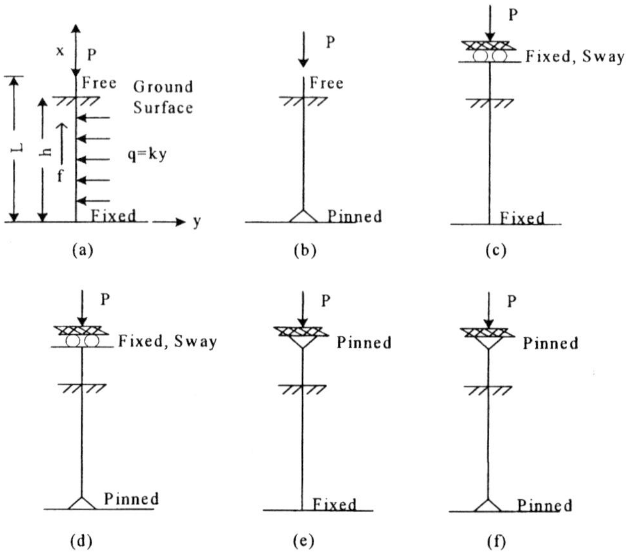

Fig. 1 Pile deflection models and boundary conditions

Assuming elastic condition for buckling under axial loads and small buckling deformation, choosing the undeflected state as a convenient datum position, the total potential energy to satisfy the condition of equilibrium is represented by

$$U+V=\frac{EI}{2}\int_0^L\left(y''\right)^2dx+\frac{1}{2}\int_0^h q(x)y\,dx-\frac{1}{2}\int_0^L p(x)\left(y'\right)^2dx \tag{7}$$

Where V = potential energy of external loads; U = summation of strain energy of system due to bending of the pile and elastic deformation of soil; EI = flexural stiffness of the pile; $q(x)$ = soil reaction; $p(x)$ = axial force of pile; y = lateral deflection of pile, which can be written as $y(x)=\sum c_n a_n(x)$, and the form varies under different boundary conditions, as shown in Table 1. $y'=dy/dx$; and $y''=d^2y/dx^2$.

Substituting Eqs (2) and (6) into Eq. (7), the following general equation is established

$$\begin{aligned}U+V=&\frac{EI}{2}\int_0^L\left(y''\right)^2dx+\frac{m_h\zeta^{1-\omega}d_0}{2}\int_0^h(h-x)^\omega y^2dx-\frac{p_0}{2}\int_0^L\left(y'\right)^2dx\\&+\frac{p_0(1-\psi)}{L}\int_0^h\int_{h-x}^h\left[\lambda_1+(\lambda_2-\lambda_1)\frac{(h-x)}{h}\right]\left(y'\right)^2dxdx\end{aligned} \tag{8}$$

According to the energy principle, $\delta(U+V)=0$ or

$$\frac{\partial(U+V)}{\partial C_i}\delta(C_i)=0 \tag{9}$$

The symbol δ denotes change caused by virtual displacement; and C_i denotes constants of deflection function. δC_i = arbitrary variation displacement. So Eq. (9)

can be written as

$$\frac{\partial(U+V)}{\partial C_i}=0 \tag{10}$$

Substituting Eq. (8) into (10), the following equation is obtained

$$\begin{aligned}&\int_0^L y'' \frac{\partial y''}{\partial c_i}dx+\beta^5\int_0^h (h-x)^{\omega} y\frac{\partial y}{\partial c_i}dx-\frac{p}{EI}\int_0^L y'\frac{\partial y'}{\partial c_i}dx\\&+\frac{2p_0(1-\psi)}{EIL}\int_0^h\int_{h-x}^h\left[\lambda_1+(\lambda_2-\lambda_1)\frac{(h-x)}{h}\right]y'\frac{\partial y'}{\partial c_i}dxdx=0\end{aligned} \tag{11}$$

Where $i=0,1,2,\ldots\ldots n; n=$ half-wave number of deflection function; and $\beta=$relative stiffness, and is defined as

$$\beta=\sqrt[5]{\frac{m_h d_0}{EI}} \tag{12}$$

Eq. (11) can also be expressed as the following form

$$[A][C]=0 \tag{13}$$

in which [A] is the coefficient matrix and can be expressed as

$$[A]=[A_1]_0^L+\beta^5[A_2]_0^h-\frac{p_0}{EI}[A_3]_0^L+\frac{2p_0(1-\psi)}{EIL}\left[[A_3]_{h-x}^h\right]_0^h \tag{14}$$

in which

$$A_1=\begin{bmatrix}\int a_1''(x)a_1''(x)dx & \int a_1''(x)a_2''(x)dx & \cdots & \cdots\\ \int a_2''(x)a_1''(x)dx & \vdots & \vdots & \vdots\\ \vdots & \vdots & \vdots & \int a_{n-1}''(x)a_n''(x)dx\\ \cdots & \cdots & \int a_n''(x)a_{n-1}''(x)dx & \int a_n''(x)a_n''(x)dx\end{bmatrix}$$

$$A_2=\begin{bmatrix}\int a_1(x)a_1(x)dx & \int a_1(x)a_2(x)dx & \cdots & \cdots\\ \int a_2(x)a_1(x)dx & \vdots & \vdots & \vdots\\ \vdots & \vdots & \vdots & \int a_{n-1}(x)a_n(x)dx\\ \cdots & \cdots & \int a_n(x)a_{n-1}(x)dx & \int a_n(x)a_n(x)dx\end{bmatrix}$$

$$A_3=\begin{bmatrix}\int\int Sa_1'(x)a_1'(x)dxdx & \int\int Sa_1'(x)a_2'(x)dxdx & \cdots & \cdots\\ \int\int Sa_2'(x)a_1'(x)dxdx & \vdots & \vdots & \vdots\\ \vdots & \vdots & \vdots & \int\int Sa_{n-1}'(x)a_n'(x)dxdx\\ \cdots & \cdots & \int\int Sa_n'(x)a_{n-1}'(x)dxdx & \int\int Sa_n'(x)a_n'(x)dxdx\end{bmatrix}$$

$$S=\lambda_1+(\lambda_2-\lambda_1)\frac{(h-x)}{h}$$

The deflection functions listed in Table 1 can be used and substituted into Eq. (11). In addition, integration can be performed to obtain a set of homogeneous equations in terms of c_i. The obtained set of homogeneous equations possesses nonzero solutions only if the determinant of the coefficient matrix is equal to zero. This equation is referred to as the characteristic equation of the system. The *n* root of this equation consists of the eigenvalues. p' is a non-dimensional parameter and is defined as

$$p' = pL^2/(\pi^2 EI) \tag{15}$$

Solutions of Eq. (11) can be calculated via the software program Mathematica, and the smallest root of the solution p'_{min} is obtained. Then the critical buckling capacity is obtained as following

$$p_{cr} = \frac{\pi^2 EI}{L^2} p'_{min} \tag{16}$$

And the equivalent buckling length, L_e can be obtained: $L_e = \sqrt{\frac{\pi^2 EI}{p_{cr}}}$.

CASE STUDY

One arch bridge was designed to use super-long rock-socketed filling pile foundation to support the bridge loads (Zou 2005), in order to investigate the bearing behavior of super-long piles and supply needed design parameters in soft soil area, a vertical loading test was carried out on the test pile, which was constructed around the site of the main pier. The 1-m diameter test pile was embedded 60m in the soils with 1.5m of free length above soil.

According to the actual supporting of the pile, the deflection function is chosen as

$$y(x) = \sum_{n=1}^{k} C_n \left(1 - \cos\frac{2n-1}{2L}\pi x\right) \tag{17}$$

In order to simplify calculation, make $\zeta = 1$, Davission and Prakash (1963) suggested that a power variation $\omega = 0.15$ was a more realistic value for clays which was postulated under undrained conditions. Taking into account the characteristics of the soil at the soft soil area in China, in this case, $\omega = 0.15$ is considered, the parameters of EI and m_h are equal to 1.47×10^6 kN.m^2 and 4000 kN/m^4, respectively, which are the measured values.

According to the method that is recommended by the Technical Code for Building Pile Foundation of China (JGJ 94-2008), the parameters $\beta = 0.345$ is obtained from Eq. (12). Obviously, the value of dimensionless length of the pile under the ground $\beta h > 4$, according to the boundary condition of test pile and requirements of JGJ 94-2008, the stability of calculated length of pile can be obtained as following

$$l_p = 0.7(l_0 + \frac{4}{\beta}) = 0.7 \times (1.5 + \frac{4}{0.345}) = 11.59 \text{ m}$$

Where l_0 = free length above soil of the pile.

Using the Euler's formula to calculate the pile buckling load as

$$p_{cr} = \frac{\pi^2 EI}{l_p^2} = \frac{3.14^2 \times 0.85 \times 1.47 \times 10^9}{11.59^2} = 91.71 \text{ MN}$$

If using the method of this paper the calculated buckling load $p_{cr} = 21.19$ MN, compared with the measured buckling load $p_{cr} = 16.72$ MN, the calculated method gave a little greater value than the measured. The main reason is that the calculation model can not fully reflect the non-linear and layered condition of soil around the pile, and also the parameters which are chosen to calculate may not represent the actual situation with a certain degree of deviation because of complexity of the test geological conditions, especially coefficient of horizontal subgrade reaction m_h. The simplified of ζ may also introduce some deviations.

The calculation results provided by JGJ 94-2008 is much larger than the measured value, so the calculated buckling load by the proposed method is more reasonable in the stability of super-long pile analysis.

CONCLUSIONS

Based on the principle of minimum potential energy, in this study, a pile buckling model was developed with the Rayleigh-Ritz method adopted to select suitable deflection functions. The analytical approach provides a general and systematic methodology for predicting the buckling loads of super-long piles with six boundary conditions and realistic variation in both soil stiffness and shaft friction.

Despite the well-known inherent limitations of a linearly elastic Winkler model for the soil, the authors' systematic analytical of solution appears to offer design predictions, which are closer to the results of pile load test reported in the literature compared with the calculated value by JGJ 94-2008. Therefore, the proposed method is considered promising for practical use.

In fact, reasonably accurate predictions of the actual buckling load are possible provided by considering the layered soil and the value of parameter, such as coefficient of horizontal subgrade reaction, shaft friction and its variation with depth.

ACKNOWLEDGMENTS

The work reported herein was supported by the Science and Technology Commission of Shanghai Municipality (Grant No.09231200900), Innovation Program of Shanghai Municipal Education Commission (10ZZ24), and the Specialized Research Fund for the Doctoral Program of Higher Education of China (Grant No. 20070247017). The authors wish to express their gratitude for the above financial support.

REFERENCES

Brandtzaeg A, Harboe E. 1957, "Buckling tests of slender steel Piles in soft, quick clay." *In: Proc., 4th Int. Conf. on Soil Mech. and Found. Engrg.*, Vol. 2:19-23.

Davisson, M. T., and Prakash, S., 1963, "A Review of Soil-pile Behavior." Hwy. Res. Rec., (39):25-48.

Davisson, M. T., and Robinson, K. E., 1965, "Bending and Buckling of Partially Embedded piles." *Proceedings 6th Int. Conf. on Soil Mech. and Found. Eng.*, Vol. 2, 243-246.

Gabr, M. A., Wang, J. J., and Zhao, M., 1997, "Buckling of Piles with General Power Distribution of Lateral Subgrade Reaction." *Journal of Geotechnical and Geoenvironmental Engineering*, ASCE, Vol. 123 (2): 123-130.

Golder H G, Skipp B O. 1957, "The buckling of Piles in soft clay." *In: Proe., 4th Int. Conf. on Soil Mech. and Found. Engrg.*, Vol. 2:35—39.

M. E. Heelis, M. N. Pavlovic and R. P. West, 2004, "The Analytical Prediction of the Buckling Loads of fully and partially Embedded Piles." *Geotechnique,* Vol. 54 (6):363-373.

Peng, X. D., 1996, "Calculation of Critical Load of Pile Considering Resistance of Soil on the Side of Pile." *China Civil Engineering Journal* Vol. 29 (5):43-48. (in Chinese)

Reddy, A. S., and Valsangkar, A. J., 1970, "Buckling of Fully and Partially Embedded Piles." *Journal of Soil Mech. and Found. Div.*, ASCE, Vol. 96 (6): 1951-1965.

Rowe, P. W., 1956, "The Single Pile Subjected to Horizontal Force." *Geotechnique,* Vol. 6 (4):70-85.

Terzaghi, K., 1955, "Evaluation of Coefficients of Subgrade Reaction." *Geotechnique*, Vol. 5:297-326.

JGJ-The People's Republic of China Industry Standards "Technical Code for Building Pile Foundations" (JGJ 94-2008).

Zou, X. J. 2005 "Theoretical and Experimental Study on the Buckling Analysis of Piles". Ph.D Thesis, Hunan University, Changsha.

Nonlinear Buckling Analysis of Piles with High-rise Pile Cap

Xin-jun ZOU[1], Bang-shun YIN[2], Ming-hua ZHAO[3], Xiao-li YANG[4]

[1]PhD, Associate Prof., College of Civil Engineering, Hunan University, Changsha 410082, China; xjzouhd@yahoo.com
[2]Master candidate, College of Civil Engineering, Hunan University, Changsha410082, China
[3]Prof., College of Civil Engineering, Hunan University, Changsha410082, China
[4]Prof., School of Civil and Architectural Engineering, Central South University, Changsha 410075, China

ABSTRACT: To study the buckling stability of piles with high-rise cap widely used in bridge and harbor engineering, the advantages as well as the insufficiency of traditional analysis methods (e.g., the method using the minimum potential energy theory and variational method by treating the pile shaft as ideal axially-compressed bars) were first discussed. Then, based on the second type of stability problem (i.e., the pile shaft is viewed as an unideal bar with initial quality defects) and the nonlinear finite element method (FEM), a new method was presented to consider the nonlinear pile-soil interaction. And influencing rules of various factors especially those initial defects on the critical buckling load P_{cr} were discussed through the numerical analysis model by ANSYS. The analysis result shows that, with the increase of the free length at the pile top from 2.5m to 5.0m, the P_{cr} will decrease to a relative extent of 32.5%, and similar changing rules are also found with the increasing degree and location of initial bending and necking along the pile shaft. Moreover, the increasing inclination degree of pile shaft will result in an obvious linear cut down of P_{cr}. Finally, the present method is applied to analyze the data from a site loading test, which found a good agreement between the calculated buckling load 18.27MN and the measured failure load 17.28MN with a relative error of 5.7%.

INTRODUCTION

In bridge or harbor engineering, pile foundations especially piles with high-rise cap are widely used to avoid underwater construction, and meet the bearing capacity and settlement requirements of the superstructures (especially wide-span bridges). For this kind of piles, the axial allowable load is often determined by the allowable pile top displacement or the buckling stability of pile shaft (Zhao 2005). When the covering soil around the pile shaft is soft or easy to get liquefied after vibration, and the unsupported or free part of pile shaft above the ground is too long, the buckling failure of pile shaft tends to take place easily, accompanied by unpredictable and serious consequences. However, in the first 20 years of the 19th century, buckling failure of piles was regarded as an important problem. From then on, researchers around the world have strived to resolve this problem (Lee 1968; Reddy 1970; Hu 1973; Zhu 1974; Zhao 1984). The findings of these researchers all verified that buckling failure of piles especially with high-rise cap in soft subsoil is possible and must be paid attention. However, the available research methods are all based on the first type of stability problem to determine the buckling critical load by eigenvalues (Liu 1998), in which the pile shaft is viewed as an ideal Euler compression bar. Many complex factors (such as pile-soil interaction, material or geometrical nonlinearity, possible initial defects of pile shaft, etc) can't be considered. Furthermore, the obtained critical load is the upper limit of real failure load and is not safe to use in engineering. Therefore, buckling stability of pile shaft should be regarded as the second type and analyzed accordingly (Zou 2007, 2008; Gaber 1994).

Unfortunately, an analytic solution considering complex factors mentioned above is hardly available at present. As a strong numerical tool in engineering, the finite element method (FEM) has found its great success in structural analysis. Therefore, the FEM is adopted in this paper to analyze those difficult problems from pile-soil nonlinear interaction (such as the stiffness disparity between the pile shaft and surrounding subsoil, sliding and cracking along pile-soil interface) by introducing pile-soil interface elements (Goodman 1968; Desai 1984).

COMMON BUCKLING ANALYSIS METHODS OF PILES

According to the elastic stability theory, two types of stability problems may be used to classify the stability behavior of structures by the changing characteristic behavior of equilibrium path during the loading process, as shown in Fig.1. For the first type of stability problem, a divarication point may be found during the equilibrium path, at which qualitative change of the equilibrium model is also observed. While for the second type, the equilibrium model won't change during the loading process. And usually, the load at the divarication point is defined as the buckling load P_{cr}, while the load at the peak point A of second type is called ultimate load P_u, which are both called by a common name, i.e., buckling critical load. Generally, the buckling load P_{cr} of first type is the upper limit of the second type. As the mechanic concept of the first type stability problem is

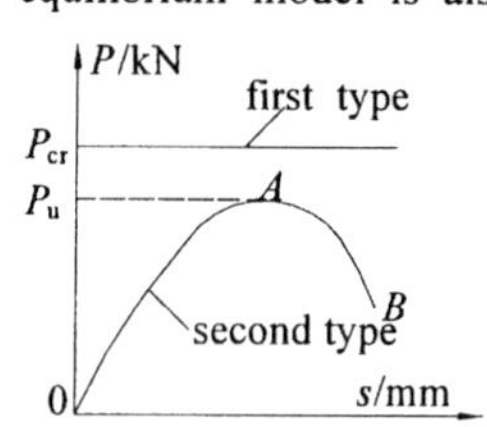

Fig.1. Classification of two types of stability problems

straightforward, the solution P_{cr} can be obtained more easily by the eigenvalue of stability equation. Common or available buckling analysis methods of piles are mostly based on the first type of stability problem, which primarily include the following two methods.

Solutions of Energy Methods

In order to analyze the buckling stability of piles by energy methods, the deflection curve of pile shaft (i.e., the lateral displacement of pile shaft at different embedded depth) should first be assumed, and then the energy equation of the pile-soil system may be set up as follows

$$\Pi = \frac{EI}{2}\int_0^l (y'')^2 \mathrm{d}x + \frac{1}{2}\int_0^h qy\mathrm{d}x - \frac{1}{2}\int_0^l P(x)(y')^2 \mathrm{d}x \tag{1}$$

where, l is the length of pile shaft, m; EI is the stiffness of pile shaft section, kN.m^2; h is the embedded depth, m; and $P(x)$ is defined as the axial force along the pile shaft, kN; q is the lateral resistance by the surrounding soil, kN/m; y is the deflection function of pile shaft with depth x.

Then, based on the minimum potential energy principle and the variational method, the buckling load P_{cr} and the corresponding stability calculation length may be derived from the Eq.(1) above. Although various boundary conditions and embedded depth of pile can be considered in energy methods, the elastic deformation energy of subsoil are mostly determined by using elastic subsoil reaction method under the condition that the coefficient of subsoil reaction keeps constant or increases linearly with the depth. In fact, due to the diversities of subsoil met in engineering practice, it's difficult to use a pre-defined model to describe the distribution of the subsoil reaction coefficients. Moreover, the side resistance along the pile shaft is often ignored or only considered by a uniform distribution. While the development and distribution mechanism for the side resistance of piles with high-rise cap tends to be very complicated, the influence by which on the buckling stability of pile shaft deserves a deeper study.

General Finite Element Method for the Buckling Analysis of Piles

In fact, the current finite element method in analyzing the buckling load P_{cr} of piles is a kind of matrix analysis method based on the plane bar system, in which the pile-soil interaction can't be considered, and only an assumed axial force distribution model and equilibrium conditions at the nodes are used to set up the iteration equation and calculate the P_{cr}. During the calculation process, as all eigenvalues (i.e., P_{cr}) will not come out by rule, necessary step by step modification should be made to the iteration matrix to get the final solution. It should be pointed out that, this kind of analysis method by getting the buckling load and mode through the eigenvalues of stiffness matrix at singularity belongs to linear buckling analysis, which neglects all those nonlinear factors and possible initial defects of the pile shaft. Furthermore, an obvious disagreement is often found between the calculated buckling load P_{cr} and the real value, and it's not safe to use the calculated P_{cr} to evaluate the stability of piles in engineering. As for the failure buckling mode, only the relative lateral displacement instead of the real absolute value of the pile shaft can be obtained, which is not sufficient to control the safety in terms of the allowable maximum displacement of the superstructures, especially those hypersensitive to displacement.

Consequently, in order to consider those necessary nonlinear factors and initial defects of pile shaft as much as possible, nonlinear numerical analysis based on the second type of stability problem should be adopted to reflect the real buckling stability of engineering piles.

NONLINEAR FINITE ELEMENT BUCKLING ANALYSIS OF PILES

Essential Theories

The buckling analysis of piles based on the second stability type comes down to solving the system of nonlinear equations mathematically. As it's difficult to have a direct solution, incremental nonlinear finite element analysis and solution of buckling eigenvalues may be combined to obtain the critical buckling load. In other words, step by step loading with a load increment of ΔP_i has to be used during the analysis process. If iteration convergence is reached at the ith incremental step, a linear buckling eigenvalue may be drawn from the tangent stiffness matrix at the start point of ith step, which can be called as nonlinear buckling analysis with the corresponding equation denoted as follows:

$$[K + \lambda \Delta K^G(\Delta u, u, \Delta\sigma)] - \Delta u = 0 \tag{2}$$

where, K is the tangent stiffness matrix set up based on the stress and strain state at the start point of ith step, which contains influences of various kinds of nonlinear factors cumulated from all former loading steps; while ΔK^G is a linear function about the load increment ΔP_i to cause buckling failure from the ith step, which is also called geometry stiffness matrix.

Then the critical buckling load may be calculated from the following formula:

$$P_{cr} = P_{i-1} + \lambda \Delta P_i \tag{3}$$

where, P_{i-1} is the load at the start point of ith step; ΔP_i is the load increment during the ith step; P_{cr} is the calculated structure failure load from the buckling load increment drawn from the ith step.

For the above nonlinear buckling analysis, it is noted that to determine the buckling eigenvalue by using the stiffness matrix containing effects of necessary nonlinear factors during loading process is more reasonable and accurate. Especially, when the buckling eigenvalue is drawn from the increment step near the real state losing stability, the calculated critical buckling load of piles will be more accurate.

Setting up of the Analysis Model

Based on the above idea, the arc length method widely discussed and used in structure nonlinear analysis as well as the large and universal finite element calculation software ANSYS were introduced to set up the nonlinear buckling analysis model of a test pile.

The test pile was 1.0 m in diameter and 60.0 m in length. 24 2-D beam elements (BEAM54) were used to represent the pile shaft. Every node has three degree of freedom (DOF), i.e., two translation DOF at x and y axis, and one rotation DOF at z axis. The elastic modulus of pile shaft concrete E had a value of 28 000 MPa. The stratification effect of the subsoil was not considered. The lateral coefficient of subsoil reaction (i.e., m) was 5.0 MN/m^4. As for the constraint condition of pile shaft ends, free boundary was considered for the pile top, while the rock-socketed pile toe was assumed fixed without any DOF.

Parameters Analysis

Based on the numerical model, a parameter analysis was carried out to judge the effect by some parameters on the critical buckling load P_{cr} of pile shaft, including the free length at the pile top l_0, the depth of initial bending z, the degree of initial bending A_0 (i.e., rise of bending arch), the inclination degree of pile shaft α, and necking of the pile shaft d_0 (i.e., the reduced diameter of pile shaft section). The influencing curves between P_{cr} and these parameters were shown in Fig.2 ~ Fig.5.

It can be seen from Fig.2 ~ Fig.5 that, the critical buckling load P_{cr} will decrease obviously with the free length l_0. For example, an increase of l_0 from 2.5 m to 5.0 m will cause a decrease of P_{cr} from 28.0 MN to 18.90 MN by a relative percentage 32.5%. Therefore, the free length of pile with high-rise cap should be noted. Similarly, the influence by initial bending of pile shaft will change with the bending degree and depth, i.e., a larger bending degree with a lower depth will cause a more disadvantageous effect on the buckling stability of pile shaft, which is similar with that

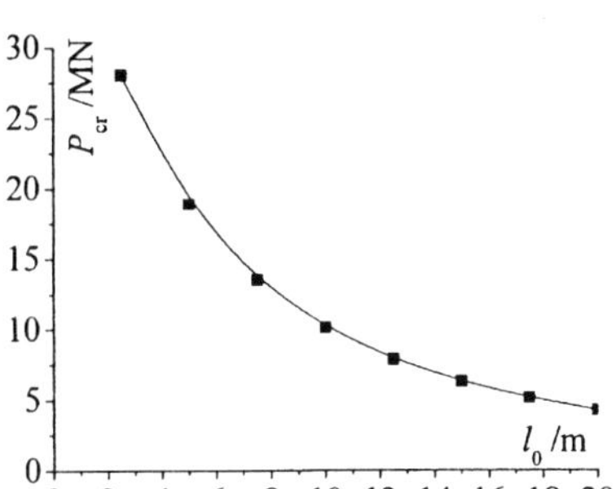

Fig. 2 Curve of P_{cr} with the free length

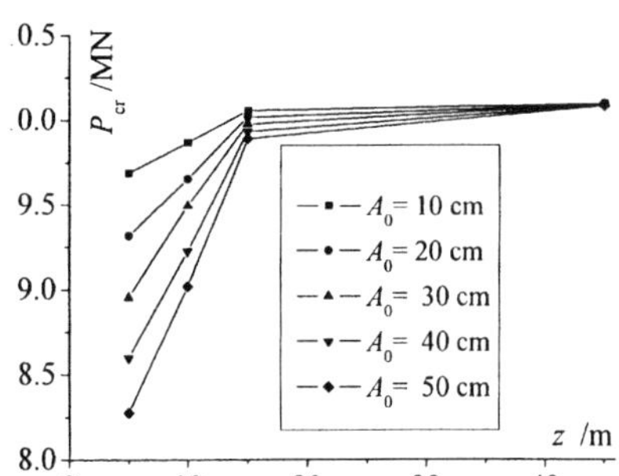

Fig. 3 Curve of P_{cr} with the initial bending

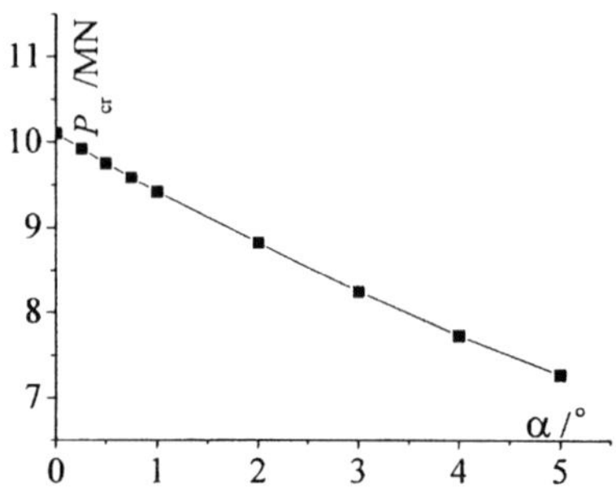

Fig. 4 Curve of P_{cr} with α

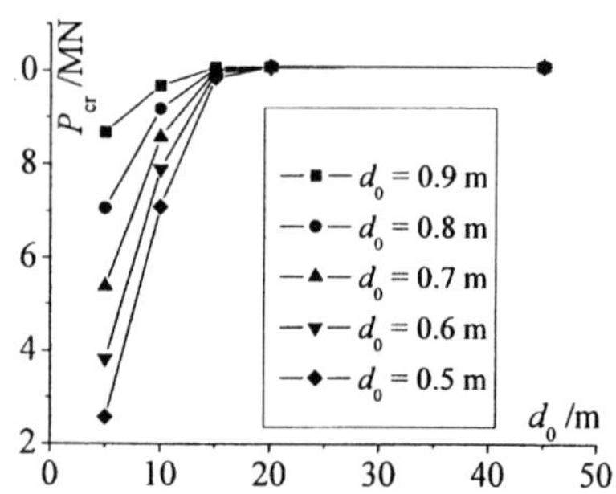

Fig. 5 Curve of P_{cr} with d_0

by the necking. A linear decrease of P_{cr} will happen with the increase of inclination degree α. Therefore, during the construction of piles with high-rise cap, precise location and rigor perpendicularity control of pile shaft should be carried out.

APPLICATIONS

The present method was applied to study the behavior of the test pile with a same size as that in the parameter analysis example. All the analysis was based on the data obtained from the field loading test. As shown in Fig.6, the counter force system for the loading test was composed of a counter force beam with a section size of 1.7 m×2.7 m and two same 1.8 m-diameter anchor piles with an uplift capacity of about 9.0 MN. A total of 9 2500 kN-capacity jacks were connected in parallel by an electric hydraulic pump to supply the load. Slowly maintained loading method was adopted. 4 dial indicators were located symmetrically to measure the pile top settlement.

During the loading course, a total number of 17 steps of load Q were applied to the pile top step by step with a step size of 1.08 MN. The final and maximum steady load on the test pile was 16.72 MN, as the corresponding pile top settlement s could nearly maintain at a value of 35.31 mm. But when the load was added up to 17.28 MN, the s increased sharply till exceeded the range of dial indicators. At the same time, the load could only be kept stable for 3 minutes. After that, an obvious offset of pile top was observed with a sharp decrease of load from 17.28 MN to 11.00 MN. Furthermore, crack appeared in the ground surface around the pile top. All these signs showed that ultimate bearing state of the test

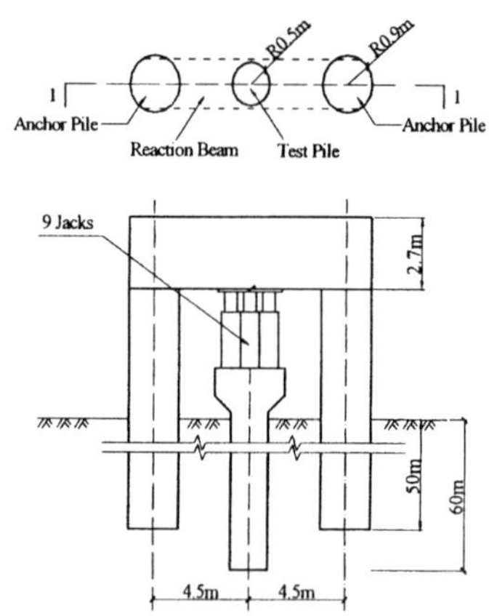

Fig.6. Sketch of the loading test

pile had reached, possibly as a result of fracture or buckling failure of pile shaft.

Therefore, nonlinear FEM was introduced to get more accurate buckling load. The analysis model was shown in Fig.7. The analysis result showed that, when the load added up to 17.65 MN, the calculated pile top settlement increased sharply (Fig.8), and couldn't reach a point of convergence after a number of repeated iterative calculation process. Then, this point may be considered as a minimum stable limit, and the corresponding load value was the buckling load of pile shaft, which showed a good agreement with a relative error of 5.6%. On the contrary, the calculated P_{cr} of linear buckling analysis was 22.91 MN, far greater than the measured ultimate load of 16.72 MN with a relative error of 37.0%.

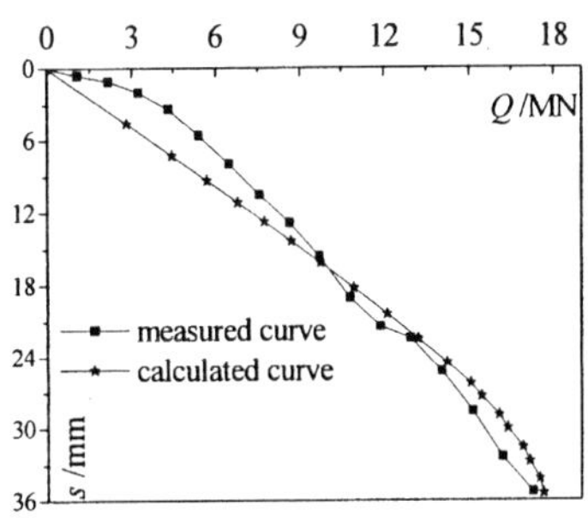

Fig. 8. Calculated and measured *Q*~*s* curves

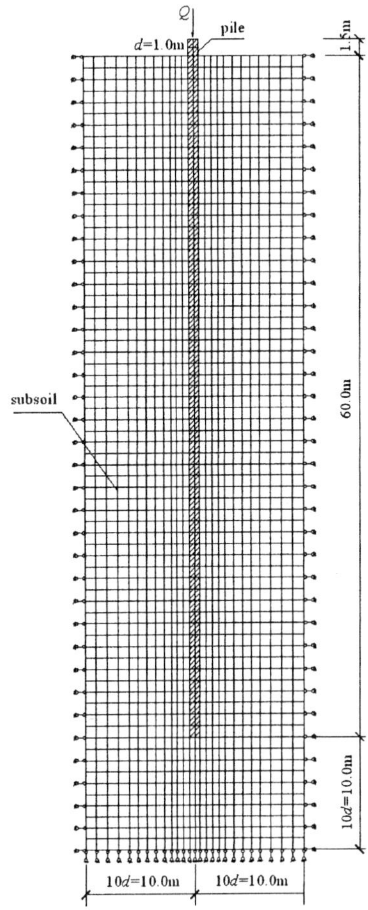

Fig.7. Analysis model by nonlinear FEM

CONCLUSIONS

1) Buckling analysis of piles with high-rise cap should be treated as the second type of stability problem instead of the first type.

2) Based on the arc length method and ANSYS, a nonlinear finite element buckling analysis is presented to judge the buckling stability of pile shaft. Parameter analysis results show that the present method is more effective than traditional methods in considering influences of non-concordant deformation, nonlinear pile-soil

interaction, and initial defects of pile shaft, such as bending, inclination or necking.

3) Application of the present method in a test pile shows that the calculated *Q-s* curve agrees well with the measured one. Furthermore, the calculated P_{cr} is very close to the measured ultimate load with a relative error of 5.6%, far better than the result by linear buckling analysis with a greater relative error of 37.0%.

ACKNOWLEDGMENTS

The authors appreciate the support by a grant from the Ph.D. Programs Foundation of Ministry of Education of China (No. 200805321042).

REFERENCES

ZHAO, Ming-hua, YANG, Ming-hui, and ZOU, Xin-jun (2005). "Vertical bearing capacity of pile based on load transfer model." *Journal of Central South University of Technology*, 12(4): 488-493.

Lee, K L (1968). "Buckling of partially embedded piles in sand." *Journal of Soil Mechanics and Foundation Division, ASCE*, 94(1): 255-270

Reddy, A S, and Valsangkar, A J (1970). "Buckling of fully and partially embedded piles." *Journal of Soil Mechanics and Foundation Division, ASCE*, 96(6): 1951-1965.

HU, Ren-li (1973). "A few problems in the calculation of side resistance by m method for bridge piles." *News Report in Railway Standard Design*, (1): 18-26.

ZHU, Da-tong (1974). "Calculation length of piles fixed in both side." *News Report in Railway Standard Design*, (5): 27-29.

ZHAO, Ming-hua (1984). "Buckling equivalent length of pile." *Engineering Mechanics*, 4(1): 94-105.

ZHAO, Ming-hua (1990). "Buckling analysis and tests of bridge piles." *China Journal of Highway and Transport*, 3(4): 47-57.

LIU, Guang-dong, and LUO, Han-quan (1998). "Stability of framed structure." Beijing: *The People's Communications Press*.

ZOU, Xin-jun, and ZHAO, Ming-hua, LIU Guang-dong (2007). "Buckling analysis of super-long rock-socketed filling piles in soft soil area by element free Galerkin method." *Journal of Central South University of Technology*, 14(6): 858-863.

ZOU, Xin-jun, ZHAO, Ming-hua, and LIU, Guang-dong (2008). "Buckling analysis of cast-in-situ rock-socketed piles with highrise pile cap under complicated subgrade reaction modulus." *Rock and Soil Mechanics*, 29(4): 963-967.

Gaber, M A, Wang, J J, and Kiger, S A (1994). "Effect of boundary condition on buckling of friction piles." *Journal of Engineering Mechanics, ASCE*, 120(6): 1392-1400

Goodman, R E, Taylor, R L, and Brekke, T L (1968). "A model for the mechanics of jointed rock." *Journal of the Soil Mechanics and Foundations Division, ASCE*, 94(SM3): 637-660.

Desai, C S, Zanman, M M, Lightner, J G, et al (1984). "Thin layer element for interfaces and joints." *International Journal for Numerical and Analytical Methods in Geomechanics*, 8(1): 19-43.

Study on Transfer Behavior of Negative Friction of Single Pile in Two-Layer Soil

Hancheng Dan[1], Liang Li [2], Hui Yao[3], Lianheng Zhao[4] and Xiaoli Yang[5]

[1]Doctoral candidate, School of Civil engineering and Architecture, Central South University, Changsha 410075, Hunan, P.R.China, danhancheng@163.com

[2]Professor, School of Civil engineering and Architecture, Central South University, Changsha 410075, Hunan, P.R.China, liliang_csu@126.com

[3]Doctoral candidate, School of Civil engineering and Architecture, Central South University, Changsha 410075, Hunan, P.R.China, jianxin_yaohui@163.com

[4]Doctoral candidate, School of Civil engineering and Architecture, Central South University, Changsha 410075, Hunan, P.R.China, zlh8076@163.com

[5]Professor, School of Civil engineering and Architecture, Central South University, Changsha 410075, Hunan, P.R.China, yxnc@yahoo.cn

ABSTRACT: This paper aims at calculating the transfer behavior of negative friction of single pile in two-layer soil in pile foundation. Based on the traditional load transfer function method, a model for calculating the transfer behavior of negative friction on a single pile in a two-layer soil profile is presented in this paper. In this model, the settlement of soil is assumed to follow bi-linear distribution; the transfer function is linear elastic and full plastic; the pile toe resistance is assumed to vary linearly with movement. By solving the model equations, the analytic solution of relative displacement and axial force of pile have been obtained. Moreover, a case history from literature is used to validate the proposed theoretical model. The prediction of negtive friction from the theoretical model agrees well with the measured value from the case history.

INTRODUCTION

A thorough understanding on the development of negative friction (NSF) of piles is essential for the analysis and design of pile foundations. Generally, NSF has two effects on pile foundations: the development of additional compressive axial force (dwondrag load) in piles and excessive pile settlement (dwondrag load). Current methods for the predictions of NSF on pile foundations fall into the following four broad categories: (I) the empirical method (Terzaghi K. and Peck R.B., 1967) (II) the load transfer method (Alonso E.E., Josa A. and Ledesma A., 1984) (III) the continuum method (Chow Y.K., Chin J.T. and Lee S.L., 1990), and (IV) the finite element method (Lee C.J., Bolton M.D. and Al-tabbaa A., 2002).

The classical method used by Johannessen and Bjerrum (Johannessen, 1965) is the effective stress method for calculating the negative friction. However, this method cannot determine the neutral plane. In 1969, based on the Mindlin solution, Poulos and Mattes (Poulos H.G. and Mattes N.S., 1969) had proposed the elastic theory for calculating the negative friction, which is only suitable for the long pile sitting on the incompressible soil layer.

In addition, during past years, different approaches had been presented to assess the response of piles (single or group) based on both complicated and simplified mathematical or numerical analyses (Tabesh A. and Poulos H.G, 2001; Liyanapathirana D.S. and Poulos H.G, 2005). Above researches were all based on the assumption that the soil is homogeneous. In this paper, the soil is divided into two layers. Based on the load transfer function of pile and considering the elastic and elasto-plastic stage of soil between the piles, the mechanical model of the load transfer mechanism of negative friction of a single pile has been established.

MODEL ASSUMPTIONS

A large number of studies have shown that the relationship between the side friction of pile and pile-soil relative displacement can be described by bilinear line model (Zhou Guo-lin, 1991). This model can be simplified in order to illustrate the negative friction model of single pile.

The soil settlement caused by heaped load is one-dimensional and follows broken double linear distribution as shown in FIG.1, where v_0 is subsidence of surface soil; v_h is the interface settlement between upper layer soil and lower layer soil, v_l is the settlement of bottom surface of soil layer. m_1 and m_2 are settlement distribution slope of upper layer soil and that of lower layer soil respectively, which can be given by

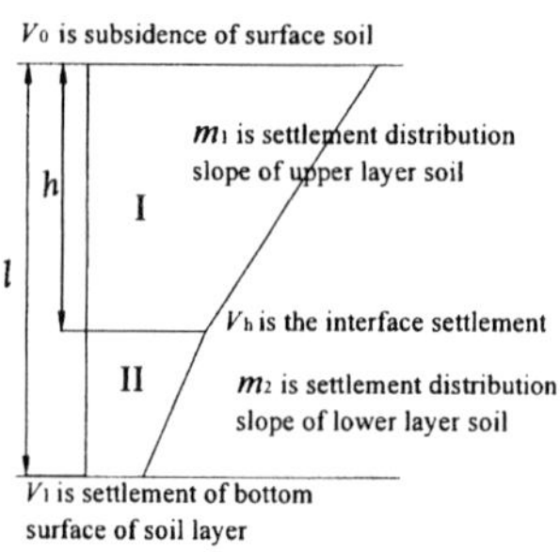

FIG.1. Distribution of soil settlement

$$m_1 = -\frac{dv}{dx} = \frac{v_0 - v_h}{h} \tag{1}$$

$$m_2 = -\frac{dv}{dx} = \frac{v_h - v_l}{l - h} \tag{2}$$

Where h is the thickness of upper layer soil [L], and l is length of pile [L].

It is assumed that the transfer function follows the linear-elastic and full plastic model relationship such as FIG.2(a) and FIG.2(b), in which f_x is the friction offered by soil per unit depth whose dimension is *KN/mm*; z_x is pile-soil relative displacement whose dimension is *mm*; z_{u1} is the elastic limit displacement value of upper layer soil around pile; k_1 is the stiffness coefficient of upper layer soil per unit depth around pile, of which the dimension is *KN/mm*; z_{u2} is elastic limit displacement value of lower layer soil around pile; k_2 is the stiffness coefficient of lower layer soil per unit depth around pile whose dimension is *KN/mm*. When z_x is greater than or equals to z_u, the friction offered by soil per unit depth around pile is constant and equals to f_{u1} or f_{u2} respectively.

As much, it is assumed that pile end bearing follows linear variation such as FIG.2(c), where R_l is pile-reaction force whose dimension is *KN*; z_x is the displacement the pile toe relivate to soil below the pile; k_3 is the stiffness coefficient of pile toe. For modeling the soil as an elasto-plastic material, it is assumed that the resistance behaves linearly elastic as long as the pressure is below R_x and is yielded with no ability to sustain additional load when it is equal or more than R_x as shown in FIG.2(c).

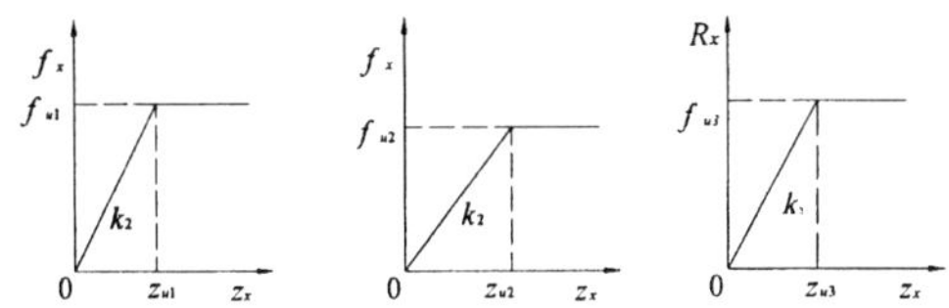

(a) (b) (c)

FIG.2. The calculation model of pile-soil system

MODEL ESTABLISHMENT

When the pile-soil relative displacement z_x is less than elastic limit displacement value z_u, the soil around pile is in elastic status. An element is selected at x depth of pile, at which the soil undergoes a settlement $v(x)$ because of consolidation, i.e. the spring for simulating the soil deformation generates a deformation $v(x)$, so the element friction is $f_x = k_1(u_x - v_x) = k_1 z_x$, where u_x is the displacement at the x section of pile.

According to the equilibrium condition, the axial force of pile N_x must agree with the equation $\frac{dN_x}{dx} = -k_1 z_x$; pile section generates elastic compression $du_x = -\frac{N_x dx}{EA}$, where A is the cross section area of the pile; E is the elastic modulus of the pile. Thus, we obtain $EA\frac{d^2u_x}{dx^2} = k_1 z_x$.

The load on the pile top section is $N_x|_{x=0} = Q$ (Q is the pile top load). In this paper, the pile top load is supposed to be zero in order to simplify the discussion and show an agreement with the general testing condition of negative friction.

$$N_x\big|_{x=0} = EA\frac{du_x}{dx}\bigg|_{x=0} = 0 \tag{3}$$

Boundary condition of pile toe is

$$N_x\big|_{x=l} = EA\frac{du_x}{dx}\bigg|_{x=l} = k_3 z_x\big|_{x=l} \tag{4}$$

And boundary condition of soil interface is

$$N_x\big|_{x=h} = EA\frac{du_x}{dx}\bigg|_{x=h} = P_1 \tag{5}$$

According to the relationship of u_x, z_x and v_x, the equation can be given by

$$\frac{du_x}{dx} = \frac{dz_x}{dx} - m_1 \tag{6}$$

Based on the Eq.(6), we have the equation which can be written as below

$$\frac{d^2 z_x}{dx^2} - \frac{k_1}{EA} z_x = 0 \tag{7}$$

By solving differential Equation (7) with the boundary conditions, we have

$$z_x^1(x) = \frac{m_1\left\{\left(\frac{P_1}{\beta_1}+1\right)\cosh(\alpha_1 x) - \cosh\left(\alpha_1 (h-x)\right)\right\}}{\alpha_1 \sinh(\alpha_1 \cdot h)} \tag{8}$$

$$N_x^1(x) = EA\left\{\frac{m_1\left\{\left(\frac{P_1}{\beta_1}+1\right)\sinh(\alpha_1 x) + \sinh\left(\alpha_1 (h-x)\right)\right\}}{\sinh(\alpha_1 \cdot h)} - m_1\right\} \tag{9}$$

where $\alpha_1 = \sqrt{\frac{k_1}{EA}}$, $\beta_1 = EA \cdot m_1$

In the low layer soil, $m_1$and α_1 are replaced by m_2 and α_2 respectively, and $\alpha_2 = \sqrt{\frac{k_2}{EA}}$.

According to the boundary conditions

$$\begin{cases} \left[\left[\frac{dz_x^2}{dx} - m_2\right]\right]_{x=h} = \frac{P_1}{EA} \\ \left[\left[\frac{dz_x^2}{dx} - m_2\right]\right]_{x=l} = \frac{k_3}{EA} z_x\bigg|_{x=l} \end{cases} \tag{10}$$

the axial force and relative displacement can be given by

$$z_x^2(x) = \frac{1}{\gamma}\left\{EA\alpha_2\beta_2 \cdot \cosh\left[\alpha_2(x-h)\right] - \lambda(\beta_2+P_1)\sinh\left[\alpha_2(l-x)\right] - EA\alpha_2(\beta_2+P)\cosh\left[\alpha_2(l-x)\right]\right\} \tag{11}$$

$$N_x^2(x) = \frac{\alpha_2 \cdot EA}{\gamma}\left\{EA\alpha_2\beta_2 \cdot \sinh\left[\alpha_2(x-h)\right] + \lambda(\beta_2+P_1)\cosh\left[\alpha_2(l-x)\right] + EA\alpha_2(\beta_2+P)\sinh\left[\alpha_2(l-x)\right]\right\} - \beta_2 \tag{12}$$

In which $\gamma=\alpha_2\cdot EA\{\lambda\cdot\cosh[\alpha_2(l-h)]+EA\alpha_2\cdot\sinh[\alpha_2(l-h)]\}$, $\lambda=\frac{k_3}{EA}$, $\beta_2=EA\cdot m_2$

And

$$P_1=\frac{\dfrac{EA\alpha_2\beta_2-\{\lambda\cdot\sinh[\alpha_2(l-h)]+EA\alpha_2\cdot\cosh[\alpha_2(l-h)]\}\beta_2}{\alpha_2\cdot EA\{\lambda\cdot\cosh[\alpha_2(l-h)]+EA\alpha_2\cdot\sinh[\alpha_2(l-h)]\}}-\dfrac{m_1}{\alpha_1}\dfrac{\cosh(\alpha_1 h)-1}{\sinh(\alpha_1 h)}}{\dfrac{m_1}{\alpha_1\beta_1\tan(\alpha_1 h)}+\dfrac{m_2}{\alpha_2\beta_2}\dfrac{\lambda\cdot\sinh[\alpha_2(l-h)]+EA\alpha_2\cdot\cosh[\alpha_2(l-h)]}{\lambda\cdot\cosh[\alpha_2(l-h)]+EA\alpha_2\cdot\sinh[\alpha_2(l-h)]}} \tag{13}$$

As well as known that neutral plane is the point at which the axial force of pile is the greatest, i.e. the derivative of axis force is equal to zero ($\frac{dN_x(x)}{dx}=0$). Therefore, we have $z_x^1(x)=0$ and $z_x^2(x)=0$, then we can obatian

$$x_1=-\frac{1}{2\alpha_1}\ln\left(-\frac{m_1\cdot P/\beta_1+m_1(1-e^{-\alpha_1\cdot h})}{m_1\cdot P/\beta_1+m_1(1-e^{\alpha_1\cdot h})}\right) \tag{14}$$

$$x_2=-\frac{1}{2\alpha_2}\ln\left(-\frac{-EA\alpha_2\beta_2\cdot e^{-\alpha_2\cdot h}+(\beta_2+P)(\lambda-EA\alpha_2)\cdot e^{-\alpha_2\cdot l}}{EA\alpha_2\beta_2\cdot e^{\alpha_2\cdot h}-(\beta_2+P)(\lambda+EA\alpha_2)\cdot e^{\alpha_2\cdot l}}\right) \tag{15}$$

Here, two x values (neutral plane depth) were obtained as shown in Eqs.(14) and (15). However, the neutral plane is unique because the neutral plane locates at the depth where the axial force is the greatest. Substitute the x_1 and x_2 into the corresponding axial force expression, and compare the axial force. If $N_x^1(x_1)>N_x^2(x_2)$, the neutral plane locates in the upper layer soil and $x^*=x_1$, while $N_x^1(x_1)<N_x^2(x_2)$, the neutral plane locates in the lower layer soil. Otherwise, $N_x^1(x_1)=N_x^2(x_2)$, the neutral plane just locates at the interface of two soil layers.

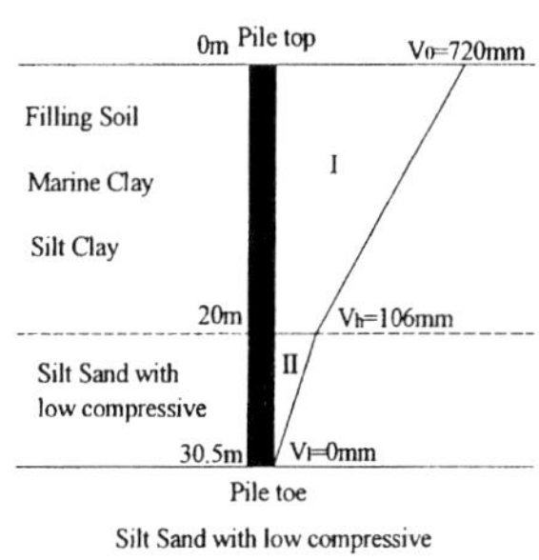

FIG.3. Distribution of soil settlement

MODEL VALIDATION

To check the validity of the model in this paper, the model is verifed against a case history which is referenced to the negative friction test of pile foundation in Hongkong (Ning Zhuang, 2003). The soil around pile from top to down are filling soil, marine clay, silt clay and silt sand with low compressive. Thus, the soil around pile can be divided into two layers (i.e. we can classify filling soil, marine clay silt clay as class "I" because of the approximately physical soil properties, while the silt sand with low compressive on which the pile toe overlays as class "II" (See

FIG.3). In addition, the distribution of soil settlement is shown in FIG.3 as well.

Moreover, the stiffness coefficient of upper layer soil and lower layer soil in elastic stage k_1 and k_2 are 2.7×10^5KN/mm and 1.8×10^6KN/mm respectively, while the stiffness coefficient of pile toe in elastic stage k_3 is 7.45×10^6KN/mm. The test pile is a steel pipe pile, whose external diameter is 609. 44mm and internal diameter is 597.44mm. The elastic modulus of steel (E_0) is 21000MPa, then the elastic modulus of steel pipe pile (E) can be calculated. In addition, other calculation parameters are provided in Table 1.

Table 1. Calculation parameters

b	α_1(m^{-1})	β_1(N)	m$_1$	α_2(m^{-1})	β_2(N)	m$_2$	λ(m^{-1})	h(m)	L(m)
3.73×10^7	0.0851	3.02×10^5	0.0081	0.2198	1.86×10^6	0.05	0.2	20	30.5

The proposed method provides the following results.

$$P_1 = -546.4KN$$

$$\begin{cases} N_{upper}^{elastic}(x) = -92205\sinh(0.0851x) + 113797\sinh[0.0851(20-x)] - 301843 \\ N_{subsoil}^{elastic}(x) = 374453\sinh[0.2198(x-20)] + 0.0065\cosh[0.2198(30.5-x)] + 264640\sinh[0.2198(30.5-x)] - 1863225 \end{cases}$$

The location of neutral plane is x^*=24.28m. The comparison of axial force of pile changing with the depth between the test results and calculation results is shown in FIG.4.

As shown in FIG.4, the calculated axial forces along the depth match the measured values well.

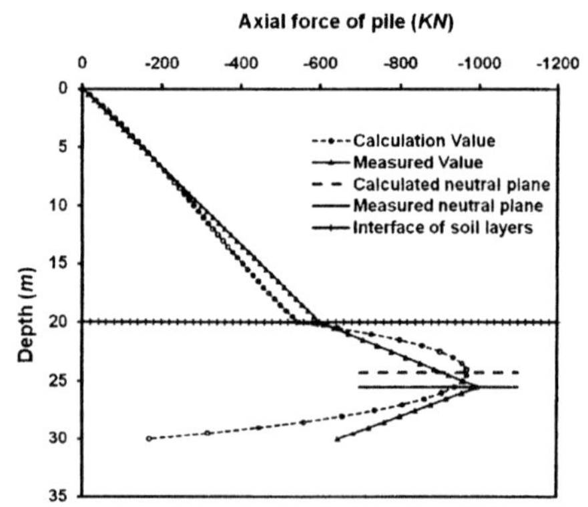

FIG.4. Comparison of predicted and measured axial forces along depth

CONCLUSIONS

In this paper, numerical models have been deduced which express the transfer behavior and distribution of negative friction of pile. Once the calculation parameters of pile and soil are determined by test and the distribution of soil layer is given as well, the plastic deformation depth of soil can be calculated by an iterative procedure and the negative friction can be obtained. Then, the proposed model and method are verified through a case history despite its simplicity. Nevertheless, the negative friction is always associated with many effect factors. Such various factors as asymmetry of heap load, hydrostatic pressure and the timeliness of friction cannot be considered completely in this paper because these factors are too complicated to allow for a closed form solution. Therefore, the model needs to be studied further in the light of future experimental and filed data.

ACKNOWLEDGEMENT

The authors appreciate the support of Ting ZHANG in spelling and grammar. In addition, the research is funded by the Central South University (No. 2009ybfz05).

REFERENCES

Alonso, E.E., Josa, A., and Ledesma, A. (1984). "Negative skin friction on piles-A simplified analysis and prediction procedure." *Geotechnique*, 34(3), 341–357.

Chow, Y.K., Chin, J.T., and Lee, S.L. "Negative skin friction on pile groups." *Int. J. Numer. Anal. Meth. Geomech.*, 14, 4–91.

Johannessen, I.J, and Bjerrum, L. (1965). "Measurements of the compression of a steel pile to rock due to settlements of the surrounding clay." *Proceeding of 6th International Conference of Soil Mechanics and Foundation Engineering*, Canada, 2, 261–264.

Lee, C.J., Bolton, M.D. and Al-tabbaa, A. (2002). "Numerical modelling of group effects on the distribution of dragloads in piles foundations." *Geotechnique*, 52(5), 325–335.

Liyanapathirana, D.S., and Poulos, H.G. (2005). "Pseudostatic approach for seismic analysis of piles in liquefying soil." *Geotech. Geoenviron. Eng., ASCE*, 131(12), 1480–1487.

Poulos, H.G., and Mattes, N.S. (1969). "The Analysis of Downdrag in End-bearing Piles." *Proceeding of 7th International Conference of Soil Mechanics and Foundation Engineering*, Mexico City, 2, 205–209.

Tabesh, A., and Poulos, H.G. (2001). "Pseudostatic approach for seismic analysis of single piles." *Geotech.Geoenviron. Eng., ASCE*, 127 (9), 757–765.

Terzaghi, K., and Peck, R.B. (1967). Soil Mechanics in Engineering Practice, 2nd ed, John Wiley & Sons, New York.

Zhou, G. L. (1991). "The analysis on Negative Friction of piles." *J. Rock and Soil Mechanics*, 12(3), 35–41.

Zhuang, N. (2003). "Simplified model and formation mechanism of negative friction of single pile." *Master Thesis, Chang'an University, Xi'an.*

Observations on Viscosity Reduction of PHPA Polymer Support Fluids

Carlos Lam[1], M. ASCE, Stephan A. Jefferis[2], and K. Gifford Goodhue, Jr.[3]

[1]Research Student, Department of Engineering Science, University of Oxford, Parks Road, Oxford, OX1 3PJ, U.K.; carlos.lam@eng.ox.ac.uk
[2]Visitng Professor, Department of Engineering Science, University of Oxford, Parks Road, Oxford, OX1 3PJ, U.K.; stephan.jefferis@eng.ox.ac.uk
[3]Research & Development Director, KB International LLC, 735 Broad Street, Suite 209, Chattanooga, TN 37409, U.S.A.; gif@kbtech.com

ABSTRACT: Polymer fluids have been used in foundation drilling for many years and yet no previous studies have been reported which assess their sensitivity to shear such as is induced by pumping equipment. A field study has been carried out on two specific polymer products under actual site conditions. It is found that there is a significant loss of viscosity when partially hydrolyzed polyacrylamide (PHPA) fluids are circulated through a centrifugal pump in a loop system. This behavior is quite distinct from that of bentonite slurries which require high shear mixing and recirculation to ensure good hydration. At first sight the sensitivity of PHPA fluids to shear may be perceived as a disadvantage but in fact if shear can be reduced so can on-site energy consumption. A reduction in fluid viscosity was also observed over a standing period between mixing and use, suggesting that a second mechanism may be contributing to the properties change. From the trial results, recommendations are given on pump selection and on the maintenance of fluid properties.

INTRODUCTION

Polymer fluids have been widely used in foundation drilling for nearly two decades in many parts of the world (Corbet et al. 1991; Ho and Tan 1996; Brown et al. 2002; Bustamante and Boato 2005). In contrast to bentonite slurries which are basically dilute clay suspensions (typically 2.5% to 6% bentonite in water), polymer fluids develop their viscosity through the hydrodynamic interaction of polymer molecules dissolved in the make-up water. The molecules have high molecular weights, typically between 14 and 17 million and a surface charge density of 30 to 45% (O'Neill and Reese 1999). It is the interaction between these charged long-chain molecules that gives rise to the solution viscosity and the other polymer properties. The most basic of these polymer fluids are partially hydrolyzed polyacrylamides (PHPAs), which are the subject of this paper. It is appropriate to note that PHPAs are by far the most popular type of polymer in the foundation industry, accounting for more than 90% of all the polymers used today.

Before considering the technical aspects of this paper, it is necessary to expand on the term 'polymer' as applied to excavation support fluids. This is essential for the proper interpretation of this paper and to the successful use of such fluids in practice. Basically, the world of polymers is vast and highly varied and it is fundamental to recognize this when considering these materials. Engineers should not treat polymers as if they were a single product and must recognize their many differences in chemical composition, physical form and properties. To treat all polymers fluids as a single product would be analogous to treating all clay slurries, such as bentonite, attapulgite, kaolin, and native clay as if they were all identical. In this paper, two polymers of very specific chemical composition are considered; both are PHPAs and yet even within this composition we find variations in their properties as will be shown later.

Although bentonite slurries and polymer fluids are two distinctly different systems, when pumped into an excavation, trench or shaft etc., they both work by exerting hydrostatic fluid pressure against the walls of the excavation to prevent groundwater infiltration and to provide a stabilizing force to the excavation. It is well known that bentonite slurries maintain this fluid pressure by forming a filter cake on the exposed soil surface so that fluid loss into the adjacent ground is reduced to low levels (Fleming and Sliwinski 1977). However, pure PHPA fluids cannot form a filter cake although they may do so if there are fine soil/clay particles suspended in the fluid. Fundamentally for fluid loss prevention, it is the rheological properties of PHPA fluids which control their flow into and through the soil pores within the surrounding formation. The rate at which the fluids are lost into the surrounding ground mainly depends their solution viscosity and the resulting viscous drag resistance achieved as the fluid migrates into the soil. Herein lies the major deficiency of PHPAs as excavation-support fluids and the underlying cause for the majority of the stabilization failures seen to date. Without proper monitoring and maintenance of the viscosity, PHPAs will fail in their role as support fluids.

Despite the importance of the viscosity of polymer fluids, little information can be found in the civil engineering literature about its sensitivity to external influences. Majano et al. (1994) report the effects of some chemical contaminants (sodium chloride, gasoline and high pH conditions) on the viscosity and the side friction of laboratory-scale model piles. In terms of physical interference such as pumping, O'Neill and Reese (1999) argue that diaphragm pumps will cause less damage to the fluids than centrifugal pumps, but they do not present any data to support this thesis. A survey of product literature produced by leading polymer suppliers reveals contradictory advice: one supplier strongly discourages the use of centrifugal pumps whilst others either allow their use or provide no guidance on the matter. To help resolve the issue, a field study was undertaken on a construction site in Glasgow, U.K. where drilled shafts (bored piles) were being installed under the support of PHPA fluids. During the work, opportunity was taken to study the loss of viscosity of clean unused PHPA fluids when subjected to the mechanical action of a centrifugal pump. Details of the test conditions are given below.

PLANT SETUP AND TEST EQUIPMENT

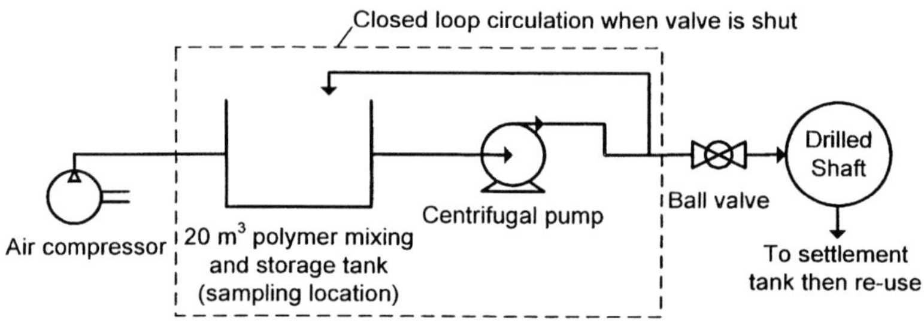

FIG 1. Schematic of Plant Setup for Test Program

Figure 1 shows a schematic diagram of the plant setup for this study. The PHPA fluids were mixed in a 20 m^3 mixing and storage tank by sprinkling the required weight of dry PHPA granules into a stream of flowing water followed by gentle agitation with compressed air lances for about an hour. A centrifugal pump with a maximum capacity of 320 m^3/h was connected between the tank and the drilled shaft. The pipe work was laid out so that the polymer fluid was recirculated back to the storage tank when the ball valve in the feed line to the excavation was shut. This meant that the pump did not need to be turned repeatedly off and on during excavation – an important aspect of plant operation as the pump and storage tank may be at some distance from the excavation.

During the test period, samples of the PHPA fluid were taken from the storage tank and tested with a Marsh funnel, which is a crude but a simple way of measuring the viscosity of support fluids in the field. The test basically involved filling the funnel with about 1.5 liter of the polymer fluid and measuring the efflux time for 946 mL (one U.S. quart) to flow out of the tip. The time in seconds was recorded as the Marsh funnel viscosity. Further details of the Marsh funnel test can be found in Darley and Gray (1988). Since the aim of the study was to assess the damage to the PHPA fluids caused by pumping, only clean fluids were used because samples taken from the drilled shafts could have been influenced by contaminants from the ground and would also contain suspended soil particles, which would affect the viscosity results.

POLYMER MATERIALS AND MIXING DETAILS

Two types of synthetic polymers were tested. Both were partially hydrolyzed polyacrylamides or PHPAs, which are copolymers of sodium acrylate (CH_2-CH-COONa) and acrylamide (CH_2-CH-$CONH_2$). The products used were in dry granular form although liquid emulsion forms of PHPA also are available. While both of the polymers tested were composed of the same two base components, it must be recognized that they may be considerably different in the ratio of these components, the length and geometry of their polymer chains, and most importantly in their performance characteristics. The suppliers of the PHPAs advised that their products were designed to work with a range of other chemical additives for enhanced shaft stability and soil-binding properties, but in this study only the base polymers were tested to reduce the number of variables.

Before addition of the PHPA granules, the pH of the make-up water was raised to between 10 and 11. The polymer suppliers recommend this pre-treatment to minimize the negative effects of water hardness and other ionic species in the water so as to maximize the viscosifying potential of the polymers. The particular pH control agent used and the optimum pH value required varied between the two PHPAs. After treatment of the make-up water, the final pH was confirmed by 4-colour indicator papers and the desired amount of PHPA granules slowly poured into a stream of water flowing into a tank under the gentle agitation with compressed air.

The PHPA concentrations used were determined following the suppliers' guidelines for the expected site conditions. The actual dosages are given in Table 1 from which it can be seen that the concentration of PHPA in Fluid B per cubic meter of water was 30% less than that in Fluid A. Although to the user there may be no obvious reason for the significant difference in the suggested dosage, it must be recognized that polymers are chemical systems formulated by the suppliers and it is therefore appropriate to respect their recommendations. By doing so, the test fluids would also have similar properties to those used on other sites with similar ground conditions.

The PHPA fluids used in the trials did not require a long hydration period for viscosity development such as is required for bentonite slurries. However, to ensure that further dissolution of PHPA solids did not affect the test results they were prepared a day in advance of use. Table 1 summarizes the data on the PHPA materials and the make-up water.

Table 1. Summary of PHPA and Make-up Water Data

PHPA polymer designation	PHPA concentration (kg/m^3)	pH adjustment agent	Water hardness[a] (mg/L)	Original and final pH of make-up water
Fluid A	0.90	alkali hydroxide	7.5 as calcium carbonate (soft water)	8 / 11
Fluid B	0.60	alkali carbonate		8 / 10

Note: [a] data obtained from Scottish Water for the central Glasgow area in year 2009.

Since this study was carried out on a working construction site, during the monitoring period the PHPA fluid was drawn off from the storage tank to support the drilling of shafts (Figure 1). For each polymer approximately 9 m^3 of fluid were used in excavations during the test period, thus at the end of each trial approximately 11 m^3 of fluid remained in the storage tank. To ensure identical test conditions between the two PHPAs, after completion of the first trial the storage tank was emptied and cleaned before the second PHPA fluid was mixed. The used fluid after concreting of the drilled shafts was routed to a separate storage tank so as not to contaminate the test fluids.

RESULTS AND DISCUSSIONS

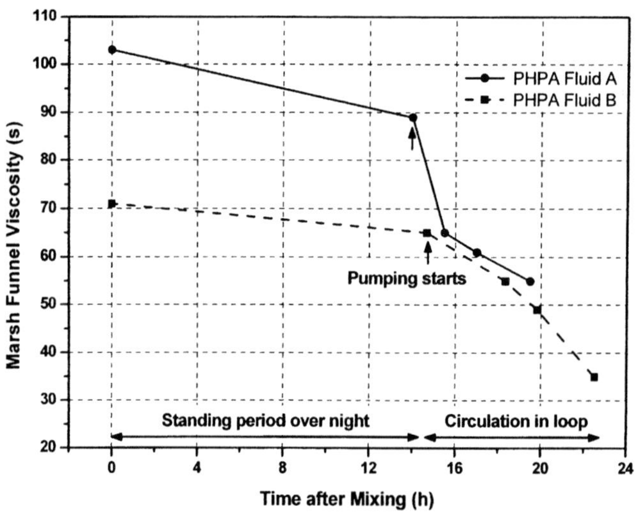

FIG. 2 Viscosity Change of PHPA Polymer Fluids

Figure 2 shows the results of the Marsh funnel viscosity tests for the two PHPA fluids for the whole test period. Interestingly, during the initial standing period between the end of mixing and the start of the pumping trial the following day, the viscosity dropped from 103 to 89 s for Fluid A and from 71 to 65 s for Fluid B, equivalent to a reduction of 14% and 9% respectively. This is a feature of PHPAs which is not always recognized, although the extra time will have allowed any undissolved PHPA granules to break up and also further disentanglement of polymer molecules to occur – both processes being expected to contribute to viscosity development. The storage tank was open at the top but as the weather during the test period was fine the fluids could not have been diluted by rainwater. Intermediate measurements during the overnight standing period were not practicable so it is not known how the viscosity reduction developed with time. This viscosity reduction has been the subject of further research, but this is beyond the scope of this paper.

For each PHPA fluid, after the initial standing period of about 14 hours, the fluid began to be drawn off for shaft support. In the shafts, when the excavation reached the bottom of the surface casing fluid was pumped to the excavation. As digging progressed, the valve on the feed line was used by site staff to add further PHPA fluid to the excavation. The fluid level was always maintained at about 1.5 m below the ground level. When the fluid was not being drawn off for use, as previously discussed, the centrifugal pump continued to run and the PHPA fluid was recirculated to the storage tank (Figure 1).

As can be seen in Figure 2, the viscosity of both PHPA fluids dropped significantly as soon as pumping started. When excavation at the Fluid A supported drilled shaft was complete, the viscosity had dropped from 89 to 55 s, which is equivalent to a reduction of 38% in viscosity. The situation with Fluid B was similar; it showed a drop from 65 to 35 s – a 46% reduction. Both PHPAs are understood to be of high molecular weight and it is believed that the reduction in viscosity was largely due to damage to the long-chain PHPA polymer molecules (molecular scission) caused by the strong mechanical action of the pump.

Towards the end of the monitoring period the rate of viscosity reduction appeared to accelerate. This is probably due to the fact that for each batch of PHPA a significant portion of the fluid had been drawn off to the excavation (~9 m^3) thus increasing the rate of recirculation of the fluid remaining in the storage tank. Finally, as we do not yet know the exact mechanism for the viscosity reduction during the standing period, it must be allowed that some of the reduction observed during the recirculation period may have been due to this mechanism. Therefore, the overall results may be the combination of effects of two totally different mechanisms.

It is also worth noting that, for both PHPA fluids, the reduction in viscosity seen during recirculation caused concerns about possible instability of the excavations. As a remedial measure, the viscosity was boosted by adding extra polymer materials directly into the holes. After completion of the monitoring work, all the remaining drilled shafts on the construction site were excavated under Fluid A with the additives recommended by the polymer supplier and were completed without any problems.

CONCLUSIONS

The viscosity of two PHPA polymer fluids has been monitored on a construction site in Glasgow, U.K. The results show that viscosity does not stay constant but tends to decrease over time even without any external interference; further research is needed to examine this behavior. The results presented here confirm the caution given in O'Neill and Reese (1999) that synthetic polymer fluids are susceptible to damage by centrifugal pumps, and in fact any pumping equipment which induces high shear stresses on the long-chain polymer molecules. Under the test conditions of this study, the viscosity of the two PHPA fluids dropped by 38% and 46% over a period of less than 10 hours. This reduction is clearly a function of the volume of the fluids in the system and the duration of the circulation.

In the light of the findings from these brief trials, it is recommended that the use of centrifugal pumps be avoided and alternatives such as diaphragm pumps be sought. If diaphragm pumps are not available in the area of work, circulation of the PHPA fluids should be minimized. The pump should be switched off when not required to supply polymer fluid to the excavation. It is appreciated, however, that this suggestion may be difficult to implement unless the pump switches off automatically when the valve to the excavation is closed. It is also clear that regardless of the final selection of pumping equipment, the condition of the PHPA fluid should be checked regularly, so that any viscosity reduction can be identified and, if necessary, the fluid properties boosted. Furthermore, it seems prudent to prepare the fluids at a slightly higher viscosity than is actually required to allow for any potential reduction in viscosity on standing.

Although the trials showed that PHPA fluids are sensitive to the mechanical action of centrifugal pumps, this is actually an advantage in terms of energy use. Bentonite slurries require high shear, i.e. high energy mixing, and continued recirculation to develop their viscosity fully (Jefferis 1982). On the other hand, PHPA fluids can be mixed by simply pouring the granules into flowing water and do not require recirculation to develop their properties. The use of PHPA fluids and possibly other types of polymer products will therefore simplify site operations and reduce on-site energy consumption once appropriate plant is the norm.

Although both PHPA fluids tested in this study showed a similar degree of viscosity reduction, it is possible that some of the newer polymer products available in the market are less susceptible to damage by shear due to different chemistries and compositions. In fact, these products claim to rely less on their fluid viscosity and more on their soil-binding ability for shaft and trench support. Details of these polymer systems can be found in several U.S. patents by Goodhue et al. (2001, 2005). Research on these newer systems is in progress and will be reported in due course.

ACKNOWLEDGMENTS

The data presented in this paper was collected as part of a research project funded by the Engineering and Physical Sciences Research Council (EPSRC) in the U.K.; grant reference EP/C537815/1. The technical and financial support of Stent Foundations Ltd. is gratefully acknowledged as is the advice on polymer systems provided by KB International LLC and CETCO Construction Drilling Products. We would also like to thank Messrs. Viv Troughton, Allan Bowers, Chris Martin and Peter Martin.for their continuing help and advice with the project.

REFERENCES

Brown, D., Muchard, M., and Khouri, B. (2002). "The effect of drilling fluid on axial capacity, Cape Fear River, NC." *Proc., DFI 27th Annual Conference on Deep Foundations*, San Diego, CA, Deep Foundations Institute, Hawthorne, NJ.

Bustamante, M. and Boato, R. (2005). "Polymer slurry in large diameter pile drilling – case histories." *Proc., 16th Int. Conference on Soil Mech. and Geotech. Engrg.*, Osaka, Japan: 2083–2086.

Corbet, S.P., Culley, D.S., Sherwood, D.E., and Cockcroft, J.E.M. (1991). "Testing and analysis of preliminary test piles in very weak chalk." *Proc., 4th Int. Conference on Piling and Deep Foundations*, Stresa, Italy: 57-63.

Darley H.C.H. and Gray G.R. (1988). *Composition and properties of drilling and completion fluids,* 5th Edition, Gulf Publishing Company, Houston, Texas, 643 p.

Fleming W.K. and Sliwinski Z.J. (1977). "The use and influence of bentonite in bored pile construction.", *Report PG3*, Construction Industry Research and Information Association, London, U.K., 93 p.

Goodhue, K.G., Jr., Holmes, M.M., Norman, C.S., and Wilkerson, J.M., III (2001). *Composition and method for a dual-function soil-grouting excavating or boring fluid*, U.S. Patent 6,248,697 B1, June 19, 2001.

Goodhue, K.G., Jr., Holmes, M.M., Norman, C.S., and Wilkerson, J.M., III (2005). *Composition and method for dual function soil grouting excavating or boring fluid*, U.S. Patent 6,897,186 B2, May 24, 2005.

Ho, C.E. and Tan, C.G. (1996). "Barrette foundation constructed under polymer slurry support in old alluvium." *Proc., 12th Southeast Asian Geotechnical Conference*, Kuala Lumpur, Malaysia: 379-384.

Jefferis, S.A. (1982). "Effects of mixing on bentonite slurries and grouts." *Proc., Specialty Conference in Grouting in Geotechnical Engineering*, New Orleans, LA, ASCE: 62-77.

Majano, R.E., O'Neill, M.W., and Hassan, K.M. (1994). "Perimeter load transfer in model drilled shafts formed under slurry." *J. Geotech. Engrg.*, 120(12): 2136-2154.

O'Neill, M.W. and Reese, L.C. (1999). "Drilled shafts: Construction procedures and design methods." *Publ. No. FHWA-IF-99-025*, Federal Highway Administration, U.S. Department of Transportation, Washington, D.C., 758 p.

An Analytical Solution to Transform O-cell Pile Test Data into Conventional Load-settlement Curve

Xiao-zhou XI[1], Long-zhu CHEN[1], Wei LIU[1,2]

1 Institute of Engineering Safety and Disaster Prevention, Dept. of Civil Engineering, Shanghai Jiao Tong University, Shanghai 200240, China
2 Dept. of Civil Engineering, Ningbo Institute of Technology, Zhejiang University, Ningbo 315100, China

ABSTRACT: In this paper, an analytical method is proposed to transform the results of Osterberg-Cell pile testing (OPT) to conventional load-settlement curve of pile, which is loaded downward at the pile top. First, a tri-linear load transfer function representing pile side resistance was used to analyze the OPT of the pile segment above the Osterberg Cell. Parameters needed for the transfer function were obtained by fitting the OPT data. The self-weight of the upper pile segment was taken into account. As the friction between soil and pile in OPT has opposite direction compared to that of the conventional test, the downward transfer function has been transformed into the upward function. The OPT data of the lower pile segment was considered as boundary conditions of the upper segment of the equivalent top-loaded pile to transform the OPT data into conventional results with the upward friction obtained above. Case histories were used to validate that the conventional pile load-settlement curve obtained using this analytical method compared well with the measured data.

1 INTRODUCTION

Self-Balanced testing method is an up-to-date static pile testing approach. In this method, specially designed loading equipment, viz. Osterberg Cell is placed at the specified location along piles, with hydraulic pipes and displacement transducer wires extended to the ground surface. Hydraulic jack provides hydraulic pressure into the loading cell, which applies loads to the test piles. In this approach, actual loading is in equilibrium with frictions between upper-segment of the pile and the soil, self-weight of pile, frictions of lower-segment of pile and pile tip resistance. The capacity of the pile can be obtained from the two load-displacement curves of the upper and the lower segment of pile. This approach was first introduced into practices by Osterberg (1989) and is referred as Osterberg-Cell load test or O-cell load test. This method has been widely used for its simple instrumentation and lower expense, especially with no needs of reaction piles or massive mass.

Since the mechanism of load transfer and soil-pile interactions in an O-cell load test is similar to that in a conventional static pile test, allowing this method considered as a reliable testing approach. However, the mode of load transfer in a self-balanced load test is different from that in a static load test consequently a certain transform approach is required to convert test data into a pile top load-settlement relationship. Empirical method and displacement harmony method are two kinds of such converting methods (GONG et al., 2002, and JIANG et al., 2006). However, the empirical method lacks sufficient theoretical explanation, and the displacement harmony method requires transmission functions for each layer of soil from field measurements. The load transfer method described herein divides the pile into several elastic elements with nonlinear springs connected in-between to simulate the load transfer mechanism between piles and soil (Seed and Reese, 1957, and CHEN et al., 1995). With static equilibrium and elastic theory, the analytical description about the behavior of piles can be derived, which requires the assumptions about loading transfer functions.

There have been quite number of research works confirmed that the hyperbolic function is suitable to simulate the loading transform. However, it was realized that the hyperbolic function is difficult to be solved and therefore was simplified into multi-linear functions (CHEN et al., 1994). This paper adopts a tri-linear model to simulate the load transfer of piles. In this method, the analytical expression of a load-displacement relationship for the upper-segment of piles is obtained with consideration of the pile self-weight. With comparisons with actual measurement, the parameters of load transfer function can be obtained. Then convert the load transfer function along the downwards direction in the self-balanced method into upwards direction. From that, the overall load-settlement curve of piles can be obtained with the actual measurements of the lower segments of piles as the boundary condition for the upper-segments of piles.

2 ANALYTICAL PREDICTION OF THE LOAD-DISPLACEMENT BEHAVIOR OF THE UPPER SEGMENT OF SELF-BALANCED PILES

2.1 Assumptions

Assuming the pile acting as a uniform cross-section linear elastic beam surrounded by single layer homogeneous soil, the equilibrium condition of the micro segment of the upper-segment of pile is shown in Fig.1. The relationship between axial force Q, self-weight G_u, perimeter u and side friction stress q_s is provided in the following equation.

$$\mathrm{d}Q(z)/\mathrm{d}z = -uq_s(W) - \gamma A \tag{1}$$

in which, q_s is function of the relative displacement between piles and soil (W), A is sectional area of pile, γ is the density of pile. Also with the assumption of pile acting as elastic bar, the following equation can be derived from the elastic stress-strain relationship of the pile.

$$-\mathrm{d}W/\mathrm{d}z = Q(z)/EA \tag{2}$$

From Equations (1) and (2), the basic relationship for the upper-segment of the pile is shown as the follows.

$$\mathrm{d}^2W/\mathrm{d}z^2 = uq_s(W)/EA + \gamma/E \tag{3}$$

In the above analytical descriptions, due to the different directions between the side frictions of piles in a self balanced pile test and in a static load test, the effects from self-weight of piles has been incorporated into the differential equations. The friction stress q_s along the pile can be described into load transfer function, a tri-linear model as shown in Fig. 2, which consists of an elastic state and two plastic hardening states. In this tri-linear model, λ_i is the shear modulus of the surrounding soil, S_{m1} is elastic displacement limit of the surrounding soil and S_{m2} is the displacement limit at the first plastic hardening.

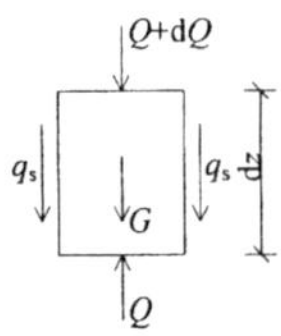

Fig. 1. Loading Diagram for the Micro Segment of Pile

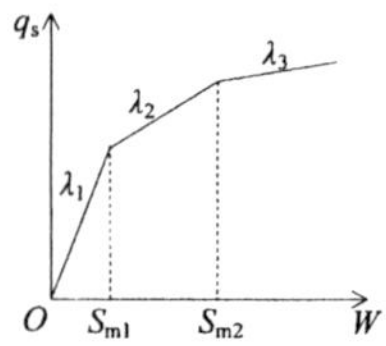

Fig. 2. Load Transfer Function between Pile and Surrounding Soil

2.2 Analytical Description for the Load-displacement Relationship of the Upper-segment of Pile

For the upper-segment of the pile, the surrounding soil is in elastic initially. With the load increased, plasticization happens in the soil near the loading cell and gradually develops upwards. After a certain load level, all surrounding soil becomes plastic in the second plastic hardening state. As shown in Fig.3, R_1, R_2, R_3 represent the height of regions of the three states, respectively. Substitute the load transfer function q_s into Equation (3), with consideration of the boundary condition at the top of pile (zero axial force) and the continuous conditions of surrounding soil in different states, the analytical description of load-displacement curve for the upper-segment of pile can be derived as follows.

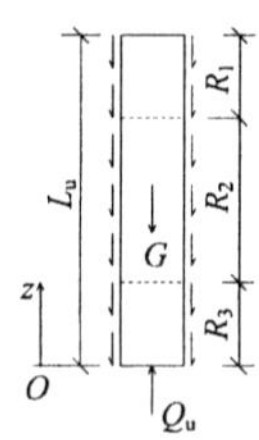

Fig. 3. Different Loading Stages for the Upper-Segment of Pile

as $\alpha_i = \sqrt{u\lambda_i / EA}$, $\beta_i = \alpha_i L_u$, $b_i = \alpha_i \operatorname{th}(\alpha_i R_i)$, ($i$=1,2,3), $m = (\alpha_1^2 - \alpha_2^2)/\alpha_2^2$, $G_u = \gamma A L_u$.

(i) All the soil surrounding the upper segment of pile keeps elastic. R_1=L_u.

$$Q_u = \alpha_1 EA \operatorname{th} \beta_1 (S_u + \gamma / E\alpha_1^2) \tag{4}$$

(ii) Part of the soil surrounding becomes plastic as the first plastic hardening state. R_1+ R_2=L_u.

$$\left.\begin{aligned} S_u &= (S_{m1} + \frac{\gamma}{E\alpha_1^2}) \frac{\alpha_1^2 \operatorname{ch}(\alpha_2 R_2) + \alpha_2 b_1 \operatorname{sh}(\alpha_2 R_2)}{\alpha_2^2} - mS_{m1} - \frac{\gamma}{E\alpha_2^2} \\ Q_u &= EA(mS_{m1} + S_u + \frac{\gamma}{E\alpha_2^2}) \frac{\alpha_1^2 b_2 + \alpha_2^2 b_1}{\alpha_1^2 + b_1 b_2} \end{aligned}\right\} \tag{5}$$

With the load increased, displacements of the surrounding soil have different behavior when

$$\frac{\alpha_1^2}{\alpha_2^2}(S_{m1} + \frac{\gamma}{E\alpha_1^2}) \operatorname{ch} \beta_2 > mS_{m1} + S_{m2} + \frac{\gamma}{E\alpha_2^2} \tag{6}$$

The soil reaches S_{m2} at the loading cell firstly, which means the beginning of the second plastic state. Otherwise, the soil reaches S_{m1} at the top of pile firstly, which means the end of the elastic state. If the surrounding soil at loading cell reaches the second hardening state firstly, with the load increased, the soil behaves as follow.

(iii) The three states exist simultaneously in the soil surrounding the upper segment of pile. $R_1+R_2+R_3=L_u$.

$$\left.\begin{aligned} S_u &= S_{m2}+(mS_{m1}+S_{m2}+\frac{\gamma}{E\alpha_2^2})[\frac{\alpha_2^2}{\alpha_3^2}\mathrm{ch}(\alpha_3 R_3)+\frac{\alpha_1^2 b_2+\alpha_2^2 b_1}{\alpha_1^2\alpha_3+\alpha_3 b_1 b_2}\mathrm{sh}(\alpha_3 R_3)-\frac{\alpha_2^2}{\alpha_3^2}] \\ Q_u &= EA(mS_{m1}+S_{m2}+\frac{\gamma}{E\alpha_2^2})[\frac{\alpha_2^2}{\alpha_3}\mathrm{sh}(\alpha_3 R_3)+\frac{\alpha_1^2 b_2+\alpha_2^2 b_1}{\alpha_1^2+b_1 b_2}\mathrm{ch}(\alpha_3 R_3)] \end{aligned}\right\} \tag{7}$$

As the continuous displacement at the boundary, equation (8) must be met.

$$(S_{m1}+\frac{\gamma}{E\alpha_1^2})\frac{\alpha_1^2\,\mathrm{ch}(\alpha_2 R_2)+\alpha_2 b_1\,\mathrm{sh}(\alpha_2 R_2)}{\alpha_2^2}=mS_{m1}+S_{m2}+\frac{\gamma}{E\alpha_2^2} \tag{8}$$

(iv) If the elastic region vanishes firstly, all the soil surrounding becomes plastic as the first plastic hardening state. $R_2=L_u$. And the equations in (iii) should be replaced by equation (9).

$$Q_u=\alpha_2 EA\,\mathrm{th}\,\beta_2(S_u+mS_{m1}+\gamma/E\alpha_2^2) \tag{9}$$

With the load increased, the soil behaves as mentioned.

(v) Part of the soil becomes plastic as the first plastic hardening state while others as the second plastic hardening state. $R_2+R_3=L_u$.

$$\left.\begin{aligned} S_u &= S_{m2}+(mS_{m1}+S_{m2}+\frac{\gamma}{E\alpha_2^2})[\frac{\alpha_2^2}{\alpha_3^2}\mathrm{ch}(\alpha_3 R_3)+\frac{b_2}{\alpha_3}\mathrm{sh}(\alpha_3 R_3)-\frac{\alpha_2^2}{\alpha_3^2}] \\ Q_u &= EA(mS_{m1}+S_{m2}+\frac{\gamma}{E\alpha_2^2})[\frac{\alpha_2^2}{\alpha_3}\mathrm{sh}(\alpha_3 R_3)+b_2\,\mathrm{ch}(\alpha_3 R_3)] \end{aligned}\right\} \tag{10}$$

(vi) All the soil surrounding becomes plastic as the second hardening state. $R_3=L_u$.

$$Q_u=\alpha_3 EA\,\mathrm{th}\,\beta_3[S_u-S_{m2}+\frac{\alpha_2^2}{\alpha_3^2}(mS_{m1}+S_{m2}+\frac{\gamma}{E\alpha_2^2})] \tag{11}$$

If $\lambda_3=0$, the second plastic hardening state becomes perfect plastic, and the equations relevant can be obtained using L Hospital rule. If $\lambda_3<0$, the first plastic hardening state becomes a softening state, and Euler's formula can be used to obtain relevant equations. As mentioned above, this tri-linear load transfer function can simulate different type of pile side resistance.

Through the analytical predictions of load-displacement for the upper segment of pile, the tri-linear load transfer function can be derived using the experimental measurements. As described in Fig. 1, the acting direction of the friction force for self-balanced piles is downwards. There is a need to verify certain parameters to convert the load transfer function to be suitable for the top-loading condition. According to Gong's paper in 2002, it can be achieved by dividing the parameters by coefficient η. For multi-layer soil, the coefficient η is calculated using the following equation.

$$\eta=L_u/\sum(L_i/\eta_i) \tag{12}$$

where L_i is the thickness of the ith layer, η_i is the correction coefficient of the ith layer. For clay and silt, η=0.8. For sand, η=0.7 and for rock, η=1.0.

3 ANALYTICAL METHOD TO CONVERT PREDICTION OF SELF-BALANCED PILES INTO LOAD-DISPLACEMENT CURVES FOR STATIC LOAD TEST PILES

3.1 Transforming Mechanism

The self-balanced pile is divided to upper and lower segments at the loading cell. Pile top load of the lower segment of pile is in equilibrium with the side friction resistance and the tip resistance of pile. The reaction force at the bottom of the upper segment of the pile is equal to the negative side friction resistance and self-weight of the pile. Therefore, the converting mechanism can be simplified into transforming the negative friction force along the upper segment of pile into the positive friction force of the static load pile. As shown in Fig.4, the loaded pile can be divided into two segments as shown in Fig.4 (b), in which the lower segment of the pile has the same transfer mechanism as the self-balanced piles. The Q_u-S_u curves of the self-balanced pile can be assumed as in accordance with that the axial loaded piles within the same region. With this assumption, the axial force and settlement caused by the top pressure and the positive side friction resistance of pile at the upper segment of pile can be calculated. As the boundary conditions, the top pressure and settlement of pile can be derived by solving the above differential equations with friction load transfer function being substituted in it.

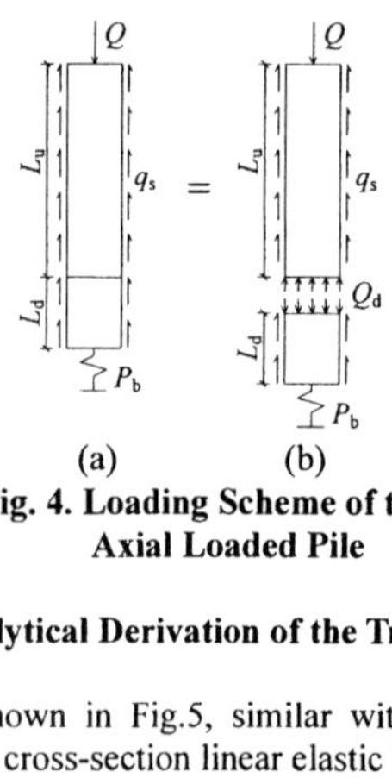

(a) (b)

Fig. 4. Loading Scheme of the Axial Loaded Pile

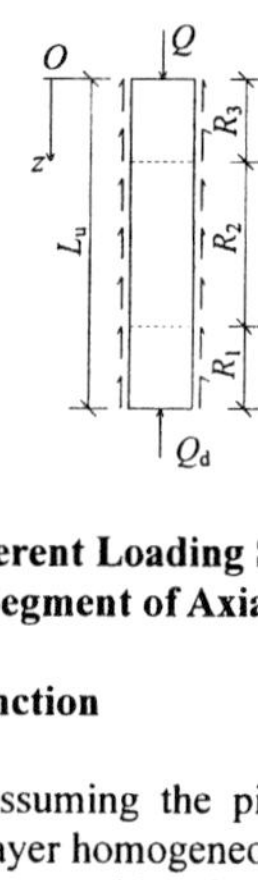

Fig. 5. Different Loading Stages for the the Upper-Segment of Axial Loaded Pile

3.2 Analytical Derivation of the Transforming Function

As shown in Fig.5, similar with section 2.1, assuming the pile acting as a uniform cross-section linear elastic bar with single-layer homogeneous soil around, the differential equation is shown as follows, without consideration of self-weight of piles (CHEN et al., 1994).

$$\mathrm{d}^2W/\mathrm{d}z^2 = uq_s(W)/EA \tag{13}$$

According to Equation (12), the transforming coefficient is re-evaluated, with corresponding updated values of parameters α_i, β_i, m. With substitution of each measurement of (Q_d, S_d) as boundary conditions, the state of the surrounding soil can be determined according to the relationship between S_d and S_{m1} or S_{m2}. The following section describes different situations, from which the (Q, S) relationship at the top of pile can be calculated according to each (Q_d, S_d) and finally obtain the load-settlement curves at the top of pile.

(1) $S_d \leq S_{m1}$

It means that the soil surrounding upper segment of pile is elastic at the location of the loading cell. Three conditions can be further considered according to the possible states of soil.

as $C_1=\alpha_1 S_d \operatorname{sh}(\alpha_1 R_1)+\frac{Q_d}{EA}\operatorname{ch}(\alpha_1 R_1)$, $C_2=\frac{\alpha_1^2}{\alpha_2}S_{m1}\operatorname{sh}(\alpha_2 R_2)+C_1\operatorname{ch}(\alpha_2 R_2)$,

$C_3=\alpha_2(S_d+mS_{m1})\operatorname{sh}(\alpha_2 R_2)+\frac{Q_d}{EA}\operatorname{ch}(\alpha_2 R_2)$

(i) All the soil surrounding the upper segment of pile keeps in elastic. $R_1=L_u$, requires $S\leq S_{m1}$

$$\left.\begin{aligned} S&=S_d\operatorname{ch}\beta_1+\frac{Q_d}{\alpha_1 EA}\operatorname{sh}\beta_1\\ Q&=\alpha_1 EAS_d\operatorname{sh}\beta_1+Q_d\operatorname{ch}\beta_1\end{aligned}\right\}\tag{14}$$

(ii) Part of the soil surrounding the upper segment of pile becomes plastic as the first plastic hardening state. $R_1+R_2=L_u$, requires $S_{m1}\leq S\leq S_{m2}$

$$\left.\begin{aligned} S&=\frac{\alpha_1^2}{\alpha_2^2}S_{m1}\operatorname{ch}(\alpha_2 R_2)+\frac{C_1}{\alpha_2}\operatorname{sh}(\alpha_2 R_2)-mS_{m1}\\ Q&=\frac{\alpha_1^2}{\alpha_2}EAS_{m1}\operatorname{sh}(\alpha_2 R_2)+C_1 EA\operatorname{ch}(\alpha_2 R_2)\end{aligned}\right\}\tag{15}$$

From $S_d\operatorname{ch}(\alpha_1 R_1)+\frac{Q_d}{\alpha_1 EA}\operatorname{sh}(\alpha_1 R_1)=S_{m1}$ to determine R_1.

(iii) The three states exist simultaneously in the soil. $R_1+R_2+R_3=L_u$, requires $S>S_{m2}$

$$\left.\begin{aligned} S&=S_{m2}+\frac{\alpha_2^2}{\alpha_3^2}(mS_{m1}+S_{m2})[\operatorname{ch}(\alpha_3 R_3)-1]+\frac{C_2}{\alpha_3}\operatorname{sh}(\alpha_3 R_3)\\ Q&=\frac{\alpha_2^2}{\alpha_3}EA(mS_{m1}+S_{m2})\operatorname{sh}(\alpha_3 R_3)+C_2 EA\operatorname{ch}(\alpha_3 R_3)\end{aligned}\right\}\tag{16}$$

R_1 can be determined from $S_d\operatorname{ch}(\alpha_1 R_1)+\frac{Q_d}{\alpha_1 EA}\operatorname{sh}(\alpha_1 R_1)=S_{m1}$,

and R_2 can be determined from $\frac{\alpha_1^2}{\alpha_2^2}S_{m1}\operatorname{ch}(\alpha_2 R_2)+\frac{C_1}{\alpha_2}\operatorname{sh}(\alpha_2 R_2)=mS_{m1}+S_{m2}$.

(2) $S_{m1}\leq S_d\leq S_{m2}$

It means that the soil surrounding the upper segment of pile becomes plastic as first plastic hardening state at the location of loading cell, which can be categorized into the following two conditions.

(i) All the soil surrounding the upper segment of pile becomes plastic as the first plastic hardening state. $R_2=L_u$, requires $S_{m1}\leq S\leq S_{m2}$

$$\left.\begin{aligned} S&=(S_d+mS_{m1})\operatorname{ch}\beta_2+\frac{Q_d}{\alpha_2 EA}\operatorname{sh}\beta_2-mS_{m1}\\ Q&=\alpha_2 EA(S_d+mS_{m1})\operatorname{sh}\beta_2+Q_d\operatorname{ch}\beta_2\end{aligned}\right\}\tag{17}$$

(ii) Part of the soil surrounding the upper segment of pile becomes plastic as the first plastic hardening state while others as the second plastic hardening state. $R_2+R_3=L_u$, requires $S>S_{m2}$

$$\left.\begin{aligned} S&=S_{m2}+\frac{\alpha_2^2}{\alpha_3^2}(mS_{m1}+S_{m2})[\operatorname{ch}(\alpha_3 R_3)-1]+\frac{C_3}{\alpha_3}\operatorname{sh}(\alpha_3 R_3)\\ Q&=\frac{\alpha_2^2}{\alpha_3}EA(mS_{m1}+S_{m2})\operatorname{sh}(\alpha_3 R_3)+C_3 EA\operatorname{ch}(\alpha_3 R_3)\end{aligned}\right\}\tag{18}$$

R_2 can be determined from $(S_d+mS_{m1})\operatorname{ch}(\alpha_2 R_2)+\frac{Q_d}{\alpha_2 EA}\operatorname{sh}(\alpha_2 R_2)=mS_{m1}+S_{m2}$.

(3) $S_d\geq S_{m2}$

It means that all the soil surrounding the upper segment of pile becomes plastic as the second plastic hardening state and requires $S>S_{m2}$

$$\left.\begin{aligned} S &= S_{m2} - \frac{\alpha_2^2}{\alpha_3^2}(mS_{m1}+S_{m2}) + [\frac{\alpha_2^2}{\alpha_3^2}(mS_{m1}+S_{m2}) - S_{m2} + S_d]\text{ch}\,\beta_3 + \frac{Q_d}{\alpha_3 EA}\text{sh}\,\beta_3 \\ Q &= \alpha_3 EA[\frac{\alpha_2^2}{\alpha_3^2}(mS_{m1}+S_{m2}) - S_{m2} + S_d]\text{sh}\,\beta_3 + Q_d\,\text{ch}\,\beta_3 \end{aligned}\right\} \quad (19)$$

4 ENGINEERING CASE STUDY

Red Mountain Manor Project

Detailed experimental data for this project has been provided in the study by LIU et al. (1999). Cast-in-place bored piles have been used, with experimental measurements from both self-balanced method and heaped loading method. Detailed description of soil is listed in Table 1. The diameter of pile is 0.6 m with concrete grade of C25. Length of self-balanced pile is 25 m with loading cell located 9 m from the bottom of the pile. The self-weight of the upper segment of pile is 100kN with elastic modulus of pile 2.8×10^4 MPa. For the heaped loaded pile, the length of pile is 28m. According to the approach as described above, the parameters can be calculated as λ_1=11837 kPa/m, λ_2=3037 kPa/m, λ_3=1081 kPa/m, S_{m1}=0.98 mm, S_{m2}=5.54 mm. The load-displacement curves for the upper segment of pile are presented in Fig.6. With transforming coefficient of friction force as 0.65, 0.7 and 0.8, the measurements from self-balanced piles have been converted into load-displacement curves for axial loaded piles, which are compared with experimental measurements as shown in Fig.7.

Table 1. List of Parameters for Different Layer of Soil

Layer	Item	Depth (m)	Density (kN/m^3)	Porosity E	Liquidity Limit I_L	Shear Parameters	
						c(kPa)	φ(°)
①-2	Plain Fill	0~3.7	19.7	0.76	0.37	32.3	16.3
②-1	Clay	2.2~15.0	17.2	1.38	0.75	20.4	11.0
②-2	Silty Clay		18.7	0.97	1.09	14.4	21.0
②-3	Silty Clay		19.5	0.79	0.91	16.4	26.2
③	Silty Clay	16.0~27.0	19.8	0.74	0.37	51.0	19.7
④	Pebble soil with coarse sand	20.2~31.0	-	-	-	-	-

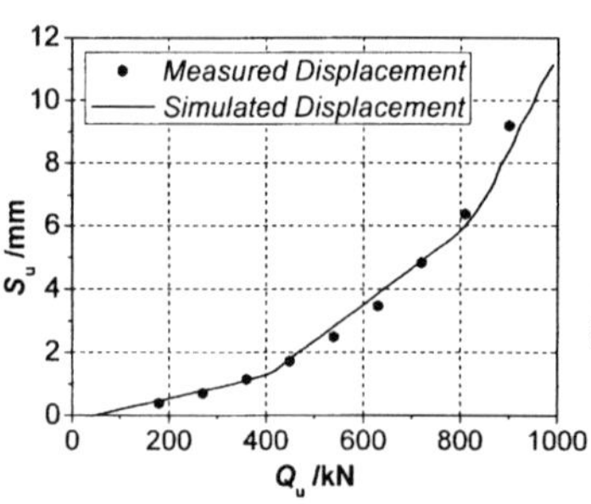

Fig. 6. Q_u-S_u Curves from Self-Balanced Piles

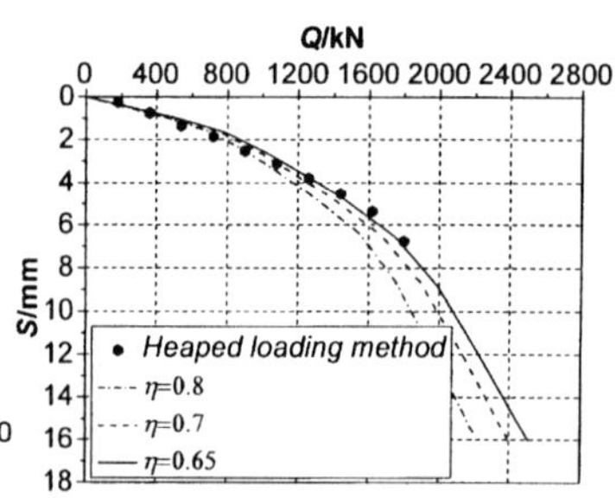

Fig. 7. Comparisons between Transforming Results with Experimental Measurements for Axial Loaded Piles

5 CONCLUSIONS

A load transfer method has been described in this paper with the assumption of the tri-linear load transfer function. The analytical expression of the load-displacement relationship for the upper segment of pile can be derived firstly with consideration of the deadweight of pile. Then, use the measured data of the lower segment of pile as the boundary condition, the results of self-balanced pile tests can be successfully converted into conventional load-settlement curve for static pile tests. The case study validated that the conventional pile load-settlement curve obtained by using this analytical method fit the measured data well. The tri-linear load transfer function is properly used to simulate various kinds of pile-soil interaction such as softening and perfect plasticity.

6 ACKNOWLEDGEMENTS

The work presented here is supported by the National Natural Science Foundation of China with the granted number 50779034. The authors wish to express their gratitude for the financial assistance. The anonymous reviewers' comments are also greatly acknowledged.

REFERENCES

Osterberg, J. (1989). New Device for load testing driven piles and drilled shaft separates friction and end bearing. *Piling and Deep Foundations*. pp. 421-427

Gong, W., et al. (2002). Theory and practice of self-balanced loading test for pile bearing capacity. *Journal of Building Structure*, 23(1), pp. 82-88.

Jiang, Y., et al. (2006). Analytical equation of loading-settlement curve under O-cell pile testing method. *Chinese Journal of Rock Mechanics and Engineering*, 25(S1), pp. 3258-3264.

Seed, H. B., and Reese L. C. (1957). The action of soft clay along friction piles. *Transactions, ASCE*, 122, pp. 731-754.

Chen, Z., and Hong, Y. (1995). Settlement Calculation of Single Pile and Piles Group, from Manual of Pile Foundation Engineering. *Beijing: China Architecture and Building Press*, pp.140-144.

Chen, L., et al. (1994). Analytical calculation of axial loading-settlement curve of piles. *Chinese Journal of Geotechnical Engineering*, 16(6), pp.30-38.

Liu, J., et al. (1999). The application of two-directional loading test on drilled shafts. *China Journal of Industrial Construction*, 29(3), pp.70-72.

Dai, G., et al. (2001). Engineering applications of a new static load testing method for piles with large bearing capacity in bridge. *Journal of Southeast University*, 31(4), pp.54-57.

Drilled Shaft Response in Piedmont Residuum Using Elastic Continuum Analysis and Seismic Piezocone Tests

Paul W. Mayne[1], M. ASCE, Fawad Niazi[2], and David Woeller[3]

[1]Professor, Geosystems Engineering Group, Civil & Environmental Engineering, Georgia Institute of Technology, Atlanta, GA 30332-0355 USA; paul.mayne@gatech.edu
[2]Graduate Research Assistant, Geosystems Engineering Group, Civil & Environmental Engineering, Georgia Institute of Technology, Atlanta, GA 30332-0355 USA: fniazi6@gatech.edu
[3]President, ConeTec Investigations Ltd., 2140 Vulcan Way, Richmond, British Columbia V6V 1J8 Canada; insitu@conetec.com

ABSTRACT: The axial load-displacement-capacity response of drilled shaft and bored pile foundations can be evaluated during the analysis and design stage using an elastic continuum framework and results from in-situ testing methods, in particular, the seismic piezocone test (SCPTù). The SCPTù is an optimal and economical means for collection of geotechnical data because the same sounding provides up to five separate measurements with depth: cone tip resistance (q_t), sleeve friction (f_s), porewater pressure (u_2), time rate of consolidation (t_{50}), and shear wave velocity (V_s). Moreover, the SCPTù provides information on soil behavior at both ends of the stress-strain-strength curves, namely the peak strength and geostatic stress state for evaluating axial pile capacity and the small-strain stiffness ($G_{max} = \rho_t \cdot V_s^2$) for the initial soil-pile deformations. Using a Randolph-type analytical elastic pile model, the approach can handle either traditional top down compression loading by dead-weight or reaction beam systems, or the more recent Osterberg cell that juxtaposes base and side resistances in opposite directions. A case study involving O-cell tests on a drilled shaft in Piedmont residuum and partially-weathered rock in Atlanta are presented.

INTRODUCTION

The axial load-displacement response and load-transfer distribution of piles can be evaluated rationally within an elastic continuum solution (Poulos & Davis, 1980). Of particular convenience, the closed-form analytical solution of Randolph & Wroth (1978, 1979) can handle rigid piles, compressible piles, and very long piles, as well as presence of a lower stiffer soil or rock layer, homogeneous or Gibson soil modulus variations with depth, and straight or belled shafts (Fleming et al. 1992).

The Osterberg load cell (O-cell) is an innovative and convenient means for mobilizing both the axial side and base resistance components of drilled shaft foundations, also termed bored piles (Fellenius 2001). The O-cell is placed in the drilled shaft foundation during the installation of the rebar cage and then concreted in-place. During load testing, the hydraulic jack is inflated using a high-pressure pump, thereby lifting the one upper shaft section while simultaneously pushing the lower segment section downward (Osterberg 2000). The results are evaluated to obtain the mobilized side and base resistances, as well as derivation of an equivalent top-down curve for the axial load-displacement-capacity response of the bored pile.

In the original setup, the O-cell was positioned at the base of the bored pile, yet can be installed at any convenient elevation within the shaft. Multiple levels of O-cells can be installed to stage-load the bored pile in sections, thereby achieving huge capacities during load testing. The analysis of individual segments from O-cell tests, as well as top-down loading, is conveniently handled using the Randolph-Wroth analytical model, as depicted in Figure 1. For clarity, the solution here is presented for rigid pile segments, yet can be easily upgraded to account for pile compressibility effects and stiffer geomaterials under the base, if desired.

O-Cell Elastic Solution

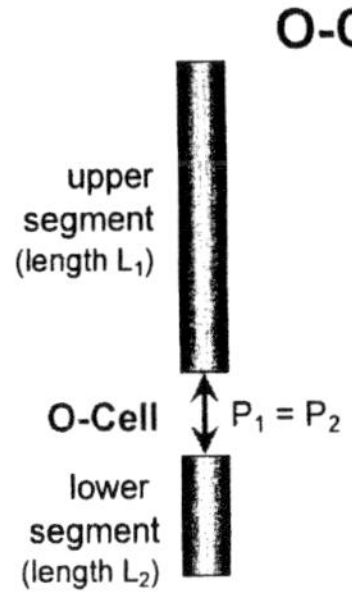

Rigid pile shaft under upward loading

$$\frac{P_1}{G_{s1} r_{o1} w_1} = \frac{2\pi}{\zeta_1} \cdot \frac{L_1}{r_{01}}$$

Rigid pile under compression loading

$$\frac{P_2}{G_{s2} r_{o2} w_2} = \frac{4}{(1-\upsilon)\xi} + \frac{2\pi}{\zeta_2} \cdot \frac{L_2}{r_{o2}}$$

P = applied force
L = pile length
r_o = pile shaft radius
E_p = pile modulus
G_s = soil shear modulus along sides
ν = Poisson's ratio of soil

w = pile displacement
$\lambda = E_p/G_{sL}$ = soil-pile stiffness ratio
$\xi = G_{s2}/G_{sb}$ (Note: floating pile: $\xi = 1$)
G_{sb} = soil modulus below pile base/toe
$\zeta = \ln(r_m/r_o)$ = soil zone of influence
$r_m = L\{0.25 + \xi\,[2.5\,(1-\nu) - 0.25]\}$

Figure 1. Osterberg cell evaluation using elastic continuum solution.

DRILLED SHAFT CASE STUDY IN PIEDMONT RESIDUUM

A case study involving O-cell load testing of a drilled shaft for a Georgia DOT viaduct next to CNN in downtown Atlanta is used to illustrate the approach. The soil borings indicated residual silts and sands grading to saprolite and partially-weathered rock of the Piedmont geology, as shown in Figure 2. Very high blow-counts were recorded below depths of 11 m. The site was also explored by a GT crew using seismic piezocone penetration tests (SCPTù), as presented in Figure 3. Equivalent *N*-values from the CPT tip resistance (q_t) at the site were obtained from the CPT - SPT

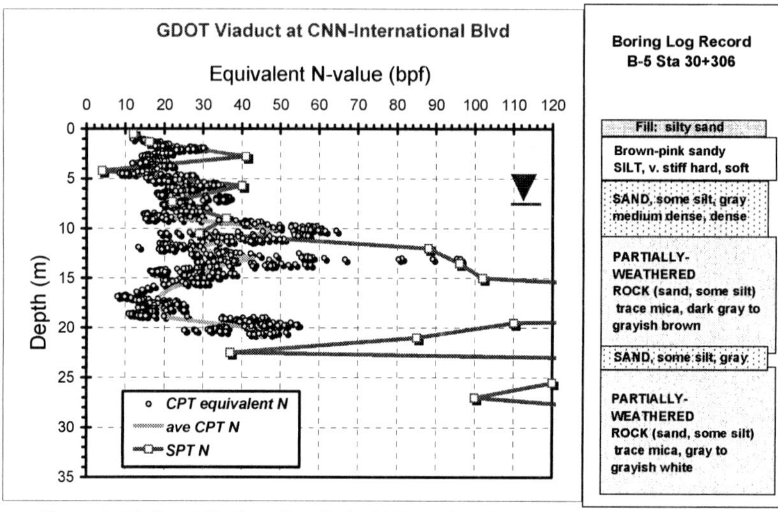

Figure 2. Soil profile from borehole SPTs and equivalent N-values from CPT.

conversion ratio q_t (bars)/N_{60} = 3, using an available correlation (Kulhawy & Mayne 1990) based on characteristic grain sizes of Piedmont soils: 0.06 mm < D_{50} < 0.15 mm. Reasonable agreement is evident between these two penetration tests for the upper 11 m in Figure 2, below which, they are quite different.

Based on the soil boring information, an instrumented drilled shaft foundation was installed with two-stages of O-cells for the load test arrangement. The shaft was constructed using the dry cased method with a final length L = 20.9 and stepped diameters: d = 1.68 m from 0 to 2.9 m; d = 1.59 m from 2.9 to 14.7 m; and d = 1.44 m from 14.7 to 20.9 m. The upper 1.4 m was cased and without concrete. A lower O-cell was situated at the base elevation, with an upper O-cell located 5.2 m above the first. As the contractor was from out of state with no prior experience in the Piedmont, difficulties were encountered in the drilling operations, resulting in a softened bottom because of a long wait before concreting operations.

SOIL PARAMETER EVALUATION

The seismic piezocone test provides up to five separate readings with depth: cone tip resistance (q_t), sleeve friction (f_s), porewater pressure (u_2), time rate of Δu decay with time (t_{50}), and downhole shear wave velocity (V_s). The evaluation of geotechnical parameters for the Piedmont geology is discussed in the following section, as based on prior calibration efforts (Mayne et al. 2000; Mayne & Brown, 2003).

The CPT-interpreted frictional angle (ϕ' and c' = 0) of Piedmont geomaterials has shown agreement with a statistical method developed by Kulhawy and Mayne (1990)

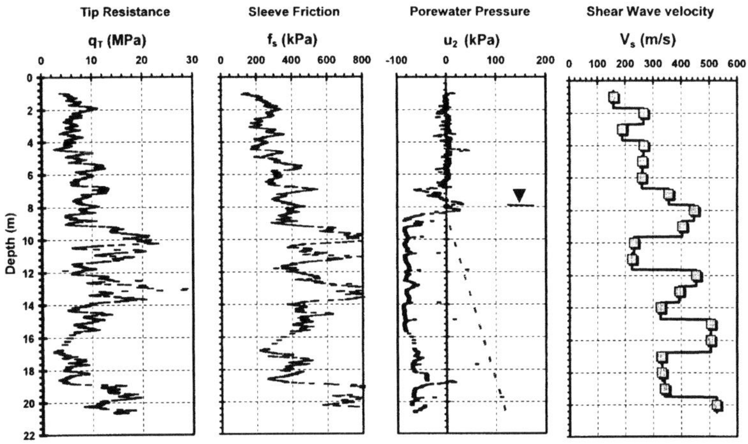

Figure 3. Seismic piezocone sounding at the GDOT International Blvd viaduct.

for sandy soils:

$$\phi' = 17.6^\circ + 11.0^\circ \cdot \log(q_{t1}) \tag{1}$$

where $q_{t1} = (q_t/\sigma_{atm})/(\sigma_{vo}'/\sigma_{atm})^{0.5}$ and σ_{atm} is a reference stress = 1 bar = 100 kPa.

The small-strain shear modulus (termed G_o, G_{max}, or G_{dyn}) is a fundamental initial stiffness relating that is obtained from the total soil mass density ρ_t and shear wave velocity (V_S):

$$G_{max} = \rho_t \cdot V_s^2 \tag{2}$$

Soils display a highly nonlinear trend in stress-strain behavior, and as such, G_{max} must be reduced to G at corresponding strain levels for the applied loads. Fahey (1998) proposed the following algorithm for the reduction factor:

$$G/G_{max} = 1 - (1/FS)^g \tag{3}$$

where $1/FS$ (reciprocal of factor of safety = P/P_{ult}) is the mobilized load and the exponent $g = 0.3 \pm 0.1$ for well behaved soils (Mayne 2007).

The drained value of Poisson's ratio (ν') ranges from 0.1 to 0.2 for all types of geomaterials at working loads up to strains of 0.1%, increasing to larger values as failure states are approached. The value for undrained loading is $\nu_u = 0.5$.

A first-order evaluation of the preconsolidation stress σ_p' (formerly designated σ_{vmax}' or P_c') for a wide variety of geomaterials can be obtained from the net cone tip resistance (q_t) as detailed by Mayne et al. (2009):

$$\sigma_p' = 0.33\,(q_t - \sigma_{vo})^m\,(\sigma_{atm}/100)^{1-m} \tag{4}$$

where corresponding values of the exponent m for intact clays, silts, silty sands, and clean sands are 1.00, 0.85, 0.80, and 0.72, respectively. For the Piedmont silts-sands, a value of m of 0.85 was used. The corresponding OCR = σ_p'/σ_{vo}'.

Additional details on soil parameter evaluation (e.g., γ_t, K_o), unit pile resistances (f_p, q_b), and capacity calculations for deep foundations are given elsewhere (Mayne 2007). In particular, the Piedmont is considered as an intermediate geomaterial and can be handled using a beta-type analysis (O'Neill & Reese, 1999).

O-CELL LOAD TEST RESULTS

The results from the capacity analysis of the various pile foundation segments by the CPT resistances and the deformation analysis from the G_{max} profile together allow an evaluation of the axial-load-displacement curves, as shown in Figure 4. These compare well with the measured O-cell performance from the staged loading. In stage 1, the lower cell was inflated to measure end-bearing only (upper cell was open). Stage 2 involved pushing the lower segment downward for shaft friction only (lower cell open). Finally, with lower cell closed, stage 3 pushed the upper pile segment upward simultaneous as the lower pile segment was forced downward. The softened toe or base is evident in the stage 1 loading of Figure 4. The pile shafts were also outfitted with vibrating wire strain gauges on the reinforcing cages, however, due to page restrictions, those data are not presented herein.

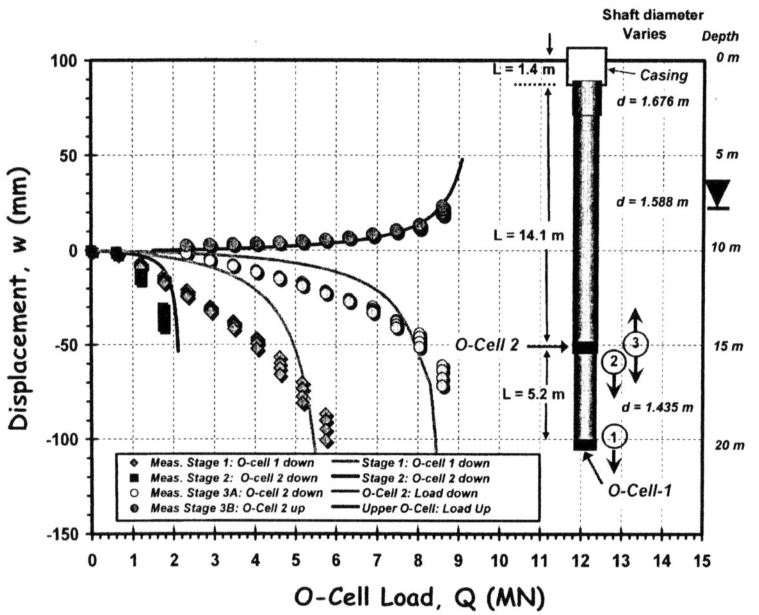

Figure 4. Measured and predicted O-cell response for GDOT drilled shaft Atlanta.

CONCLUSIONS

Elastic continuum provides a rational framework for the axial evaluation of drilled shaft foundations, loaded either by conventional top-down systems or the recent O-cell arrangement. Seismic piezocone test provides an optimal method to collect the many types of necessary geotechnical data for the analysis. A case study in Piedmont residuum is presented to show the general use.

ACKNOWLEDGMENTS

The authors thank Tom Scruggs (GDOT), Dr. Alec V. McGillivray (Berkel) and Guillermo Zavala (Catholic Univ. - Lima) for their efforts on this project.

REFERENCES

Fahey, M. (1998). Deformation and in-situ stress measurement. *Geotechnical Site Characterization*, Vol. 1 (Proc. ISC-1, Atlanta), Balkema, Rotterdam: 49-68.

Fellenius, B H. (2001). The O-cell: an innovative engineering tool. *Geotechnical News Magazine*, Vol. 19 (2): 32-33.

Fleming, W G K, Weltman A J, Randolph M F and Elson W K. (1992). *Piling Engineering*, 2nd Edition, Blackie/Halsted Press, London: 122-128.

Kulhawy, F.H. & Mayne, P.W. (1990). *Estimating Soil Properties for Foundation Design*, Rept. EL-6800, Electric Power Res. Inst., Palo Alto: 306 p.

Mayne, P.W. (2007). Cone penetration testing – a synthesis of highway practice. *NCHRP Synthesis 368*, Transportation Research Board, Washington, DC: 118 p.

Mayne, P.W., Brown, D., Vinson, J., Schneider, J.A., and Finke, K.A. (2000). Site characterization of Piedmont residual soils at the NGES, Opelika, Alabama. *National Geotech. Experimentation Sites*, (GSP 93), ASCE, Reston/VA: 160-185.

Mayne, P.W. and Brown, D.A. (2003). Site characterization of Piedmont residuum of North America. *Characterization and Engineering Properties of Natural Soils*, Vol. 2, Swets & Zeitlinger, Lisse: 1323-1339.

Mayne, P.W., Coop, M.R., Springman, S., Huang, A-B., and Zornberg, J. (2009). State-of-the-Art Paper (SOA-1): Geomaterial behavior and testing. *Proc. 17th ICSMGE,* Vol. 4 (Alexandria), Millpress/IOS Press, Amsterdam: 2777-2872.

O'Neill, M.W. and Reese L.C. (1999). *Drilled Shafts: Construction Procedures and Design Methods*, FHWA-IF-99-025, ADSC-IAFD, Dallas, TX: 758 pages.

Osterberg, J.O. (2000). Side shear and end bearing in drilled shafts. *New Technological and Design Developments in Deep Foundations*, GSP No. 100 (Proc. GeoDenver), ASCE, Reston, Virginia: 72-79.

Poulos, H.G and Davis E.H. (1980). *Pile Foundation Analysis & Design*. Wiley & Sons, New York, 1980: 379 pages.

Randolph M.F. and Wroth C.P. (1978). Analysis of deformation of vertically loaded piles. *Jour. Geotechnical Engineering Division* (ASCE) 104 (GT12): 1465-1488.

Randolph M.F. and Wroth C.P. (1979). A simple approach to pile design and the evaluation of pile tests. *Behaviour of Deep Foundations*, STP 670, ASTM, West Conshohocken, Pennsylvania: 484-499.

Axial Capacity of Drilled Shafts in Clay Over Gravel Formation

San-Shyan Lin[1], Jen-Cheng Liao[2], and Kun-Jui Wang[3]

[1]Professor, Dept. of Harbor and River Eng., National Taiwan Ocean University, Keelung, Taiwan 20224 (email: sslin@mail.ntou.edu.tw)

[2]Associate Research Fellow, Taiwan Construction Research Institute, Taipei County, Taiwan.

[3]Engineer, Trinity Foundation Engineering Consultants, Taipei, Taiwan.

ABSTRACT

This paper presents six case histories of load tests on reverse circulation method installed drilled shafts, with or without tip grouting, under axial load. Shaft displacement readings based on both telltale/extensometer and rebar strain gage are presented first. Subsequently, effect of loading cycle on shaft elastic modulus estimation is evaluated. The load versus displacement relationships at the shaft head, the t-z curves of the clay layer and of the gravel layer and the tip resistance versus displacement relationships are other important concerns and are compared in this paper. In addition, revised adhesion factor α and the parameter β based on the pile load test results are proposed in the paper.

INTRODUCTION

In some area of Taipei City, over tens of meters thick of gravelly soil layer sandwiched between top silty clay layer and bottom bedrock can often be found. Increasing use of rock socket drilled shafts supporting buildings in Taipei and how to estimate side friction resistance of shaft through the gravelly soil layer have been major concerns for local geotechnical engineers. In addition, effect of tip grouting on skin friction resistance through gravelly soil layer is another concern.

Bearing capacity of drilled shafts in gravel formation has drawn geotechnical engineers' attention in the past decade, such as the available researches by O'Neill and Reese (1999) and by Rollins et al. (2005). These researchers conducted a series of uplift tests on drilled shafts in soil profiles ranging from uniform sand to sandy gravel to evaluate side friction of drilled shafts in gravel soils. These researches were performed on drilled shafts embedded in gravel soils without overburden.

To compare the axial performance of drilled shafts through gravel layer, six instrumented drilled shafts with and without tip grouting installed at eastern Taipei area were selected and evaluated based on load test results. The summary of the test shaft geometries and measured capacities is provided in Table 1. All the drilled shafts were installed by the same contractor. The modified U-shape grouting device (Lin et al. 2000) was pre-installed on three pre-selected shafts for tip grouting purpose. Telltale/extensometer and rebar gage were installed in all test shafts to obtain load transfer in side and tip resistances. The telltales were extended to several selected depths along the length of the shafts to measure the movements of the telltale rods relative to the head of the drilled shafts. The forces in the shafts was obtained at selected depths by multiplying the axial stiffness of the test shaft (cross-sectional area at the selected depth times composite modulus of the shaft) by the strain estimated at the depth of interest (O'Neill and Reese 1999). Similarily, the forces in the shafts were estimated by multiplying the axial stiffness of the shafts by the strain values measured using the sister bar strain gages.

In this paper, the forces in the shafts estimated using telltales were compared with those estimated using rebar strian gages. Subsequently, effect of loading cycle on elastic modulus estimation is studied. The load versus settlement curves at the shaft head and the t-z curves along shaft are also main concerns and are presented in the paper. Finally, the adhesion factor α and the parameter β based on the pile load test results are also proposed for drilled shafts installed in similar soil condition and construction methods in the region.

SITE LOCATION AND GROUND CONDITIONS

The six shafts were installed at five locations within the test site located in Eastern Taipei, as shown in Fig. 1. In addition to top 3m of backfill material, typical soil conditions at the site consists of ~50m of low plasticity clay layer and underlying ~15m of gravelly soil layer, which contains about 42 to 70 percent gravel, 28 to 42 percent sandy soil and other fine soils. Beneath the gravel layer, sandstone bedrock was encountered. Typical subsurface soil physical and strength properties are presented in Table 2. Fig. 2 shows the undrained shear strength of the top silt clay layer along depth.

Table 1 Test Shaft Geometry and Measured Capacity.

	Diameter (m)	Length (m)	Tested loading (MN)	Resistance (MN)		Settlement (mm)		Loading Cycle	Tip grouting (m^3)	resistance/ tested load (%)	
				Side	Tip	Head	Tip			Side	Tip
Case 1	1.2	76.0	36.3	35.9	0.4	70.8	6.8	2	-	99	1
Case 2	1.5	65.0	37.7	35.8	1.9	33.8	8.7	1	2.7	95	5
Case 3	1.5	56.0	26.5	22.6	3.9	29.0	9.3	3	6.6	85	15
Case 4	1.2	59.0	24.5	23.7	0.8	17.4	0.3	5	3.2	96	3
Case 5	2.5	70.3	55.9	54.8	1.1	17.4	7.0	1	-	98	2
Case 6	2.0	70.7	39.2	39.2	0.0	17.4	0.0	1	-	100	0

Table 2 Soil Properties of the Test Site.

No.	Depth (m)	Soil type	SPT-N	W_n (%)	e	r_t (kN/m^3)	W_L (%)	I_p (%)	Es (MPa)
1	0~3	Fill (SF)	3~ >50	23~ 36.9	0.89~ 0.97	17.7~ 18.2	-	-	-
2	1.7~3 2.6	Silty clay (CL)	2~7	33~ 41	0.99~ 1.34	16.1~ 18.3	35~ 44	14~ 21	16~ 56
3,4	29.1~ 53	Silty clay or sandy silt (CL-ML, SM)	13~ 37	24~ 30	0.82~ 0.9	17.7~ 18.8	33~ 36	13~ 15	56~ 80
	30.7~ 53	Silty clay with some gravel layer(SM/ML/GW)	10~ >50	20.2~ 33.6	0.66~ 1.2	17.9~ 19.6	-	-	40~ 100
5	49.4~ 66.6	Sandy gravel (GM/GW)	24~ >50	15~ 32	0.7~ 0.9	18.6~ 21.6	-	-	200
6	54.6~	Bed rock(SS/SH/MS)	>50	11~ 13		19.6~ 22.9	-	-	200

Note:
γ_t :Total Density ; W_n : Natural Water Content ; W_L: Liquid Limit ; I_P: Plasticity Index ; e: Void Ratio ; Es: Elastic modulus; ( ) :average; SPT-N: blows/30 cm.

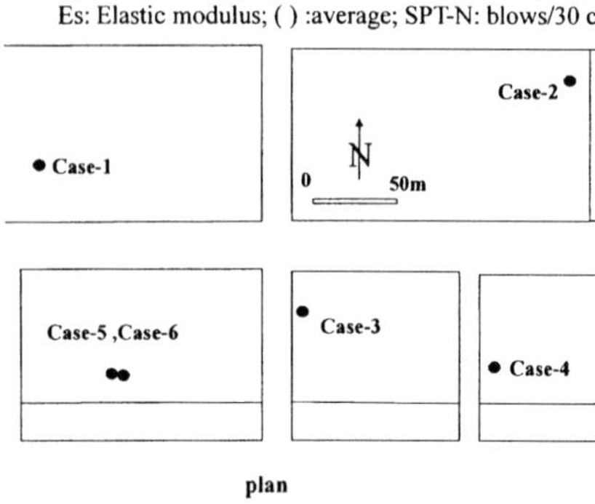

Figure 1 Site Plan and Locations of Test Shafts.

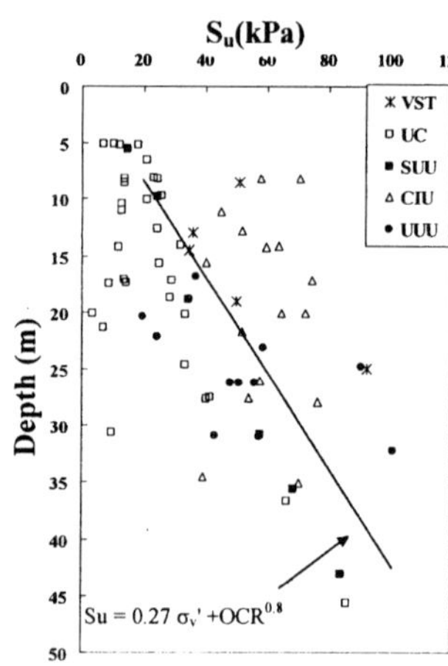

Figure 2 Undrained Shear Strength of the Top Clay Layer Along Depth.

CONSTRUCTION

The reverse circulation method using polymer slurry to stabilize the drilled hole was used for the drilled shaft installation. The polymer slurry was pumped into the shaft borehole while the slurry-soil mixture was discharged from the centrally placed air-lift riser pipe to allow desanding of the slurry. All drilled shafts were concreted using tremie method. Four 5.08cm PVC pipes were attached to the rebar cage of each shaft for crosshole sonic logging integrity testing. To evaluate the total load carried at different depths along the shaft, both rebar strain gages and telltale/extensometer were installed at several selected depths of each shaft (Fig. 3). Four rebar strain gages were attached to the rebar cage at each level and were protected by covers.

TIP GROUTING

A modified U-shape cleaning/grouting device was pre-installed at the tip of selected test shafts. A high-pressure water jet was used to clean undesirable material from the shaft base prior to tip grouting. After the shaft concrete reached the required strength, high-pressure water was used to push down the sediment or weak material below the tip of the shaft to form a void. Using the drilling rig, a drilling rod was lowered through the pre-installed sonic logging tube to the bottom of the shaft to loose the soils so that they can be flushed up through the grouting tubes. The water was pressurized at between 14.7 and 19.6 MPa. In the mean time, while maintaining a rotation speed of 6 to 8 RPM, the rod was lowered at a speed of 100mm/min within about 30cm or deeper of the shaft base. The water was circulated at least twice by jetting in each shaft to be tip grouted.

Subsequently, low-pressure washing at 5 MPa was used to flush the loose debris to the ground surface with returning water. Washing was stopped once the clean water was returned over a given period of time.

After the base sediment was cleaned, the grout, at a pressure of 2 to 5 MPa, was injected through one of the sonic logging pipes. Only two pipes were used for grouting each time, the other two holes were packed off. The grout used was composed of water and cement at a water/cement ratio of 1:1. Once the grout returned from the other grouting pipe, injection of grout continued under the same pressure for about five minutes or until a pressure of reaching 4,900 kPa was maintained to finish the grouting process. The 4,900 kPa pressure was set based on the experience of local engineers. Typically, the excess water from the washing process at the shaft bottom will return prior to the grout return after beginning of tip grouting. Detail information of the tip grouted shafts is shown in Table 3.

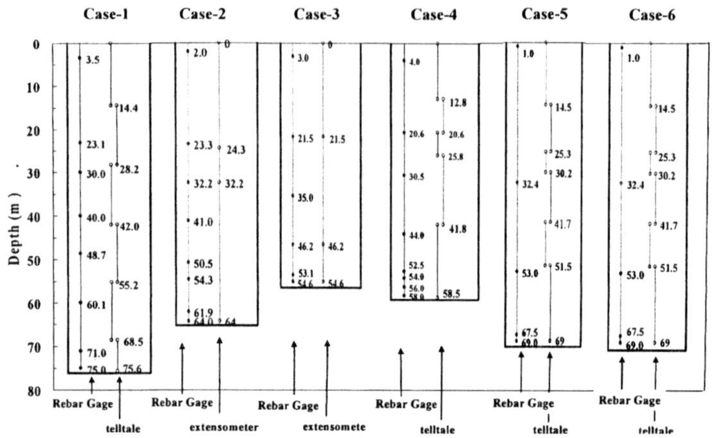

Figure 3 Selected Depths of the Telltale/extensometer and Rebar Gage.

Table 3 Quantity of Grout used for Tip Grout of Cases 2 to 4.

Cases	Shafts Diameter (m)	Shafts Length (m)	Tip grouting range (m) Start	Tip grouting range (m) End	Washing or grouting pressure (MPa) 1st washed	Washing or grouting pressure (MPa) 2nd washed	Washing or grouting pressure (MPa) grouting	Grouting volume (m^3)	Grout volume/ washed volume
2	1.5	65.0	65.0	65.3	14.7	5.0	2~5	2.66	1.7
3	1.5	56.0	56.0	59.0	19.6	5.0	2~5	5.43	0.9
4	1.2	59.0	59.6	60.6	19.6	5.0	2~5	3.20	1.6

SHAFT LOAD TESTS AND TEST RESULTS

Settlement readings were taken immediately after each load increment and at one minute and then at every two minutes before increasing the load. When the maximum load was reached, the loading was maintained until the settlement rate was less than 0.25 mm/hr before unloading. Then, the maximum load was unloaded at four decrements.

In order to estimate frictional resistance along the shaft, the strains at different depths of the shaft and the modulus of the concrete and the steel are required. The axial load is the strain multiplied by cross sectional area of the pile and the elastic modulus as shown in the following equation

$$P = \varepsilon \cdot (A_c \cdot E_c + A_s \cdot E_s) \quad (1)$$

where $P =$ axial force of the considered cross section; $A_c =$ cross sectional area of concrete; $A_s =$ cross sectional area of steel; and $E_s =$ elastic modulus of steel.

Variation of strain along shaft can be obtained from either telltale/extensometer or rebar strain gage readings. Depths considered for telltale/extensometer and rebar strain gage readings of each shaft are shown in Fig. 3. Comparing to the rebar strain gage, telltale or extensometer is considered to be easier to install and also cheaper. The difference of the readings from telltale/extensometer and rebar strain gage is given in Fig. 4. Apparently, the readings obtained from either telltale or extensometer is higher than that from the rebar strain gages. None of the available researches discussed which reading data is more accurate. Local engineers believe that the data obtained from rebar strain gage is more reliable and is hence used in the following shaft friction resistance estimation. However, the reason why rebar strain gage reading is more reliable than that of telltale/extensometer still needs further study.

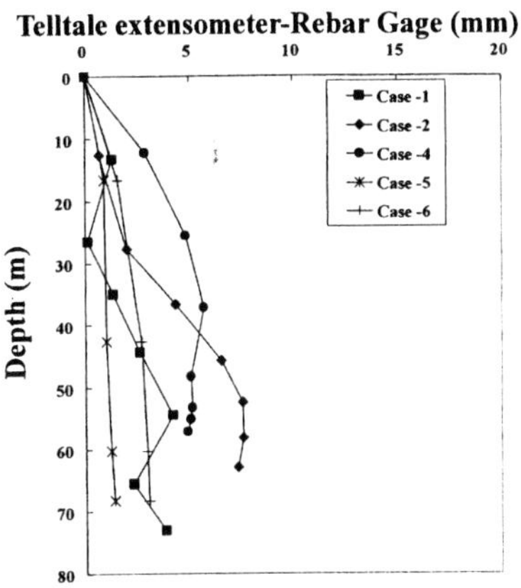

Figure 4 Difference of Readings between Telltale/extensometer and Rebar Strain Gage.

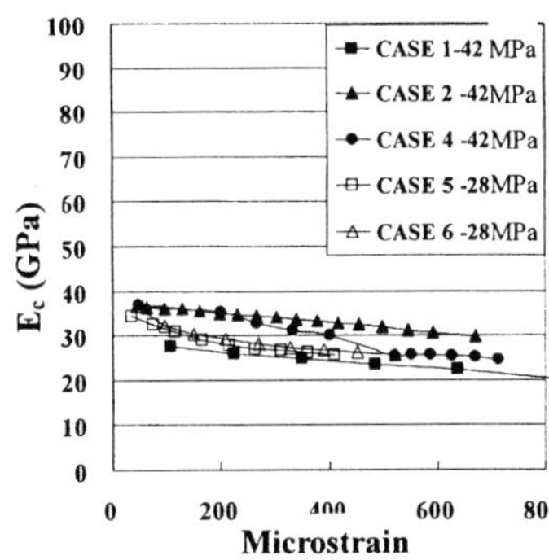

Figure 5 Concrete Modulus versus Measured Strain.

Strain reading of the rebar strain gage located nearest to the pile head is often used for concrete modulus estimation; the modulus calculated for each increment affected by shaft resistance is negligible and hence the calculated tangent modulus is close to the actual one, as suggested by Fellenius (2001). In addition, similar to the observation by Fellenius (2001), the modulus values are also strain dependent as shown in Fig. 5 for the studied cases. Furthermore, it is observed in this study that the modulus values are also test loading cycle dependent as shown in Fig. 6. Increasing the number of loading cycle, the modulus value decreases with increasing strain. Selected modulus value versus strain curve for Case 4 is also presented in Fig. 6. The method for concrete secant modulus determined from tangent modulus suggested by Fellenius (2001) is adopted in this paper.

The normalized t-z curves of the silt clay layer, gravel layer and the bedrock layer are given in Figs. 7 to 10, respectively. Ultimate resistance has been reached at silty clay layer but not for gravel soil and bedrock. Strain softening was also observed for tip grouted cases of 2 to 4 at lower silty clay layer. For gravelly soil layer, the side friction resistance corresponding to the measured displacement of 15mm is considered as the ultimate value as suggested by Woo (2001).

For the cases without tip grouting, the ultimate skin frictional resistance of silty clay layer versus SPT-N is shown in Fig. 11, in which the f_s = N, 3.3N, and 5N are the suggested correlations in Canadian and Japanese highways as well as Japanese buildings design codes, respectively. Higher resistance was observed than that of the

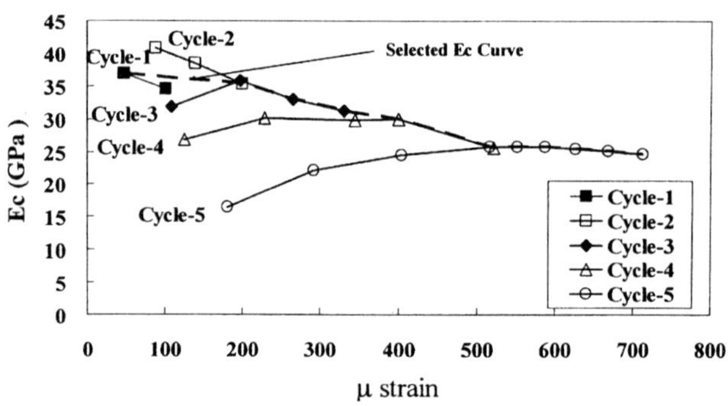

Figure 6 Effect of Loading Cycles on Concrete Modulus.

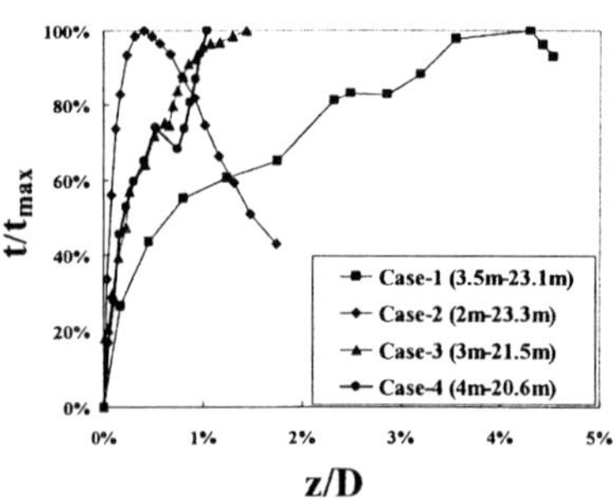

Figure 7 Normalized t-z Curves of the Upper Silt Clay Layer.

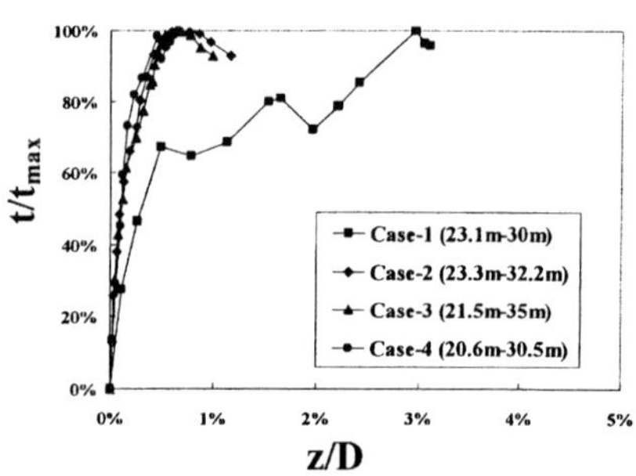

Figure 8 Normalized t-z curves of the lower silt clay layer.

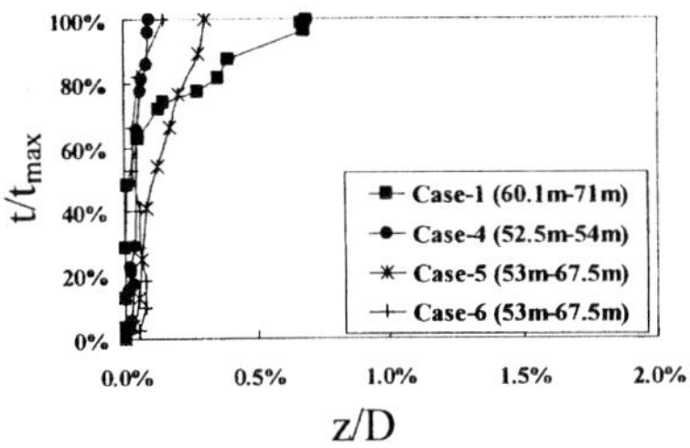

Figure 9 Normalized t-z Curves of the Gravelly Soil Layer.

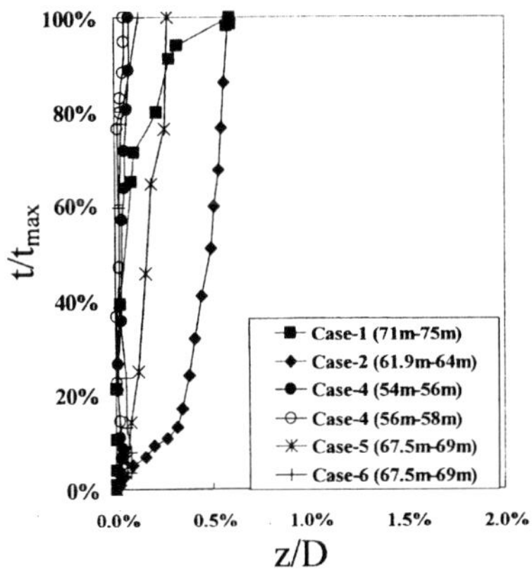

Figure 10 Normalized t-z Curves of the Bedrock.

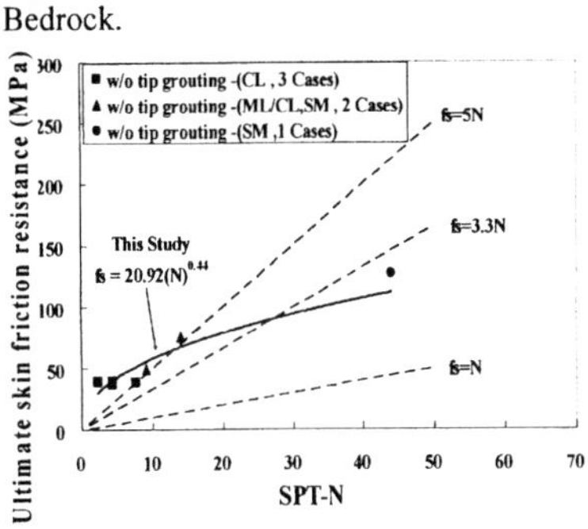

Figure 11 Relation between Ultimate Skin Frictional Resistance versus SPT-N Values.

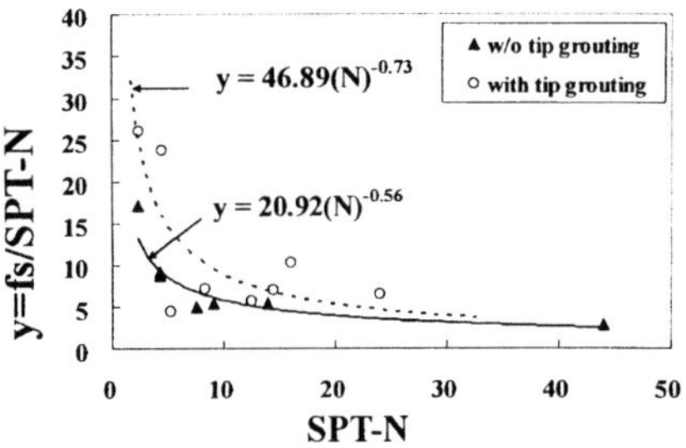

Figure 12 Relation of $f_s / SPT-N$ versus SPT-N Values.

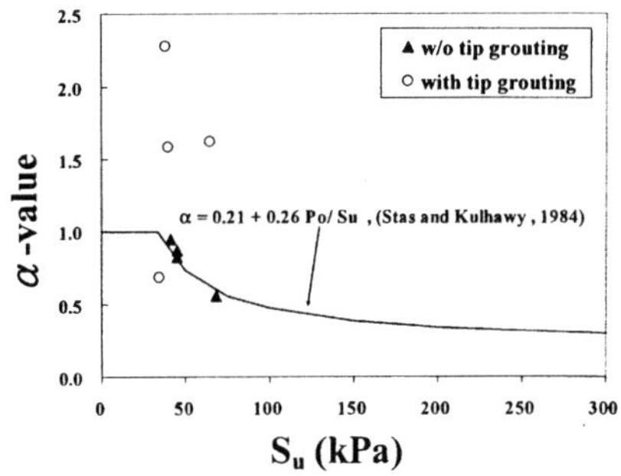

Figure 13 Back-calculated α-value versus Undrained Shear Strength.

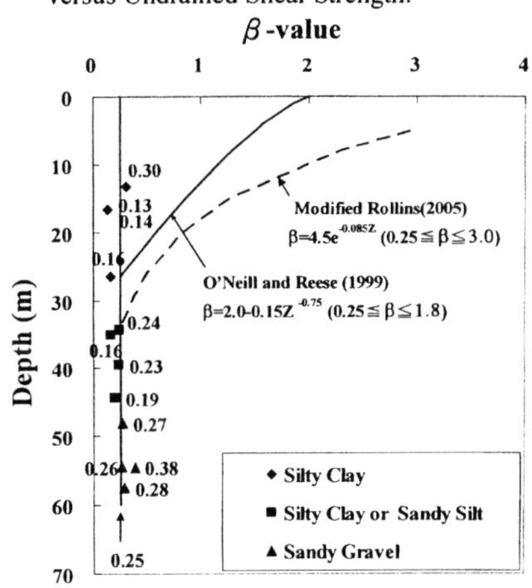

Figure 14 Variation of β-value with Depth.

suggested design code values under lower SPT-N value. The ratios of the ultimate skin frictional resistance to the SPT-N value for the cases with and without tip grouting are shown in Fig. 12. The $f / SPT-N$ ratio appears to be decreasing with increasing of SPT-N values. Apparently, the cases with tip grouting also increase their skin friction resistance. For the SPT-N value less than 50, the best fitted results for the cases with and without tip grouting can be expressed using Eqs. (2) and (3), respectively.

$$f_s / SPT-N = 46.89(N)^{-0.73} \tag{2}$$

and

$$f_s / SPT-N = 20.92(N)^{-0.56}, \tag{3}$$

For the top silt clay layer, the α values obtained from the pile load test results are shown in Fig. 13. For the cases without tip grouting, the α value well fits the α versus undrained strength relationship of the clay as recommended by Stas and Kulhawy (1984). However, three of the four α data for the shafts with tip grouting treatment are significantly higher than the recommendation by Stas and Kulhawy (1984).

The β values along depth from the pile load test results are shown in Fig. 14, in which the recommended values by O'Neill and Reese (1999) and by Rollins et al. (2005) are also given for comparison. Since only uplift load tests were conducted by Rollins et al. (2005), the Rollins et al. (2005) curve was modified based on the suggestion by Rollins et al. (2005) that the shaft resistance in tension could be 12-25 percent smaller than that in compression. 25m and 35m are the limit depth of gravel layer considered by O'Neill and Reese (1999) and by Rollins et al. (2005), respectively. Based on the above studied cases, suggested design parameters are summarized in Table 4.

Table 4 Summary of Design Parameters Suggested Based on the Studied Cases.

Soil type	Ultimate skin resistance, fs (kPa)	α method	β method	SPT-N method	Z (mm)	z/D (%)
Silty clay(CL)	37~39	0.7~0.98	0.13~0.3	5N~17N	3~22	0.4~1.4
Silty clay or sandy silt	49~80	-	0.16~0.24	3.4N~5.3N	-	-
Silty sand with some gravel	114~127	-	0.26~0.27	2.9N~3N	-	-
Sandy gravel(GM/GW)	149~188	-	0.28~0.39	1.6N~2.4N	>30	1.7<z/D

CONCLUSIONS

Based on the results of the six case histories discussed in this paper, the following conclusions are drawn:

1. In addition to the effects of the measured strain and measured depth, the elastic modulus of the shaft concrete is also affected by the test loading cycles.
2. Side frictional resistance for the clay and the gravel layer are 37 to 39 kPa and 149 to 188 kPa, respectively.
3. Comparison on the shaft test results for the shafts with and without tip grouting show tip grouting improves mainly the shaft frictional resistance.
4. Back-calculated α value of the clay layer is between 0.7 and 0.95.
5. Back-calculated β value of the clay layer and gravel layer is 0.13 to 0.3 and 0.28 to 0.38, respectively.
6. By using SPT-N value for side friction resistance estimation, the ultimate skin resistance is bounded between 3N (kPa) and 5N (kPa)

REFERENCES

Fellenius, B.H. (2001). "From strain measurements to load in an instrumented pile" Geotechnical News Magazine, 19(1), 35-38.

Lin, S.S., Lin, T., and Chang, L.T. (2000). " A case study for drilled shafts base mud treatment" in New Technological and Design Developments in Deep Foundations, GSP No. 100, ASCE, pp. 46-58.

O'Neill, M.W. and Reese, L.C. (1999). Drilled Shafts: Construction Procedures and Design Methods Publication No. FHWA-IF-99-025, Department of Transportation, Washington D.C., USA.

Rollins, K.M., Clayton, R.T., Mikesell, R.C., and Blaise, B.C. (2005). "Drilled shaft side friction in gravelly soils" Journal of Geotechnical and Geoenvironmental Engineering, ASCE, 131(8), 987-1003.

Stas, C.V., and Kulhawy, F.H. (1984). "Critical evaluation of design methods for foundations under axial uplift and compression loading" Report El-3771, Electric Power Research Institute, Palo Alto, California, USA.

Woo, S.M. (2001). "Design and Construction Problem of Large Diameter long piles based on pile load test results" in Proceedings of Pile Talk Symposium, Taipei, Taiwan, B1-B37.

Evaluation of Axial Capacity of Post Grouted Drilled Shafts

Emad Farouz[1], M. ASCE, Mike Muchard[2], M. ASCE, and Ke Yang[1], M. ASCE

[1]CH2M HILL, 15010 Conference Center Drive, Suite 200, Chantilly, VA 20151; efarouz@ch2m.com
[2]Applied Foundation Testing, Inc., Tampa, Florida, USA; mmuchard@testpile.com

ABSTRACT: A design phase load test program was performed at the Broadway Viaduct bridge site to evaluate load carrying capacity and constructability of post grouted drilled shafts. Three strain gauge instrumented test drilled shafts were installed in medium dense sand. Two drilled shafts included sleeve port and plate base grouting apparatus and were post grouted. The third drilled shaft was un-grouted and served as a control drilled shaft for comparison of the un-grouted end bearing. All drilled shafts were 1.52 m (5 feet) in diameter. The control drilled shaft was 16.76 m (55 feet) long. Post grouted drilled shafts were 16.76 m (55 feet) and 20.12 m (66 feet) long. The three drilled shafts were tested including integrity testing using the crosshole sonic logging method on each shaft, post grouting, and statnamic load testing. The test results were used to evaluate the axial capacity increase and an existing design method for post grouted drilled shafts.

CE Database subject headings: Drilled Shafts; Grouting; and Load Test.

INTRODUCTION

Post grouted drilled shafts can mobilize higher end bearing capacity under the same permissible settlement compared to conventional drilled shafts. Additionally, the construction quality and capacity of each drilled shaft can be verified by pressurizing the drilled shaft tip during post grouting operation, which provides additional quality assurance and increases the confidence in the load carrying capacity. Because of these advantages, post grouted drilled shafts had been used successfully worldwide for nearly four decades. However, they have not been widely used in the United States. This may be attributed to lack of knowledge about the effectiveness of post grouting and successful local practice utilizing post grouted drilled shafts. There is a need to provide practitioners local case studies to illustrate the advantages of post grouted drilled shafts.

A design phase load test program was performed at the Broadway Bridge Viaduct site in Council Bluffs. The main objectives of the load test program are:

- to evaluate the increase of capacity of post grouted drilled shafts embedded in medium dense sand; and

- to evaluate existing design methodology for post grouted drilled shafts.

The test program entailed construction of three strain gauges instrumented test drilled shafts. Two shafts included sleeve port and plate base grouting apparatus and were post grouted. The third shaft was un-grouted and served as a control shaft for comparison of the un-grouted end bearing. All shafts were 1.52 m (5 feet) in diameter. The control shaft was 16.76 m (55 feet). Post grouted shafts were 16.76 m (55 feet) and 20.12 m (66 feet).

POST GROUTED SHAFT BACKGROUND

Post grouting of drilled shafts has been successfully used in the United States since 2002 (Dapp, et al. 2007). A post grouted drilled shaft design methodology has also been developed based on research funded by the Florida Department of Transportation (Mullins et al. 2001, and Mullins et al. 2006).

The post-grouting process entails: (1) installation of a grout distribution system during conventional cage preparation that provides grout tube-access to the bottom of the shaft reinforcement cage, and (2) after the shaft concrete has cured, injection of high pressure grout beneath the toe of the shaft. During post grouting, many parameters, including the grout pressure, upward shaft movement, and grout volume injected, are measured and recorded to ensure quality.

There are a number of grout distribution systems in use (Dapp et al. 2007). A sleeve port with plate above was used at this project. This system is sometimes referred to as a tube-manchette. Sleeve port systems use a rubber tube over a perforated pipe section. See Figures 1 and 2.

Figure 1. Tube-a-manchette underside of grout plate

Figure 2. Grout plate affixed to toe of cage

GENERALIZED SOIL CONDITIONS

A full geotechnical investigation was performed for design of the bridge. In addition, SPT borings were performed at each of the three test shaft locations as part of the load test program. For the purpose of this paper a description of soils encountered at each specific test shaft location are provided. These three borings were drilled to a depth of 21.34 m (70 feet) below the ground surface. In general, the upper 6.1 m (20 feet) in all three borings consisted of soft to stiff silty clay (CL). SPT N values in this upper silty clay ranged from 3 to 14 blows per 30.5 cm. The water table was around 4.88 to 5.18 m (16 to 17 feet) in depth below ground surface. The upper silty clays were underlain by fine sand and silty fine sand (SP-SM) to depths of 16.76 to 18.29 m (55 to 60 feet). SPT N values in the sands ranged from 6 to 42 blows per 30.5 cm. The fine silty sands gave way to more coarse grained sand that extended to the boring bottom of 21.34 m (70 feet). In the TS-1 boring a soft clay layer was encountered between the bottom of the fine sand at 16.76 m (55 feet) and the coarse sand at 18.29 m (60 feet).

TS-1 which was the 19.81 m (65 feet) post grouted shaft was tipped in the coarse sand while TS-2 (16.76 m post grouted shaft) and TS-3 (ungrouted) were tipped in the fine grained sandy soils. These medium dense granular soils were very conducive to improvement by base grouting. The average SPT N values within 3 diameters from tip of the three drilled shafts were approximate 16 blows per 30.5 cm.

DRILLED SHAFT INSTRUMENTATION AND CONSTRUCTION

A summary of test shaft geometry and type are provided in Table 1. Test shaft geometries were designed such that ultimate capacities could be reached within the constrains of the 2,000 ton Statnamic load testing device.

Table 1 Summary of Test Drilled Shafts

Test Shaft ID	Diameter (m)	Planned Embedment Length (m)	Post-Grouted
TS-3	1.524	16.76	No
TS-2	1.524	16.76	Yes
TS-1	1.524	19.81	Yes

Shaft reinforcement was the same for all three test shafts and consisted of 23 - #32 mm longitudinal bars with #16 mm shear hoops on 0.3 m centers. A 7.62 cm (3 inches) clear spacing governed the steel hoop diameter. Five 5 cm (2 inches) steel crosshole sonic logging (CSL) access tubes were also installed on the cage. In addition, a post grout distribution system was affixed to the toe of the cage utilizing 2.54 cm PVC grout access tubes. Furthermore, sister bar strain gages were installed at four vertical levels in all three test shafts. The upper three level strain gages were 0 m, 5.49 m, 10.06 m below ground surface. The lowest level strain gages were placed 0.6 to 0.76 m (2 to 2.5 feet) above the grout plate.

Shaft construction was performed using the "wet method". The general procedure included placing a short upper permanent casing 1.22 to 1.52 m (4 to 5 feet) in length to both stabilize the near surface soils and provide a clean and level surface for load testing. A track mount drill rig was used with various drilling tools, auger and cleanout bucket to excavate the soils. Water and polymer slurry admixture drilling fluid was introduced early on to stabilize the walls of the excavation. Once the required depths were attained, the shaft bottoms were mechanically cleaned using a standard clean-out bucket and inspected for cleanliness by sounding with a weighted tape. The upper 1.22 to 1.52 m of each shaft was constructed 1.68 m (66 inches) in diameter using a permanent casing that extended approximately 0.3 m (1 foot) above grade. The nominal shaft diameter for the remaining depth was 1.524 m (60 inches).

A pump truck slick line was used to place the concrete while maintaining the discharge 3 to 4.6 m (10 to 15 feet) below the rising concrete head at all times.

POST GROUTING TEST SHAFTS

Drilled shaft post grouting equipment is specialized but the grout mix is simple. A typical post grout pump consists of three components including a colloidal mixer, agitated holding tank, and a hydraulic actuated piston type pump capable of high pressures. The maximum target grout pressure for this project was 4,137 kPa (600 psi) based on preliminary design calculations. The grout consists of Type I or II Portland cement with a design water/cement ratio of 0.5. However, the WC ratio can readily be lowered (not below 0.40) during the grouting process to reduce large volume takes in more open soil conditions.

The general procedure consisted of first flushing the grout lines with fresh water until clear water return was observed. This was done shortly after the completion of the shaft construction and again just prior to post grouting. Each base grout apparatus included two separate U-Tubes for redundancy. Grouting was performed on one of the U-Tube systems than the other. The grout was injected at low flow rates to build pressure slowly. Once the design pressure was achieved, the grout pressure was allowed to bleed off to atmospheric pressure (sometimes referred to as an unlocked condition).

During grouting, the grout pressure, volume, shaft displacement, and embedded strain gages were monitored with a data acquisition system. To accurately measure upward shaft displacement, an independent reference beam was set up over the test shaft to attach the LVDT's. Back up displacement measurements were made with a surveyor's level. Grout pressure was measured with an electronic pressure transducer and with a Bourdon oil filled dial gage. Grout volume measurements were made manually by recording the levels in the holding tank.

TEST SHAFT POST GROUT DATA

The maximum grout pressure achieved was 4,551 kPa (660 psi) for both shafts. The grouted volume roughly equates to a grout cylinder beneath the shaft base of equal diameter to the shaft and 0.3 m (1 foot) in height. A summary the grouting data is provided in Table 2.

Table 2 Summary of Post Grouting Results

Drilled Shaft	Maximum Upward Displacement (mm)	Permanent Upward Displacement (mm)	Maximum Verified Side Resistance (kN)	Maximum Verified End Bearing (kN)	Maximum Unit End Bearing (kPa)	Total Verified Capacity (kN)
TS-1	2.1	1.6	2,789	3,390	1,858	6,178
TS-2	1.8	1.5	2,713	3,212	1,762	5,925

The strain gages within the shaft were monitored during the grouting. The average forces at each gage level in the shaft were also calculated. The verified upward side shear resistance corrected for buoyant weight is shown in Table 2 along with the verified end bearing and the total verified capacity during the grouting. The upward side shear resistance is below ultimate due to the small upward displacement experienced during grouting.

AXIAL STATNAMIC LOAD TEST RESULTS

Statnamic load testing was performed in general accordance with ASTM 7383-08. The testing device is shown in Figure 3. The derived static capacities from the Statnamic tests were determined using the Segmental Unloading Point Method (SUP) (Mullins et al, 2002). The Segmental Unloading Point Method (SUP) discretizes a foundation into segments. The number of segments and their lengths are defined by the locations of the embedded strain gages. This allows the standard or Modified Unloading Point Method (UPM) to be applied to each segment. Then the total derived static response is calculated as the sum of the derived static response from the individual segments. Rate Effect Factors (REFs) were used in conjunction with the SUP method as suggested in the national Cooperative Highway Research Program (NCHRP) Project: NCHRP 21-08.

The derived static load versus displacement for each pile is presented in Figure 4. A limit state of 2.54 cm (1.0 inch) of movement at the shaft top was specified as the failure criteria for this project.

Figure 3 2000 ton Statnamic test device

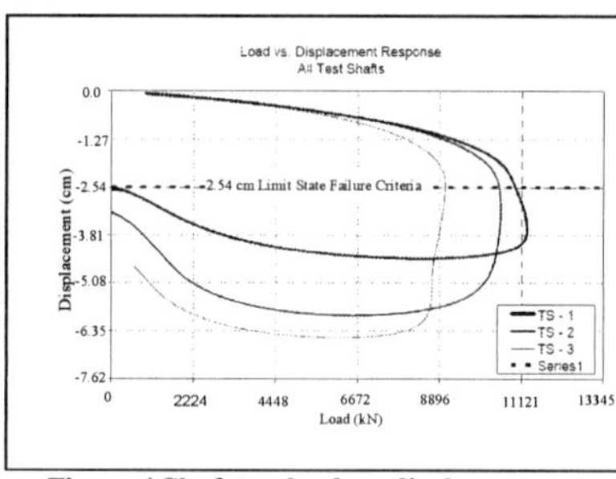

Figure 4 Shaft top load vs. displacement

The measured static capacities of the three test drilled shafts using statnamic load tests are summarized in Table 3. It can be seen that the ultimate side resistance of the drilled shaft TS-1 is approximately 580 kN (130 kips) greater than the average ultimate side resistance of the drilled shafts TS-2 and TS-3 due to 3.05 m (10 feet) longer shaft length. Based on the close match of the ultimate side resistance of drilled shafts TS-2 and TS-3, end post grouting did not affect the side resistance as expected. The average standard penetration test N-values within the 3 test drilled shaft diameter of the three drilled shafts tip are almost same, i.e. 16 blows per 30.5 cm. Therefore, the end bearing of TS-3 is comparable to the end bearing of TS-1 and TS-2 prior to post grouting. The increase of the end bearing due to post grouting varied from 1397 to 1575 kN (314 to 354 kips) with an average value of 1486 kN (334 kips).

The ultimate (maximum mobilized) unit side resistance of the top 6 m (20 feet) of the clay layer varied from 81 to 86 kPa (1.7 to 1.8 ksf). The ultimate (mobilized maximum) unit side resistance of the sand layer below the clay layer varied from 81 to 134 kPa (1.7 to 2.8 ksf) increasing with depth.

Table 3 Summary of Ultimate Capacities for Test Shafts

Drilled Shaft	Drilled Shaft Diameter (m)	As-built Drilled Shaft Embedment Length (m)	Type	Ultimate Total Capacity @ 2.54 cm Settlement (kN)	Unit Ultimate End Bearing (kPa)	Ultimate End Bearing (kN)	Ultimate Side Resistance (kN)
TS-1	1.524	19.96	Post grouted	10, 964	1,532	2,793	8,171
TS-2	1.524	16.89	Post grouted	10,506	1,628	2,971	7,535
TS-3	1.524	16.70	Not grouted	9,043	766	1,397	7,646

EVALUATION OF THE EXISTING DESIGN METHOD FOR POST-GROUTED DRILLED SHAFT

The end bearing of post grouted drilled shafts were estimated using method proposed by Mullins et al. (2006). The side resistance and end bearing of un-post grouted drilled shafts were estimated using the method by O'Neill and Reese (1999). The following step-by-step procedure for predicting axial capacity of post grouted drilled shafts was employed per Mullins et al. (2006):

1. Calculate the ungrouted end bearing capacity at 5% Diameter (D) displacement, $q_{pUltimate}$.
2. Calculate the ultimate side shear resistance, F_s, for the total length of embedded shaft.
3. Divide the ultimate side shear resistance by the cross sectional area, A, of the shaft to determine the maximum anticipated grout pressure, GP_{max}

$$GP_{max} = \frac{F_s}{A} \tag{1}$$

4. Calculate the grout pressure index (GPI), which is defined as a nondimensional ratio of the applied grout pressure to the ungrouted end bearing at a displacement of 5%D, as the ratio of the maximum anticipated grout pressure (Step 3) to the ungrouted unit tip resistance (Step 1).

$$GPI = \frac{GP_{max}}{q_{pUltimate}} \tag{2}$$

5. Establish the maximum permissible displacement, which is the desired displacement at the mobilized (ultimate) capacity, as a ratio of the shaft diameter in percentage, %D.
6. Determine the tip capacity multiplier, TCM, given the grout pressure index (Step 4) and the permissible displacement (Step 5) using the following equation.

$$TCM = 0.713(GPI)(\%D)^{0.364} + \frac{\%D}{0.4(\%D)+3.0} \tag{3}$$

7. Estimate the grouted unit tip resistance as the product of the tip capacity multiplier (Step 6) and the ultimate ungrouted end bearing capacity (Step 1)

$$q_{grouted} = (TCM)(q_{pUltimate}) \tag{4}$$

8. The total mobilized (ultimate) capacity, R_{total}, of the post-grouted drilled shaft at the permissible displacement (Step 5) can be calculated as:

$$R_{total} = q_{grouted}(A) + F_s \quad (5)$$

Using above procedure, the predicted drilled shaft capacities of the test drilled shafts and the ratio of predicted capacity over measured capacity are provided in Table 4. The predicted ultimate end bearing capacity was based on displacement of 5% shaft diameter for the non-post grouted drilled shaft.

Table 4 Predicted Capacities

Drilled Shaft	Predicted Ultimate Side Resistance (kN)	Predicted Ultimate End Bearing (kN)	Predicted Unit Ultimate End Bearing (kPa)	Predicted Ultimate Total Capacity (kN)	Post Grouted	Predicted / Measured Ultimate Side Resistance	Predicted / Measured Ultimate End Bearing	Predicted / Measured Ultimate Total Capacity
TS-1	6005	5676	3112	11681	Yes	73%	203%	107%
TS-2	4546	4573	2509	9119	Yes	60%	154%	87%
TS-3	5062	1308	718	6370	No	66%	94%	70%

It can be seen from Table 4, the ultimate side resistance was under predicted by 27% to 40 %. The predicted end bearing capacity of the non grouted drilled shaft (TS-3) closely matches the measured value. The post grouted drilled shaft ultimate end bearing capacity was over predicted by 54 to 103%. Fortunately, due to the under estimate of side resistance, the total capacity of post grouted drilled shafts was estimated to be 87 to 107% of the measured total capacity, which is considered quite well from design view point.

CONCLUSIONS

A design phase load test program was successfully carried out at the Broadway Bridge Viaduct site in Council Bluffs, Iowa. The following are considered major conclusions from the load test program:

1. The end bearing improvement expressed as a capacity multiplier ranged from 200 % to 213 % (Table 3) for the post grouted shafts in comparison to the un-grouted shaft.
2. The end bearing of post grouted drilled shafts was over predicted by 54 to 103%. Due to under prediction of side resistance, the estimated total capacity of post grouted drilled shafts using the method by Mullins et al. (2006) in combination with the method for non-post grouted drilled shaft by O'Neill and Reese (1999) closely matched the measured values.
3. Based on the measured side resistance at commensurate depths, the increase in capacity from post grouting is equivalent to a 2.74 m (9 feet) long drilled shaft if the same factor of safety (or resistance factor) is used for post grouted and non-post grouted drilled shafts. However, for projects without a load test, a factor of safety of 2 and 2.5 may be used for post-grouted and non-post grouted drilled shafts, respectively, since the post grouting validates the resistance of each drilled shaft. Assuming factor of safety of 2.5 is to be used for non post grouted drilled shafts without a load test, the increase of end bearing from post-grouted drilled shaft and the use of factor safety of 2.0 can reduce the drilled shaft length by approximately 3.4 m (11 feet) for this project.

4. Post grouting of drilled shaft tips not only increases usable end bearing capacity but also provides heightened quality assurance. Its inherent ability to provide a measure of axial capacity for each shaft is the basis for this quality assurance mechanism. No other quality assurance tool provides, with such practicality, an assurance of capacity in all drilled shafts. This could be particularly useful in non-redundant foundation systems.
5. Based on the conclusions #3 and #4 above, for drilled shafts utilizing post grouting and using load resistance factor design (LRFD) method a resistance factor of 0.70 may be justified versus the typical values of 0.40-0.60. For allowable stress design method, the use of factor of safety (FS) of 2 may be adequate for post grouted drilled shafts, where a FS of 2.5 is typically used considering that no static load test is performed. If such values are adopted by code officials, then the use of post grouted drilled shafts will be become even more cost effective.
6. The reduction in drilled shaft length by using post grouted shafts will save construction time, which indirectly will result in construction cost savings.
7. Future load test programs may consider multi-cycle post grouting to evaluate the potential of further improvements for end-bearing mobilization.
8. Based on experience on various post grouted drilled shafts including this load test program, it appears that the larger the design load the more cost effective the post grouting will be.

ACKNOWLEDGEMENTS

The authors gratefully acknowledge the contributions of Mr. Bob Stanley, Iowa DOT Mr. Curtis Monk, FHWA, Geotechnical Services, Inc. for providing SPT borings and concrete testing services. Thanks are also extended to Longfellow Drilling for all their efforts installing the shafts and field support during the post grouting and load testing. Mr. Phil King of SynchroPile is also thanked.

REFERENCES

Mullins, G., Lewis, C., and Justason, M.D., 2002. "ADVANCEMENTS IN STATNAMIC DATA REDUCTION TECHNIQUES". ACSE Geotechnical Special Publication 116, pp. 915-928.

Mullins, G., Dapp, S., Frederick, E., Wagner, R. "Post GROUTING DRILLED SHAFT TIPS – PHASE I – FINAL REPORT" Prepared by the University of South Florida for Florida Department of Transportation, December, 2001.

Mullins, G., Winters, D., and Dapp, S. (2006). "Predicting End Bearing Capacity of Post-Grouted Drilled Shaft in Cohesionless Soils." Journal of Geotechnical and Geoenvironmental Engineering, Vol. 132, No. 4. pp. 478-487.

Dapp, S. D., Muchard, M., Brown, D. A. (2007) "EXPERIENCES WITH BASE GROUTED DRILLED SHAFTS IN THE SOUTHEASTERN UNITED STATES". DFI 2007 Conference, Amsterdam.

O'Neill, M. and Reese, L. (1999). "Drilled Shafts: Construction Procedure and Design Methods", FHWA-IF-99-025, Federal Highway Administration.

Numerical Simulation for Wave Propagation in Pile Integrity Test

Abdolhosein Haddad

Civil Engineering Faculty, Semnan University, Semnan 35195-363, Iran, ahadad@semnan.ac.ir

ABSTRACT: During the construction of deep foundations there are different circumstances that are likely to affect the final shape of piles that tend to affect the designed depth, cross section and material quality. For quality control assurance of these elements pile integrity test (PIT) is often used. PIT as a method that involves the impact of a hammer on the head of the pile generating a low strain compression wave that propagates along the pile and afterwards is reflected at the bottom of the pile, or where there are changes of impedance related with variations in cross section or density of the material. In this paper a numerical solution is presented for the problem of one dimension wave propagation along a concrete pile. The problem consider a longitudinal element with a piece wise varying section, friction forces along the element and a reaction force at the tip, with a transient stress pulse applied at the top. This problem is representative of low strain pile integrity tests. Different alternatives for the solution of the problem were considered, and the spectral analysis method was chosen and implemented using Matlab. The test was modeled for several cases, comparing the results with actual PIT measurements. The parameters for the analytical model were obtained from numerical models of soil-pile interaction.

INTRODUCTION

Integrity testing is a widely used method to check the integrity of concrete piles by means of a low strain shock wave. The reflections of the pile toe or discontinuities give information about the integrity of the pile shaft. The popularity of this method is due to by the speed and the cost, especially since the equipment prices have dropped and portability and performance have increased.

Although integrity testing is the most widely used by geotechnical engineers to determine the continuity and estimate the shape of the pile embeded in soil (Rausche 1992; Massoudi, 2004 and Haddad, 2008), in some cases the signal interpretation is not an easy task because of soil-pile interaction effects, variations in density of the material and major section changes near the top of the pile. Because of these interpretation limitations, a study and understanding of the physical phenomena was carried out to see what happens in the pile reviewing the wave propagation one-dimensional theory in rods. In the present article this theory was used using an analytical solution of the problem, considering the effects of external forces along the pile shaft to simulate the PIT test through a software application.

ONE DIMENSIONALWAVE PROPAGATION

By the nature of the impact, the material and geometry, it is reasonably assumed that the wave propagation occurs in one dimension, and therefore is described by the following partial differential equation (Koshlyakov, 1964):

$$\frac{\partial^2 u}{\partial t^2} = c^2 \frac{\partial^2 u}{\partial x^2} \tag{1}$$

Where u= displacement function; t= time; c=wave propagation velocity and x= position. The wave propagation velocity c can be estimated using the formulation of the one-dimensional differential equation in an elastic media according to the linear elastic modulus E and the material density ρ by the expression:

$$c = \sqrt{\frac{E}{\rho}} \tag{2}$$

To adjust the equation (1) to more realistic pile conditions, the friction forces (spring) and energy dissipation (damper) are incorporated into the equation as shown in Eq.(3):

$$\rho A \frac{\partial^2 u}{\partial t^2} = EA \frac{\partial^2 u}{\partial x^2} - Ku - \eta \frac{\partial u}{\partial t} \tag{3}$$

Where E=linear elastic modulus of pile material; A=cross-section pile area; K=soil friction effect constant; and η= material damping effect constant.

FUNDAMENTAL SOLUTION

The Eq.(1) can be solved in different ways (Doyle, 1989). It can be solved either by analytical or numerical methods. To avoid precision limitations inherent of numerical solutions, this paper considers an analytical solution. The following mathematical methods were reviewed:

1. Method of characteristics.
2. Variables separation method
3. Laplace transformation method
4. Spectral analysis methodology

Depending on the boundary and initial conditions of the system, one can obtain similar solutions by the method of separation of variables and Laplace transformation using harmonic solutions. However, due to the presence of retardants forces raised in the Eq.(3) and the shape of the stress pulse, the solution by these methods gets quite challenging.

Similarly, it was studied the Eq.(1) can be solved using the method of characteristics, which provides a suitable solution for the study of propagation of the pulse along the pile, but the analysis increased in complexity considering the presence of external forces related with soil and pile material effects. Taking into account these limitations, the implemented analytical solution for Eq.(3) was the spectral analysis methodology using Fourier transform, considering that it has the following advantages:

- Removal of time variation allows the partial differential equation be an ordinary equation and can be integrated directly.
- Easy to incorporate to the partial differential equation the effects of external forces due to the soil-pile shaft interaction and pile damping, without making the solution complex to solve.
- Easy to model the solution by a left and right traveling wave, taking into account the system boundaries conditions.
- Easy to include the discontinuities and cross section changes to more specific problem solution.

As well as stated by Doyle (1989), "It should be pointed out that while the method uses a computer, it is not a numerical method in the usual sense, because the analytical description of the waves is still retained", and therefore, it is considered as an analytical solution. For the numerical solution the Fast Fourier Transform (FFT) is used.

In Fig. 1 it is observed that the process of the basic algorithm to solve theEq.(3) consists firstly on converting the input signal H(t) on its spectral form through the use of the Fast Fourier Transform (FFT).

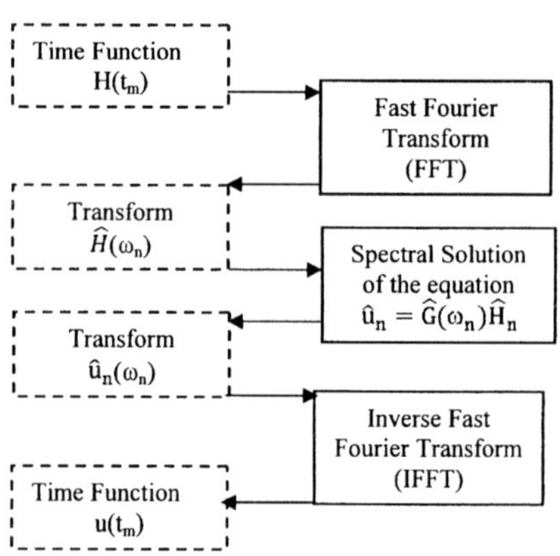

FIG 1. Basic methodology of wave propagation spectral analysis. (Doyle, 1989).

The transformed solution is obtained evaluating the product of the force function $\widehat{H}_n$ and the transfer function $G(k_{mn})$ for each frequency, as shown in Eq.(4):

$$\hat{u}_n = \hat{H}_n G(K_{mn}) \tag{4}$$

Finally, the solution is reconstructed in the dominion of time using the Inverse Fast Fourier Transform (IFFT). Eq.(3) has the following spectral representation:

$$EA\frac{\partial^2 \hat{u}}{\partial x^2} + [\omega^2 \rho A - K - i\omega\eta]\hat{u} = 0 \tag{5}$$

Eq.(5) has the following solution:

$$\hat{u}(x) = Ae^{-ikx} + Be^{+ikx} \tag{6}$$

where:

$$k = \pm\left[\omega^2 \frac{\rho}{E} - \frac{K}{EA} - i\omega\frac{\eta}{EA}\right]^{1/2} \tag{7}$$

and A, B are the indeterminate amplitudes for each frequency. When time variation is evaluated, two waves were considered in the solution, one travels in positive direction and the other travels in the negative direction. That is:

$$u(x,t) = \sum Ae^{-i(kx-\omega t)} + \sum Be^{+i(kx+\omega t)} \tag{8}$$

The values of A, B, k, etc., are calculated for each frequency ω and depend on initial and boundary conditions.

WAVES REFLECTION AND TRANSMISSION

Any change in pile cross-section or in material properties, produces new waves. The one-dimensional spectral analysis considers an incident wave A_1, a reflected wave B_1 and one transmitted wave A_2. Therefore, for two pile sections with different impedance values, the value of the respective displacement would be:

$$\hat{u}_1 = A_1 e^{-ik_1 x} + B_1 e^{+ik_1 x}$$
$$\hat{u}_2 = A_2 e^{-ik_2 x} \tag{9}$$

The wave propagation analysis considers continuity of force and displacement in the discontinuity or impedance change, obtaining the solution for the reflected wave coefficients B_1 and transmitted wave A_2 in the following way:

$$B_1 = \frac{Z_1 - Z_2}{Z_1 + Z_2} A_1 \tag{10}$$

$$A_2 = \frac{2Z_1}{Z_1 + Z_2} A_1 \tag{11}$$

Where impedance Z_i for each pile section, is defined by the equation:

$$Z_i = \frac{E_i A_i}{c_i} \tag{12}$$

Where E_i , A_i and c_i are linear elastic modulus, cross-section and wave propagation velocity for each pile section respectivily.

SOLUTION IMPLEMENTATION

The analytical solution described previously, was implemented in a graphical software application using Matlab facilities. There are two pile section considered with independent properties and compatibility of strains and stresses in its interface and spectral transfer functions are applied for each pile section in order to obtain the problem solution. It was considered that pile integrity test analysis of two sections were sufficient for a practical test interpretation. The effect of a single major change determines the answer in the test, and the important effort is to identify and interpret properly. PILEGRAPH software simulates the pile integrity test and allows analysis of piles that include a decrease or increase in impedance at any position along the pile shaft. Likewise, the program allows the analysis of soil properties around the pile shaft and at the tip including material damping of the pile. In order to verify the results consistency of the software, three modeling cases were made: the case of a uniform pile section and cases of pile cross section reduction/increment, as shown in Figs. 2,3 and 4 respectively.

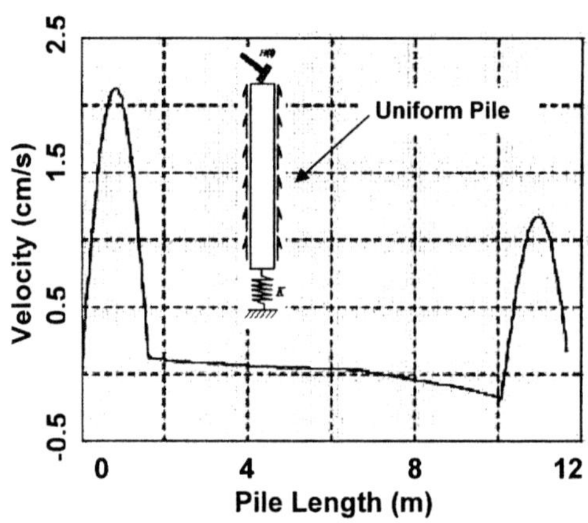

FIG. 2. Result of pile integrity test for a uniform pile simulated by PILEGRAPH

Figure 2 shows a pile without integrity problems, this result is typical for a pile with proper construction and good soil conditions.

A typical result of a reduced impedance pile, which presents wave reflection with an increase of speed at that point and the corresponding reflection at the pile bottom is shown in Fig.3. A case like this occurs when there are problems that don't allow the construction of the pile with the designed diameter, and therefore is evaluated in a special way in order to determine how it can affect the loading capacity of the pile.

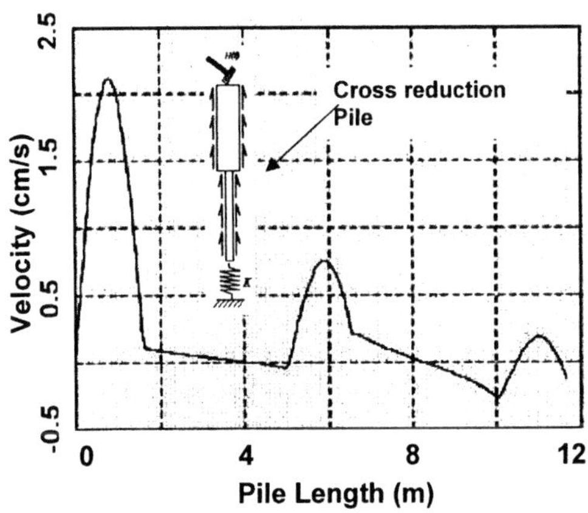

FIG. 3. Result of pile integrity test for cross reduction pile simulated by PILEGRAPH

Similarly, Fig. 4. shows a pile with increased impedance and tip reflection.

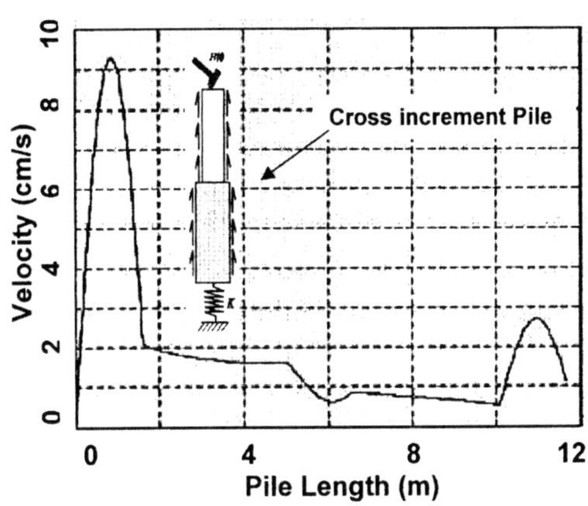

FIG. 4. Result of pile integrity test for cross reduction pile simulated by PILEGRAPH

VALIDATION WITH PIT EXPERIMENTS

PILEGRAPH can be used as a result interpretation tool of signals obtained on a field test that otherwise can only be evaluated in a qualitative manner. It allows parameter quantification of such as following:

1. Impedance change by the detected defect
2. Effects of surrounding soil and pile material in wave propagation.
3. Maximum impact force given by the hammer
4. Force time contact (characteristic wave length)

As an illustration of the applicability of the model, Fig. 5 presents the results of a PIT test carried out in a site located sought part of Iran together with PILEGRAPH modeling results. From the analysis it was possible to determine the following test parameters:

- The defect is located at a depth of 18.0m
- The diameter reduction range from 0.5m to 0.2m
- The force magnitude due to the hammer impact at the tip was 6.5 kN
- The initial pulse duration is 0.0015 seconds
- The soil stiffness around the pile shaft are $3x10^7$ N/m
- The minimum value of soil stiffness at the tip for positive velocity reflection is $1x10^{11}$ N/m
- The calibrated damping value of the signal in the pile is approximately $2x10^4$ kg/s

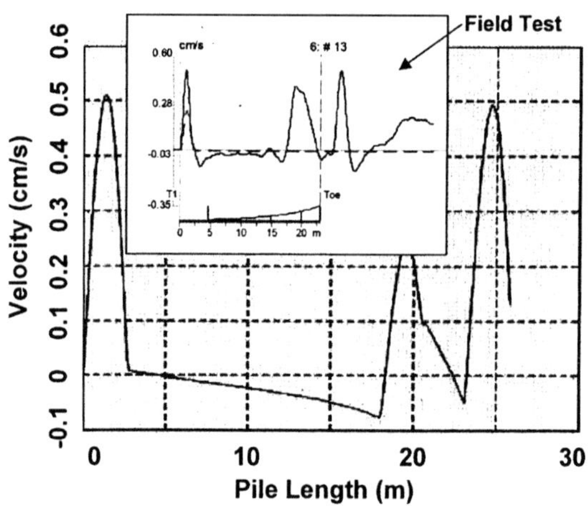

FIG. 5. Result of a field pile integrity test and PILEGRAPH simulation

CONCLUSIONS

In this paper the use of one-dimensional wave propagation theory in integrity testing of concrete pile was discussed. Because of its slenderness, it can be considered that only compression waves travel along the pile. Among the possible analytical solutions for one-dimensional wave propagation equation, the spectral analysis methodology was selected, since it allows the solution simplification of the partial differential equation incorporating external forces effects, as well as the introduction of discontinuities and section changes to more specific cases, without compromising the estimation of the solution. The applied spectral analysis for the wave equation allows a solution of the partial differential equation in terms of the position variable so that, using the Fast Fourier Transform, the initial pulse can be incorporated into the solution. When operating the transformed pulse in the spectral form of the wave equation using transfer functions, one can get the solution in the frequency domain and then apply the inverse Fast Fourier transform to obtain the solution of the partial differential equation in terms of time and position variables. Such solution was used to perform different types of analysis and simulations. PILEGRAPH allows the user to solve the one-dimensional wave propagation equation applied to pile conditions using a simple graphical user interface. The comparison between the real PIT signal and the simulated graph allow to get quantitative information on the approximate defect diameter, assessment of the possible surrounding soil and material damping effects, maximum force of contact and time contact for such force. This assessment is not possible through traditional interpretation of the evidence based only on qualitative analysis. Therefore, the software developed using the spectral methodology is useful as a tool for extending the applicability of the PIT test and reduce its subjectivity.

REFERENCES

Doyle, J.F. (1989). ''Wave Propagation in Structures''. New York: Springer-Verlag.

Haddad, A. and Kheyrodin, A. (2008) "Dynamic measurements of precast concrete jointed piles in a layered soil." Eighth International Conference on the Application of Stress-Wave Theory to Piles. Lisbon. Portugal, IOS Press

Koshlyakov, N.S. (1964). ''Differential Equations of Mathematical Physics'' Amsterdam: North Holland. Publishing Company.

Likins, G. (2000) ''Recent Advances and proper Use of PDI low Strain Pile Integrity Testing.'' Application of Stress-Wave Theory to Piles. Rotterdam, Balkema.

Massoudi, N. (2004). ''Non-Destructive Testing of Piles using the low Strain Integrity Method." Fifth International Conference on Case Histories in Geotechnical Engineering. NewYork,

Rausche, F. (1992). ''Pile Integrity Testing and Analysis." Barends, F.B.J. Application of Stress-Wave Theory to Piles. Rotterdam: Balkema

Construction Technologies of Super-long Pile Foundation in Hangzhou Bay Bridge

Junneng Ye

Ningbo urban rail transit project construction headquarters, Ningbo 315010, China; yjn75@139.com

ABSTRACT: Due to the complicated construction condition, i.e. formidable natural conditions and complicated geological conditions and the large scale of bay bridges, the design and construction of Hangzhou bay bridge have inevitably encountered numerous obstacles, among which the construction of large scale pile foundation is a crucial one. In this paper, the key technology and countermeasures of construction of large scale pile foundation in rich shallow-buried gas in offshore zone of the south shore beach area are introduced. Deformation of the steel casing pipe and countermeasures project was studied in the pile construction of south and north channels bridge piers. In addition, such technologies as installation of cathode protection, large-sized pile-sinking barge performance and pile-sinking hammer model, are studied in steel-pipe pile foundation construction of water area. The results will serve as a theoretical foundation and guidance for successful construction, operation safety and sufficient durability of Hangzhou bay bridge, and also as a reference for design and construction of other bay bridges in China.

INTRODUCTION

The Hangzhou bay bridge, which links Haiyan, Jiaxing City to Cixi, Ningbo City, in Zhejiang Province, is 36 km long and is considered the world's longest sea-spanning structure in the world. The bridge is composed of south channel bridge, north channel bridge, and non-navigation opening bridges.

Two kinds of steel pipe pile and cast-in-place bored pile are used in the bridge foundations. A total of 5513 steel pipe piles, 1.5m and 1.6m in diameter, 71~89m in length, are used in the approach bridges of water area. 3550 cast-in-place bored piles of 1.5~2.8m in diameter, 125m in the maximum of length, are constructed in the approach bridges of the south and north tidal-flat areas, the approach bridge of region of high piers, the north channel and the north channel bridge. The scale of steel pipe pile and cast-in-place bored pile used on this project is rare in our country.

The Hangzhou bay, located at one of the three strong tide bays, has large tides with high velocities, and has such formidable natural conditions as varied meteorology, scour, shallow gas and so on. In addition, the site geological conditions of Hangzhou bay bridge are complex. The distribution of soft and harden stratum is interlaced. The embedded depth of the piles is deep, and the construction period spans periods of typhoon and monsoon. All of these factors create great difficulty in the construction of pile foundations. This paper discusses several of the many construction technologies of pile foundations studied and adopted in the Hangzhou bay.

ALL-WEATHER GPS SURVEY CONTROL TECHNOLOGY

As the width of sea surface is 32km in the location of bridge, the external interference factors are complicated. So the GPS survey control technology is adopted in the vertical control survey and in the control orientation of steel pipe pile offshore. The pile-sinking survey control system of steel pipe pile is composed of a GPS external preparation system and a GPS global positioning system on the pile-sinking barge. To ascertain the location precision, the uniform GPS-RTK reference station system is built in the first stage network control point of the south bank, the north bank and the B flat roof in the sea.

The key GPS technologies are: (1) the series of GPS reference station is established. The reference station receives the satellite signal. The location instrument on the piling barge and the cooperating point can receive the satellite signal and the localization signal sent out by the reference station, thus the real time position of the location instrument is determined. (2) The establishment of engineering survey coordinate system. As the Beijing coordinate system of 1954 year is used in the design, the bridge calibrated by the 1954 coordinate system is transformed to that calibrated by the engineering survey coordinate system. Lofting is then carried out by using the engineering survey coordinate system which can assure the correctness of pier and top structure lofting. (3) Fitting height with high precision GPS. The transitional curved surface fit is created and the measurement error of GPS fitting height in sea is less than 3cm. This satisfies the requirement of construction elevation control. (4) GPS positioning system for offshore piling. Compared with the general method, the GPS system can realize automatic position of steel pipe pile, and improve the work efficiency.

CONSTRUCTION TECHNOLOGY OF CAST-IN-PLACE BORED PILE

(1) Cofferdam construction

There are 388 bored pile caps in tidal flat of south bank. The caps have a thickness of 2.5m and diameter of 8.5~9.8m. The bottom surface elevation of pile caps is – 1.00m at the nearshore zone, and – 0.5m at far off-bank. Before the construction of the pile caps, the elevation of tidal flat is +2.0m~ – 3.0m, the average high water level is +2.52m, and the average low water level – 2.12m. Since there exists adverse geological and meterological phenomena such as tide, typhoon surge, liquefiable sandy soil stratum and so on, many difficulties in construction can occur, for example, mud leakage, sinkage of steel casing, shrinkage cavity, hole collapse, verticality over standard occur in the construction.

The use of steel sheet pile cofferdams was the original method for construction of pile caps. As there is sandy loam with incipient velocity at the tidal flat surface and the stability is very poor, great local scour occurred in the bed surface during the construction, resulting in large cofferdams. As a result, it was difficult to cut-off water infiltration by using steel sheet pile cofferdam. Therefore, the steel sheet pile cofferdam was abandoned and boxed cofferdam is utilized in the construction of the bored pile caps.

(2) Shallow gas in tidal flat of south bank

Shallow gas, with a methane content of up to 96%, is present in the local soft stratum in the south bank tidal flat, mainly located in the depth from 40m to 60m. The rapid release of shallow gas can bring violent disturbance and remove great quantities of soil and water, which lead to quick conditions and collapse of soil. Therefore, the shallow gas can affect the safety of construction.

In the Hangzhou bay bridge engineering, some countermeasures were adopted to prevent the problem of shallow gas. 1) The advanced controllable exhaust technology was executed at the enrichment area of shallow gas before the construction of bored pile and the fragmentary exhaust bore was arranged at both sides of pier. 2) When drilling into the shallow gas layer, such measures as slow decompression drilling, avoiding lifting the drill pipe quickly, increasing the ratio of slurry and adjusting the water head of slurry dynamically were adopted to avoid the hole wall collapse caused by the emission of shallow gas. In addition to the deeper gas, there is surface layer gas in the muddy loam and clay layer in the zone of one or two piers. Although, the surface layer gas has a little pressure, the distribution is wide. So significant hole wall collapse takes place because advanced controllable exhaust technology is not suit able in this area. As a result, steel casing is lengthen to 12~14m to avoid the problem of hole wall collapse.

(3) Steel casing technologies in main channel bridge pier

In the engineering, the diameter of steel casing is 3.1m, thickness 18mm, and material Q235A (with a few Q345C). A reinforcement hoop is set at the bottom of the steel casing. The pile is driven to the embedded depth of 25~30m by using a vibratory hammer. There are 38 bore piles with diameter of 2.8m in the D13 pier foundation, south channel bridge (shown in Fig.1), among which 8 steel casings experienced deformations driving installation as listed in Table 1.

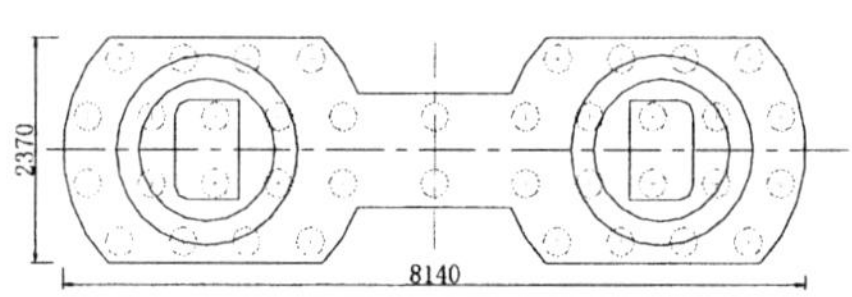

FIG.1 Layout chart for bored pile of south channel bridge pier

The reasons for deformation of the steel casing are summarized as following: (1) Steel casing subjects to great driving force and resistance during installation. The driving force acted on steel casing is 3500~6400kN. The wall of steel casing is so thin that the ratio is just 1/172(thickness is 18 mm and diameter is 3.1m). (2) During the pile driving, the hoop stress of steel casing is great because a part of soil pushes into the steel casing and a majority of soil pushes out in the bottom of steel casing. The earth pressure in the steel casing is greater than that calculated according to initiative earth pressure formula and reversely at the out of steel casing.

Table 1. Steel Casing Deformation Statistics of D13 Pier Foundation

Type	Material	Numbering	Elevation Sticking (m)	HB (m)	Structure model		Reinforcement hoop model	
					Thickness (mm)	Welding type	Thickness (mm)	Height (mm)
I	Q235A	1	-40.0	5.0	20	straight welding	12	3.0
I	Q235A	2	-38.70	6.3	20	straight welding	12	3.0
I	Q235A	18	-42.30	2.7	20	straight welding	12	3.0
I	Q235A	19	-39.3	5.7	20	straight welding	12	3.0
I	Q235A	20	-43.50	1.5	20	straight welding	12	3.0
I	Q235A	24	-39.0	6.0	20	straight welding	12	0.8
I	Q235A	15	-38.5	6.5	20	straight welding	12	0.8
II	Q235C	29	-42.0	3.0	18	spiral welding	14	0.8

HB is Height along the steel casing bottom.

The survey value of earth pressure at the out of steel casing is 110MPa, and is greater than passive earth pressure. A hoop stress test at the bottom of steel casing was carried out on the No. 18 steel casing. The distribution of test point is shown on Fig. 2. The test result of hoop stress, as shown Table 2, is 38~96MPa. Whereas, the calculated value is 50MPa~71MPa according to buckling crucial hoop stress of non linear thin walled tube.

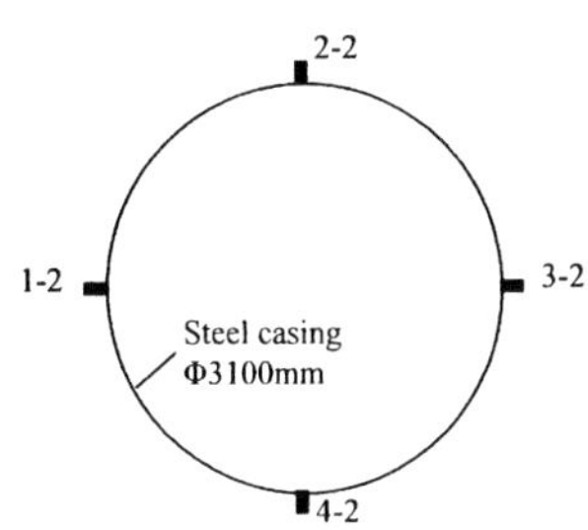

FIG. 2 Distribution of test point in No.18 steel casing

As the diameter of steel casing is great and embedded depth is relative shallow, the electric method is used to cut the steel casing and the jack is used to correct the deformation under water which is an efficient, simple method to deal with the deformation of steel casing.

Table 2 Stress Test Result of No.18 Steel Casing

Test point No.	Hoop stress peak/MPa	Axial stress peak/MPa
1-2	-96.4	-21.5
2-2	-77.2	-13.5
3-2	-96.0	-8.0
4-2	+37.8	-22.8

CONSTRUCTION TECHNOLOGY OF STEEL PIPEPILE IN WATER AREA

(1) Cathode protection system in the antisepsis

According to the natural division, geological and hydrological condition and section of bridge in Hangzhou bay, different foundation distribution of steel pipe piles is adopted. Furthermore, Hangzhou bay is located at subtropical monsoon climate area. As there is high chlorine ion content in the seawater and air, Hangzhou bay is a strong corrosion area. Therefore, the antisepsis of steel pipe pile is especially important. A comprehensive anti-corrosive scheme of high performance epoxy coating, cathodic protection and reserving corrosion allowance is used in the steel pipe pile.

In the sediment corrupt seawater environment of Hangzhou bay, the traditionally cathodic protection, i.e. welding anode block, can satisfy the requirement of this engineering. The bracelet type anode is used in the combination form of cathodic protection, a new installation method of sacrificial anode, i.e. fastness using clasp under water, feed welding above water, , and multi-position traction location method is used to lead the anode positioning (as shown in Fig.3).This assures the integrity of the high performance coating and can rapidly complete the installation of the welding and antisepsis in the zone of zero-zero visibility.

FIG.3 Installation of cathode protection system

FIG.4 Haili 801# pile-sinking barge

(2) Large-sized pile-driving equipment

In the difficult conditions of wind, billow and flow, to complete the driving of large diameter, long steel pipe pile, the use of advanced piling equipment is important. Based on a calculation analysis executed on the pile driving barge, the height of pile holder in the barge should be more than 90m and the lifting weight than 100t. Therefore,

6 pile driving barges were put into use in the pile installation: Sea pile 8#, Haili801#, Hanggong pile 7#, Sanhang pile15#, , among which the Haili801#, as shown in Fig 4, is the most advanced, multifunction, full boost circle, lifting pile driving barge. All pile-sinking barges are equipped with GPS global positioning system, large energy hydraulic hammer and diesel hammer.

The embed depth of steel pipe pile is 57~74m in this design, and the steel pipe pile must impenetrate the fine sand layer of 20m. The SPT blow count of fine sand layer is greater than 50 and ultimate bearing capacity is greater than 15000kN. Therefore, a pile hammer with high rated energy should be adopted in the pile driving engineering. As the ultimate bearing capacity of the pile with a diameter of 1.6m reaches 20000kN and exceeds the design value 15000kN. The D-128 and DELMAG150-42 diesel hammer was used to install a part of short pile with a diameter of 1.5m. As a result, two S-280 hydraulic hammers, one D-180 and two D-160 diesel hammers were used to complete installation. The quality of the executed pile driving was satisfactory with an average dissymmetry of 15.7cm that was within the allowable deviation of D/4. In addition, the pile inclination satisfies the demand of the criterion.

CONCLUSIONS

Such construction technologies as all-weather GPS survey control technology, shallow gas in tidal flat of south bank, steel casing technologies in main channel bridge pier, cathode protection system in the antisepsis and large-sized pile-sinking equipment of pile foundation were adopted in the Hangzhou bay bridge as described in this paper. Accordingly, numerous construction obstacles have been solved in construction of pile foundation in Hangzhou bay bridge, and may contribute to the design and construct of other bay bridges in China as a reference.

REFERENCES

Wang Delong, He Xubin, Wu Jianzhong. (2006). "Handling of deformation of steel casings for large diameter bored piles." *Bridge Construction*, Vol. (1): 66-68

Lu Zhongda.(2006). "Key technologies for Hangzhou bay bridge." *China civil engineering journal*, Vol. 39(6): 78-82

Liu Shaotang, Xiao Haihong , Zhao Zhanyang. (2005). "GPS pilling positioning system for Hangzhou bay major bridge." *J. southwest univer. Sci. Tech.*, Vol. 20(4): 42-45

Wang Changjiang.(2005). "The application of GPS-RTK in pile foundation construction survey of Hangzhou bay cross-sea bridge." *J. Shijiazhuang Railway institute*, suppl, Vol. 18:72-75

Model Test research on the Compaction Effect of Jacked Pile in Layered Soil

Yu-nong LI[1,2], Jing-pei LI[1,2], Shu-tao ZHANG[1,2]
1.Key Laboratory of Geotechnical and Underground Engineering of Ministry of Education, Tongji University, Shanghai ,200092 China;
2. Department of Geotechnical Engineering, Tongji University, Shanghai, 200092 China
Corresponding author (E-mail:liyunong.nong@hotmail.com)

ABSTRACT: During the penetration process of jacked pile installation, the soil around pile will move laterally and heave. The existing structures and facilities were often damaged due to the soil movement resulted from jacked pile installation. To reduce those accidents, it is important to study the law of the lateral soil movement and upward heave due to compaction. This paper describes results from a series of experiments that investigated the radial soil displacement due to jacked piles in layered soils. A number of jacked piles were installed in a test chamber filled with dry sand and rubber powders, which was used to simulated soft soil. Closed-ended piles with 40-mm diameters were used. The experiments were carried out to study how radial displacement and vertical displacement changed as the pile penetration depth varied. We determined how the displacement value changed at different soil interfaces during the pile pressing process. The results revealed the soil displacement behavior produced by different position and penetration depth of the pile. We found that a lag effect existed between the maximum displacement and the pressed pile depth. The experimental results allow for a better understanding of the inherent mechanism of compaction of the jacked pile during the penetration process. These results are important for estimating the influence of pile installation and guiding jacked pile design.

INTRODUCTION

With the increasing precision of construction technology and jacked pile machinery in recent years, as well as improvement of urban environmental protection requirements, the jacked pile has been welcomed by design and construction organizations. The jacked pile is widely used in areas with soft soil and in densely populated cities. Due to good results, it is becoming the main pile type for foundation design and construction.

However, the jacked pile is a displacement pile, which, in theory, can cause pile tip spherical cavity expansion and pile body cylindrical cavity expansion. During the penetration process, the shallow soil layer moves upwards and the lower soil moves laterally, leading to the generation of radial pressure and vertical uplift force towards adjacent embedded piles. Thus, bending, tilting, horizontal displacement and other adverse consequences may occur. These effects may also negatively impact adjacent buildings. Therefore, an understanding of the pile sinking mechanism and the compaction effect of the jacked pile have important practical significance for estimating the construction impact of pile sinking and may help guide the design of future pile projects.

Many scholars have widely researched the pile sinking compaction effect of jacked pile. Notable contributions to our improved understanding of jacked pile behavior have been achieved through slip line theory (Mayerhof 1961; Koumoto 1982), cavity expansion theory (Cao et al. 2002; Lou Zhanyou et al. 2006), the strain path method (Lou Zhanyou 2004; Einav et al. 2005), numerical analysis (Zhang Mingyi et al. 2003; Guerrero et al. 2005) and test method (Zhang Mingyi et al. 2000; Xu Jianping et al. 2000). However, due to the complexity of penetration process of pile pushing and combination of soil layer, little information is available concerning the inherent effect of the mechanism of compaction during the penetration process of jacked piles.

In this study, a laboratory model scale jacked pile test was developed to examine the behavior of the jacked pile during the penetration process in layered soils. The results of this study explain the previous experimental restriction of test piles being unable to continuously penetrate the soil.

EXPERIMENTAL TEST FACILITY AND PROCEDURE

Test chamber

The pile tests were performed in a 1.0-m long, 0.6-m wide and 1.2-m tall chamber with 3-mm thick walls, which was placed within a mild steel chamber (see Fig. 1). The front of the chamber was made of tempered glass in order to visually examine the model test. The chamber contained three drain valves at the bottom of the three walls. Standard drainage consolidation techniques were utilized.

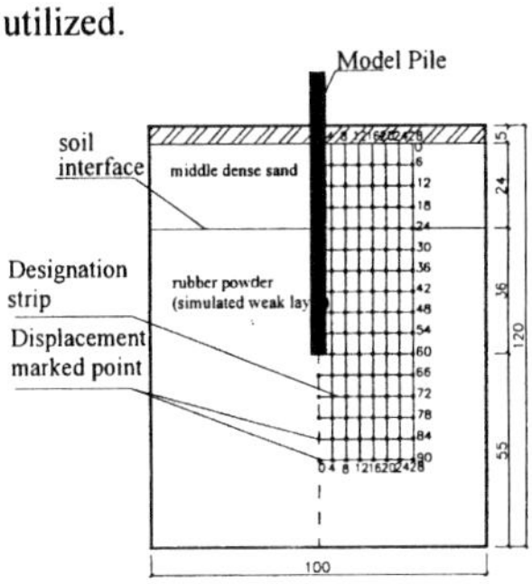

FIG. 1. Pile testing chamber / (cm)

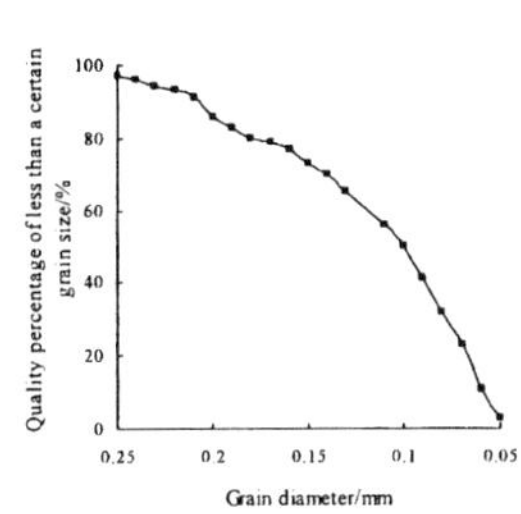

FIG. 2. Grain size distribution of prepared sand

Soil Sample Preparation

In order to simplify the complexity of model, in this paper, the effect of pore water pressure is ignored. In the model jacked pile tests, the rubber powder of grain size 50 mesh was used to simulate weak soil. The soil sample used for testing was layered within the chamber. The sand samples were placed by sprinkling dry sand from a height of 20 cm and by controlling the weight of the each soil sample layer to ensure that the overall porosity and density of the soil sample was kept at a nearly constant level. This procedure yielded a uniform sand sample. A number of tests were conducted to obtain the average index properties of the sand, which are shown in Table 1.

Table 1. Some Relevant Properties of the Sand used in the Present Study

Property	Value
Specific gravity	2.69
Void radio	0.793
Relative density	40%
Friction angle	32°

The tests were conducted in accordance with the recommendation of the China Standard (GB/T50123—1999). The grain size distribution curves of the prepared sand specimens were shown in Fig. 2. Finally, after completion of the soil sample preparation, the soil samples were stood for 2 days (see Fig. 1).

Model Jacked Pile

According to the test results described by Kishida H (1965), the radial displacement range of sand during pile jacking is approximately 7D (D is the diameter of pile) distance from the pile axis. To minimize the boundary effect from the chamber walls, a close-end half model aluminium pipe pile that was 40 mm in external diameter and 30 mm in internal diameter was used for this series of tests. The model piles were fabricated to be 80 cm in total length, with an embedment of about 60 cm.

Penetration System

The penetration device was refitted with a CLD-3-type hand static penetrometer, which was fixed on the top of the chamber with a U-shaped clasp (see the Fig.3). The installation of the model pile was carried out at a constant penetration rate of about 0.2 cm per second.

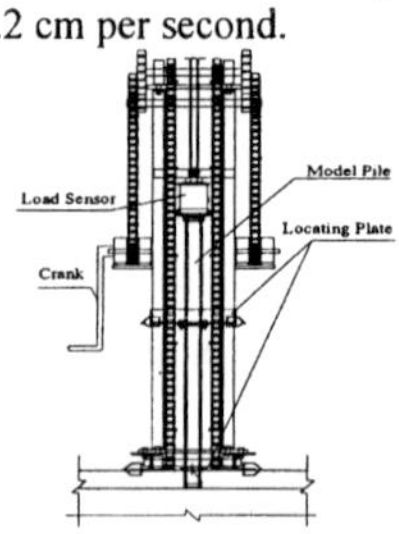

FIG. 3. The penetration device

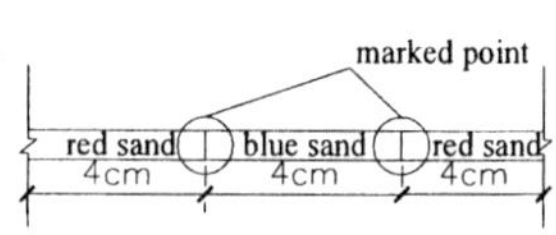

FIG. 4. The designation strip

Displacement Observation Marks

In order to observe the displacement of the pile during pile jacking process, the designation strip was marked with red and blue fine sand every 6 cm of depth from the soil surface (see Fig. 1). The red and blue fine sand laid alternately every 4cm along the radial direction in each designation strip (see Figs. 1 and 4).

Tests were conducted to observe and analyze the displacement field by image analysis methods using a high-resolution digital camera that captured photographs every 5 s.

▪EXPERIMENTAL RESULTS

The results obtained from this series of tests are presented and discussed in the following sections.

Compaction Effect during Pile Penetration Process

Figs. 5 (a) and (b) show the curves of the vertical displacement (U_Z) of uniform sand and layered soil at depth of 6 cm along the radial direction(r), respectively, in the process of pile jacking. The changes in the two curves are approximately the same. The values of vertical displacement change smaller with increasing the penetration depth (H). For uniform sand, (1) The maximum soil heave is located at $2D$ from the pile axis; (2) The value of soil heave was about 1.8 mm, which is approximately 4.5% of the pile diameter. Beyond this range, the soil heave along the radial direction gradually decreased. For the limitation of the test conditions, the value of surface uplift was not obtained, but it should be greater than that at depth of 6cm. The test results are similar to the results obtained by Cooke (1973), who used a testing pile that was 5 m in length and 16.8 cm in diameter. The test by Cooke (1973) indicated that the maximum uplift is about 6%D, and appeared at a location $2D$ from the pile axis.

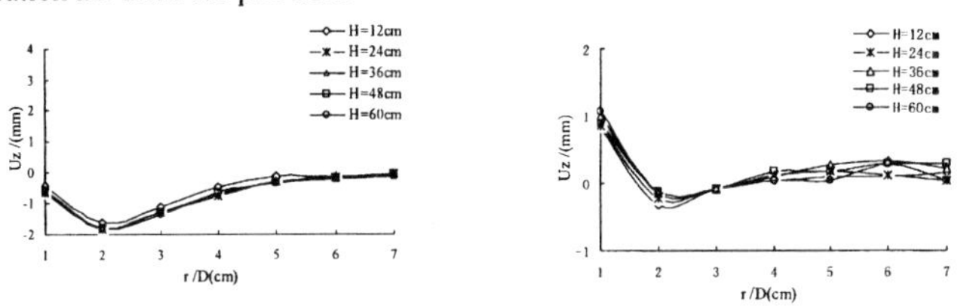

(a) Uniform sand (b) Sand stratum overlying the weak soil layer

FIG. 5. Vertical displacement at Z=6cm during pile jacking

Compared with the test in uniform sand, the value of the vertical displacement from the test in layered soil is greater. It is mainly due to the value of vertical displacement of the weak soil layer is greater than that of the uniform sand when the pile sinking.

The variations of the radial displacement (Ur) of uniform sand and layered soil at depth of 24cm during the installation of the model pile are shown in Fig. 6. The changes of radial displacement in the two curves showed exponential decrease with increasing of the radial distance. For sand stratum overlaid with the weak soil layer, the value of radial displacement increased rapidly when the pile end crosses the sand and weak soil interface. The value of radial displacement at soil interface (i.e. Z=24cm) from the test in layered soil is smaller than that from the test in uniform sand.

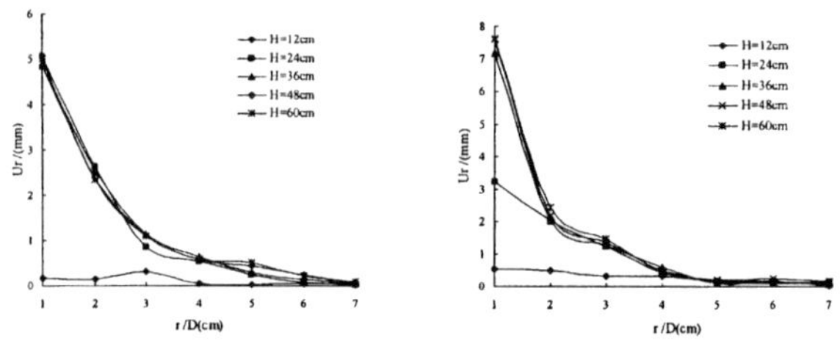

(a) Uniform sand (b) Sand stratum overlying the weak soil layer

FIG. 6. Radial displacement at *Z*=24cm during pile jacking

The radial displacement curve of uniform sand and layered soil at x=2D (where x is the distance from the pile axis) during pile penetration is shown in Fig. 7.

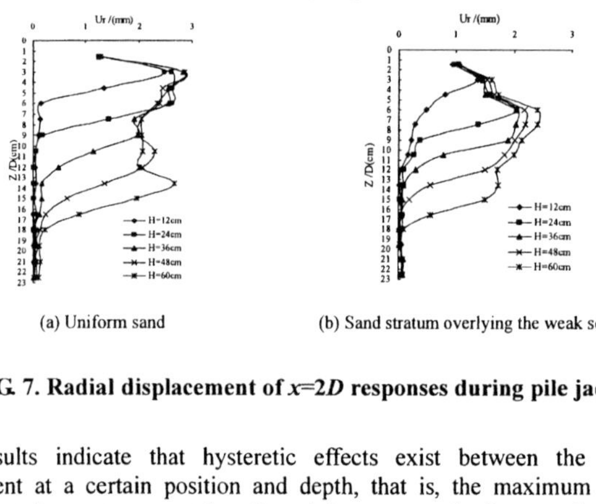

(a) Uniform sand (b) Sand stratum overlying the weak soil layer

FIG. 7. Radial displacement of *x*=2*D* responses during pile jacking

The results indicate that hysteretic effects exist between the maximum displacement at a certain position and depth, that is, the maximum horizontal displacement does not occur at the pile end, but at a certain distance above the pile end and the distance vary with penetration depth. This result was previously described by J.H. Hwang (1994) in Taiwan was 10D. The apparently radial affected zone ranges from 5D to 7D below the pile tip, and the result based on the observations by J.H. Hwang (1994) was 8D. The soil compaction displacements of layered soil reach the maximum value at soil layer interface, but the value is smaller than that of uniform sand with increasing of the penetration depth. Therefore, the size of the soil compaction displacement may be affected by the penetration resistance of the jacked pile and compression modulus of soil sample.

In addition, due to the spatial distribution mode of soil displacement, the influences of soil heave and subsidence are not ignored. From the angle of theoretical analysis of compaction effect of jacked pile, a spatial analysis method should be adopted. Fig. 8 shows the vertical displacement curve of uniform sand and layered soil at $x=2D$ during pile penetration.

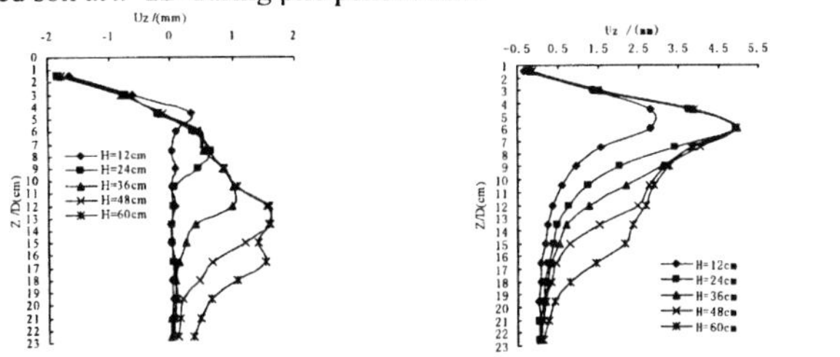

(a) Uniform sand (b) Sand stratum overlying the weak soil layer

FIG. 8. Vertical displacement of $x=2D$ responses during pile jacking

The results show that the changes of the vertical displacement are similar to that of the radial displacement. For the layered soil, the maximum appears at the soil interface, and the value of the vertical displacement is greater than that of the uniform soil.

Compaction Effect Analysis after Pile Jacking

Fig. 9 shows a soil displacement isogram of the soil around the pile after the end of pile jacking. It can be seen that it has obvious soil heave for the uniform sand in the range of $5D$ from the soil surface, but the soil heave did not appear in the layered soil. The movement is mainly lateral downward movement under the interaction of the normal pressure and the shaft resistance of the pile. Furthermore, the deformation of most of the zone for the uniform sand is mainly the radial compaction, but it is mainly the vertical compaction for the layered soil.

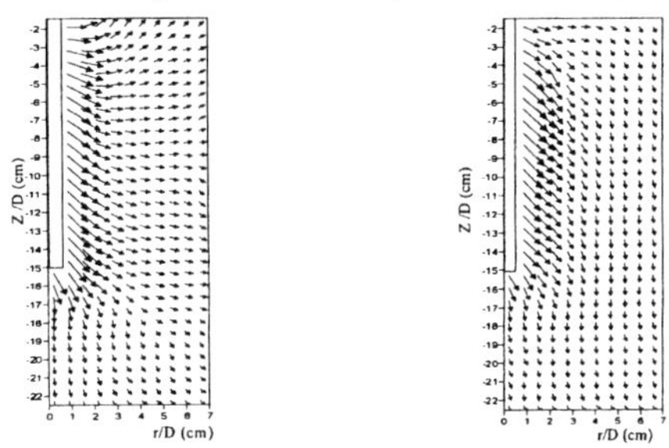

(a) Uniform sand (b) Sand stratum overlying the weak soil layer

FIG. 9. Radial displacement responses after pile jacking

Fig. 10 and Fig. 11 show a distribution curve of the radial displacement down through the soil at different distances from the pile axis and a radial displacement curve when the pile jacking ended, respectively.

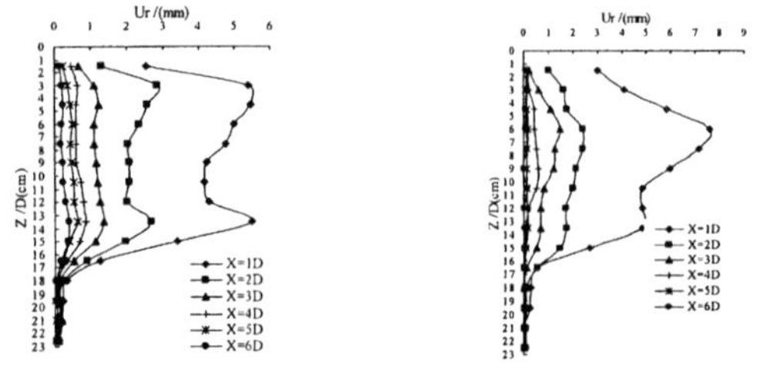

(a) Uniform sand (b) Sand stratum overlying the weak soil layer

Fig. 10 .Profiles of soil radial displacement with depth

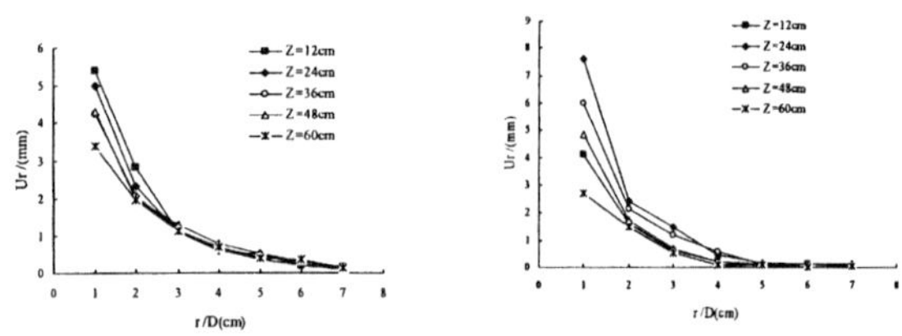

(a) Uniform sand (b) Sand stratum overlying the weak soil layer

FIG. 11. Soil lateral displacement distribution

The results indicate that the compaction effect is most evident in the range of 4-12*D* from the ground surface. The radial displacement of the other areas is relatively small, which agrees with the syllogism of the compaction effect proposed by Yu (2000). In addition, the displacement values decrease with increasing distance from the pile axis. The influence range is 5-7*D* for the uniform sand, but it is relatively small for the layered soil, which is only 4*D*. Application of traditional plane-strain cavity expansion theory yielded a displacement value of 10*D*, which is greater than the test result. This lack of agreement is due to the plane strain problem, in which one cannot obtain the spatial distribution mode of soil displacement and neglect the influences of soil heave and subsidence.

CONCLUSIONS

The results obtained from this series of tests are presented and discussed in the following section.

(1) According to the displacement variation at different depths during the pile installation stage, soil deformation can be divided into three different regions. Region 1 is 2-3D from the soil surface. The main feature of this region is its generation of surface uplift. Region 2 is the radial compaction zone. In region 3, the change of lateral displacement is significant. It is mainly defined as the spherical cavity expansion disturbance area;

(2) During the pile installation stage, the maximum compaction

displacement and the penetration depth cause the hysteresis effect. The maximum value of radial displacement appears at several times the diameter position above the pile tip but not at the pile tip;

(3) The radial compaction zone increases with increasing depth during installation at a range between 5D to 7D, which is smaller than the zone estimation obtained by use of traditional plane-strain cavity expansion theory.

(4) When the pile was penetrated in the situation of the sand overlying the weak soil, the radial displacement value is smaller than that of the uniform sand, which is concerned with the nature of the soil sample and penetration resistance of the pile.

REFERENCES

Mayerhof, G G. (1961). "The ultimate bearing capacity of wedge-shaped foundations." *Proc.5th Int. Conf. on Soil Mechanics and Foundation Engineering*, Paris, pp. 105-109.

Koumoto T., Kaku K. (1982). "There dimensional analysis of static cone penetration into clay." *Proc. 2nd Europe Symp Penetration Test*. Amsterdam, Netherland Press, pp.635~640.

Cao L F., Teh C I., Chang M F. (2002). "Analysis of undrained cavity expansion in elasto-plastic soils with non-linear elasticity." *International Journal for Numerical and Analytical Methods in Geomechanics*, 26(1), pp. 25-52.

Luo Zhan-you, Xia Jian-zhong, Gong Xiao-nan. (2006). "Unified solution for the expansion of spherical cavity in strain-softening materials with different elastic module in tensile and compression." *Engineering Mechanics*, 23(4), pp.22-27.

Luo Zhan-you. (2004). "Research on compaction effect and construction measures of Jacked pile." Hangzhou: Zhe Jiang University.

Einav I., Randolph M F. (2005). "Combining upper bound and strain path methods for evaluating penetration resistance." *International Journal for Numerical Methods in Engineering*, 63(14), pp. 1991-2016.

Zhang Ming-yi, Deng An-fu, Gan Teng-jun. (2003). "Displacement penetration method used for numerical simulation to jacked pile." *Rock and Soil Mechanics*, 24(1), pp.113-117.

Guerrero S L., Vallejo L E. (2005). "DEM analysis of crushing around driven piles in granular materials." *Geotechnique*, 55(8), pp. 617-623.

Zhang Ming-yi, Deng An-fu. (2000). "Experimental Study on Jacked Precast Piles in Layered Soil." *Chinese Journal Geotechnical Engineering*, 22(4) pp.490–492.

Xu Jian-ping, Zhou Jian, Xu Zhao-yang, Zhang Chun-yu. (2000). "Model test research on pile driving effect of squeezing against soil." *Rock and Soil Mechanics*, 23(1), pp. 235-238.

Kishida H, Meyerhof G G. (1965). "Bearing capacity of pile group under central and eccentric load in sand." *Proc. 5th ICSWFE.*, pp. 270-274.

Cooke R W. Price G. (1973). "Strain and displacements around friction pile." *Proc. 8th Int. Symp. on Mechanics of Foundation.* Houston, Texas University.

J.H. Huang, Neng Liang and Cheng-Hsing Chen. (2001). "Ground response during pile driving." *J. Geotech. Geoenviron. Eng.*, 127(11), pp.939-949.

Yu, H .S. (2000). "Cavity expansion methods in geomechanics." The Netherlands, Klower Academic Publishers, pp.275-276.

Numerical Simulation Study on Use of Groove in Controlling Compacting Effects of Jacked Pile

Zhanyou Luo[1], Jiangzhong Xia[2], Xiaonan Gong[3]

[1]Vice Professor, Zhejiang University of Science and Technology,318 Liuhe Road, Hangzhou 310023, China; PH (86-571)8507-0518; FAX (86-571) 8507-0518; email: Lzy0395@163.com
[2]Professor, Zhejiang University of Science and Technology,318 Liuhe Road, Hangzhou 310023, China; PH (86-571)8507-0500; FAX (86-571) 8507-0500; email: xia.jz@163.com
[3]Professor, Zhejiang University, 20 Zheda Road, Hangzhou 310027, China; PH (86-571)8795-2077; FAX (86-571)8502-1310; email: xngong@hzcc.com

ABSTRACT: Groove of preventing the compacting effect was an effective construction measure of decreasing compacting effects of jacked pile. An appropriate finite element model adopting the Drucker-Prager failure criterion and finite deformation theory was proposed according to the actual process of jacked pile. This model can simulate the interaction of pile and soil and continuous penetration. Displacement fields on the measure of groove were studied through the finite element model, and the simulation results were compared. The results show depth and situation of groove are important influencing parameters of the compacting effect of jacked pile. The compacting soil displacement of jacked pile can obviously be decreased due to the use of groove. Therefore, when the building and underground utilities need to be protected, the depth of grooves preventing the compacting displacements should be larger than that of protected surroundings.

INTRODUCTION

The jacked piles compared with the driven piles have been widely used in urban areas due to noiseless and vibrationless installation. However, displacement piles jacked are adverse to surrounding environment because of compacting induced displacement (Hagerty, 1971; Huang, 2001; Yang, 2006). In order to avoid the damage to environment some measures are taken to decease the compacting effects of jacked piles. In many practical engineering the measured results showed that the compacting effects of jacked piles due to some construction measures such as prebored hole and groove were different from those conditions in which no measure was taken. It is therefore necessary to study the effect of the construction measure to protect surrounding buildings (Sagaseta, 2001; Luo, 2006).

Grooves referring to a surface trench adjacent to the pile are an effective measure to prevent the compacting effect of jacked piles for environment surroundings. Because of the complexity of installation of jacked pile, the study on the compacting effect under the construction measure of grooves was few. The analytical solution based on the cavity expansion theory (CEM) or strain path method (SPM) cannot obtain the horizontal and vertical displacement fields of whole pile-depth (Carter, 1986). Furthermore the CEM or SPM solution cannot represent the interface friction between pile and soil. So the analytical solutions may result in large errors. However the finite element method can effectively overcome theses difficulties (Mabsout, 1995). In this paper, the finite element model based on the interaction of pile-soil and continuous penetration was established. The compacting effects of jacked pile due to depth and situation of groove were analyzed according to the model.

FINITE ELEMENT MODEL

Basic Assumption

Due to the complexity of installation of jacked piles, some modeling issues include the following engineering properties of soil, finite deformation, contact problem and constitutive relation of soil and pile. In order to effectively simulate the installation of jacked piles some assumptions were made as follows:

(1) The model was analyzed based on the total stress method.

(2) The two-dimensional axisymmetric space was adopted according to the actual penetration process.

(3) Constitutive relation of reinforced concrete pile was linear elastic and four-node axisymmetric element for pile was adopted.

(4) Soil was modeled as continual elastic-plastic relation and the Drucker-Prager failure criterion, taking into consideration the hardening of material nonlinearly and geometry nonlinearly.

(5) Face-face friction contact elements were adopted for the interface of pile and soil. The friction type of contact elements belongs to coulomb theory.

Finite Element Model

Figure 1 shows the finite element model of pile-soil system. The diameter of the pile is 0.4m and its length is 20m. The tip of the pile is assumed parabolic in shape. The diameter of clay soil domain is taken as 50 times the one of the pile and its depth is 100 times of the diameter of the pile. The soil surface is free. Lower and right surfaces are fixed boundary. Because of the Continuum of soil, it is necessary to set a narrow opening before pile installation in order to make the pile penetration process become early and avoid the large mesh distortion. The simulation result of compacting soil displacements cannot be affected owing to the small size of narrow opening. If the section of pile is rectangle, the section can be transformed into the Circular section according to the Equivalent area.

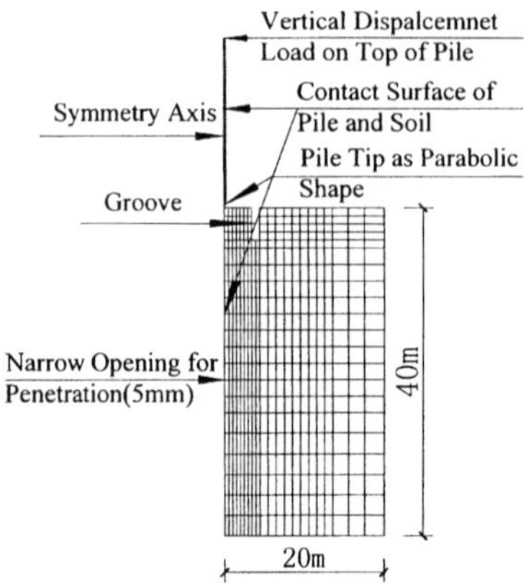

Figure 1. Finite element model

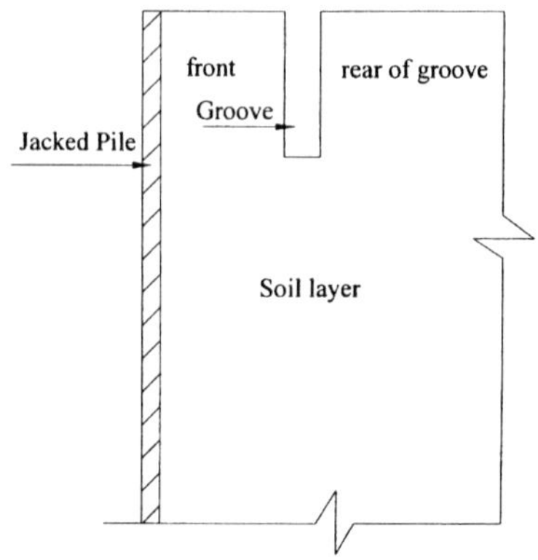

Figure 2. Sketch of Groove Location

ANALYSIS OF COMPACTING DISPLACEMENTS

Simulation Parameters

Parameters of pile and soil. In order to obtain the effective results, it was necessary to assume soil is homogeneous. The parameters of pile and soil are shown in Table 1.

Table 1. Parameters of pile and soil

Parameters	soil	pile
Modulus of Elasticity, E(MPa)	2.0	2000
Poisson's ratio, ν	0.40	0.30
The density of concrete, ρ (kg • m^{-3})	1890	2500
Cohesion, C(kPa)	15.0	
Internal friction angle, Φ (°)	15.0	
Friction coefficient, μ	0.20	0.20
Maximum Shear strength, τ (kPa)	20.0	20.0
Contact stiffness, K(kN • m^{-1})	8000	

Parameters of grooves. Before the pile was penetrated, it was effective for the Grooves to decrease the lateral displacements and heaves of shallow soil and differential settlement of surrounding building and underground utilities. Because the grooves were used to protect shallow foundation and underground utilities, the depth of groove should be less than 4 m. The width of groove was 1-2 m. Figure 2 shows the sketch of the groove location. The parameters of grooves are shown in Table 2.

Table 2. Parameters of groove

Distance between pile and groove (m)	Parameters of groove			
	Depth (m)			Width (m)
2.0	1.0	2.0	4.0	1.0
4.0	1.0	2.0	4.0	1.0
6.0	1.0	2.0	4.0	1.0
No groove				

Compacting Displacements Affected by the Depth of Groove

Figure 3 shows the horizontal and vertical displacements between groove and pile. The horizontal and vertical displacements increase with the increasing depth of groove.

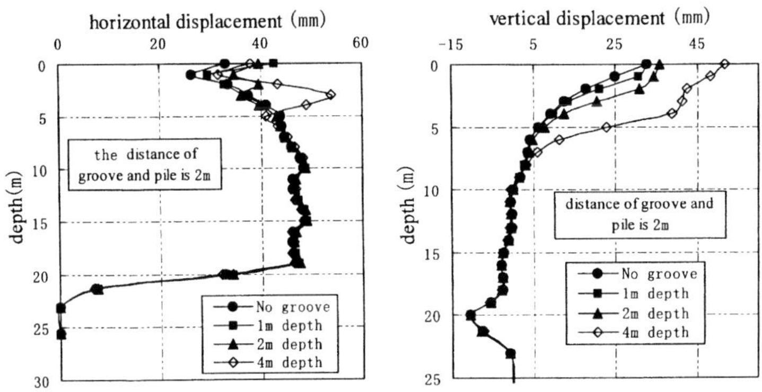

Figure 3. Displacement field (measured distance from pile is 0.8 m, groove front)

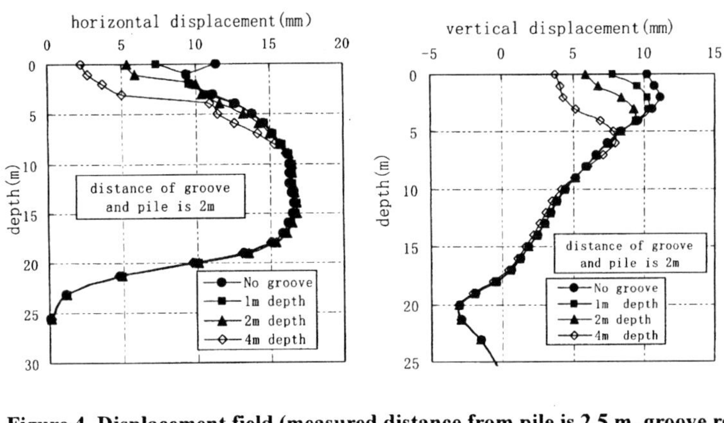

Figure 4. Displacement field (measured distance from pile is 2.5 m, groove rear)

Figure 4 shows the horizontal and vertical displacements of groove rear. The horizontal and vertical displacements decreased with the increasing depth of groove. The result shows the soil displacements of groove rear would effectively be decreased due to the action of groove. However the difference of displacements between action of groove and no groove is small. Therefore, when the building and underground utilities need to be protected, the depth of grooves preventing the compacting displacements should be larger than that of protected surroundings.

Compacting Displacements Affected by the Distance between Groove and Pile

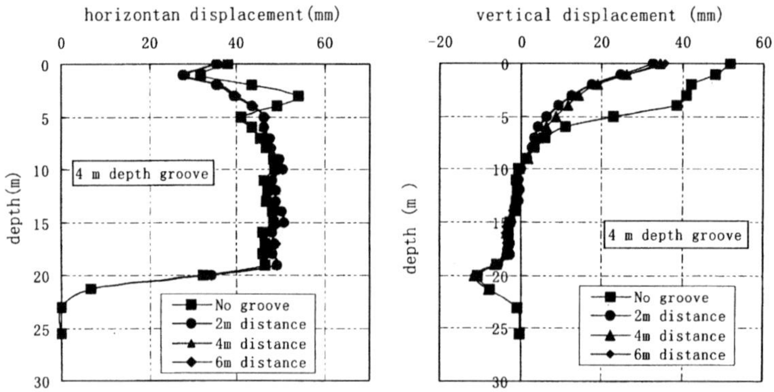

Figure 5. Displacement field (measured distance from pile is 0.8 m,groove front)

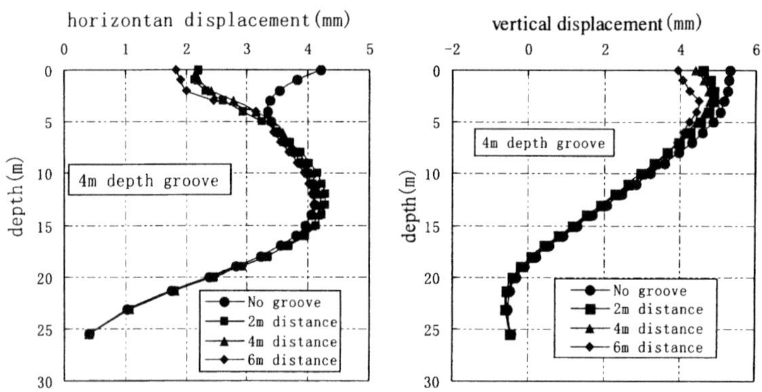

Figure 6. Displacement field (measured distance from pile is 7.4 m,groove rear)

Figure 5 shows the horizontal and vertical displacements of groove front. When the distance between groove and pile was 2 m, the horizontal and vertical displacements increased with the increasing distance between groove and pile. However, when the distance between groove and pile became larger than 4 m, the difference between action of groove and no grooves may be small. Therefore, the compacting displacements would be affected easily with the decreasing of distance between groove and jacked pile.

Figure 6 shows the horizontal and vertical displacements of groove rear. The horizontal and vertical displacements decreased with the increasing of distance between groove and pile due to action of groove. However the difference of displacements between action of groove and no groove was small. Therefore, the compacting displacements would not be affected easily with the increasing of distance in back of groove.

CONCLUSION

The relationship of compacting soil displacements and the depth of groove and distance between groove and jacked pile were analyzed according to the finite element model. Main conclusions are written as follows.

(1) Groove can obviously affect the compacting soil displacement both groove front and groove rear. The compacting soil displacement of groove ahead will become larger than that without of groove. And the compacting soil displacement of groove rear will become less than that without of groove.

(2) When the building and underground utilities need to be protected, the depth of grooves preventing the compacting displacements should be larger than that of protected surroundings.

ACKNOWLEDGMENTS

The authors appreciate the support of the National Natural Science Foundation of China (No.50708097) and the science and technology of Zhejiang Province (No. 2009C31160).

REFERENCES

Carter, J.P., Booker, J.R., and Yeung, S.K.(1986). "Cavity expansion in cohesive frictional soils. " *Geotechnique*, 36(3), 345-358.

Hagerty, D.J., and Peck, R.B.(1971). "Heave and lateral movements due to pile driving. " *J.Soil Mech. And Found. Div.*, ASCE, 97(11), 1513-1532.

Huang, J.H., Liang, N. and Chen, C.H.(2001). "Ground response during pile driving." *J.Geotech.Geoenviron.Eng.*, 127(11), 939-949.

Luo Z.Y., Zhu X.R., Wang L.F.(2006). "Field studies on effect of jacked pile on adjacent buildings and roads in clay. " *Foundation Analysis and Design*(ASP153), ASCE Special Publication,153,195-199.

Mabsout, M.E., Teese, L.C. and Tassoulas, J.L.(1995). "Study of pile driving by finite-element method. "*Journal of Geotechnical Engineering*, ASCE, 121,535-543.

Sagaseta, C., Whittle, A. J.(2001). "Prediction of ground movements due to pile driving in clay. " *J.Geotech.Geoenviron.Eng.*,127(1), 55-65.

Yang, J., Tham, L.G., Lee, P. K.K., Chan, S. T., and Yu, F.(2006). "Bahaviour of jacked and driven piles in sandy soil. " *Geotechnique*, 56(4),245-259.

Effects of Cyclic Frequency and Pre-Loading on Behaviour of Plate Anchors

S. P. Singh [1] and S. V. Ramaswamy [2]

[1]Professor, Department of Civil Engineering, National Institute of Technology, Rourkela – 769 008, Orissa, India, Email; spsingh@nitrkl.ac.in, singh_p_suresh@yahoo.co.in
[2]Former Professor, Department of Civil Engineering, Anna University, Chennai – 600 025, Tamilnadu, India,

ABSTRACT: This paper reports the response of embedded circular plate anchors to varying frequencies of cyclic loading. It also highlights the effect of pre-loading on anchor movement and post-cyclic monotonic pullout behavior using model circular plate anchor, buried at embedment ratio of six in soft saturated clay. The frequencies of loading cycle are found to have considerable effect on anchor movement. For a given duration of loading, high frequency cycle causes more anchor movement than that caused by low frequency cycles. Pre-loading causes a reduction of anchor movement in subsequent loading stages. Anchors subjected to cyclic loading and then monotonic pullout, show stiffer load-displacement behaviour at an initial stage compared to anchors not subjected to any cyclic loading. Pre-loading causes a reduction of anchor movement in subsequent loading cycles. When anchors are recycled at a load ratio level less than the pre-cycling load, the anchor movement in the recycling phase is very much reduced, but if the recycling is done at a higher load ratio level, the effect is not that pronounced, and the anchors behave as if they were not subjected to any cycling load in the past. The relative post-cyclic stiffness of anchors for the present test conditions varies from 1.169 to 1.327.

INTRODUCTION

Anchors form an important component of many civil engineering structures. The primary function of these anchors is to transmit upward forces to the soil at certain depth below the ground. In an offshore environment, these anchors are subjected to dynamic loadings caused by waves, wind, ocean current and tides in addition to sustained pulls. Additional complications of loading conditions are caused by cyclonic storms, which could last for several hours inducing different patterns of cyclic loading. The time period of loading cycles could vary depending upon the wave cycles prevalent in that locality and are also strongly influenced by the cyclonic storms and other environmental factors. During the in service life, the anchors are expected to be subjected to a number of cyclonic storms of varying intensity and duration. An attempt has been made to study the response of embedded circular plate anchors to these types of loadings. Although a real-time loading is best to evaluate the performance of anchors in the field, considerable limitations exist in replicating these loading patterns in laboratory model tests. Therefore, in the present work, uniform rectangular loading cycles are used to study the effects of time period of loading cycles and pre-loading on anchor behaviour.

REVIEW OF EARLIER WORK

Ponniah and Finally (1983) reported the long-term behaviour of circular (50 mm dia) plate anchors subjected to sinusoidal loading of 10 sec time period. Based on the test results it was reported that anchors did not fail when the load (SLRL±CLRL), cycled up to 50 ± 20% of the drained ultimate pullout capacity. With recycling the failure load increased to 70 ± 20% of the drained anchor capacity. The short term cyclic behavior of a deep circular (50 mm dia) plate anchor in soft cohesive soil was reported by Datta et al. (1990). The principal parameters studied were the influence of mean load and the cyclic amplitude on the permanent anchor movement and post-cyclic static pullout capacity. Based on the experimental results they proposed that the plate anchors should be designed for a load of 1/3rd of its static pullout capacity to take into account the effects of cyclic loading. Singh and Ramaswamy (2002) have studied the behaviour of plate anchors in soft saturated soil under sustained-cyclic loading to highlight the relative influence of static load ratio level (SLRL) and cyclic load ratio level (CLRL) on permanent anchor movement as well as post-cyclic monotonic load-deformation behaviour of deep anchors. The movement of plate anchor was reported to be governed primarily by the amplitude of cyclic loading rather than the static load ratio level. Anchors subjected to cyclic loading and then monotonic pullout load showed stiffer load-settlement behaviour than anchors not subjected to any cycling loading. Singh and Ramaswamy (2008) reported the response of embedded circular plate anchors to varying frequencies of cyclic loading (0.30 and 0.45, with cyclic time periods of 2 sec, 6 sec, 12 sec and 24 sec). The frequency of loading cycles was shown to have considerable effect on the movement of anchors. For a given duration of loading, high frequency cycles caused more anchor movement than that caused by low frequency cycles. However, a marginal loss of anchor capacity up to an extent of 12% was observed due to cyclic loading. Literature review shows that the available information on the behavior of plate anchors under cyclic loading is limited. Little attempt has been made to understand the influence of cyclic time period and pre-loading on the behavior of plate anchors.

EXPERIMENTAL PROGRAMME

The experimental program undertaken in the present study was broadly divided into two distinct phases. In the first phase, the anchors were subjected to two different cyclic load ratio levels (CLRL) i.e. 0.30 and 0.45, with cyclic time periods of 2 sec, 6 sec, 12 sec and 24 sec. The cumulative anchor movements with loading cycles were recorded. After completion of 1000 loading cycle the anchors were pulled out at a rate of 5 mm/min to find out the post-cyclic monotonic pullout behaviour. In the second phase, the anchors were subjected to a given cyclic load ratio level, thereafter, the anchors were subjected to re-cycling load levels less than, equal to, and greater than the load applied in first cyclic phase. In between the first phase of cycling and re-cycling phase an unloading period of 22 hrs was allowed. Re-cycling at load ratio levels of 0.30, 0.45 and 0.60 were done for anchors subjected to a pre-cyclic load ratio of 0.45; whereas, load ratios of 0.15, 0.30 and 0.45 are used for anchors subjected to pre-cyclic load ratio of 0.30. In each stage of loading (i.e. pre-loading phase or re-cycling phase), the anchors were subjected to 500 rectangular loading cycles of 12 sec time period. This time period of loading cycles was based on the prevailing wave conditions along the Indian east coast. At the end of re-cycling, monotonic pullout tests were conducted to find out the post-cyclic pullout capacity of anchors. All the above tests were conducted using a circular plate anchor of 80 mm diameter, buried at a depth of 480 mm in saturated clay with an average moisture content of 57.3% (I_c = 0.40). A highly plastic commercial clay which exhibited a LL = 75% and PI = 44% was used in the present study. It contained 96.5% fines (< 75 micron) and a 3.5% coarse fraction. The X-ray diffraction pattern showed the presence of illite, kaolinite, chlorite & vermiculite clay minerals with quartz. The soil was classified as 'CH' as per the Indian Standard soil classification system (IS: 1498-1970).

Preparation of Test Sample

The pulverized clay was thoroughly mixed with required amount of water by hand kneading and stored in airtight containers. Care was taken to remove the entrapped air during the mixing operation. The wet soil was again remixed after 2 days and stored in airtight plastic containers for another 7 to 8 days before being used. This procedure was followed to ensure proper moisture equilibrium in the soil sample. The wet soil was placed in the test tank in small quantities by hand and patted uniformly. Because of the low consistency of the soil used, no problem was faced to fill the test tank using this method. The cylindrical test tank used for the model tests had an internal diameter of 560mm and 710mm deep. The diameter of the test tank was seven times bigger than that of the least dimension of the anchor in order to avoid boundary effect of the test tank on the anchor. After filling the test tank to the base level of the anchor, the anchor with the connecting rod was placed and the filling operation continued till required embedment depth (480 mm) was achieved. The test tank was kept undisturbed for 22 hours before the load being applied. The average unit wet weight and the undrained unit cohesion of clay in the test tank was 16.08 kN/m^3 and 4.70 kPa, respectively.

Cyclic Loading Test

One-way vertical cyclic pullout load on anchor the imparted was by using a pneumatic loading apparatus. This consisted of (i) an air compressor with a reservoir (ii) pressure regulator with indicator (iii) double acting pneumatic power cylinder with 40 mm bore diameter and 150 mm stroke length (iv) three-way solenoid valve and (v) an electronic timer capable of operating the solenoid valve in the frequency range of 1/24 to 1 Hz. The cyclic loading on the anchor was imparted by the piston of the double acting pneumatic power cylinder which was connected to the anchor rod by a flexible wire through a system of frictionless pulleys. The piston of the pneumatic power cylinder was actuated by regulated compressed air, passing through a solenoid valve system controlled by an electronic timer. A schematic diagram of the cyclic loading set-up used is shown in Fig. 1.

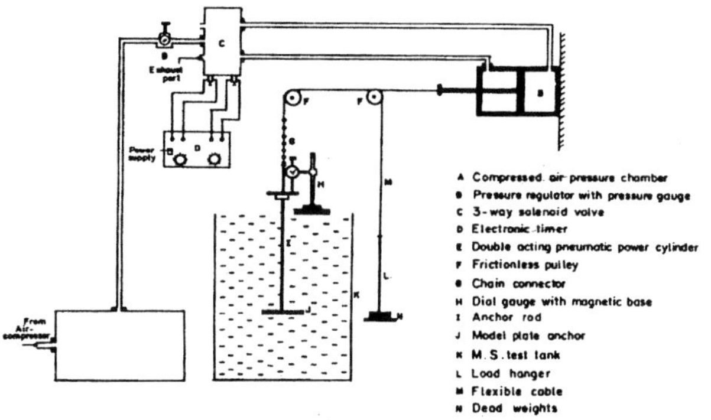

FIG. 1. Schematic layout of the cyclic loading test set-up.

Post-Cyclic Monotonic Pullout Test

The post-cyclic monotonic pullout tests were carried out by using a strain controlled test set-up. After completion of the cyclic loading, the anchors were monotonically pulled out at a rate of 5 mm/min. The pullout resistance of anchors at required displacement levels was measured using a tension proving ring. Both cyclic loading tests and post-cyclic monotonic pullout tests were carried out without eliminating adhesion and suction force.

RESULTS AND DISCUSSIONS

Effect of Time Period of Loading Cycles on Movement of Anchors

Typical variations of anchor movement with number of loading cycles are illustrated in Fig. 2 for CLRL of 0.45. The movement of the anchor was found to increase continuously with cycles of loading. However, the rate of increase of anchor movement tended to slow down with an increase of loading cycles. It is further noticed that the rate of movement of anchors with number of loading cycles tended to stabilize earlier when subjected to lower frequency cycles. For anchors subjected to 2 sec loading cycles, the movement was found to be almost linear until the cyclic loading was stopped (i.e. after 1000 cycles of loading). Rao (1988) found similar results from the undrained triaxial tests conducted on marine clays and reported that at high frequency of loading, the built-up cyclic strain per load cycle was slow and more or less uniform, whereas at low frequency loading it was quite high and the movement tended to stabilize at a smaller number of loading cycles. The movement of the anchor during cyclic loading was related to the development and dissipation of excess pore water pressure. The cyclic pore water pressure increased during the cyclic loading phase and subsequently dissipates under cyclic unloading. The dissipation of excess pore water pressure in the soil mass above the anchor results in the upward movement. The movement of an anchor after 1000 loading cycles with 24 sec period is 1.75 times that of anchors subjected to a period of 2 sec for a CLRL of 0.45. This value is 1.93 for CLRL of 0.60. But, it was to be noted here that the loading duration for 1000 cycles with a 2 sec period was 33.33 minutes compared to 400 minutes for a 24 sec period. The curves in Fig. 3 indicated a non-linear relationship between the movement of anchor and duration of loading. Initially, the rate of anchor movement was high, which tended to stabilize with time. For a given duration of time, cyclic loadings of lower time period produced higher anchor movement than movements caused by cycles of higher time period.

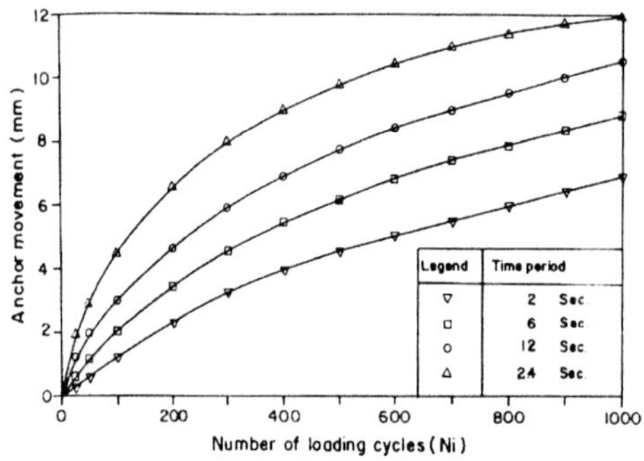

FIG. 2. Movement of anchors with number of loading cycles at CLRL of 0.45

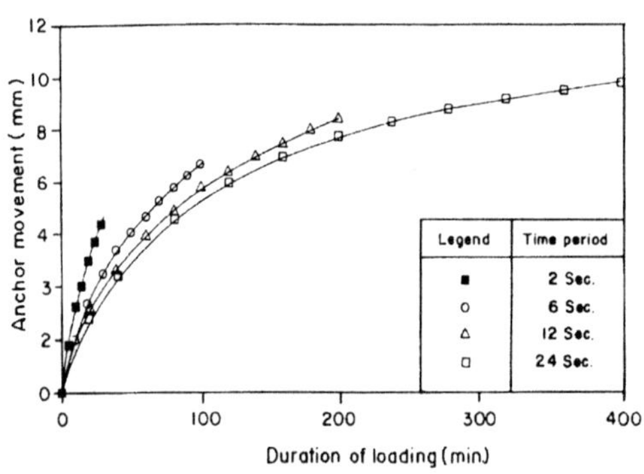

FIG.3. Movement of anchors with loading time at CLRL of 0.45

Effect of Pre-Loading on Movement of Anchors

For studying the effect of pre-loading on anchor movement in subsequent loading stages, two sequences of cyclic loading were adopted. In the first phase, the anchors were cycled 500 times at CLRL of 0.30 or 0.45. In the next phase i.e. after an unloading period of 22 hours the anchors were again cycled at a CLRL value less than, equal to or more than that of the first stage of loading. This is referred to as re-cycling. The re-cyclic load ratio levels of 0.15, 0.30 and 0.45 were used for anchors cycled at load ratio level of 0.30. Whereas, re-cycling load ratio levels of 0.30, 0.45 and 0.60 are used for anchors cycled at CLRL of 0.45. The movements of anchor with number of loading cycles in re-cycling phase were compared with the movements registered during the cycling phase with same intensities of loading (Fig. 4). When anchors are cycled at load ratio level of 0.30 and further recycled at load ratio levels of 0.15, 0.30 and 0.45 the anchor movements obtained after 500 cycles of loading in re-cycling phase were 25.7%, 52.3% and 72.8% of anchor movements those were obtained in the initial cyclic stage with CLRL of 0.15, 0.30 and 0.45 respectively. Similarly, the anchor movement obtained in re-cycling stage for load ratio levels of 0.30, 0.45 and 0.60 with pre-loading intensity of 0.45 was 31.3%, 58.8% and 86.6% of the movement registered in cycling stage with comparable loading intensities. This showed that when anchors were re-cycled at a load ratio level less than the pre-cycling load, the movement of anchor in re-cycling phase were very much reduced, but if the re-cycling was done at a higher load ratio level, the effect was not that pronounced, and the anchors behaved as if they were not subjected to any cycling load. The movement during cyclic loading was related to the development and dissipation of pore water pressure. The cyclic pore water pressure increased during the loading phase of a cycle and subsequently dissipates. The dissipation of excess pore water pressure from the soil just above the anchor created a locally consolidated soil mass with comparatively higher shear strength. For low cyclic load amplitudes the anchor movement was arrested by the stiffer soil mass, formed above the anchor during the initial few cycles of loading. At higher cyclic load levels, this phenomenon also occurred but due to substantial movement of the anchor in each cycle of loading, a localized consolidated soil zone could not be formed as in each cycle, the anchor moved upward through undisturbed soil. This resulted in a decrease in anchor movement in the re-cycling phase. However, if the CLRL in re-cycling phase was higher than of the pre-cycling phase, the effect is not pronounced, and the anchors behaved as if they had not been previously subjected to any cycling load.

Effect of Pre-Loading on Post-Cyclic Pullout Behaviour

Typical pullout load displacement behaviour of anchors subjected to a pre-cyclic load ratio level of 0.30 and subsequently re-cycled at load ratio levels of 0.15, 0.30 and 0.45 is shown in Fig. 5. Load-displacement curve for an anchor, which has not been subjected to any cyclic loading, is also plotted in both this figure for comparison. For all these cases of cyclic loading, there was a marginal reduction in peak pullout load, while the initial stiffness of the anchor increased compared to the anchor which was not subjected to any cyclic loading. For a given pre-cyclic load the peak pullout load decreases when the re-cycling load ratio level increases, whereas, the relative cyclic stiffness increases. The relative stiffness is the ratio of initial stiffness of anchors subjected to cyclic loading to that of the anchor without any cyclic loading. The relative post-cyclic stiffness was found to vary from 1.169 to 1.282, and the degradation factor varied from 0.946 to 0.981. The increase of initial stiffness of anchors is related to the local consolidation of the soil above the anchor during cyclic loadings and depends upon the intensity and duration of cyclic loading. Datta et al. (1990) also reported the increase of initial stiffness of anchors in remoulded clay and attributed this to the strain hardening of soil caused by repeated loading.

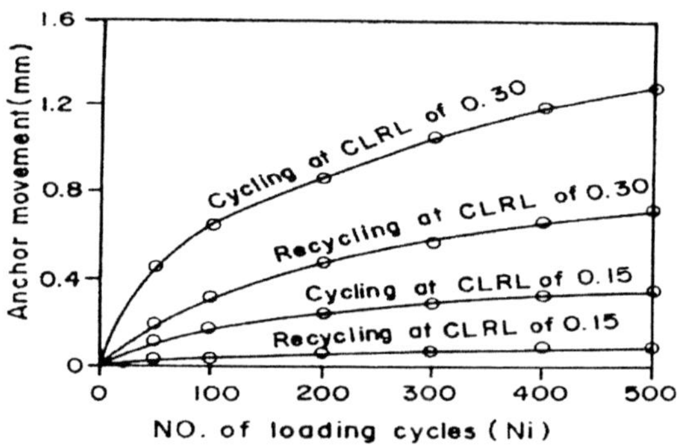

FIG. 4(a). Effect of pre-cycling on anchor movement (with pre-cycling load ratio level of 0.30 and re-cycling load of 0.15 and0.30)

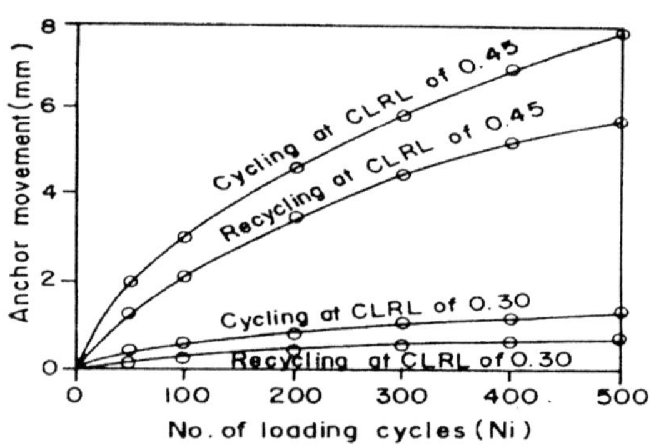

FIG. 4(b). Effect of pre-cycling on anchor movement (with pre-cycling load ratio level of 0.30 and re-cycling load of 0.30 and0.45)

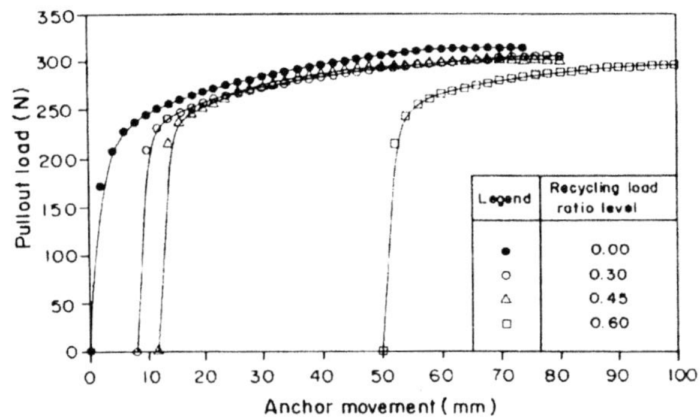

FIG.5. Post-cyclic pullout load-displacement curves for anchors subjected to pre-cycling load ratio level of 0.45

CONCLUSION

Based on the results of the present investigation it was concluded that the magnitude of anchor movement was primarily governed by the amplitude and frequency of cyclic loading and for a given duration of loading, high frequency cycle caused more anchor movement than that caused by low frequency cycles. Pre-loading reduced anchor movement in subsequent loading stages. When anchors were re-cycled at a load ratio level less than the pre-cycling load, the movements in the re-cycling phase were very much reduced, but if the re-cycling was done at a higher load ratio level, the effect was not highly evident, and the anchors behaved as if they were not subjected to any cycling load in the past. Anchors subjected to cyclic loading and then monotonic pullout load showed stiffer load-displacement behaviour than anchors not subjected to any cyclic loading. For the present test conditions (remoulded clay) a marginal loss of anchor capacity up to 12% was observed due to cyclic loading. This was believed to be due to the loss of embedment depth during cyclic loading. However, marine clays which exhibit strength on account of their in-situ structure, may experience degradation of the soil structure during cyclic loading and thus experience subsequent loss of anchor stiffness as well as anchor capacity.

REFERENCES

Bemben, S.M. and Kupferman, M. (1975). "The vertical holding capacity of marine anchor flukes subjected to static and cyclic loading". *Proc. Offshore Technology Conf. OTC* 2185:363-374.

Bouazza, A. and .Finlay, T.W. (1990). "Uplift capacity of plate anchors buried in two layered sand". *Soils and Foundations*. Vol.36(4):57-70.

Datta, M., Gulhati, S.K. and Achari, G. (1990). "Behavior of plate anchors in soft cohesive soils under cyclic loading". *Indian Geotech. J.* Vol.20(3):206-224.

Ponniah, D.A. and Finlay, T.W. (1983). "Cyclic behaviour of plate anchors". *Canadian Geotech. J.* Vol.25:374-381.

Rao, A.V.N. (1998). "Strength behaviour of marine clay subjected to repeated loading". *Ph.D. Thesis. IIT, Madras,* Chennai, India.

Singh, S.P. and Ramaswamy, S.V. (2002). "Response of plate anchors to sustained – cyclic loading". *Indian Geotechnical Journal*. Vol.32(2):161-172.

Singh, S.P. and Ramaswamy, S.V. (2008). "Influence of frequency on the behaviour of plate anchors subjected to cyclic loading". *Journal of Marine Georesources & Geotechnology*, Vol.26(1):36-50.

Displacement Field of an Uplifting Anchor in Sand

Mingliang Liu[1], Jinyuan Liu[2], M.ASCE, P.E., P.Eng., and Hongmei Gao[3]

[1]Ph.D. Candidate, Geo-Institute, Hohai University, Nanjing, China 210098; mliu@hhu.edu.cn
[2]Assistant Professor, Ryerson University, Toronto, Canada M5B 2K3; jinyuan.liu@ryerson.ca
[3]Ph.D. Candidate, Geo-Institute, Hohai University, Nanjing, China 210098; hongmei54@163.com

ABSTRACT: This paper presents a scaled-model test on sand displacement field during anchor uplifting. An in-house data acquisition system is developed to acquire the force, the anchor displacement, and soil images during anchor uplifting. The sand displacement is calculated using the digital image correlation (DIC) method from two images: one is taken at the initial stage and the other is at the peak pullout load moment. The failure plane is identified by locating the maximum shear strains deduced from the displacement field. Two distinctively different shear bands are observed in loose and dense conditions: 1) in loose sand, the shearing bands start at the edges of the anchor plate and converge and form a bell above the anchor; 2) while in dense sand these shear bands extend outward to the ground surface with an inclination angle with the vertical of approximately ½ to 1/3 the friction angle of soil.

INTRODUCTION

Anchors are often used for suspension bridges and television and transmission towers to resist the uplift forces caused by wind. With the development of ocean petroleum industry, anchors are also used to resist the uplift forces caused by ocean waves. Many methods have been used to study the uplift behavior of anchors, including field tests (Giffels *et al.* 1960; Sutherland 1988), model tests (Balla 1961; Murray and Geddes 1987; Ilamparuthi *et al.* 2002) and numerical methods (Rowe and Davis 1982; Sakai and Tanaka 2007). Various prediction methods have also been proposed to calculate the peak uplift resistance based on the anchor behavior observed in the model tests and assumed soil failure surfaces (Meyerhof and Adams 1968; Vesic 1971; Merifield *et al.* 2006). However, the predictions from these methods vary in a wide range. The inaccurate predictions are due to the lack of full understanding of the failure mechanism.

This paper is to study sand displacement field during anchor uplifting and investigate the influence of sand density on the failure mechanism of an anchor. The digital image correlation (DIC) method is used to measure sand movement from images. Two different soil densities are used in these tests: one is loose and the other is dense.

EXPERIMENTAL SET-UP AND PROCEDURES

Experimental Set-up

A test set-up is developed in this study, which consists of a camera, a loading frame, a load cell, a linear variable differential transformer (LVDT), a Plexiglas mould, and a computer. The camera is PL-B741E model camera with a resolution of 1280 x 1024 pixels from *PixeLink*. It is controlled by the PC through an in-house developed driver using Matlab® Simulink commands. A load cell (the loading capacity of 890 N) and an LVDT (a linear strike of ±25 mm) are used to measure the anchor pullout force and displacement. The data acquisition system consists of NI-MIO-16XE-50(NI 6011E) PC card and SCB-68(68-PIN) shielded connector block.

The Plexiglas mould has dimensions of 500 mm (length) x 300 mm (width) x 500 mm (depth). A semi-circular anchor with a diameter of 50.8 mm and a thickness of 5 mm is used in this study. A 1 m long threaded steel rod with a diameter of 5 mm is used to connect the anchor. The rod is connected to the load cell and then to the loading frame through an adaptor. Load is applied vertically through a screw mechanism. The anchor is lifted along a guide slot upward by manually rotating a handle. The guide slot guarantees the anchor to move vertically during testing.

Soil Properties

The sand is classified as SP uniformly graded according to the Unified Soil Classification System. Typical soil properties are shown in Table 1. The angles of friction, measured from direct shear tests according to ASTM D3080, are 30.8° for loose sand and 43.3° for dense sand.

Sample Preparation

The sample density is controlled by the pulverization and tamping methods. As for loose samples, first, a 50 mm thick sand bed is placed at the bottom of the mould. Second, the semi-circular anchor is set on the sand bed and aligned vertically through the guide slot and horizontally against the front window of the Plexiglas box. Third, the upper sand is prepared by the pulverization method to the required embedment depth. Fourth, the rod is then connected to the loading frame for tests.

As for dense samples, the same steps are followed except that in the third step sand is compacted by tamping layer by layer until the final height is reached. Based on the density measurements, these methods ensure a consistency in soil preparation.

The anchor embedment depth is 150 mm for both loose and dense sand, which corresponds to an embedment ratio H/D of 3, where H is the anchor embedment depth and D is the anchor diameter.

Test Procedure

First, the camera is set at approximately 300 mm away with its optical axis perpendicular to the front window of the Plexiglas mould. Since images are sensitive to the changes in ambient light, the light source for illuminating the sand surface is the only light source left on during the test. Second, the data acquisition and image capture drivers are activated in the PC. Third, the camera is set in auto data acquisition mode with one frame per second rate. Fourth, the anchor is lifted up by manually rotating the handle while the images and data are acquired by the computer. The rate of uplifting is controlled at approximately 3 mm/min while the load and displacement acquired through the load cell and the LVDT with corresponding time are saved automatically to the computer. Fifth, the test is terminated until an apparent failure rupture is observed in the sample or no additional loading can be taken by the anchor. Sixth, the images and a log file with corresponding time for each image are transferred to a desired folder into the computer.

Table 1. Physical and geotechnical properties of sand used in the tests

Soil parameters	Loose (CL)	Dense (CD)
Uniformity coefficient, Cu	1.27	1.27
Coefficient of curvature, Cc	1.25	1.25
Soil particle size, d_{50} (mm)	1.15	1.15
Max dry unit weight, γ_{max} (kN/m3)	16.8	16.8
Min dry unit weight, γ_{min} (kN/m3)	14.6	14.6
Sample dry unit weight, γ (kN/m3)	15.07	16.31
Relative density, D_R (%)	23.8	80.1
Peak angle of friction, (degrees)	30.8	43.3

DIGITAL IMAGE PROCESSING METHOD

The digital image correlation (DIC) method is a classic pattern recognition technique, where two images are compared to obtain the relative displacement between them. DIC is also called Particle Image Velocimetry (PIV) in the fluid dynamic field. DIC has been recently used to study soil deformations by many researchers (Gudehus and Nubel 2004; Cheuk *et al.* 2008). The cross-correlation function is often used in DIC to obtain the displacement of two images. The discrete form of standard cross-correlation function is as follows:

$$C(\Delta x, \Delta y) = \frac{1}{MN}\sum_{m=0}^{M-1}\sum_{n=0}^{N-1} f(m,n)g(m+\Delta x, n+\Delta y) \qquad (1)$$

Where M and N are the dimensions of the interrogated images, f and g are the intensities of two images being interrogated. The correlation function given above is sensitive to the average intensity of images. Therefore, the zero normalized cross-correlation function is normally used in the analysis.

In this paper, the PIVview2c software is used to calculate soil displacement field. This software has features to allow users to select the window size, cross-correlation algorithm, peak function, etc. More details can be found in PIVTEC (2006). Unless noted, the features used in this research are the final window size of 32 X 32 pixels, the multiple-correlation algorithm, and the multi-grid interrogation method.

RESULT ANALYSIS

Loading Capacity of an Uplift Anchor

The load vs. displacement curves for both loose and dense conditions are shown in Fig. 1. In dense sand, the anchor uplift resistance shows a rapid increase with increasing displacement at the initial stage up to approximately 1 mm and then the rate of load increase gradually reduces with displacement until the peak pullout resistance is reached. After arriving at the peak, the uplift resistance decreases gradually with increasing displacement until the anchor fails at approximately 12 mm.

In loose sand, there is a significant post-peak stage, where the pullout resistance stays at the peak load with a displacement of more than 5 mm. There is no softening phase noticed after the peak. The peak resistance in loose sand is almost half that of dense sand, while the displacement at the peak load is about 6 mm compared to 3 mm in dense sand.

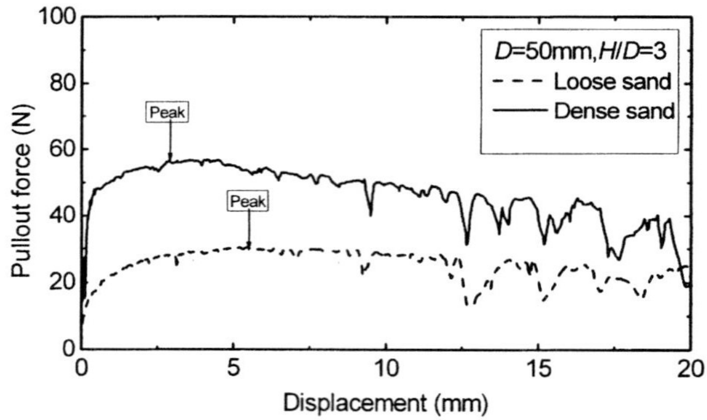

FIG. 1. The load vs. displacement curves for the anchor

Displacement Field around an Uplifting Anchor

The displacement fields for both loose and dense conditions at the peak pullout resistance are shown in Fig. 2. In loose sand, soil deforms mainly within a bell-shaped zone, above the anchor without propagating to the soil surface, as delineated in Fig. 2a. Beyond this zone, there is minimal displacement observed from the images.

In dense sand, a whole block above the anchor is mobilized up to the soil surface compared to the bell-shaped zone in loose sand. There is an appreciable shearing zone on each side of this block, as shown in Fig. 2b.

Strain Field around an Uplifting Anchor

Strains can be calculated using the predicted displacements per Roscoe *et al.* (1963). The failure in soil is normally known as shear failure. The failure plane can be studied by identifying the maximum shear strains. The contours of shearing strains for both loose and dense conditions are shown in Fig. 3. The shearing strain field clearly shows that two shearing bands start at the right and left edges of the anchor plate. In loose sand, these shear bands converge and form a bell above the anchor. In dense sand, however, these shear bands extend outward to the ground surface with an inclination angle of approximately 15° with the vertical.

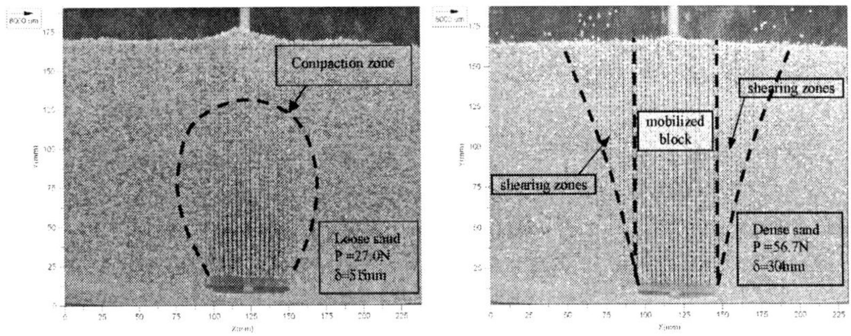

a) Loose sand b) Dense sand

FIG. 2. The displacement fields at the peak pullout resistance

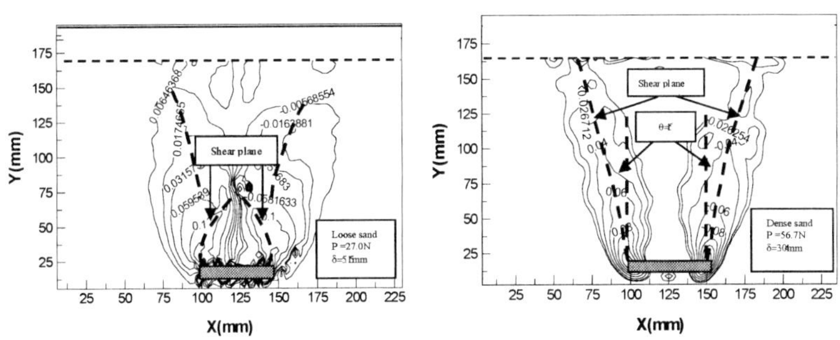

a) Loose sand b) Dense sand

FIG.3. Contours of the shear strain fields

CONCLUSIONS

The displacement fields around a scaled semi-circular anchor are obtained from two images captured during anchor lifting. The soil density has significant influence on the failure mode of the anchor in sand. Soil failure planes are studied by locating the maximum shear strains deduced from the displacement field. Two distinctively failure planes are observed in the tests. In loose sand, the shearing bands start at the edges of the anchor plate and converge and form a bell above the anchor. In dense sand, however, these shear bands extend outward to the ground surface with an inclination angle with the vertical of approximately $1/2\phi$ to $1/3\phi$, where ϕ is the friction angle of soil. Based on this study, it would overestimate the anchor pullout capacity in loose sand based on a failure plane extending to the ground surface observed only in dense sand.

No prototype was attempted to be simulated in this study, since there are so many factors influencing the anchor pullout capacity, including anchor geometry, embedment depth, soil particle size, moisture content, cohesion, overburden, etc. There is also limitation that the soil nonlinear behavior cannot be properly simulated in scaled model tests due to a lower stress level than the prototype. More detailed analyses considering similarity law and overburden will be investigated in the future.

ACKNOWLEDGMENTS

The authors want to thank China Scholarship Council for the scholarships awarded to the first and third authors during their one-year study at Ryerson University. The authors also acknowledge the Seed Fund awarded to the second author by Ryerson University.

REFERENCES

Balla, A. (1961). "The resistance to breaking out of mushroom foundations for pylons." *In* Proceedings of the 5th International Conference on Soil Mechanics and Foundation Engineering, Paris, France, 569-576.

Cheuk, C. Y., White, D. J., and Bolton, M. D. (2008). "Uplift mechanisms of pipes buried in sand." *Journal of Geotechnical and Geoenvironmental Engineering*, 134(2), 154-163.

Giffels, W. C., Graham, R. E., and Mook, J. F. (1960). "Concrete cylinder anchors." *Electrical World*, 154, 46-49.

Gudehus, G., and Nubel, K. (2004). "Evolution of shear bands in sand." *Geotechnique*, 54(3), 187-201.

Ilamparuthi, K., Dickin, E. A., and Muthukrisnaiah, K. (2002). "Experimental investigation of the uplift behaviour of circular plate anchors embedded in sand." *Canadian Geotechnical Journal*, 39(3), 648-664.

Merifield, R. S., Lyamin, A. V., and Sloan, S. W. (2006). "Three-dimensional lower-bound solutions for the stability of plate anchors in sand." *Geotechnique*, 56(2), 123-32.

Meyerhof, G. G., and Adams, J. I. (1968). "The ultimate uplift capacity of foundations." *Canadian Geotechnical Journal*, 5, 225-244.

Murray, E. J., and Geddes, J. D. (1987). "Uplift of anchor plates in sand." *Journal of Geotechnical Engineering*, 113(3), 202-215.

Roscoe, K. H., Arthur, J. R. F., and James, R. G. (1963). "Determination of strains in soils by X-ray method." *Civil Engineering (London)*, 58(685), 1009-1012.

Rowe, R. K., and Davis, E. H. (1982). "the behaviour of anchor plates in sand." *Geotechnique*, 32(1), 25-41.

Sakai, T., and Tanaka, T. (2007). "Experimental and numerical study of uplift behavior of shallow circular anchor in two-layered sand." *Journal of Geotechnical and Geoenvironmental Engineering*, 133(4), 469-477.

Sutherland, H. B. (1988). "Uplift resistance of soils." *Geotechnique*, 38(4), 493-516.

Vesic, A. S. (1971). "Breakout resistance of objects embedded in ocean bottom." *American Society of Civil Engineers Proceedings, Journal of the Soil Mechanics and Foundations Division*, 97(9), 1183-1203.

Uplift Capacity of Pile Anchors in Cohesionless Soil

V. B. Deshmukh[1], D. M. Dewaikar[2] and Deepankar Choudhury[3], M.ASCE

[1]Research Scholar, Department of Civil Engineering, Indian Institute of Technology Bombay, Powai, Mumbai – 400076, INDIA. Email: z6404808@iitb.ac.in
[2]Professor, Department of Civil Engineering, Indian Institute of Technology Bombay, Powai, Mumbai – 400076, INDIA. Email: dmde@civil.iitb.ac.in
[3]Associate Professor, Department of Civil Engineering, Indian Institute of Technology Bombay, Powai, Mumbai – 400076, INDIA. Email: dc@civil.iitb.ac.in

ABSTRACT: An analysis of uplift capacity of pile anchors in cohesionless soil is proposed in which Kötter's equation is employed to obtain the distribution of soil reaction on the axisymmetric failure surface, assumed to be the frustum of a cone with a characteristic angle of inclination with the pile-soil interface. A closed form solution for the uplift capacity of pile anchor is obtained with no requirement of any charts or tables. The results are compared with experimental data for a total 28 cases, ranging from loose to dense cohesionless soil, with maximum embedment ratio of 40, vis-à-vis available theoretical solutions. The proposed method leads to the predictions that are very close to the experimental results in 75% cases and this demonstrates the successful application of Kötter's equation for the estimation of the uplift capacity of pile anchors.

1.0 INTRODUCTION

Some theoretical and experimental investigations have been reported over the last few decades on the behavior of pile foundations under uplift forces. On the theoretical side, relatively less information (Meyerhof, 1973; Das, 1983; Chattopadhyay and Pise, 1986 and Shanker et al. 2007) is available for the estimation of uplift capacity of pile anchors. In general, these methods need the use of graphical charts for choosing the uplift coefficient, K_u which depends on the soil friction angle and pile slenderness ratio. Results of several scale model laboratory tests (Awad and Ayoub, 1976; Chaudhari and Symons, 1983; Das and Seeley, 1975 and Das, 1983) on shallow and deeply embedded piles are also available. Some field tests results of piles embedded in sand under uplift forces have also been reported (Downs and Chieurzzi, 1966; Ireland, 1957 and Vesic, 1970).

Kötter's (1903) equation has been successfully used by several researchers in the estimation of uplift capacity of circular plate anchors. Balla (1961) assumed the failure surface to be the part of a circle and used Kötter's (1903) equation to describe the distribution of shearing stresses on the failure surface. However, normal stresses on the failure surfaces were ignored and only the components of shear stresses resolved in vertical direction were considered. Matsuo (1967) used a curved failure surface (log spiral with a tangent) to develop an approximate theoretical expression for the uplift capacity of circular shallow laid anchor plates. Ghali and Hanna (1994) developed a theoretical method using log spiral failure surface to predict the uplift capacity of a single plate anchor embedded up to shallow, transition and deep depths.

Meyerhof (1973) estimated the net pull-out resistance of a pile anchor assuming that under axial pull, failure took place at the pile-soil interface. In case of driven piles for a given value of ϕ, the uplift coefficient K_u increases with increase in slenderness ratio H/d, (where H is the pile depth and d is the diameter) of the pile up-to a maximum value, beyond which, it remains constant. The analysis is valid for slender ratios up-to 15. Das (1983) assumed the failure surface along the periphery of the pile which is similar to that of Meyerhof's (1973) method. Based on the model laboratory test results, he reported that, unit skin friction at the soil–pile interface increased linearly with depth up-to a critical embedment ratio and remained constant beyond it. Chattopadhyay and Pise (1986) assumed a curved failure surface passing through the surrounding soil mass. The uplift capacity is presented in terms of net uplift capacity factor, A_1, which depends on the H/d ratio, ϕ, δ and α which is the inclination of failure plane with the horizontal. Shanker et al. (2007) used the method similar to that of Chattopadhyay and Pise (1986) except that, the failure surface was assumed to be a truncated cone with edge passing through the pile tip. For estimating uplift capacity, an axisymmetric solid body of revolution of soil along the pile was assumed. Based on limit equilibrium condition, the gross and net uplift capacities were proposed.

2.0 PROPOSED METHOD

In the present investigations, a semi-analytical method is developed using Kötter's (1903) equation for the estimation of net uplift capacity of a pile anchor in cohesionless soil. The failure surface is assumed in the form of the frustum of a cone that originates from the pile tip at an angle α with respect to horizontal and extends to the ground surface as shown in Fig.1. The net uplift capacity is evaluated by considering vertical equilibrium of soil mass in the frustum of the cone. The analysis is performed with maximum embedment ratio H/d = 40 and predictions obtained by the proposed method are compared with available experimental results. Kötter's (1903) equation which is valid for plane strain condition was successfully used for the analysis of a retaining wall by Dewaikar and Halkude (2002) and evaluation of bearing capacity factor, N_γ by Dewaikar and Mohapatro (2003). The equation when integrated over a plane or curved failure surface, gives the magnitude of resultant soil reaction, p at a point. This reaction is then integrated over the failure surface that is generated under axisymmetric conditions. The angle, α depends upon the factors like δ and ϕ. In case of strip, rectangular and circular anchors, Meyerhof and Adams (1968) suggested that, α could vary between (90-ϕ/3) to (90-2ϕ/3). Based on this observation and with some trials, the following expression for α is adopted in the present analysis.

$$\alpha = 90 - \left(\frac{\delta}{\phi}\right)^6 \tan^{-1}\left(\frac{\phi}{2}\right) \tag{1}$$

For a perfectly smooth pile (δ = 0), the failure surface coincides with the soil-pile interface (α = 90^0) and for a perfectly rough pile (δ = ϕ) the angle becomes, $\alpha = 90^0 - \tan^{-1}(\phi/2)$.

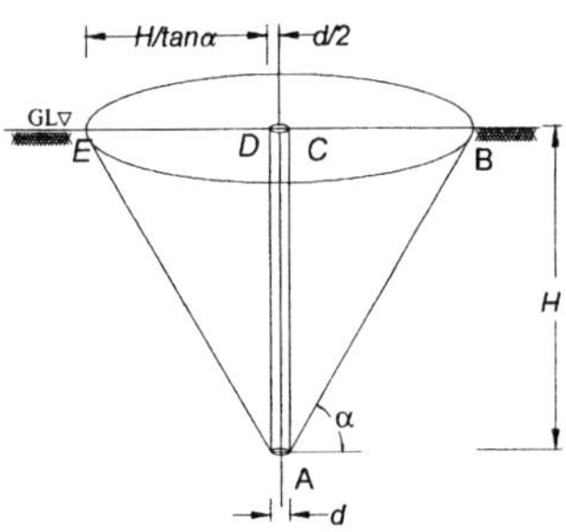

FIG. 1. Geometry of a pile anchor with axisymmetric failure surface

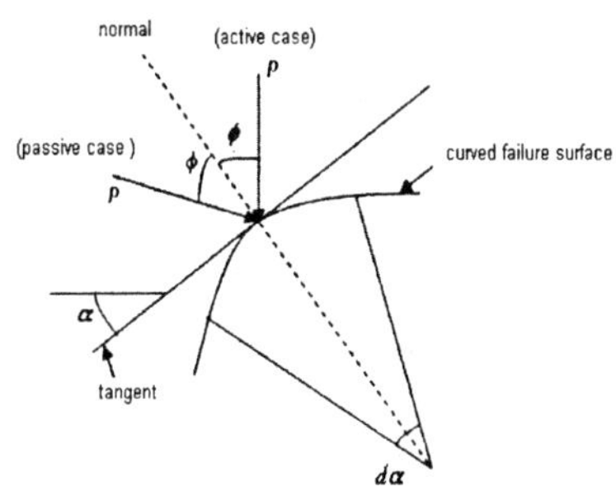

FIG. 2. Kötter's (1903) equation for a curved failure surface

Kötter's (1903) equation basically refers to the distribution of reactive pressure on a failure surface (Fig. 2). In cohesionless soil medium in passive state of equilibrium, it has the following form.

$$\frac{dp}{ds} + 2p\tan\phi\,\frac{d\alpha}{ds} = \gamma\sin(\alpha + \phi) \tag{2}$$

Where, dp = differential reaction pressure on the failure surface, ds = differential length of failure surface, ϕ = angle of soil internal friction, $d\alpha$ = differential angle and α = angle of failure plane formed by inclination of tangent at the point of interest with the horizontal.

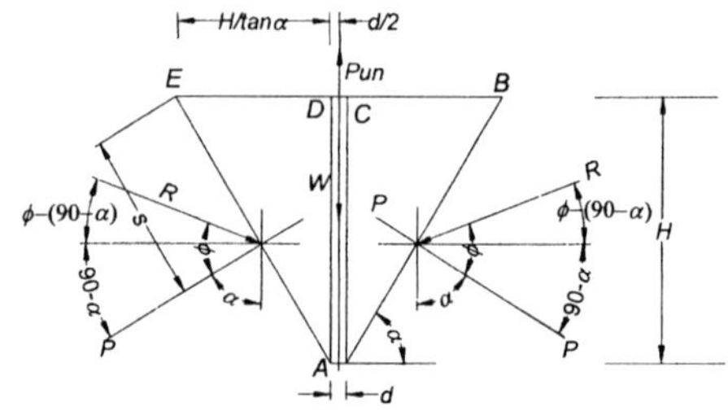

FIG. 3. Free body diagram of the failure wedge under plane strain condition

In Fig. 3, AB is shown as a part of failure wedge, ABC of a pile anchor in plane strain condition. The various forces are the net uplift force P_{un}, weight of ABC failure wedge W and soil reaction R on the failure plane AB.
For a plane failure surface, $d\alpha/ds$ = 0 and integrating with the substitution of boundary condition that, pressure p is zero at point B, which corresponds to s = 0, Eq.2 simplifies to

$$p = \gamma \sin(\alpha + \phi) s \tag{3}$$

In the above equation, reaction p has the unit of kN/m^2.

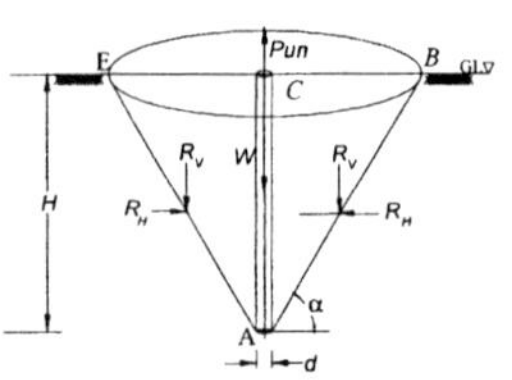

FIG. 4a. Free body diagram of the pile anchor with axisymmetric failure surface

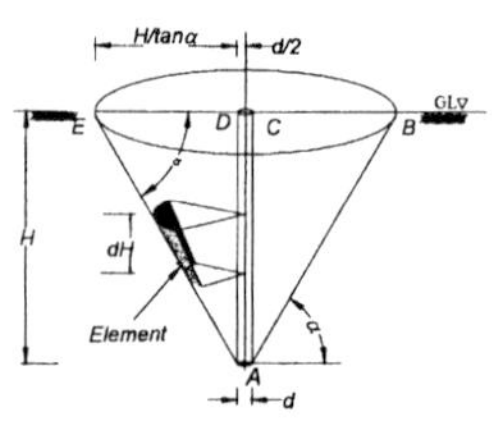

FIG. 4b. Element of the Axisymmetric failure surface

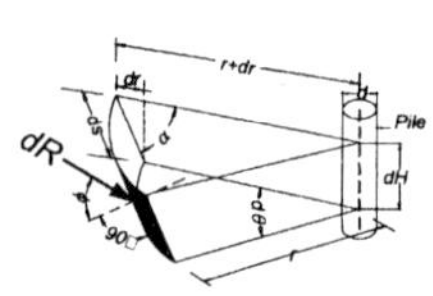

FIG. 4c. Enlarged element with forces

During uplift of a pile embedded in cohesionless soil, a frustum of a conical failure surface develops in the soil as shown in Fig. 4a. The uplift force is resisted by the vertical component R_v of resultant soil reaction and weight, W of soil. If p is the corresponding differential reaction pressure on the element, then dR which is the elemental soil reaction on the elemental area, $r.d\theta.ds$ is expressed as $dR = p.dA$ where, $dA = r\,d\theta\,ds$. From Fig. 4c, $ds = dr/\cos\alpha$, so that,

$$dR = rd\theta\,\frac{dr}{\cos\alpha} \tag{4}$$

Substitutions for p and dA give

$$dR = \gamma\sin(\alpha+\phi)s\,\frac{rd\theta\,dr}{\cos\alpha} \tag{5}$$

From Fig. 3, the following value of s is substituted in Eq.5.

$$s = \left[\left(\frac{H}{\tan\alpha}+\frac{d}{2}\right)-\left(r+\frac{dr}{2}\right)\right]\frac{1}{\cos\alpha} \tag{6}$$

The differential reaction dR is then obtained as

$$dR = \frac{\gamma\sin(\alpha+\beta)}{\cos^2\alpha}\left[\left(\frac{H}{\tan\alpha}+\frac{d}{2}\right)r - r^2\right]dr\,d\theta \tag{7}$$

The resultant soil reaction, R on half part of the axisymmetric solid body of revolution is evaluated as

$$R = \int_{\theta=0}^{\theta=\pi}\int_{\frac{d}{2}}^{\left(\frac{d}{2}+\frac{H}{\tan\alpha}\right)} dR \tag{8}$$

Resolving, R in the vertical direction, R_v is obtained as $-R\cos(\alpha+\varphi)$. Then R_v is expressed as

$$R_v = -\int_{\theta=0}^{\theta=\pi}\int_{\frac{d}{2}}^{\left(\frac{d}{2}+\frac{H}{\tan\alpha}\right)} dR\cos(\alpha+\phi) \tag{9}$$

Substituting for dR and after performing integration the following result is obtained.

$$R_v = -\frac{\gamma\pi\sin(\alpha+\phi)\cos(\alpha+\phi)}{6\cos^2\alpha}\left[\left(\frac{H}{\tan\alpha}+\frac{d}{2}\right)^3+\frac{d^2}{4}\left(d-3\left(\frac{H}{\tan\alpha}+\frac{d}{2}\right)\right)\right] \tag{10}$$

The net weight of axisymmetric solid body of revolution is divided into three parts; viz. W_1 as the weight of inverted circular cone, W_2 as weight of small inverted cone below pile tip and weight, W_3 of the pile. Then, the net weight, W of axisymmetric solid body revolution is evaluated as

$$W = \gamma\pi\left[\frac{1}{3}\tan\alpha\left(\frac{d}{2}+\frac{H}{\tan\alpha}\right)^3 - \frac{d^3\tan\alpha}{24} - \frac{d^2 H}{4}\right] \tag{11}$$

Referring Fig.4a and taking vertical equilibrium of forces, the net uplift capacity, P_{un} is obtained as $P_{un} = 2R_v + W$. Substituting for R_v and W from Eqs. 10 and 11 respectively and with some simplifications, the following expression is obtained.

$$P_{un} = \frac{\gamma\pi\tan\alpha}{12}\left\{4C_4^3 - d^2 C_5 - \frac{4\sin 2(\alpha+\phi)}{\sin 2\alpha}\left[C_4^3 + \frac{d^2}{4}(d-3C_4)\right]\right\} \tag{12}$$

Where,

$$C_4 = \left(\frac{d}{2}+\frac{H}{\tan\alpha}\right),\ C_5 = \left(\frac{d}{2}+\frac{3H}{\tan\alpha}\right) \text{ and } \alpha = 90 - \left(\delta/\phi\right)^6 \tan^{-1}\left(\phi/2\right)$$

The above simple expression represents the net uplift capacity of a pile anchor in cohesionless soil. It is readily amenable to hand calculations with no resort to charts or graphs.

3.0 COMPARISON WITH EXPERIMENTAL RESULTS

The comparisons of predictions by theoretical methods (proposed method; Meyerhof, 1973; Das, 1983; Chattopadhyay and Pise, 1986; Shanker et al., 2007 and conventional method) with experimental results of Das (1983, series, ii) are presented in Fig. 5. It is seen that, up to H/d = 16, predictions of the proposed method are in a good agreement with experimental results and beyond this embedment ratio, they deviate by more than 50%. With the exception of conventional method, other methods also show good agreements with experimental results.

In Fig. 6 the results of proposed method, Meyerhof's (1973) method, methods proposed by Das (1983), Chattopadhyay and Pise (1986), Shanker et al. (2007) and the conventional method are compared with the experimental results reported by Chattopadhyay and Pise (1986). It is observed that, predictions made by the proposed method underestimate the uplift capacity to up to H/d = 16 with a reasonably good agreement for H/d of 16 to 24 and a closer agreement for $H/d > 24$. The predictions by Meyerhof (1973) and Shanker et al. (2007) show comparatively better agreement with experimental results in the range of above embedment ratios, whereas methods proposed by Das (1983) and Chattopadhyay and Pise (1986) underestimate the experimental results for $H/d > 20$.

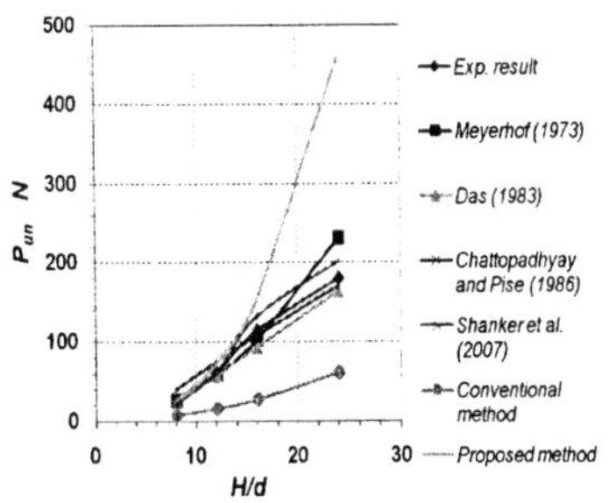

FIG. 5. Comparison of theoretical methods with exp. results of Das (1983) ($\phi = 34^0$, δ=30.5^0)

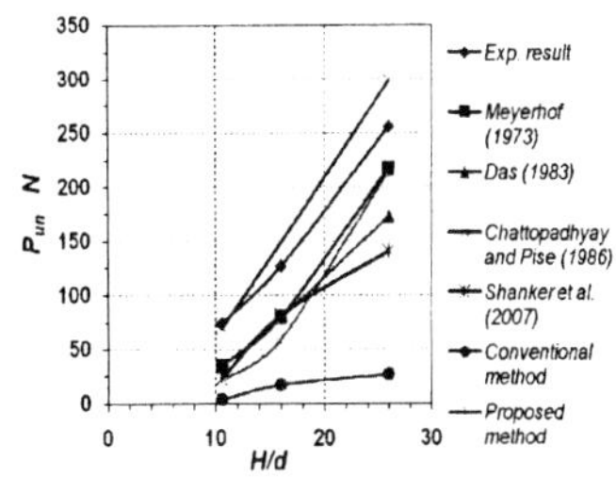

FIG. 6. Comparison of theoretical methods with exp. results ($\phi = 41^0$, δ=34^0) of Chattopadhyay and Pise (1986)

The experimental results of Dash and Pise (2003) are compared with predictions of theoretical methods as shown in Fig. 7. It is observed that the results of proposed method are nearer to the experimental results ($\phi = 30^0$, δ=21^0) up to $H/d < 16$ and beyond which they deviate by 50%. The predictions by some of the other theoretical methods (Meyerhof, 1973; Chattopadhyay and Pise, 1986 and Das, 1983) underestimate the experimental results, whereas the method proposed by Shanker (2007) marginally overestimates the experimental results.

From Fig. 8 it is seen that, some of the theoretical methods (proposed method; Shanker et al., 2007 and Meyerhof, 1973) show a good agreement with experiment results of Shanker et al. (2007). But the theoretical methods proposed by Das (1983) and Chattopadhyay and Pise (1986) underestimate the experiment results beyond $H/d > 30$.

The comparison of net uplift capacity predictions by the theoretical methods is also made with experimental results (plotted as the ideal line) as shown in Figs. 9(a) and (b). In Fig. 9a, the results of theoretical predictions (proposed method, Das, 1983; Meyerhof, 1973 and the conventional method) are shown in comparison with the ideal line that is plotted using the experimental results of Das (1983, series (ii) and (iii)), Chattopadhyay and Pise (1986), Dash and Pise (2003) and Shanker et al. (2007). Similarly, in Fig. 9b a comparison of the results obtained by the other theoretical methods (proposed method; Chattopadhyay and Pise 1986 and Shanker et al., 2007) with the ideal experimental line is shown.

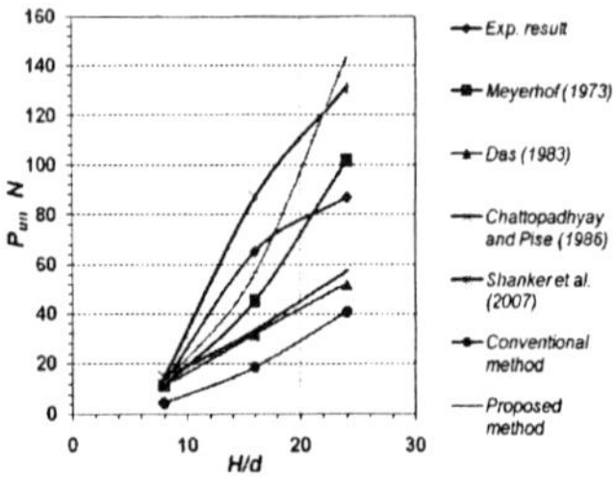

FIG. 7. Comparison of theoretical methods with experimental results ($\phi = 30^0$, $\delta = 21^0$) of Dash and Pise (2003)

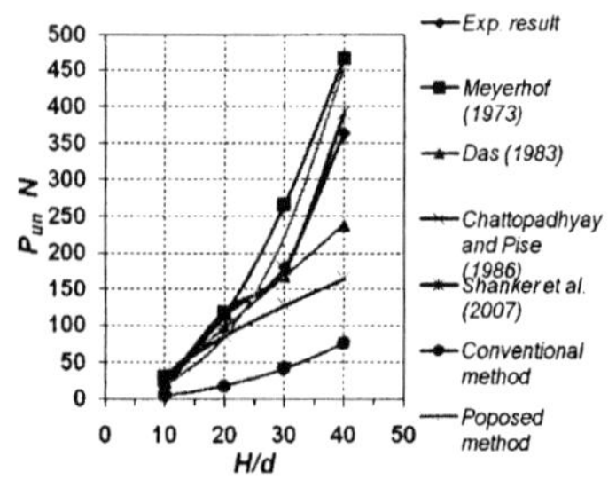

FIG. 8. Comparison of theoretical methods with experimental results ($\phi = 38^0$, $\delta = 26^0$) of Shanker et al. (2007)

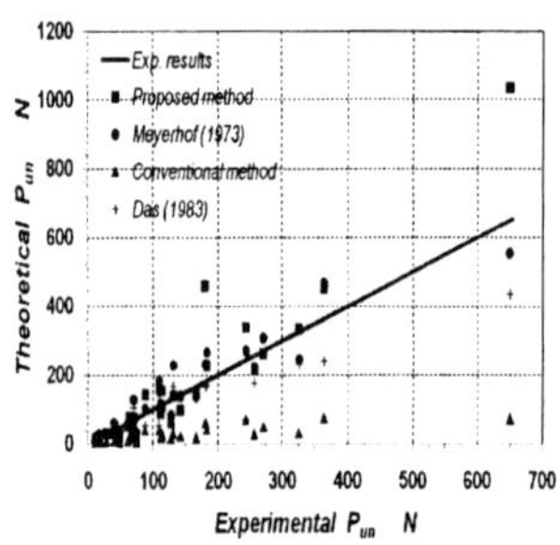

Fig. 9a. Comparison of P_{un} by theoretical methods with experimental results

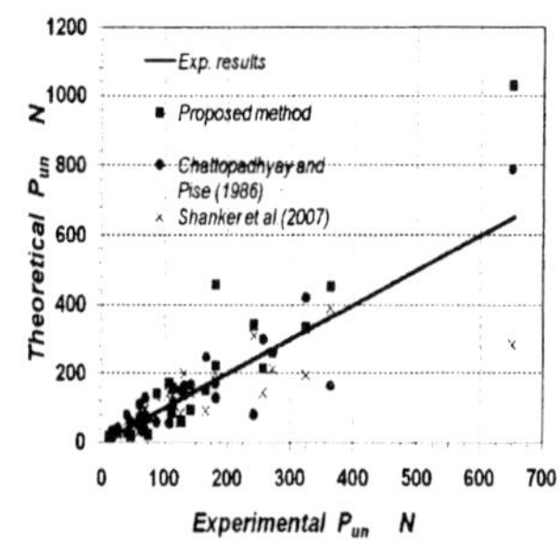

Fig. 9b. Comparison of P_{un} by theoretical methods with experimental results

The comparison of theoretical predictions of Das (1983) shows underestimation in most of the cases (Fig. 9a). This method shows a good performance in 78% of the cases. It requires the use of Meyerhof's (1973) charts for choosing an appropriate value of the uplift coefficient. The comparison of theoretical method of Chattopadhyay and Pise (1986) shows some results close to the experimental results. For H/d >30, this method makes underestimated predictions. It gives a good performance in 67% of cases; but one needs to use several charts to select the value of net uplift factor. Meyerhof's (1973) method generally shows a good agreement in 85% cases. In some cases at higher values of ϕ, the predictions are overestimated. This method shows better predictions in most of the cases; but requires the use of a chart for choosing the uplift coefficient.

The method proposed by Shanker et al. (2007) shows a good agreement in 78% cases when compared with experimental results, for comparatively smooth piles (Fig. 9a). For rough piles the predictions are overestimated. This method involves an assumption on the co-efficient of earth pressure at rest. The predictions of conventional method overestimate the pile capacity on the unsafe side. The proposed method of analysis also considers a frustum of conical failure surface. It leads to the predictions that are closer to experimental values in 75% cases. It shows a good performance for $H/d < 20$. Although this method involves an approximation with respect to the application of Kötter's (1903) equation, it is improved with a proper selection of angle, α as per Eq. (1). The integration involved is fairly simple to obtain a closed form solution for the net uplift capacity (Eq. 12), which is amenable to hand-calculations without resort to any graphs or tables. Kötter's (1903) equation lends itself as a powerful tool in this analysis.

4.0 CONCLUSIONS

The proposed semi-analytical method is simple and provides a closed form solution for the net uplift capacity of a pile anchor. It shows a better performance for $H/d < 20$ in medium to dense cohesion-less soils with a marginally inferior one for $H/d > 20$ in loose cohesion-less soils. It is amenable to hand-calculations with no resort to any graphs or charts. The application of Kötter's (1903) equation is successfully demonstrated for the axisymmetric condition. No assumptions are required regarding the coefficient of lateral earth pressure in the analysis and the predictions show a close agreement with the experimental results.

REFERENCES

Awad, A. and Ayoub, A. (1976). "Ultimate uplift capacity of vertical inclined piles in cohesionless soil." *Proc. of the 5th Int. Conf. on Soil Mechanics and Foundation Engineering*, Budapest, Vol.1:221–227.

Balla, A. (1961). "The resistance to breaking out of mushroom foundations for Pylons." *Proc of the 5th Int. Conf. on Soil Mechanics and Foundation Engineering*, Paris, France, 1, 569–576.

Chattopadhyay, B.C. (1994). "Uplift capacity of pile groups." *Proc. of XII ICSMFE,* New Delhi, India, 539–542.

Chattopadhyay, B.C. and Pise, P.J. (1986). "Uplift capacity of pile in sand." *Journal of Geotechnical Engineering*, ASCE, Vol. 112(9), 888–904.

Chaudhuri, K.P.R. and Symons, M.V. (1983). "Uplift resistance of model single piles." *Proc. of the Int. Conf. on Geotechnical Practice in Offshore Engineering*, Austin. ASCE, Tex: 335–355.

Das, B.M. (1983). "A procedure for estimation of uplift capacity of rough piles." *Soils and Foundations*, Vol. 23(3): 122-126.

Das, B.M. and Seeley, G.R. (1975). "Uplift Capacity of buried model piles in sand." *Journal of Geotechnical Engineering*, ASCE, Vol. 101(7): 711-715.

Dash B.K. and Pise, P.J. (2003). "Effect of compressive load on uplift capacity of model piles." *Journal of Geotechnical and Geoenvironmental Engineering*, ASCE, Vol. 129(11): 987–992.

Dewaikar, D.M. and Halkude, S.A. (2002). "Seismic passive/active thrust on retaining wall-point of application." *Soils and Foundations,* Vol. 42(1): 9–15.

Dewaikar, D.M. and Mohapatro, B.G. (2003). "Computation of Bearing Capacity Factor Nγ—Terzaghi's Mechanism." *International Journal of Geomechanics*, ASCE, Vol. 3(1): 123-128.

Downs, D.I. and Chieurzzi, R. (1966). "Transmission tower foundation." *Journal of Power Division,* ASCE, 92(2): 1–114.

Ghaly, A. and Hanna, A (1994). "Ultimate pullout resistance of single vertical anchors." *Canadian Geotechnical Journal*, 31: 661–672.

Ireland, H.O. (1957). "Pulling tests on piles in sand." *Proc. of the 5th Int. Conf. on Soil Mechanics and Foundation Engineering.* London, Vol. 2: pp. 43-45.

Kötter, F. (1903). "Die Bestimmung des Drucks an gekrümmten Gleitflächen, eine Aufgabe aus der Lehre vom Erddruck. Sitzungsberichte der Akademie der Wissenschaften, Berlin, 229–233.

Matsuo, M (1967). "Study of uplift resistance of footings: I." *Soils and Foundations*, 7(4): 1–37.

Meyerhof, G.G. and Adams J.I. (1968). "The ultimate uplift capacity of foundations." *Canadian Geotechnical Journal,* Vol.5 (4): 225-244.

Meyerhof. G.G. (1973). "Uplift resistance of inclined anchor and piles." *Proc. of the 8th Int. Conf. on Soil Mechanics and foundation Engineering,* Moscow, Vol. 3: 167-173.

Shanker, K., Basudhar, P.K. and Patra, N.R. (2007). "Uplift capacity of single piles: predictions and performance." *Geotechnical and Geological Engineering,* Vol.25: 151–161.

Vesic, A.S. (1970). "Tests on instrumented piles, Ogeechee River site." *Journal of Soil Mechanics and Foundation, Division*, ASCE, Vol. 96(2): 561-584.

Study on Bearing Behaviors of Single Axially Loaded SDCM Pile

Xin Zhao[1], Mai Wu[2], Shunwei Chen[3] and Dandan Kong[4]

[1] Associate Professor, School of Civil Engineering, Hebei University of Technology, Tianjin 300132, P.R.China; zhaoxintj@126.com
[2] Associate Professor, School of Civil Engineering, Hebei University of Technology, Tianjin 300132, P.R.China; wumaitj@126.com
[3] Lecture, School of Civil Engineering, Hebei University of Technology, Tianjin 300132, P.R.China; chshwei@126.com
[4] Associate Professor, School of Civil Engineering, Hebei University of Technology, Tianjin 300132, P.R.China; ddkong2008@126.com

ABSTRACT: Stiffened Deep Cement Mixed (SDCM) pile technique has provided a new approach for treating soft soil. The main focus of this paper is the bearing behavior and capacity prediction of single SDCM pile. Static load tests data of 67 piles collected from six different districts in China are analyzed in this paper. The bearing capacity, load-settlement curve and failure modes of SDCM piles are studied. According to the bearing mechanism of SDCM pile, a new approach to predict the capacity, *Sectioning and Summarizing* method, is proposed. To calculate the capacity, one SDCM pile should be divided into two segments: the upper segment with core pile and the lower segment without the core pile. The total shaft resistance of SDCM pile is the summation of the two segments. Finally the dominate factors influencing the pile resistance are analyzed. The study results show that the capacity of SDCM pile is mainly affected by the parameters of the core pile. The resistance provided by the pile shaft including core pile was 1.0~2.9 times that of bored and cast-in-place concrete pile. The proposed *Sectioning and Summarizing* method can be used to estimate the capacity of the SDCM pile and get more logical result than the currently used methods.

INTRODUCTION

SDCM pile is a new kind of composite pile as shown in Fig.1. Immediately after construction of deep cement mixed (DCM) pile, a concrete or reinforced concrete core is driven or placed into the centre of the DCM pile. Thus, two-layer structure of SDCM pile is formed. Two parts of the composite pile can work together to bear and transfer the vertical load effectively. The cement-soil pile forms the outer layer to surround and brace the stiffer core while the stiffer core is of higher stiffness and it serves to resist compressive stress along the pile shaft. The outer layer of cement-soil has a large external surface that can effectively transfer the applied load to the surrounding soil. When the load of the superstructure is transferred to the top of the composite pile, the stiff core takes most of the load and gradually transmits it to the cement-soil via the interface between the stiff core and cement-soil. Furthermore, the cement-soil transfers the load to the surrounding soil through the pile-soil interface. Thus, the SDCM pile combines the advantages of concrete pile and DCM pile.

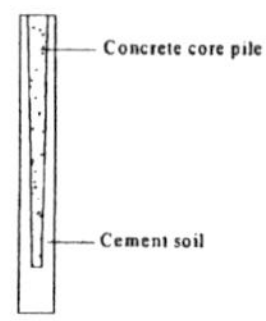

Fig.1. Schematic of SDCM pile

Recently, researchers in different areas in China studied bearing behavior of SDCM pile by means of in-situ load test. Some common conclusions were drawn while some different opinions still exist.

By performing field load tests in Tianjin, Ling et al. (2001) found that the capacity of SDCM pile is 1.36~1.54 times that of bored and cast-in-place concrete pile and the load-settlement curves of SDCM pile demonstrated plunge failure mode; Chen et al. (2003) reported similar test results in Kunming. The results presented by Liu et al. (2001) indicated that SDCM piles have similar bearing behavior to bored concrete piles while the unit shaft resistance of SDCM piles was 50%~70% higher than that of concrete piles. Unlike Ling, Liu found that the load-settlement curves of SDCM piles exhibited gradual rise character. Instrumented pile load test conducted by Dong et al. (2001) indicated that over 90 percent of the total load was carried by the pile shaft. According to the test result, Dong also recommended that capacity of SDCM pile can be predicted by adopting the same parameters as bored concrete pile.

From the above investigation results, some conclusions can be drawn as follows:

(1) Addition of the high stiffness core pile increases the deformation modulus and the shaft resistance of the pile.

(2) Bearing capacity of SDCM pile is derived mainly from shaft resistance, particularly if the core pile is designed to be shorter than the DCM pile.

(3) The shaft resistance of SDCM pile is equal or higher than the bored and cast-in-place pile.

However, some problems still exist to be solved. Firstly, for the reason that the load-settlement relation is the reflection of the bearing mechanism and failure mode of the pile, it is essential to further study the load-settlement character of the testing pile. Secondly, a more important problem is how to logically predict the capacity of SDCM pile. The existing design codes for SDCM pile have specified different calculation principles to predict the capacity of single SDCM pile. DB29-102-2004 of Tianjin city recommends that the unit shaft resistance of SDCM pile can adopt 1.3~1.7 times that of the precast concrete piles; while DB13(J)50-2005 of Hebei province suggests that the corresponding parameters can directly adopt those of the bored and cast-in-place concrete pile.

Presently, more projects employ SDCM pile to treat soft soil in China. It is urgent to find a more accurate and reliable approach to predict the bearing capacity of SDCM pile. In this paper, 67 testing pile results collected from 14 different projects located at six cities are analyzed, the bearing mechanism of SDCM pile are discussed. According to the bearing mechanism and test results, a new approach to predict the capacity of a single SDCM pile, *Sectioning and Summarizing method,* is presented. Finally, the leading factors affecting the pile resistance are analyzed.

ANALYSIS OF STATIC LOADING TEST RESULTS

The parameters of 67 test piles are shown in Table 1. All the piles were tested by slow maintained static load method specified by *Chinese Technical code for testing of building foundation piles* (JGJ106-2003). Accordingly, the ultimate resistance of the pile termed Q_u was determined using the following criterions: (1) when plunge failure occurs, the failure load is taken as the ultimate resistance; (2) for the pile with gradual rise load-settlement curve, the load at the settlement of 40mm is taken as Q_u.

Table 1. Parameters of Testing Piles

Project title/pile quantity	Main soil layer	Diameter of DCM (mm)	Length of DCM/core pile (m)	Core pile type	Cross section of core pile		Average value of Q_u (kN)
					Shape	Side length or diameter (mm)	
TD1/4	Clay and silty clay	700	11.5/7.0	Precast	Rectangular	443~100	1125
TD2/6		500	10.0/6.0	Precast	Circular	400~100	736
TD3/6		500	8.5/5.0	Precast	Circular	300~100	650
KM1/6[a]	Clay	500	8.0/3.5	Precast	Rectangular	180~100	580
KM2/1[a]		500	8.5/6.0	Precast	Rectangular	240~100	860
GD1/2	Silty caly	600	14.0/13.5	Precast	Rectangular	270	2070
GD2/1		600	13.5/13.5	Precast	Rectangular	270	2070
GD3/4		600	14.0/13.0	Precast	Rectangular	276~120	1450
YC1/6	Clay	500	8.0/5.0	Precast	Circular	180~100	500
XZ1/1	Silty clay	500	12.2/10.0	Precast	Circular	300~115	1200
TD4/5	Silty caly	600	12.0/9.5	Precast	Rectangular	240~100	960
DL1/3	Muddy clay	600	12.0/11.0	Precast	Rectangular	252~120	780
JD1/3	Silty clay	500	10.5/9.0	Precast	Rectangular	233~100	1073
JY1/1[b]	Silt	600	10.7/9.5	Precast	Rectangular	220~150	800
JY2/1[b]		600	10.5/10	Precast	Rectangular	220~150	720
JY3/1[b]		600	7.8/7.0	Precast	Rectangular	220~150	720
SH1/2[b]	Muddy clay	700	18.0//17.0	Precast	Rectangular	220~150	1320
SH2/2[b]		700	24.0/23.0	Precast	Rectangular	250~150	2070
LM1/4	Silty clay	600	12.8/12.8	Cast-in -place	Rectangular	350	1528
SJ1/2	Loessal silty clay	620	6.0/3.0	Cast-in -place	Circular	350	1575
SJ2/2		620	6.0/4.0		Circular	350	1885
SJ3/2		620	6.0/5.0		Circular	200	1550
SJ4/2		620	6.0/5.0		Circular	300	1765
SJ5/2		620	6.0/5.0		Circular	350	1950
XT1/1[c]	Silt and silty clay	900	12.0/10.0	Cast-in -place	Circular	500	2080
XT2/1[c]		900	12.0/11.0		Circular	500	2080

a: Reported by Chen et al (2003); b: Reported by Dong et al (2001);c: Reported by Liu et al. (2001).

Load-settlement curve obtained from static load test is the main basis to define the capacity of a pile. The curve also reflects the bearing mechanism and failure mode of testing pile. Typical load-settlement curves of the tested piles listed in Table.1 are plotted. The curves of SDCM piles with precast and cast-in-place core pile are plotted in Fig. 2 and Fig. 3 respectively.

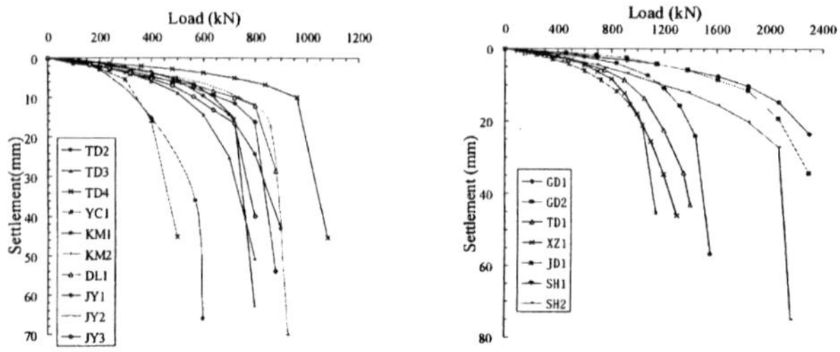

Q_u<1000kN Q_u>1000kN

Fig. 2. Load-settlement curves of SDCM pile (Precast core pile)

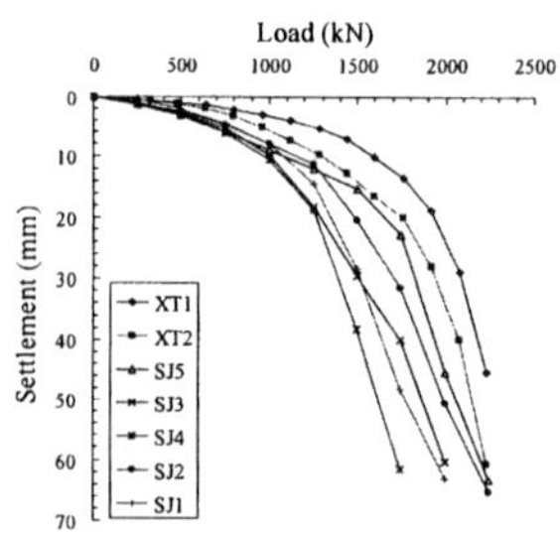

Fig. 3. Load-settlement curves of SDCM pile (Cast-in-place core pile)

According to the shape character of the end segment, the load-settlement curves can be commonly divided into two types: "plunge failure" and "gradual rise". As a whole, most load-settlement curves of SDCM pile exhibit "gradual rise" mode, which implies that the SDCM pile has good interface friction character with surrounding soil.

However, as shown in Fig. 2, several testing piles with precast core pile exhibited plunge failure mode. Aiming to avoid this unexpected failure mode in engineering practice, it is essential to find the main reason resulting in the failure. Take pile SH2 for instance, the core pile was long and slender with slenderness ratio (length over diameter) of about 115. And due to the slenderness of the core pile, the pile shaft may fail to bear prior to the failure of the surrounding soil in the following two modes: (1) the core pile will be broken when the shaft load of core pile exceeding the compression strength of the reinforced concrete; (2) the core pile slipped on the DCM due to the lesser side friction between the core pile and the cemented soil. Correspondingly, large settlement occurred at no or small increase of load. Testing pile SH1 demonstrated a similar failure mode to SH2. In engineering practice, the method to avoid the plunge failure mode is to adopt appropriate slenderness of the core pile.

However, adopting a larger diameter core pile is not adequate to avoid plunge failure. SDCM pile consists of two parts: the core pile and the cemented soil surrounding and bracing the core pile, and proper parameters should be defined to assure the effective transfer of load between the core pile and the cemented soil. Experiment results presented by Wu et al. (2004) proved that the total frictional value (or the cohesive force) between the two parts was mainly affected by two factors: the total side area of the core pile and the compression strength of the cemented soil. So, if the cohesive force is smaller than the applied load on the head of core pile, slippage or plunging failure of core pile will occur. For example, testing pile KM1 employed a very short core pile of 3.5m in length. The cohesive force provided by the cemented soil is inadequate. When the load applied on the core pile exceeded the frictional resistance bearing the core pile, the core pile plunged into the cemented soil.

Another important problem is how the parameters of core pile affect the capacity of SDCM pile. To find the answer, five pairs of testing pile, SJ1 to SJ5, are installed and tested by the author. The parameters of testing piles are listed in Table 1. Ten test piles are divided into two groups. In group A, SJ1, SJ2 and SJ3 have the same diameter and varied length. In group B, SJ3, SJ4 and SJ5 have the same length and varied diameter. The test results were presented in Fig. 4.

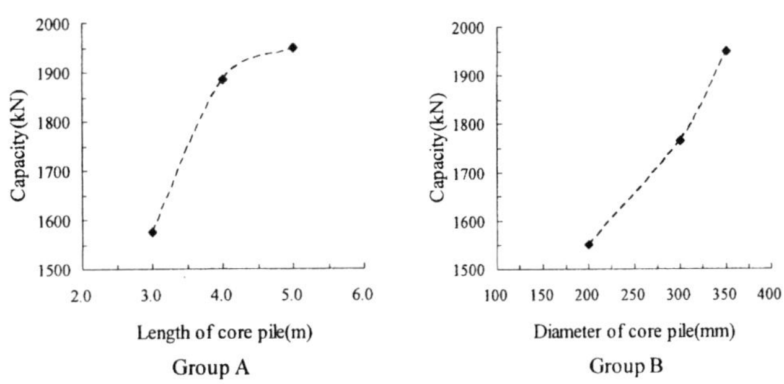

Fig. 4. Effect of core pile parameters on capacity of SDCM pile

As shown in Fig. 4, the parameters of core pile affect the capacity of SDCM pile notably. The capacity increases with the growth of length or diameter of core pile. Obviously, to predict the capacity of SDCM pile without regarding the variety of parameters is illogical and inaccurate. We must care more about the core pile parameters when designing and constructing SDCM piles.

SECTION AND SUMMARIZING METHOD

Currently, the capacity of SDCM pile is predicted as follows: (1) taking the SDCM pile as a pile with uniform material with the diameter and length of DCM pile; (2) calculating the capacity by using the shaft and toe resistance parameters recommended by codes. In other words, for the same soil, the recommended unit shaft resistance is identical regardless of the difference of the core pile. By using this approach, the calculated capacity of pile SJ1 to SJ5 will be identical. Apparently this predicting mode is illogical. Therefore, on the basis of bearing mechanism and field tests result, a new approach termed "*Section and Summarizing*" method is proposed to calculate the capacity of SDCM pile. The bearing capacity can be calculated according to the following steps.

(1) Dividing the whole SDCM pile into two segments as shown in Fig. 5: the upper segment with core pile (L_1 in length) and the lower segment without core pile (L_2 in length).

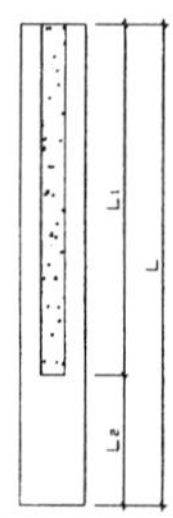

Fig. 5. Dividing of the pile shaft

(2) Calculating the total shaft resistance of the two segments respectively. The total shaft resistance of SDCM pile is the summation of the two parts.
(3) Calculating the total toe resistance of SDCM pile.
(4) Summarizing the total shaft and toe resistance as the capacity of the pile.
Thus, the capacity of SDCM pile, Q_u, can be calculated by the following equation:

$$\begin{aligned} Q_u &= Q_{su} + Q_{pu} \\ &= Q_{su1} + Q_{su2} + Q_{pu} \end{aligned} \quad (1)$$

Where Q_{su} means the total shaft resistance of SDCM pile, $Q_{su} = Q_{su1} + Q_{su2}$, where Q_{su1} means the total resistance of segment L_1, Q_{su2} means the total resistance of segment L_2 and can be calculated by adopting parameters of DCM pile; Q_{pu} means the total toe resistance of SDCM pile and can be predicted by using the common method.

In equation (1), all resistances can be estimated except Q_{su1}, the total resistance of L_1. Theoretical analysis by Dong (2001) has shown that the increase of compression stiffness is the main factor influence the value of Q_{su1}, therefore we now introduce the index of compression stiffness ratio, expressed as η, to analyze the influence of compression stiffness on the shaft resistance, where

$$\eta = \frac{A_p E_p + A_{cs} E_{cs}}{A_{cp} E_{cs}} \tag{2}$$

Where A_{cp} means the section area of DCM pile; A_p means the average area of core pile, E_p means the elastic modulus of core pile; A_{cs} means the section area of cemented soil, where A_{cs}=A_{cp}-A_p; E_{cs} means the elastic modulus of cemented soil. In equation (2), the numerator represents the compression modulus of the composite pile section while denominator represents that of the DCM pile with no core pile.

As mentioned above, SDCM pile exhibits similar bearing behavior to cast-in-place concrete pile, so another index, β_s, is also introduced. This index means the ratio of Q_{su1} to Q_{sug}, thus the total pile shaft resistance Q_{su1} can be obtained by the following equation.

$$Q_{su1} = \beta_s \cdot Q_{sug} \tag{3}$$

Where Q_{sug} means the total shaft resistance of bored concrete pile with the same diameter and length. The value of Q_{sug} can be calculated according to site condition and relevant codes. To calculate Q_{su1} by using equation (3), we must determine the value of index β_s. For the reason that compression stiffness is the key factor influencing the exertion of shaft resistance, we now try to find the relationship between the two indexes β_s and η. By calculation and analysis of the testing result of piles presented in Table.1, the relation of η and β_s is presented in Fig. 6.

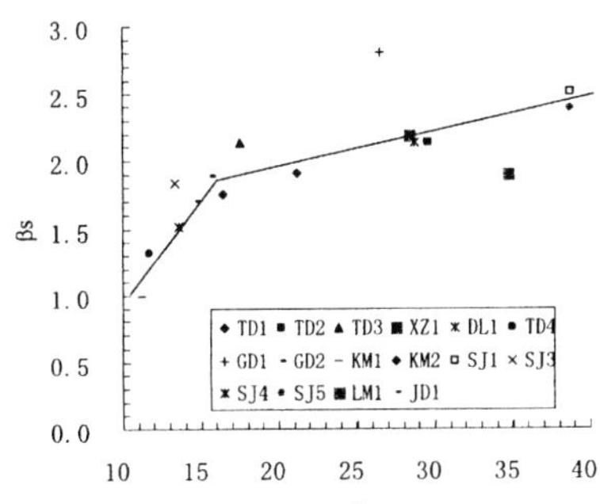

Fig. 6. η~ β $_s$ curve

As shown in Fig. 6, the ratio β_s ranges from 1.0 to 2.9 corresponding to varied η value, and the η~β_s relation can approximately be fitted into two segments of lines with different slopes. When the index η is smaller than 16, β_s increases rapidly with the growth of η and can be estimated using the following equation:

$$\beta_s = 1.0 + 0.145 \times (\eta - 10) \qquad \eta = 10 \sim 16 \tag{4}$$

After the value of η exceeds 16, the index β_s increases slowly with the growth of η and can be estimated using the following equation:

$$\beta_s = 1.87 + 0.0254 \times (\eta - 16) \qquad \eta = 16 \sim 40 \tag{5}$$

In engineering practice, the bearing capacity of SDCM piles can be predicted by using equation (1) to (5). Of course the predicting results should be verified by field test before construction of the piles.

CONCLUSIONS

(1) For proper design, the load-settlement curve of SDCM pile exhibits "gradual rise" style and no plunge failure should appear.

(2) Addition of the core pile is the source of high capacity of SDCM pile. The resistance provided by the pile shaft including core pile was 1.0~2.9 times that of bored concrete pile.

(3) In design of SDCM pile, length and diameter of the core pile must be selected carefully to satisfy the capacity requirement.

(4) *Sectioning and Summarizing* method can be used to estimate the capacity of the SDCM piles. It is considered more logical than the existing methods.

REFERENCES

Dong, P., Chen, Z., and Qin, R. (2001). "Use of concrete-cored DCM pile in soft ground". *Chinese Journal of Geotechnical Engineering*, Vol. 24 (2):204-207

Chen, Y., and Rao, Y. (2003). "Experimental researches of the concrete core mining Pile in Kunming Gu Duicun village". *Journal of Kunming Metallurgy College*, Vol. 19 (1):1-6

Ling, G., An, H., Xie, D., and Wang, E. (2001). "Experimental study on concrete core mixing pile." *Journal of building structure*, Vol. 22 (2):92-96

Liu, J., and Liu, J. (2001). "Experimental study on dry-bored composite pile under groundwater." *Chinese Journal of Geotechnical Engineering*, Vol. 23 (5):536-539

Wu, M., Zhao, X. and Wang, E. (2004). "Experimental study on stiffened DCM pile in laboratory" *Industrial Construction*, Vol. 34 (11):45-48

Experimental Study and Engineering Practice of Rammed Concrete Pile

Mai Wu[1], Xin Zhao[2], Dandan Kong[3] and Yanfeng Li[4]

[1] Associate Professor, School of Civil Engineering, Hebei University of Technology, Tianjin 300132, P.R.China; wumaitj@126.com
[2] Associate Professor, School of Civil Engineering, Hebei University of Technology, Tianjin 300132, P.R.China; zhaoxintj@126.com
[3] Associate Professor, School of Civil Engineering, Hebei University of Technology, Tianjin 300132, P.R.China; ddkong2008@126.com
[4] Lecturer, School of Civil Engineering, Hebei University of Technology, Tianjin 300132, P.R.China, liyanfengwindy@163.com

ABSTRACT: Rammed concrete pile is the development of columnar-hammer rammed and expanded technique. Due to the twice compaction effect during installation process and high compression strength of pile shaft, rammed concrete pile has the advantages of high bearing capacity and relatively low cost. The main purpose of this paper is to study the installation procedure, the bearing behavior of rammed concrete pile and compaction effect on the surrounding soil. The indoor tests results indicate that the rammed concrete specimens take only 14 days, only half of the common concrete, to achieve the design strength, which is favorable to shorten the construction period of foundation. Additionally, the compression strength of rammed concrete is about 10 to 50 percent higher than the regular concrete with the same cement content ratio. Field tests results show that rammed concrete pile offers greater capacity than driven and cast-in-place pile. Finally a case history is introduced that further verified the compaction effect of the rammed concrete pile to the surrounding soil.

INTRODUCTION

Rammed and expanded pile technique is employed widely to treat soft soil in foundation engineering. A columnar drop hammer with diameter of 300~500mm, length of 2~6m and weight of 10~80kN is utilized to drive the pile and ram the fillings to form the pile (Wang et al., 2008). The "twice compaction" effect on surrounding soil during both the pile driving and the aggregate ramming procedures greatly improves the side resistance of the pile as well as the bearing capacity of the soil. Generally, the rammed and expanded technique has the following advantages: (1) suitable for various soil conditions; (2) be convenient to find abnormity of the strata during the driving of the pile; (3) the pile material is convenient to obtain and the cost is low, particularly it is beneficial to environment protection if building rubbish is adopted as the fillings.

However, as the pile was constructed with discrete material of low strength, the bearing capacity of the pile is relatively low and its application range of the technique is limited. In order to develop the rammed and expanded technique and further increase the bearing capacity of treated foundation, semi-dry concrete of higher strength was adopted to form the pile. Usage of rammed concrete can increase the bearing capacity of the pile while inheriting the advantage of twice compaction effect of the existing technique. To some extent the rammed concrete pile is similar to Franki pile (Tomlinson,1994). An identical feature of the two techniques is the twice mechanical compaction effect on surrounding soil during both driving and ramming process. However, the following differences exist between the two techniques： (1) rammed concrete pile is employed as reinforcement member in composite foundation while Franki pile acts as part of piled foundation; and as a result, no reinforcement cage is required for rammed concrete pile; (2) in terms of bearing properties of single pile, Franki pile is considered to be end-bearing pile due to its enlarged base, while rammed concrete pile behaves as frictional pile deriving most of its carrying capacity from the side resistance.

The focus of this paper is the bearing behavior of rammed concrete pile and the treated foundation. The experiments being carried out include two parts: the indoor experiments to test the mechanical properties of the rammed concrete and field tests to study the bearing behavior of a single pile. A case history is presented finally to investigate the compaction effect of the rammed concrete pile to the surrounding soil.

INSTALLATION OF RAMMED CONCRETE PILE

Multi function piling rig (Fig. 1) consisting of a mast, leaders and two winches on a rotatable carriage is used to support, guide and lift the piling tube and the heavy hammer. The inner diameter of the commonly used tube is 400mm and the outside diameter of the hammer is about 350mm. The piling tube can be omitted when the soil is stiff cohesive materials and no soil collapse is expected during the construction process. However, for saturated soil strata under ground water level, the piling tube is essential to prevent the collapse of the hole during the whole construction procedure. Limited by the length of the mask, it is difficult to fill the shaft material from the top of the tube. To solve this problem, several filling openings were designed and cutted along the piling tube. The openings are closed (Fig.2.(a)) during driving and extracting of the tube to prevent the side soil from entering the tube. Thus it becomes convenient to fill the aggregate into the tube on the ground surface by using these openings as shown in Fig. 2.(b).

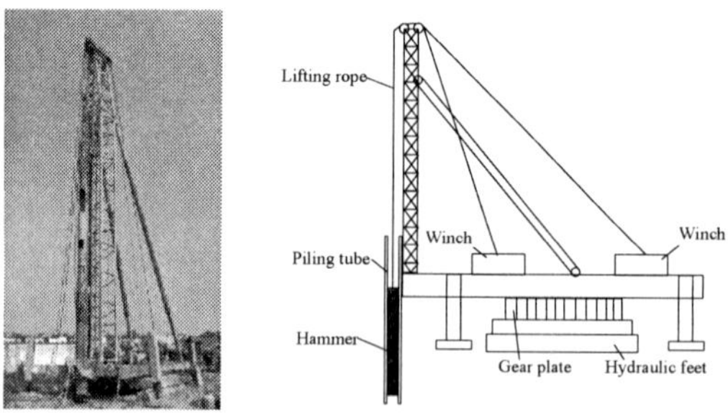

Fig. 1. Piling rig

(a) (b)

Fig. 2. Filling shaft material through openings

As mentioned above, the piling tube must be utilized when installing the rammed concrete pile under ground water level. Similar to Franki pile, the whole construction procedure is illustrated in Fig.3. However, three aspects of distinctions exist in the installation procedure of rammed concrete pile in comparison with Franki pile: (1) the semi-dry concrete was filled into the tube through the openings instead of top of the tube; (2) no enlarged base is formed though a small amount of brick or gravel can be filled and rammed into the bearing soil layer to increase the toe resistance of the pile; (3) no reinforcement cage was inserted.

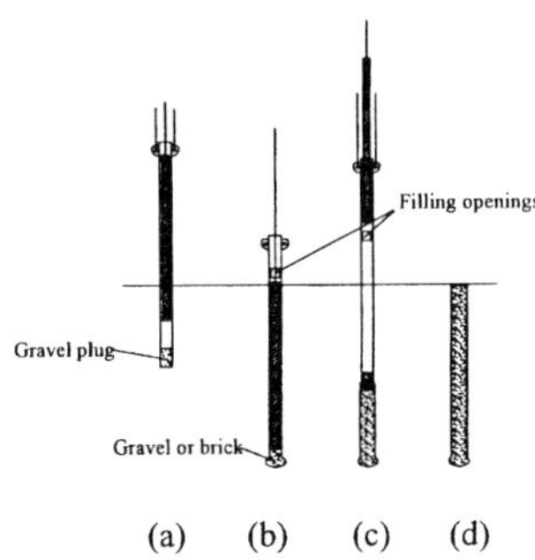

(a) Driving piling tube (b) Filling concrete in piling tube

(c) Ramming concrete in shaft (d) Completed pile

Fig. 3 Construction sequence of rammed concrete pile

COMPRESSION STRENGTH OF RAMMED CONCRETE

To study compression behaviors of rammed concrete, sixteen groups of cylinder specimens were made and tested. The specimens of 120mm height and 152mm diameter were made by using the heavy compaction apparatus as shown in Fig. 4.

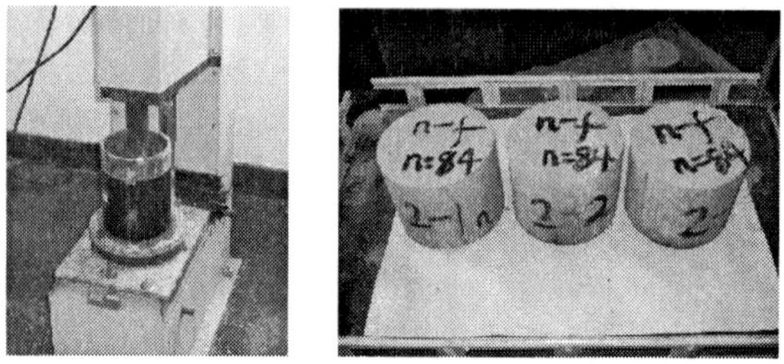

Fig. 4 Heavy compaction apparatus and specimens

To investigate the effect of curing conditions on the strength development of rammed concrete, specimens were cured in standard conditions, in water and in soil respectively. Fig. 5 shows the compression strength to curing age relationships of rammed concrete, for which the cement content ratio (the weigh ratio of cement to total aggregates) is 0.15. It is clearly shown that all specimens obtained most of their ultimate strength during the first 14 days, and 14 days later the strength remained approximately unchanged until the curing age of 60 days. The test results indicate that the curing age of rammed concrete is only half that of casted regular concrete, which is favorable to shorten the construction period of the foundation engineering.

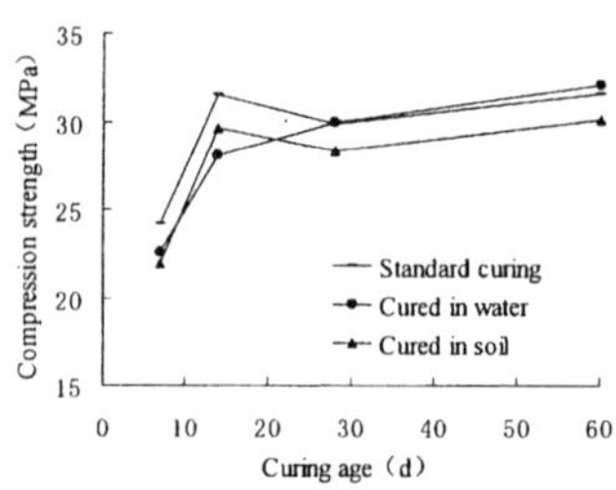

Fig.5 Curing age-compression strength curves

Casted regular concrete specimens were also made and tested to be compared with rammed concrete. The comparison result is illustrated in Fig. 6. With the same cement content, the strength of rammed piles is notably greater than that of the casted concrete. For instance, when the cement content ratio is taken as 0.15, the strength of rammed concrete is about 30MPa while the strength of casted concrete is only about 20MPa. With the increasing of cement content ratio, the strength difference between the two types of concrete decreases to about 10 percent as the cement content ratio is taken as 0.25. Therefore, it can be concluded that the rammed concrete cost fewer cement than common casted concrete to achieve the same strength.

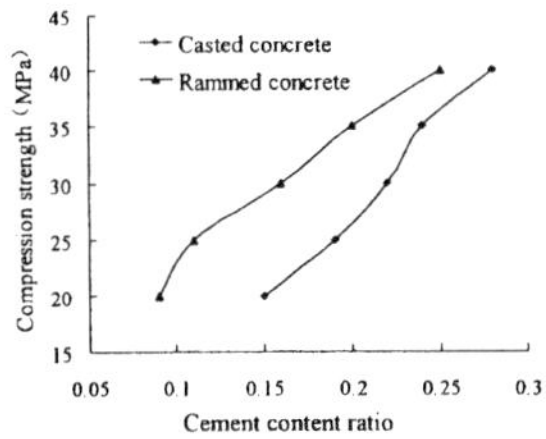

Fig.6 Cement content-compression strength curves

BEARING BEHAVIOR OF RAMMED CONCRETE SINGLE PILE

Field tests were performed in Cangzhou City in eastern China to investigate the bearing behavior of rammed concrete pile. The main stratum surrounding the test pile is plastic clay with the moisture content varying from 25.4 to 33.0 percent. The stratum beneath the pile toe is compacted silt with natural void ratio of 0.76. One rammed pile and one driven and cast-in-place concrete pile were installed and tested. The penetrations of the two piles were both 7.0m. The two holes were rammed in the same manner. The compression strength of the cast-in-place concrete pile was 20MPa and diameter of this pile is about 420mm. The rammed concrete pile is constructed according to the sequence shown in Fig. 3 and the diameter of the pile is about 450mm. 30 days later the piles were tested and the load-settlement curves are plotted in Fig. 7.

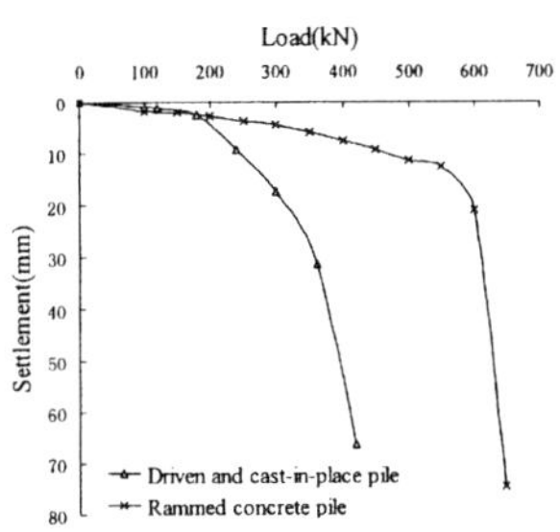

Fig.7 *Q*~s curves of static loading test

The two piles were tested by slow maintained static load method. According to *Chinese technical code for ground treatment for buildings* (JGJ79-2002), if plunge failure occurs, the plunge resistance should be taken as the ultimate resistance. Thus, the ultimate capacity of the two piles can be defined as 600kN and 360kN respectively. The capacity of the rammed concrete pile is about 1.67 times that of the driven and cast-in-place pile. The rammed concrete pile derived its high capacity from the twice compaction effect during both driving and ramming of the pile. It is a regret that no strain or stress testing apparatus was installed in the pile, and further study is expected to determine how the capacity increase is distributed.

CASE HISTORY

The Filling Station project of Fanshan Phosphorite Corporation is located at Zhangjiakou City in north China. The project comprises four 23m-high towers of which the bearing capacity of the foundation should large than 400kPa. The soil profile at the location of the filling station consisted of 15.5 m depth of collapse loess with the moisture of 6 percent and void ratio of 0.86. The bearing capacity of the nature foundation is about 160kPa. To eliminate the collapse behavior of the soil layer and increase bearing capacity of the foundation, rammed concrete pile method was designed and constructed. Seven piles were constructed and tested before formal construction and the plane arrangement of the seven piles is illustrated in Fig.8. The average pile diameter of the piles is about 450mm and the penetration is 12.0m. Because the ground water level is below the pile toe, the piling tube was not used during the construction procedure. Twenty-eight days after the construction, static loading test of composite foundation was carried out. The testing area including pile 4# is rectangular in shape with the length of 1.35m and width of 1.15m as shown in Fig. 8.

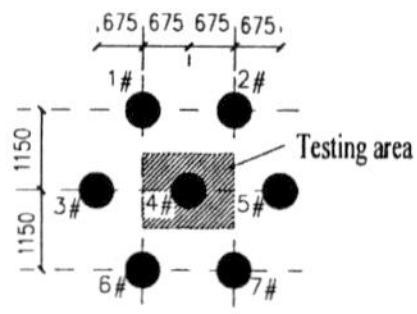

Fig. 8 Plane arrangement of testing piles (unit:mm)

The pressure-settlement curve of the testing area is shown in Fig.9. According to *Chinese technical code for ground treatment for buildings*, the bearing capacity of the composite foundation is taken to be the pressure that achieves the settlement of about 0.6 percent of the pile's diameter, thus the bearing capacity of the tested composite foundation is about 500kPa that satisfied the design requirement.

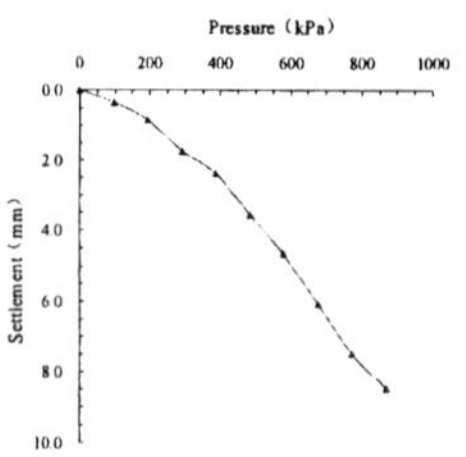

Fig. 9 Pressure-settlement curve of composite foundation

Another aim of the foundation treatment of this project is to compact the surrounding soil and to eliminate the collapsible character of the loess. In-situ tests were carried out and soil specimens of different depth were obtained and tested in laboratory. Table.1 shows the comparison results of related parameters before and after the construction of the piles. Apparently the surrounding soil was compacted with the dry density increasing while the void ratio decreasing. The collapse coefficient reduced from 0.031 to 0.006. According to *Chinese technical code for building construction in collapsible loess regions* (GB50025-2004), when the collapse coefficient is smaller than 0.015, the soil can be considered safe to structure. Thus two aims of the project are both achieved by adopting the rammed concrete pile. Additionally, it is abnormal that the compression modulus decreased after being treated. This may be caused by the testing error or something else.

Table. 1 Comparison of testing index before and after treatment

Testing index	Moisture content (%)	Dry density (g/cm^3)	Void ratio	Compression modulus (MPa)	Collapse coefficient	Bearing capacity (kPa)
Pre-treatment	6.0	1.43	0.86	22.1	0.031	160
Post-treatment	5.5	1.61	0.66	19.7	0.006	200
Improving rate	-8.5%	13%	-23%	-10%	-80%	25%

CONCLUSIONS

The following conclusions can be drawn from the study presented in this paper.

(1) To install the rammed concrete pile, the piling tube must be utilized under ground water level. It is convenient to fill shaft material by using the filling openings aligned along the longitude of the piling tube.

(2) The compression strength of rammed concrete specimens is about 10 to 50 percent higher than the regular concrete with the same cement content ratio.

(3) Due to the twice compaction effect, the rammed concrete piles offer greater capacity than driven and cast-in-place concrete pile.

(4) The case history shows that construction of rammed concrete pile can effectively eliminate the collapsible character of the loess and greatly increase the bearing capacity of composite foundation.

REFERENCES

China Building Research Institute (2002), Chinese technical code for ground treatment for buildings (JGJ79-2002). China Architecture and Building Press, Beijing, 200p.

Li, Y. (2009). Study on pile material's engineering characteristics of rammer concrete piles thrust-expanded in column-hammer. Hebei University of Technology, Tianjin, 69p.

Shaanxi Building Research Institute (2004), Chinese technical code for building construction in collapsible loess regions (GB50025-2004). Architecture and Building Press, Beijing, 158p.

Tomlinson, M. J. (1994). Pile design and construction practice, Taylor & Francis Group, Oxon, 411p.

Wang, E., and Liu, X. (2008). "Study on consolidating mechanism of piles trusted-expanded with Column-hammer". *Building Science*, Vol 24(9): 63-67

Centrifuge Study on Ultra-deep Foundation Pit Excavation in Soft Ground

Qianwei Xu[1], Xianfeng Ma[2], Hehua Zhu[3], Xiurun Ge[4], Yan Ping[5], Fangfang Jin[6].

[1]School of Naval Architecture, Ocean and Civil Engineering, Shanghai Jiaotong University, Shanghai 200030, People's Republic of China; xuqwtju@126.com
[2]Key Laboratory of Geotechnical and Underground Engineering of Ministry of Education, Tongji University, Shanghai, 200092, People's Republic of China; xf.ma@tongji.edu.cn
[3]Key Laboratory of Geotechnical and Underground Engineering of Ministry of Education, Tongji University, Shanghai, 200092, People's Republic of China; zhuhehua@tongji.edu.cn
[4]School of Naval Architecture, Ocean and Civil Engineering, Shanghai Jiaotong University, Shanghai 200030, People's Republic of China; xiurunge@public.wh.hb.cn
[5]Lovat Trading (Shanghai) Co. Ltd., Shanghai 200003, People's Republic of China; ping.yan@lovat.com.cn
[6]Department of Civil Engineering, University of Shanghai for Science and Technology, Shanghai 200093, People's Republic of China; jff_001@163.com

ABSTRACT: In order to investigate the behaviors of the support systems and the associated ground deformations, centrifuge model test is adopted to simulate the excavation of ultra-deep foundation pit in soft ground. In this test, the geometrical scale is 1:120, and the soil layer thickness, excavation depth and retaining wall height are 60 cm, 31.6 cm and 54.2 cm respectively. The model ground consists of two types of soil, including silty clay and sandy silt. The aluminum alloy plate and aluminum alloy pipes are used to simulate reinforced concrete diaphragm wall and braces respectively. The retaining wall deformation, bending moment, brace axial force, earth pressure around the retaining wall, ground surface settlement, etc. are measured and discussed in detail in this paper. This can be used to comprehensively assess the stability of pit excavation support system. The findings from the centrifuge study are particularly useful for practical engineering.

INTRODUCTION

With buildings being higher, larger and heavier, and the continuous construction of municipal facilities, more and more deep excavation projects are emerged in urban areas. Due to the impact of surrounding environment, engineering geology and hydrological conditions, varieties of complex issues are often encountered in the construction of deep excavation, such as great subsidence of ground surface, large deformation of retaining wall and failure of braces, etc. Therefore, many scholars and engineers (Thorley 2002, Liu 2005, Leung 2007, and Hashash 2008) studies the issues related to deep excavation. Although great deals of research results and practical experiences have been achieved in this aspect, there are still many engineering problems needed to be solved. To solve these problems, the most effective way is to carry out an in-situ full-scale test, but it is difficult and expensive. In contrast, laboratory model test is relatively simple, economical, operable, and highly repetitive. As a physical modeling technology, centrifuge model test is an important testing method and has been widely used in various geotechnical engineering fields, such as embankment (Peiris 2008), slope (Ling 2009), pile foundation (Brandenberg 2005), retaining wall (Madabhushi1 2004), and tunnel engineering (Nomoto 1999). In order to examine the behaviors of diaphragm wall deformation, brace axial force, earth pressure and ground surface settlement, centrifuge model test is employed to simulate an ultra-deep excavation in soft ground. The results of this study are presented in this paper.

PRINCIPLES OF CENTRIFUGE TEST

Centrifuge modeling is a technique to simulate the mechanical response of full-scale structures in reduced-scale physical models. In a centrifuge model test with model downsized to 1/N of a full size feature, the model will be accelerated to Ng centrifuge force field to replicate the self-weight stress field (where g is acceleration of gravity) when the test is performed. Under the premise of same material and same boundary stress condition, the prototype and the corresponding centrifuge model will have the same stress state, strain and failure mechanism. More importantly, it can reproduce the characteristics of the prototype.

CENTRIFUGE

The centrifuge test reported here was conducted on the geotechnical centrifuge of Tongji University, with an effective radius of 3.0 m. The centrifuge payload capacity is 150g-ton with a maximum acceleration of 200g for static tests and 100g for dynamic tests. The test system is composed of a set of data acquisition system, a high-speed camera, and a closed circuit television monitoring equipment. The effective volume of the test chamber used in the test is 0.9 m (length) × 0.7 m (width) × 0.7 m (height).

CENTRIFUGE MODEL

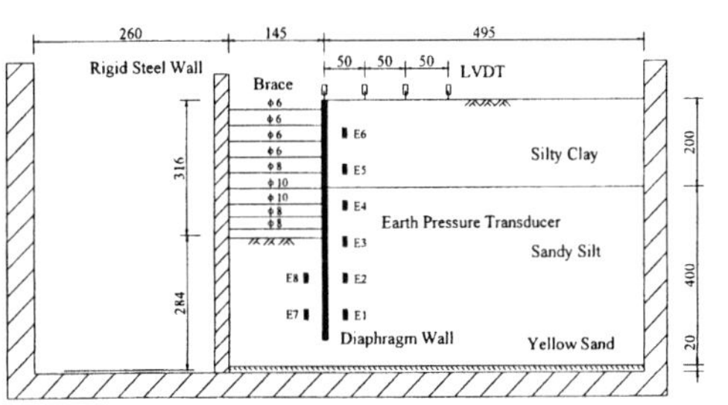

FIG. 1. Elevation view of braced deep excavation (unit: mm).

A 1/120 scale model was created for testing in the geotechnical centrifuge facility.

The plane-strain test chamber which amounts on the centrifuge arm was used to house the model and provide the appropriate boundary conditions for the test. The geometry of the model deep excavation is shown in Figure 1. A diaphragm wall with 9 level braces is adopted as excavation support of the foundation pit. In the reduced scale model, the soil layer thickness, excavation depth and retaining wall height are 60cm, 31.6cm and 54.2cm respectively, which represent a prototype of 72m, 38m and 65m accordingly when tested at 120g. Two soil layers, a 200mm thick silty clay layer and a 400mm thick sandy silt layer, were placed in the test chamber. Additionally, in order to form double drainage condition, a 20mm thick yellow sand layer is laid at the bottom of the chamber.

MODEL PREPARATION

In deep excavations, the reinforced concrete diaphragm wall and braces are mainly used as bending component and compression component respectively. The conversion relations from a prototype to a model can be calculated according to the following equations.

$$E_m I_m = \frac{E_p I_p}{N^3} \quad (1)$$

$$E_m A_m = \frac{E_p A_p}{N^2} \quad (2)$$

where E=elastic modulus (F/L^2); I=moment of inertia per unit width (L^4/L); A=area of section (L^2); EI=bending stiffness per unit width (F·L^2/L); EA=compression stiffness (F); N=scaling factor; m=model; and p=prototype.

At a scaling factor of 120, it is nearly impossible to meet the requirement of bending stiffness and compression stiffness by using prototype concrete material in the test, therefore, aluminum plate or epoxy fiber plate are often utilized to simulate prototype concrete components in centrifuge studies. In this test, a 7.5mm thick aluminum plate was selected to simulate the retaining wall, and various aluminum alloy tubes were chosen to simulate the braces as shown in Figure 1.

The soil in the model was made in the following steps. The soil taken from a construction site is firstly dried, and then crushed into powder. After that, it is saturated as flowing and plastic slurry, and finally consolidated in the centrifuge chamber at 120g. When the consolidation is completed, we measured the soil layer thickness and calculated the dry density so as to compare with that of prototype. Table 1 shows the mechanical properties of the prototype soil. Among them, the soil shear strength is got by consolidated undrained triaxial test.

Table 1. Mechanical parameters of soil stratum

Soil Name	Water Content (%)	Density (kg/m^3)	Undrained Cohesion (kPa)	Undrained Friction Angle(°)	Compression Modulus (kPa)
Silty clay	35.1	1730	19	20.1	4.21
Sandy silt	28.7	1950	4	35	14.32

INSTRUMENTATION

As shown in figure 2, the diaphragm wall was instrumented with 8 pairs of strain gauges protected by a thin epoxy layer. Each pair of strain gauges is configured to form a half-bridge circuit to measure retaining wall deformation at the corresponding elevation during the model test based on the small deformation hypothesis.

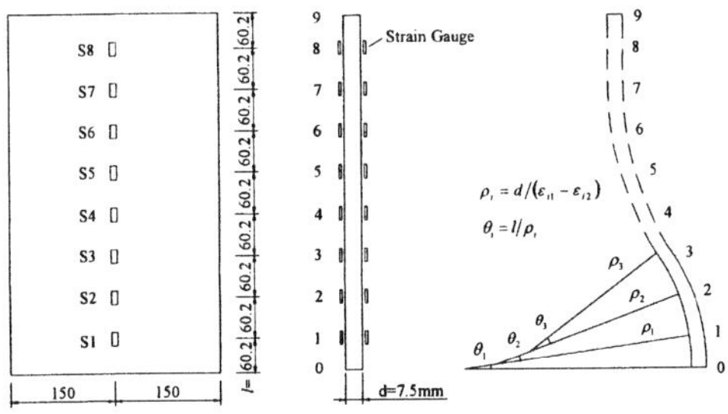

FIG.2. Arrangement of strain gauges (unit: mm).

To investigate the variation of earth pressure on both sides of the diaphragm wall, eight earth pressure transducers with dimension of ϕ 7mm × 3mm are placed in the soil. Two transducers were located below the final excavation surface as shown in Figure 1.

Four displacement transducers (LVDTs) installed in the soil are used to measure the settlements of the ground during the test as shown in Figure 1.

CENTRIFUGE TESTING TECHNOLOGY OF EXCAVATION

Staged construction was employed to install 9 levels of braces. At each stage, the excavation and installation of a brace were performed under normal gravity field after the centrifuge test is stopped. After that, the test chamber is re-accelerated to the predetermined value so that the responses of soil and retaining structure can be measured. The 'start–stop–dig–support–restart' procedure was repeated until 9 construction steps were finished.

TEST RESULTS

Retaining Wall Deformation

Figure 3 shows the retaining wall deformations at various excavation steps and the deformation values have been converted to prototype values. Due to malfunction of data acquisition system, the test results for the second and third excavation steps were not obtained. In Figure 3, negative value on the horizontal coordinate means that the retaining wall deflects towards the excavation. As shown in Figure 3, the deflection of the wall increases gradually with the increase of excavation depth. Meanwhile, the maximum deformation depth moves downwards with the increase of excavation depth. The deformation began to change obviously from step 4 and reached the maximum value of 8.3cm at step 9. From step 7 to step 9, the value of maximum deflection of the wall is almost same, but the maximum deflection depth was getting deeper. In this test, the brace was not jacked against the retaining wall, thus only compression deformation of the brace is expected. Therefore, the top of retaining wall presents the character of deforms outwards when it bends.

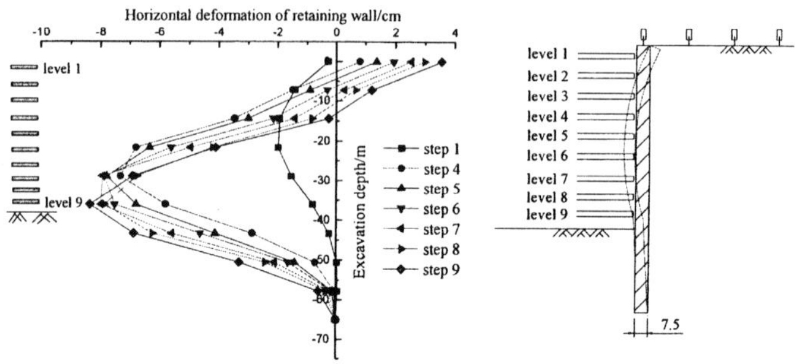

FIG. 3. Horizontal deformation of retaining wall.

Bending Moment of Retaining Wall

Figure 4 shows the bending moment envelopes of the retaining wall at various excavation steps. The bending moment is positive when the wall deflects towards the excavation; otherwise it is negative. The bending moment reached the maximum value of 3700kN-m at the depth of about 28m in step 7; at the same time, the maximum deflection of the retaining wall is reached.

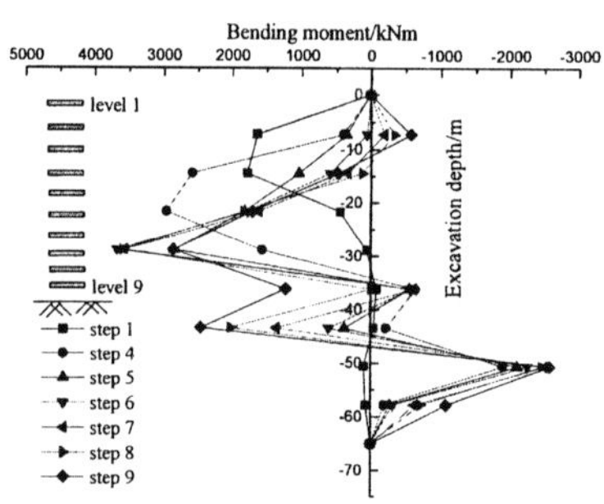

FIG. 4. Bending moment envelope of retaining wall.

Axial force of braces

The axial forces of braces at various test steps are shown in Figure 5. It can be seen that the axial force of each level brace experiences a gradual increase initially and then decreases slowly after reaching the maximum value. The axial forces of five out of nine braces reached the maximum values at step 8. The maximum axial force of approximate 3500 kN was reached in the sixth level brace.

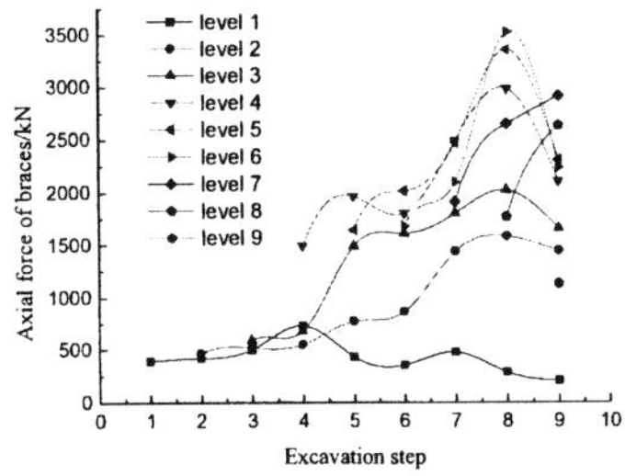

FIG. 5. Variation diagram of brace axial force.

Variations of Earth Pressure

Figure 6 shows the change of earth pressures from six earth pressure transducers E1 through E6 along the retaining wall at different steps. Figure 6 presented a phenomenon that the earth pressure of the wall segment above 28 m was increasing with progress of excavation. On the other hand, the earth pressure below depth 28 m was decreasing with progress of excavation. The increase of earth pressure above 28 m was due to the fact that the top of the bended retaining wall deforms towards the retained soil and resulted in higher contact pressure. The lower part of the retaining wall deforms towards the foundation pit (excavation) and resulted in decrease of wall-earth contact pressure. The decrease of earth pressure is most obvious at the depth of -45m. The change of earth pressure is also consistent with the deformation of retaining wall shown in Figure 3.

Figure 7 shows the change of earth pressures from earth pressure transducers E7 and E8 compared to at-rest in the passive zone in front of the retaining wall at different steps. As can be seen in Figure 7, with the progress of excavation, the earth pressure at the deep depth of -55.4m increased gradually; while the earth pressure at -45.8 m increased slowly initially and then decreased dramatically starting at step 6. This quick reduction of earth pressure at depth -45.8 m may be caused by the unloading effect of excavation. Therefore, the earth pressure in passive area is influenced not only by the squeezing of soil but also by the unloading of soil at excavation bottom.

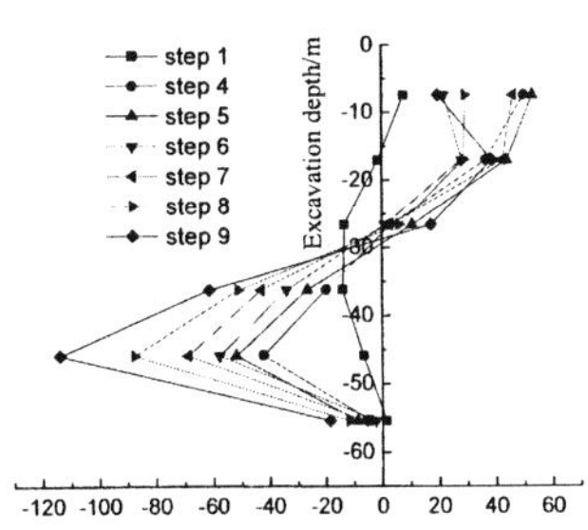

FIG. 6. Change of earth pressure behind retaining wall.

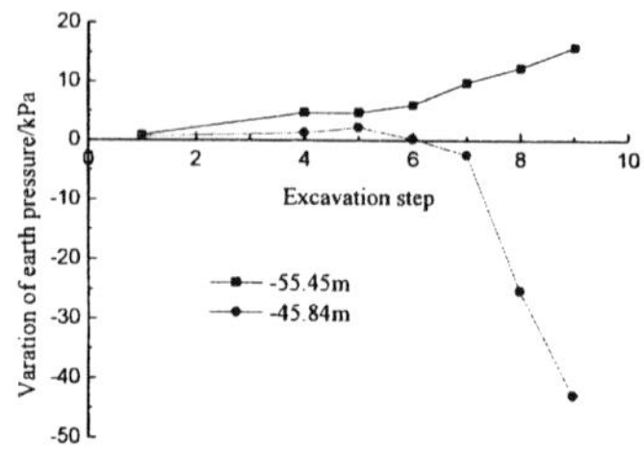

FIG. 7. Variation of earth pressure increment before the retaining wall.

Ground Surface Settlement

Figure 8 shows ground surface displacements at various excavation steps. It can be found that with the progress of excavation, ground surface subsides gradually and reaches the maximum value of 11cm at step 9. Secondly, the surface settlement generally reached maximum values at a distance 6 m away from the wall. Thirdly, the settlement increased significantly starting from step 7. This coincides with the large deformation of retaining wall at the same time.

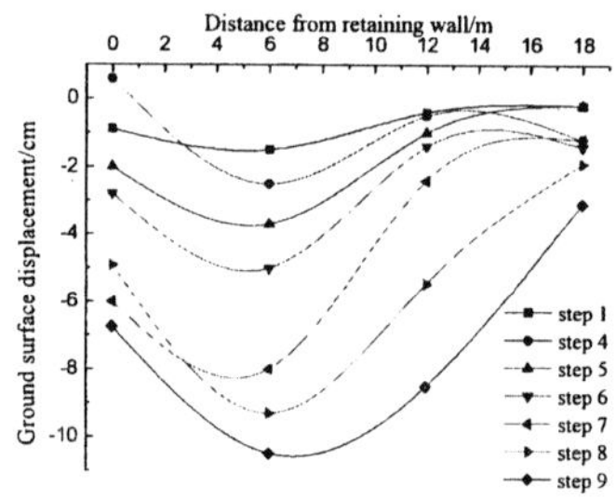

FIG.8. Subsidence of ground surface.

CONCLUSIONS

This paper investigates the behavior of a 38m ultra-deep excavation in soft ground by a centrifuge test. The main findings obtained are summarized as follows.

(1) During ultra-deep excavation, the deformation of retaining wall increases gradually and the maximum deformation position also moves down accordingly.

(2) The maximum values of bending moment and deformation of retaining wall occurs nearly at the same depth. In this test, the maximal value of bending moment appears at the seventh level brace.

(3) In ultra-deep excavation, the axial force of each level brace experiences a gradual increase first and then gradually decreases after reaching the maximal value. It is also influenced by the retaining wall deformation.

(4) The earth pressure behind the retaining wall is mainly influenced by the

deformation of retaining wall. Beneath the final excavation surface, the earth pressure in front of the retaining wall is influenced by both the retaining wall deformation and the unloading effect.

(5) As for ultra-deep excavation, if the excavation depth is relatively shallow, the maximum subsidence of ground surface is approximately equal to the maximum value of retaining wall lateral deflection; but when the excavation depth is greater, the maximum subsidence of ground surface would increase dramatically, and it is much higher than the maximum value of retaining wall lateral deflection.

ACKNOWLEDGMENTS

This research project is supported by national science foundation for post-doctoral scientists of China and science foundation of the key laboratory of geotechnical and underground engineering of ministry of education (Tongji University).

REFERENCES

Thorley, C. B. B., and Forth, R. A. (2002). "Settlement due to Diaphragm Wall Construction in Reclaimed Land in Hong Kong." Journal of Geotechnical and Geoenvironmental Engineering, ASCE, Vol. 128, No. 6, 473-478.

Liu, G. B., Ng, Charles W. W., and Wang, Z. W. (2005). "Observed Performance of a Deep Multistrutted Excavation in Shanghai Soft Clays." Journal of Geotechnical and Geoenvironmental Engineering, ASCE, Vol. 131, No. 8, 1004-1013.

Leung, E. H. Y., and Ng, C. W. W. (2007). "Wall and Ground Movements Associated with Deep Excavations Supported by Cast in Situ Wall in Mixed Ground Conditions." Journal of Geotechnical and Geoenvironmental Engineering, ASCE, Vol. 133, No. 2, 129-143.

Hashash, M. A., Osouli, A., and Marulanda, C. (2008). "Central Artery/Tunnel Project Excavation Induced Ground Deformations." Journal of Geotechnical and Geoenvironmental Engineering, ASCE, Vol. 134, No. 9, 1399-1406.

Peiris, L. M. N., Madabhushi, S. P. G., and Schofield, A. N. (2008). "Centrifuge Modeling of Rock-Fill Embankments on Deep Loose Saturated Sand Deposits Subjected to Earthquakes." Journal of Geotechnical and Geoenvironmental Engineering, ASCE, Vol. 134, No. 9, 1364-1374.

Ling, H. I., Wu, M., and Leshchinsky, D., et al, (2009). "Centrifuge Modeling of Slope Instability." Journal of Geotechnical and Geoenvironmental Engineering, ASCE, Vol. 135, No. 6, 758-767.

Madabhushi, S. P. G., and Zeng, X. (2004). "Simulating Seismic Response of Cantilever Retaining Walls." Journal of Geotechnical and Geoenvironmental Engineering, ASCE, Vol. 133, No. 5, 539-549.

Brandenberg, S. J., Boulanger, R. W., and Kutter, B. L., et al. (2005). "Behavior of Pile Foundations in Laterally Spreading Ground during Centrifuge Tests." Journal of Geotechnical and Geoenvironmental Engineering, ASCE, Vol. 131, No. 11, 1378-1391.

Nomoto, T., Imamura, S., and Hagiwara, T. et al., (1999). "Shield Tunnel Construction in Centrifuge." Journal of Geotechnical and Geoenvironmental Engineering, ASCE, Vol.125, No.4, 289-300.

Numerical Analysis on Ground Behavior for Construction of Displacement Pile

Makoto Suzuki[1], H.M. Shahin[2], Teruo Nakai[3]

[1] Deputy Director, Institute of Technology, Shimizu Corporation, 3-4-17, Etchujima, Koto-ku, Tokyo 135-8530, Japan; PH (81-3)3820-5500; FAX (81-3)3643-7260; e-mail: makoto.suzuki@shimz.co.jp
[2] Associate Professor, Nagoya Institute of Technology, Gokiso-cho, Showa-ku, Nagoya 466-8555, Japan; PH (81-52) 735-5485; FAX (81-52) 735-5485; e-mail: shahin@ nitech.ac.jp
[3] Professor, Nagoya Institute of Technology, Gokiso-cho, Showa-ku, Nagoya 466-8555, Japan; PH (81-52) 735-5485; FAX (81-52) 735-5485; email: teruo.nakai@nitech.ac.jp

ABSTRACT: Recently, a construction method for installing a precast pile nearby an existing driven pile has been developed considering the drainage of buried soil excluding the embedded soil of the ground. This type of displacement pile may affect the behavior of the surrounding ground. This study examines the influence of the displacement pile on the behavior of the surrounding ground with axisymmetric elastoplastic finite element analysis. In the analysis, increment of volumetric strain is applied to simulate the construction of the displacement pile. The behaviors of the ground consisting different types of soil characteristics such as cohesive soils, sandy soils and mixing of different soils are discussed.

INTRODUCTION

Ground improvement and cast in place pile, etc. are often used for the reinforcement of a bridge foundation to avoid the influence of the construction on the existing structures as much as possible. Considering the environmental problems, the processes of the formation of soil and slurry and the countermeasures against groundwater contamination in the vicinity of the water sources have become social demands during the construction work in recent years. Here, a construction method for installing a precast pile nearby an existing driven pile has been developed considering the drainage of buried soil excluding the embedded soil of the ground. This type of displacement pile may change the behavior of the surrounding ground.

The ground around the pile heaves along with penetration of displacement pile depending on the soil condition. In the case of the cohesive soil layer consolidation settlement may occur due to the stress increase in the ground. If the permeability of the cohesive soil layer is low, effective confining stress does not change much rather excess pore water pressures develop as the pore water pressure is not fully dissipated during the construction. Gradually the excess pore water pressure dissipates which results a primary consolidation.

This study examines the influence of the displacement pile on the behavior of the surrounding ground with axisymmetric elastoplastic finite element analysis. The finite element analyses are carried out with FEMtij-2D software in which subloading tij model is used as a constitutive model for the soil. In the analysis, increment of volumetric strain is applied to simulate the construction of the displacement pile.

OUTLINE OF CONSTITUTIVE MODEL USED IN ANALYSIS

An elastoplastic constitutive model for soils, named subloading t_{ij}-model (Nakai & Hinokio, 2004), is used in finite element analyses.

To take into consideration automatically the influence of the intermediate principal stress, the model is formulated using modified stress t_{ij} (Nakai and Mihara, 1984), which is defined as

$$t_{ij} = a_{ik}\sigma_{kj} \tag{1}$$

Here, a_{ij} is the non-dimensional symmetric tensor whose principal values are given by the following direction cosines of the normal to the spatially mobilized plane.

$$a_1 = \sqrt{\frac{I_3}{I_2\sigma_1}},\quad a_2 = \sqrt{\frac{I_3}{I_2\sigma_2}},\quad a_3 = \sqrt{\frac{I_3}{I_2\sigma_3}} \tag{2}$$

$$\left(I_2 \text{ and } I_3 : \text{2nd and 3rd invariants of } \sigma_{ij}\right)$$

The yield function is given by the following equation as a function of mean stress t_N and stress ratio X based on t_{ij} concept:

$$\begin{aligned} f &= \ln t_N + \varsigma(X) - \ln t_{N1} \\ &= \ln\frac{t_N}{t_{N0}} + \varsigma(X) - \left(\ln\frac{t_{N1e}}{t_{N0}} - \ln\frac{t_{N1e}}{t_{N1}}\right) \\ &= \ln\frac{t_N}{t_{N0}} + \varsigma(X) - \left(\frac{1+e_0}{\lambda-\kappa}\varepsilon_v^p - \frac{\rho}{\lambda-\kappa}\right) = 0 \end{aligned} \tag{3}$$

where $\varsigma(X) = \dfrac{1}{\beta}\left(\dfrac{X}{M^*}\right)^{\beta}$ (4)

As shown in Figure 1, t_{N1} and t_{N1e} measure the size of subloading surface and normal yield surface, respectively, and ρ is the state variable which represents current soil density. Here, the yield surface (subloading surface) not only expands but also shrinks for the current stress to lie always on the surface. The plastic strain increment is split into two components – a component $d\varepsilon_{ij}^{p(AF)}$ obeying an associate flow in modified stress t_{ij} space, and an isotropic compression component $d\varepsilon_{ij}^{p(IC)}$ (Nakai and Matsuoka, 1986).

$$d\varepsilon_{ij}^p = d\varepsilon_{ij}^{p(AF)} + d\varepsilon_{ij}^{p(IC)} \tag{5}$$

where

$$d\varepsilon_{ij}^{p(AF)} = \Lambda \frac{\partial f}{\partial t_{ij}} = \frac{df_\sigma - \frac{1}{t_{N1}}\langle dt_N \rangle}{\frac{1+e_0}{\lambda-\kappa}\left(\frac{\partial f}{\partial t_{ii}} + \frac{G(\rho)}{t_N}\right)} \frac{\partial f}{\partial t_{ij}} \tag{6}$$

$$d\varepsilon_{ij}^{p(IC)} = \frac{a_{ii}}{a_{ii}+G(\rho)} \cdot \frac{\lambda-\kappa}{1+e_0} \frac{1}{t_{N1}} \langle dt_N \rangle \frac{\delta_{ij}}{3} \tag{7}$$

$G(\rho)$ is a monotonically increasing function, which satisfies the condition $G(0)=0$ and is given in the form of

$$G(\rho) = a \cdot \rho^2 \qquad (a: \text{ material parameter}) \tag{8}$$

This model requires only a few unified material parameters, but can describe properly the following typical characteristics of soils: (1) influence of intermediate principal stress on the deformation and strength of soil is considered by using t_{ij} concept; (2) influence of stress path on the direction of plastic flow is considered by splitting the plastic strain increment into two components; (3) influence of density and/or confining pressure is considered by referring the subloading surface concept by Hashiguchi (1980).

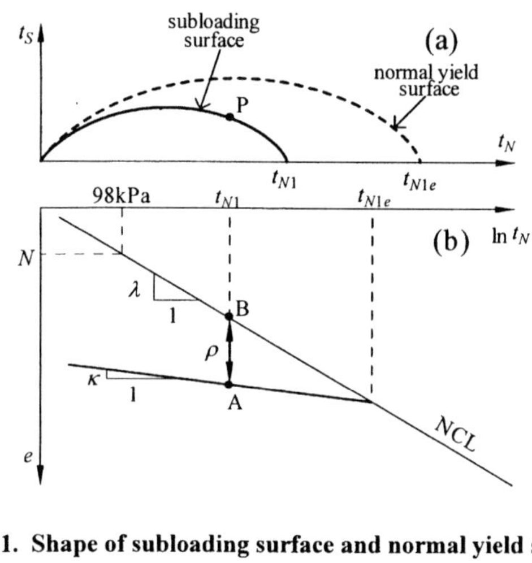

FIG. 1. Shape of subloading surface and normal yield surface.

SIMULATION TECHNIQUE FOR THE CONSTRUCTION OF DISPLACEMENT PILE

The pile installation is simulated by applying volumetric strain increment in each element corresponding to the pile element as shown in Figure 2. At first, the volumetric strain of the targeted element is being increased by 100% at step 1. Next, the property of this element is changed from the soil (elastoplastic element) to the pile (elastic element) at step 2, and at the same time the volumetric strain of the element lying underneath the previous one is being increased by 100%. These two steps are repeated till the tip of the displacement pile reaches to the desired depth.

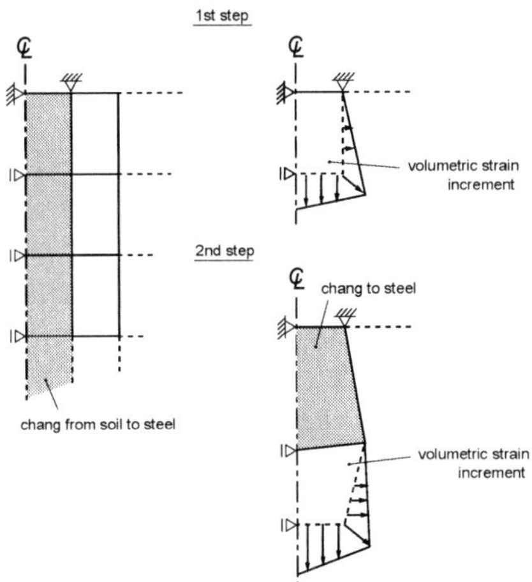

FIG.2. Installation techniques of displacement pile

NUMERICAL MODEL

Axisymmetric elastoplastic finite element analysis with subloading t_{ij} model is carried out to investigate the ground behavior in the surroundings of the displacement pile. Figure 3 shows a typical model ground with 24 meter depth having three different soil layers. Groundwater level is assumed at the ground surface to perform coupling analysis considering soil-water interaction effect. The top and bottom surfaces are assumed as drained boundary and the right side is also assumed as drained boundary.

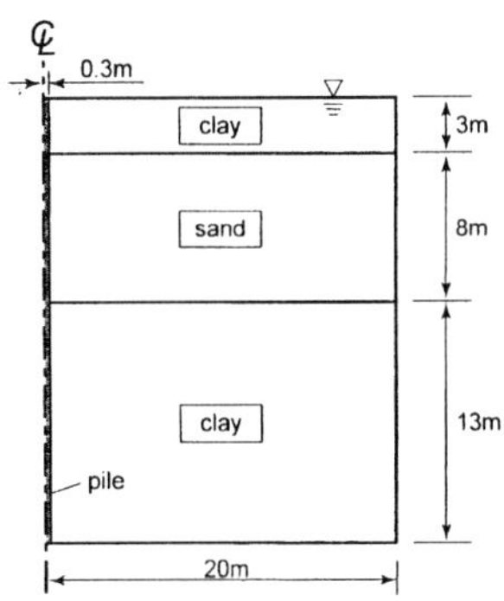

FIG.3. Model ground

Table 1 shows the parameters of clay and sand used in the finite element analysis. Fujinomori clay is used as clayey soil and Toyoura sand is used as sandy soil.

Table 1. Values of material parameters of soils

Parameter	Clay	Sand	
λ	0.1039	0.0700	
κ	0.0099	0.0045	
N=e_{NC} at p=98kPa & q=0kPa	0.922	1.100	Same parameters as Cam clay model
R_{CS}=$(\sigma_1/\sigma_3)_{CS\,(comp.)}$	3.20	3.20	
ν_e	0.2	0.2	
β	1.5	2.0	Shape of yield surface
α	500	30 500	Influence of density and confining pressure
k (m/min)	$1.0\text{x}10^{-6}$	$1.0\text{x}10^{-4}$	

CONSTRUCTION TIME FOR THE PILE INSTALLATION

Figure 4 shows the construction time for installing the displacement pile into the ground. The pile is inserted into the ground up to 8 meter depth working 4 hours in a day. Therefore, the total length (24m) of the pile is inserted into the ground by the time period of 3 days.

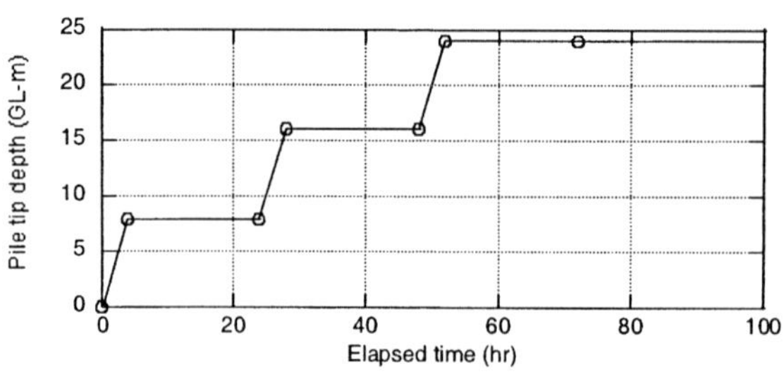

FIG.4. Construction time for the pile installation

DISPLACEMENT OF GROUND SURFACE

Figure 5 shows vertical displacements of the ground surface at different hours. Time required for installing the pile was 52 hours including the consolidation period. Therefore, the vertical displacements beyond this time (after 52 hours) represent the consolidation settlement. It is seen that the soil surrounding the pile is much consolidated because of the effective stress increment. It is revealed that the settlement due to the consolidation dominates in the ground.

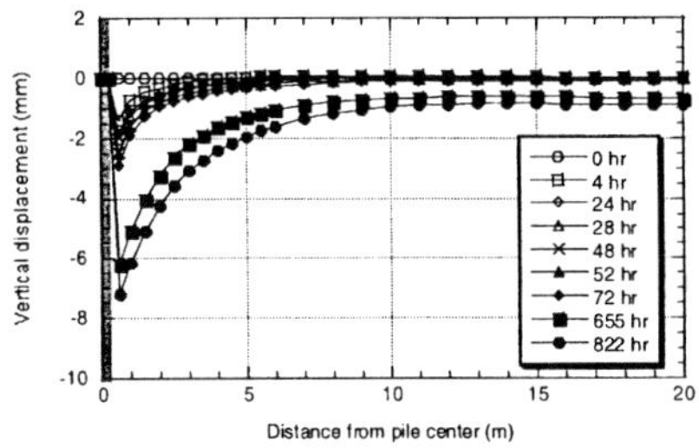

FIG.5. Vertical displacement of ground surface

The histories of vertical displacements at the points which are located at 1.0 and 3.0 meter from the pile center are shown in Figure 6. The left graph shows the data for more than 100 days and the right one shows the data during the first 4 hours. It is found that the ground surface heaves first, and undergoes a consolidation settlement due to the pile penetration. However, the heaving of surface is not so severe as compared to the total settlement.

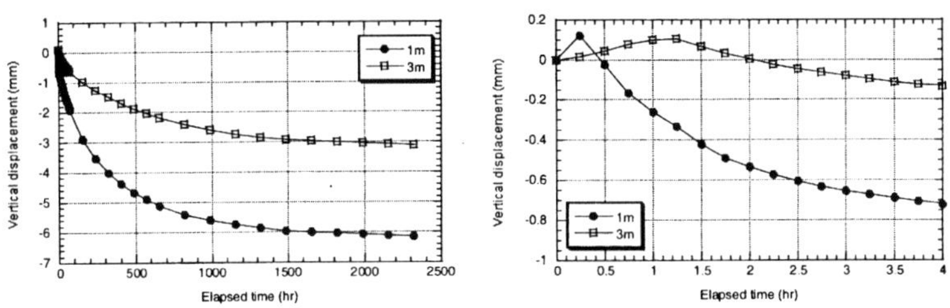

FIG.6. Histories of vertical displacements at 1.0 and 3.0 meter from the pile

DISPLACEMENT OF UPPER SURFACE OF SAND LAYER (GL-3m)

Figure 7 shows the vertical displacement at GL-3m in the ground. The left figure shows the distribution of the vertical displacements at different hours. The right figure is a close up look of the left figure taking the data of the first 4 hours. It is seen that the interface between the clay and sand layers heaves due to the consolidation of the overlying cohesive soils when the pile tip approaches to the interface. However,

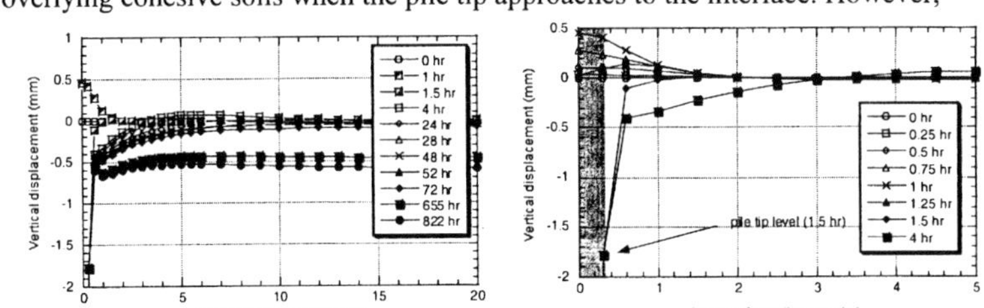

FIG.7. Vertical displacement of the interface between the clay and sand layers

the interface, adjacent to the pile, rapidly changes to the state of subsidence when the pile tip passes the face beyond 1.5 hours.

PRINCIPAL STRESS DISTRIBUTIONS

Figure 8 shows the principal stress distributions of the ground at 4 and 52 hours, respectively. The pile elements are modeled as elastic material which produces large stresses in the simulation. The effective stresses of soil elements do not change much even near the pile. It is also found that the shear deformation in the ground surrounding the pile is not much significant except a little rotation of the principal stress.

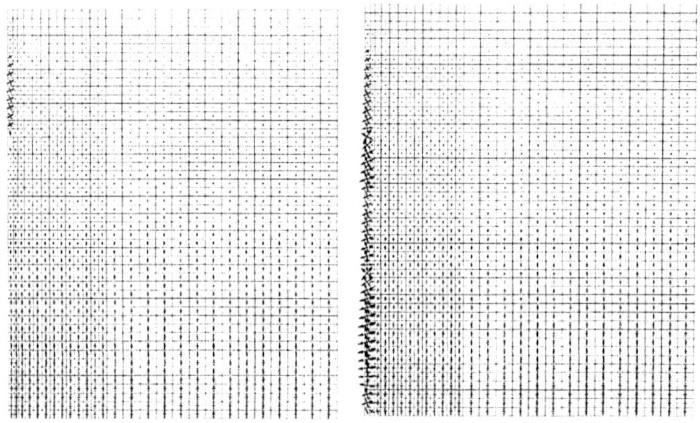

4 hours (pile tip GL-8m) 52 hours (pile tip GL-24m)

FIG.8. Principal stress distributions

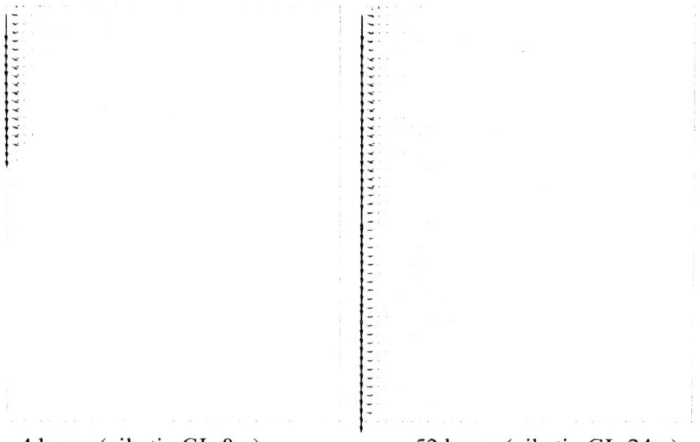

4 hours (pile tip GL-8m) 52 hours (pile tip GL-24m)

FIG.9. Displacement vectors of the ground

DISPLACEMENT VECTORS OF THE GROUND

Figure 9 shows the displacement vectors of the ground at 4 and 52 hours, respectively. A significant subsidence is seen at the pile center for applying the volumetric strain increment in the pile element. The displacement of the ground near the pile tip propagates to the horizontal and downward directions. However, with the time the displacement of the ground surrounding the pile propagates to the horizontal and upward directions.

CONCLUSIONS

The influence of the displacement pile on the ground behavior has been examined by axisymmetric elastoplastic finite element analysis. Here, the pile installation is simulated by applying volumetric strain increment in the finite element analysis. It is revealed that the ground heaves first due to the pile penetration, and it undergoes to the consolidation settlement with time. However, the heaving of the ground is not so significant as compared with the total settlement in this case. It is understood that the construction process of the displacement pile can be simulated well by the volumetric strain increment in the finite element analysis.

REFERENCES

Nakai, T. and Hinokio, M. (2004). "A simple elastoplastic model for normally and over consolidated soils with unified material parameters." Soils and Foundations, 44(2), 53-70.

Nakai, T. and Mihara, Y. (1984). "A new mechanical quantity for soils and its application to elastoplastic constitutive models." Soils and Foundations, 24(2), 82-94.

Nakai, T. and Matsuoka, H. (1986). "A generalized elastoplastic constitutive model for clay in three-dimensional stresses." Soils and Foundations, 26(3), 81-98.

Hashiguchi, K. (1980). "Constitutive equation of elastoplastic meterials with elastio-plastic transition." Jour. of Appli. Mech. ASME, 102(2), 266-272.

Numerical Simulations on a Tube-installed Displacement Pile in Sand Using Stress-Controlled Method

Chuang Yu[1] and Guangyin Du[2]

[1] Associate professor, Doctor, College of Architecture and Civil Engineering, Wenzhou University, Zhejiang Wenzhou, China, 325035; ac-yc@wzu.edu.cn
[2]Associate professor, Institute of Geotechnical Engineering, Southeast University, Jiangsu Nanjing, China, 210096; guangyin@seu.edu.cn

ABSTRACT: During the installation of a displacement pile the surrounding soil is pushed away and distorted. For a driven pile the process is even more complex, due to vibrations caused by the installation process. In engineering practice the installation effects are accounted for empirical or analytical design method. For more complex geotechnical constructions, especially in cases with complex load combinations or strong interaction with neighboring structures, the use of the finite element method is common, in order to reach an optimal and economical design. However, when the modeling of displacement piles is attempted in finite element code, the installation phase is often not taken into account. Considering effects of pile installation is a challenging task for numerical simulations of displacement piles. This paper presents 2-D axysimmetric finite element analysis of one mini-pile in sand. For initializing the effects of installation, stress-controlled cavity expansion is considered. It is shown that realistic stress fields can be obtained by stress controlled cavity expansion up to the appropriate value of K, which is later pressure parameter (later soil pressure to vertical stress). The method is considered as a feasible method for engineering practice to simulate the effects of installation.

INTRODUCTION

The bearing capacity of a driven pile in sand can not be determined very well by theoretical analyses , because the stresses acting against the pile, and the mechanical properties of the severely disturbed pile/soil contact zone, are not known. A practical design method must therefore be based upon a semi-empirical approach, calibrated against results from representative pile load tests. A numerical simulation using stress-controlled method is applied to investigate the bearing capacity of a displacement pile in sand.

To simulate a displacement pile penetration properly, one needs an advanced numerical algorithm, which takes into account large strains (Chopra 1992). In addition to that, dynamic effects due to pile driving or pile vibration also need to be taken into account. Moreover, such a large strain analysis needs an advanced numerical approach that is able to cope with large deformation and density changes. Although, large strain numerical analyses have been applied for pile penetration by researchers (Mabsout 1999). The results of large strain numerical analyses are still often deviating from the experimental data (Dijkstra et al. 2006). Moreover, due to the lack of availability of the code for large strain numerical algorithms, the complexity of the analysis and excessive computational time and cost, a large strain numerical analysis is from the point of view of practicing engineers not yet popular. Therefore we need to introduce an equivalent procedure for the installation, aiming to reproduce the same initial stress configuration after pile installation process.

The installation-induced stress history of a soil element adjacent to a displacement pile in sand involves (i) loading to failure at a stress level comparable to the cone resistance q_c as the pile tip approaches the soil element, (ii) a reduction in stress by approximately two-orders of magnitude as the pile tip passes through, and (iii) cyclic shearing along the pile shaft, leading to contraction of the interface zone and 'friction fatigue' (Poulos 1989, Randolph 2003, White 2005).

As for the initial stress conditions are concerned, the installation will tend to increase the radial stress at the pile wall and the surrounding soils. No doubt that the major effect of a particular displacement pile installation procedure is the resulting stress field around the pile. Installation of displacement piles creates an increase of the radial stress around the pile. The expression of $p_r' = K \cdot \sigma_{vo}'$ (e.g. Kulhawy 1984) is often used in practice, where p_r' is the radial pressure applied on the cavity wall, σ_{vo}' is the initial effective overburden stress and the lateral pressure parameter K is a constant which depends on the soil, the diameter of the pile and the installation procedure. If direct empirical data on K is missing, its value can be back-analyzed from a pile loading test. Even with growing computer power and advanced numerical modeling, pile load tests remain of utmost importance as advanced numerical models need field calibration.

This study is restricted to tube-installed displacement piles. The effects of installation are simulated with stress-controlled cavity expansion method. The aim of this study is to find a method to account for the effects of installation and demonstrate a feasible way of FE-displacement pile analysis to determine the bearing capacity.

FIELD LOAD TESTS OF DISPLACEMENT PILES

Several small diameter piles (diameter is 18 cm), which are called HSP piles with length of about 6.25m and were installed in Wijchen, the Netherlands. The soil at the site consists of well packed fluvial layers of medium fine to coarse sand and gravel, which were deposited during the Pleistocene era (Vermeer and Schad, 2005). The soil is considered as medium dense to loose sand. The piles are installed by jacking a steel tube at a constant speed. Once the tube has reached the required depth, the tube is withdrawn and high slump concrete is pumped continuously into the cavity, as shown in Figure 1a. The tube withdrawal can only start when a predetermined minimum concrete pressure has been reached.

Due to this installation process, the radial stress around the pile increases. The increased radial stress is commonly expressed as $K \cdot \sigma'_{vo}$ in engineering practice. In technical literatures, the value of K ranges between 0.5 to 5 depending on the density of the soil and the displacement magnitude (Lancellotta, 1995). In this study the K value will be back calculated from the measured load-settlement data.

In order to determine the bearing capacity of the piles, several load tests have been performed using static pile load test. Figure 2 presents the load-settlement measurements data of two test piles, which include the creep phases, where the initial and end settlements at maintained load steps were recorded.

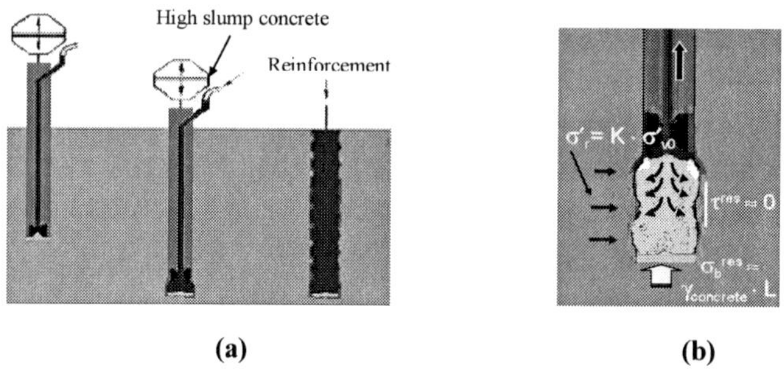

FIG. 1. Pile Installation and stress field

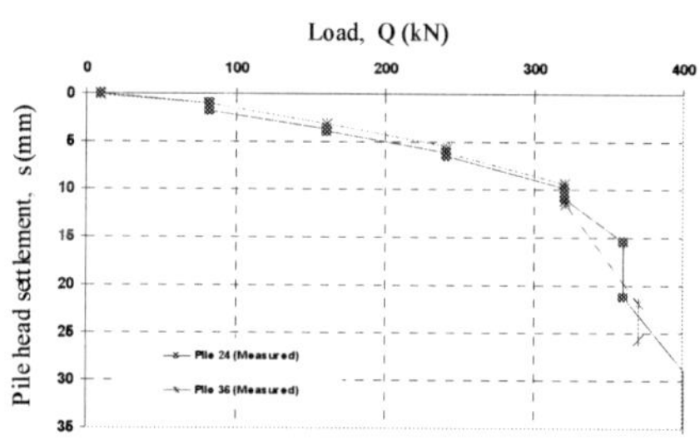

FIG. 2. Load-settlements curve of pile load tests (van der Stoel et al. 2005)

FINITE ELEMENT ANALYSIS OF A DISPLACEMENT PILE
FINITE ELEMENT MODEL

In order to simulate such tests, finite element calculations are conducted using PLAXIS code. PLAXIS (Version 8.2) is a finite element code for soil and rock analyses, originally developed for analyzing deformation and stability in geotechnical engineering projects (PLAXIS, 2002). PLAXIS is a 2-dimensional FEM model package. Several constitutive models of varying complexity and areas of application are included (Hardening model and HS-Small type models). The models most commonly used are the Mohr–Coulomb model and the Soft Soil model developed for modeling the elasto-plastic behavior within the framework of soil mechanics. However one advanced model called HS-Small is used in this simulation(Benz 2006). The model takes into account stress and strain level dependent stiffness of the soil. It is a small strain enhancement of the previously developed Hardening Soil model, which is described by Schanz et al. (1999). The code PLAXIS calculates the strains, stresses and failure states of soil. It permits full automatic mesh generation based on the triangulation principle.

The simulations of this paper produced were performed using axisymmetric mesh with the domain of 8m radius and 13m height (as shown in Figure 3). 6-noded triangular elements with second order interpolation for the displacements and three Gauss points for the numerical integration were used. The mesh is refined along the shaft and at the bottom of the pile where the stress gradient is high. For the contact between pile and soil, 3-pairs-nodes elements with 3 Gauss integration points were used.

In numerical analysis, material behavior of the soil and pile are represented by material constitutive models. The material behavior of the test pile is considered to be linear elastic whereas the soil material follows the so-called "HS-Small" constitutive model. The advantage of using the interface elements has been shown by Wehnert and Vermeer (2004) that it reduces significantly the sensitivity of the load-settlement calculation results to mesh refinement. Moreover, on using the interface element, unrealistic high stress concentration around a sharp corner can be avoided. In addition to that, in this simulation the interface friction angle is lower than the soil friction angle and the interface cohesion should always close to zero to account for the remolded soil around the pile.

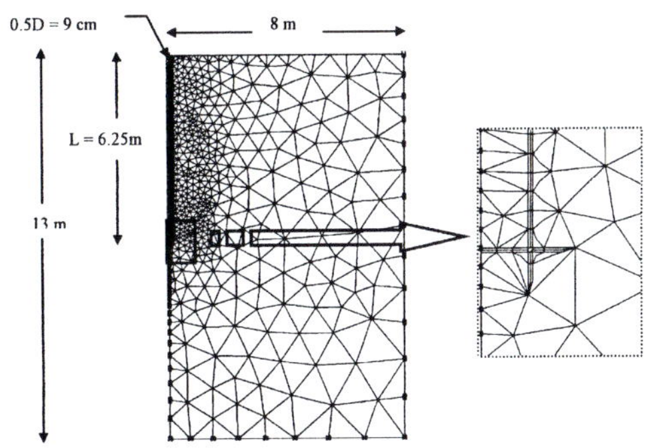

FIG.3. Geometry and FE-mesh

Material properties for the linear elastic pile and HS-Small soil model that are used in calculation are listed in Table 1.

Table 1. Material Properties

Parameters	γ_{unsat} / γ	E_{50}^{ref}	E_{oed}^{ref}	E_{ur}^{ref} / E	m	c′	φ′	ψ	ν_{ur} / ν	G_0^{ref}	$\gamma_{0.7}$	Tensile strength
Unit	kN/m³	MPa	MPa	MPa	-	kPa	°	°	-	MPa	-	kPa
Soil (HS-Small)	17 / -	15	11	45/-	0.5	0.1	36	6	0.2/-	62.63	$0.5\cdot10^{-4}$	0
Pile (Elastic)	- / 23.5			- / 15000					- / 0			

PILE INSTALLATION SIMULATION

The major effect of a displacement pile is obviously the increase of radial stress around the pile. The increase of radial stress is here simulated using stress-controlled cylindrical cavity expansion. The method is performed by imposing radial pressure with distribution of $K \cdot \sigma'_{vo}$ on the cylindrical cavity wall to generate the radial stress field in the soil (Figure 4a). This process is however not straightforward. The simulation procedure is described as follows.

The initial in-situ stresses are generated according to K_0 condition, with the lateral earth pressure at rest estimated by means of Jaky's formula $K_0 = 1\text{-}\sin \varphi'$. After that, the cylindrical cavity expansion is performed. In this phase, soil stiffness is assumed to be linearly increasing with depth. Elements which represent pile, are removed to create a cavity along the centre line and subsequently a radial stress is imposed to the cavity wall. This radial stress increases with depth according to $p'_r = K \cdot \sigma'_{vo}$, where K is taken to be 3.2 for the present calculation. This value is obtained from back analysis of simulating the measured load settlement curve. In addition to the prescribed radial stress, at the bottom of the cavity, a vertical stress of $\sigma'_v = \gamma_{unsat} \cdot L$ is applied, where L is the pile length to set the initial equilibrium. Finally, the pile elements are activated and the imposed radial stress and vertical stress are removed. The displacements at this phase are set back to zero while the generated stress field around the pile remains. Figures 4b show the stress field after material replacement and reset of displacement to zero. Radial stress σ'_r is the stress acting parallel to the direction of pile radius and hoop stress σ'_θ is the circumferential stress acting perpendicular to the direction of pile radius.

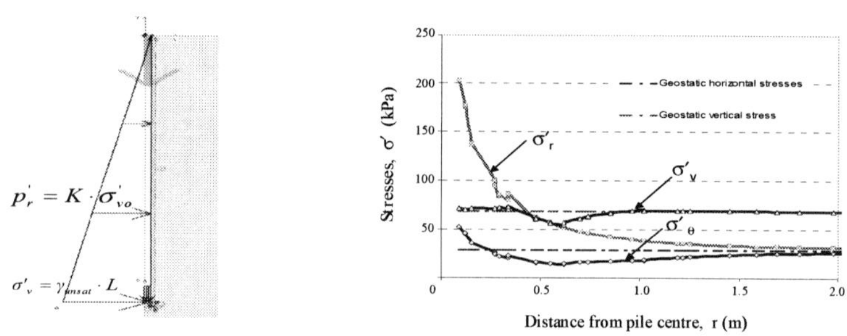

(a) (b)

FIG.4. (a)stress controlled method (b) stresses after material replacement and reset displacement to zero at 4m depth

PILE LOADING

Pile loading is performed using displacement controlled method. A displacement of 35mm as recorded in the load tests is prescribed at the pile head. The calculated load-settlement curve is shown in Figure 5a. Considering the pile resistance components (Figure 5a), it is shown that the load is mainly taken by the skin friction. A pile with this type of behavior is generally known as skin friction pile.

The distribution of radial stress in the interface as shown in Figure 5b has a curved shape which was also shown by O'Neil and Reese (1972) and Vesic (1970). The distribution of interface shear stress takes similar form to the radial stress as it is a function of the radial stress. In practice, K is often assumed to be constant with depth and taken at the ultimate pile resistance (after loading), for example as described by Lancellotta (1995), although Figure 5b shows that this is not the case. For such situation, it is possible to introduce an average K-value. It can be back calculated as follows:

$$K_{average} = \int_0^L p_r' dz \,/\, \int_0^L \sigma_{vo}' dz$$

The integral of the radial stress p_r' can be obtained by numerical integration while the integral of initial vertical stresses σ_{vo}' can be evaluated analytically to give $0.5 \cdot \gamma_{unsat} \cdot L$. In this analysis $K_{average}$ at the end of loading is found to be 2.94. Hence, K drops from K = 3.2 after installation down to $K_{average}$ = 2.94 for a settlement of 35 mm.

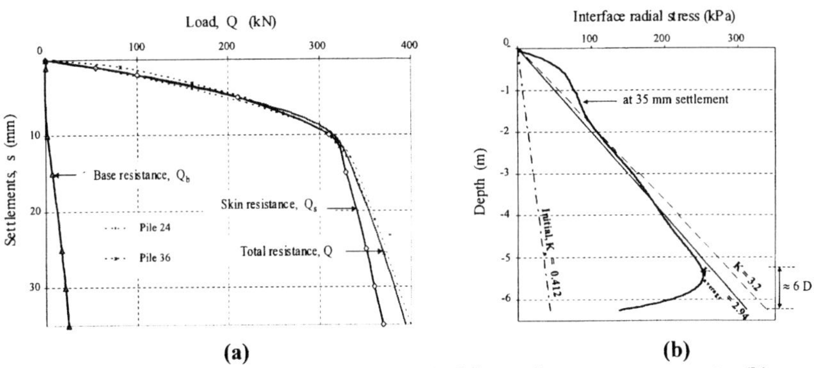

FIG.5.Pile loading to 35mm settlement: (a) Pile resistance components (b) Interface radial stress

CONCLUSION

A feasible method to simulate the installation effects of a cast-in-place displacement pile in sand using finite element analyses with the so-called stress controlled method has been proposed. On using the stress controlled method, the computed load-settlement curves keep good agreements with measured ones. During calculation process, the K value is back analyzed from the measured load settlement data, therefore a pile load test is important. Realistic stress fields after cavity expansion and after pile loading are obtained. Due to the relatively simple approach proposed for capturing the installation process of a displacement pile, it is author's opinion that the method will be accepted in engineering practice by more calibrations of load test results.

The paper provides a useful study to better understand the mechanism of friction resistance for pressure grouted piles, but its application in practice is a little limited due to the many variables involved and the load test results. Since each site is unique, and since pile design and installation procedures will vary between projects, pile load tests will be needed to confirm calculated pile capacities. With this method we can predict the behavior of bearing capacity of more piles using limited data.

ACKNOWLEDGMENTS

The authors appreciate the support of the Qianjiang Talent Program of Zhejiang Province(No. 2009R10047).

REFERENCES

Benz, T. (2006). "Small Strain Stiffness of Soils and Its Numerical Consequences" .Institute for Geotechnical Engineering, University of Stuttgart.

Brinkgreve, R. B. J. editor. (2002). "PLAXIS, 2D Version 8" . AA. Balkema.

Chopra, M. B., Dargush, G. F. (1992). "Finite Element Analysis of Time-Dependent Large-Deformation Problems" *Int. J. for Numer. Analyt. Meth. Geomech.*, 101-130.

Dijkstra, W. J., Broere, van Tol, A. F. (2006). "Numerical Investigation into Stress and Strain Development around a Displacement Pile in Sand". In Schweiger (Ed.), Numerical Methods in Geotechnical Engineering, Taylor & Francis Group, London, 595-600.

Kulhawy, F. H. (1984). "Limiting Tip and Side Resistance: Fact Or Fallacy" *.Proc., Symp. on Design and Analysis of Pile Found.*, ASCE, New York, 80-98.

Lancellotta, R.(1995)."Geotechnical Engineering" A.A.Balkema, Rotterdam, Brookfield.

Mabsout, E.M. (1999). "Pile Driving by Numerical Cavity Expansion" . *International Journal for Numerical and Analytical Methods in Geomechanics*, Vol. 23(11): 1121-1140.

O'Neil, M. W., Reese, L. C. (1972). "Behaviour of Bored Piles in Beaumont Clay". *Journal of the Soil Mechanics and Foundation Division* 98, No. SM2:195-213.

Poulos H.G. (1989). The mechanics of calcareous sediments. Jaeger Memorial Lecture, Aust. Geomechanics, Special Edition: 8-41

Randolph M.F. (2003). "Science and empiricism in pile design". Géotechnique, Vol. 53(10): 847-875

Schanz, T., Vermeer, P.A. and Bonnier, P.G. (1999). "The Hardening Soil Model: Formulation and Verification". Beyond 2000 in Computational Geotechnics-10 years of plaxis, Balkema, Rotterdam.

Van der Stoel, A. E. C., Klaver, J. M., van Dijk, S. (2005). "Test HSP Wijchen" Rapport no 04079a, Voorbij Funderingtechnieken BV, Amsterdam.

Vermeer, P. A., Schad. H. (2005). "HSP Pile Loading Tests in Wijchen" .Inspection Report, Institut for Geotechnical Engineering, University of Stuttgart.

Vesic, A. S. (1970). "Test on Instrumented Piles, Ogeechee River Site". *Journal of the Soil Mechanics and Foundation Devision* 96, No. GT2, 561-584.

Wehnert, M., Vermeer, P.A. (2004). "Numerical Analyses of Load Tests on Bored Piles" *Proceedings 9th Symposium on Numerical Models in Geomechanics* (NUMOG IX), Ottawa, Canada, A.A. Balkema Publishers, Leiden, 505-511.

White, D.J.(2005). "A general framework for shaft resistance on displacement piles in sand". *Proc. Int. Symp. on Frontiers in Offshore Geotechnics*, Perth: 697-703

Investigation on the Requirements regarding the Minimum Embedded Length of Monopiles

Yu-Shu Kuo[1], Martin Achmus[2] and Khalid Abdel-Rahman[3]

[1]Assistant Professor, Department of Civil Engineering, National Pingtung University of Science and Technology, 91201, Pingtung, Taiwan; kuo@mail.npust.edu.tw
[2]Head of Institute, Professor, Institute of Soil Mechanics, Foundation Engineering and Waterpower Engineering, Leibniz University of Hannover, 30167, Hannover, Germany; achmus@igbe.uni-hannover.de
[3]Senior Lecturer, Dr.-Ing., Institute of Soil Mechanics, Foundation Engineering and Waterpower Engineering, Leibniz University of Hannover, 30167, Hannover, Germany; khalid@igbe.uni-hannover.de

ABSTRACT: Monopiles are open-ended steel pipe piles with diameters of usually 3 to 5m, which are used as foundation structures for offshore wind energy towers. Such foundations are subject to intensive cyclic horizontal loading due to wind and wave loads. In the practical design different requirements regarding the minimum embedded pile length are used in order to limit the lateral deformation under static and in particular cyclic loads. Such are the "critical length", "vertical-tangent" and "zero-toe-kick" criteria. For piles of very large diameter and thus high stiffness like monopiles these criteria lead to very large and hardly realistic required lengths. In this study, the different design criteria are compared with respect to piles of very large diameter. By means of the stiffness degradation method, a method which makes the determination of displacement accumulation with cyclic loading possible, the different criteria are assessed. With that, the suitability of adopting the mentioned design criteria for monopile foundations is discussed.

INTRODUCTION

Pile foundations have been used in many existing offshore wind farms in Europe. The large-diameter monopile is one of the most popular foundation structures adopted. Monopiles are open-ended steel pipe piles with diameters of usually 3 to 5m. The lateral loading comes from wind and wave load, is thus intensely cyclic and leads to an accumulation of permanent deformation of the pile. Due to the extreme sensitivity of wind energy converters to the lateral deformation, a sufficient embedded pile length is needed to limit the lateral deformation within the tolerance under cyclic loads.

For a laterally loaded pile, there are some suggestions presented in the existing references for the minimum required embedded pile length. A critical length $L_{c,const}$ is first found in Hetenyi (1946). The embedded pile length beyond $L_{c,const}$ can be considered as infinite long. The horizontal deformation of a single pile is not influenced by the pile length when the pile length exceeds $L_{c,const}$. This critical length $L_{c,const}$ can be presented in terms of elastic pile length L_o by the suggestion of Hetenyi (1946) as

$$L_{c,const} \cong 4L_o \qquad (1)$$

The elastic pile length is affected by the relative stiffness of the soil and the pile and is presented in term of constant modulus of subgrade k as:

$$L_o \cong \left[E_p I_p / k\right]^{\frac{1}{4}} \qquad (2)$$

Herein E_p is the Young's modulus of pile and I_p is the moment of inertia.

By considering a linearly increasing subgrade modulus with depth due to Titze (1970), $L_{c,const}$ can be determined for different loading conditions as the case with a pure lateral load H or a pure moment M as:

$$L_{c,const} \cong \begin{cases} 4L_o & \text{(only H)} \\ 3.5L_o & \text{(only M)} \end{cases} \qquad (3)$$

The elastic pile length determined with the coefficient of horizontal subgrade modulus n_h of Titze (1970) is formulated as:

$$L_o \cong \left[E_p I_p / n_h\right]^{\frac{1}{5}} \qquad (4)$$

where the coefficient of horizontal subgrade modulus n_h can be presented by modulus of subgrade k as:

$$n_h = k/z \qquad (5)$$

A similar critical length as the results of Hetenyi (1946) is also obtained by the analysis of Titze (1970) with a constant modulus of subgrade.

To consider the variation of soil stiffness with depth, $L_{c,const}$ can be also determined by the formulation of Randolph (1981) in iterative calculation steps as:

$$L_{c,const} = D\left(E_p / G_c\right)^{\frac{2}{7}} \qquad (6)$$

wherein D is the pile diameter, E_p is the Young's modulus of pile and G_c is the average value of G^* over the length of the pile less than $L_{c,const}$. G^* can be determined from the shear modulus of soil G_s and the Poisson's ratio of soil ν_s as:

$$G^* = G_s(1 + 0.75\nu_s) \qquad (7)$$

Some suggestions regarding the required embedded pile length are also given in the Guidelines of Germanischer Lloyd (2005) to minimize the risk of accumulated deformation. The critical pile length ruled by the "zero-toe-kick" criterion as well as rigid clamping criterion and "vertical-tangent" criterion (Fig.1) are determined due to the deformation responses of pile tip, and mainly considered to limit the deflection of a pile under cyclic lateral load [Faber & Klose (2006)]. The pile length necessary to reach the zero-toe-kick criterion is defined here as $L_{c,ztk}$, and the pile length to reach the vertical-tangent criterion is defined here as $L_{c,vt}$. The zero-toe-kick criterion is presented by Germanischer Lloyd in the late 1990s for oil and gas platform foundations and is defined as the embedded pile length that ensures two zero deflection points below the soil surface. The distance from soil surface to the second point with zero deflection of pile is named as clamping length and is suggested as a required length for a laterally-loaded pile in Lesny & Wiemann (2005). For the monopile with a large diameter, the zero-toe kick criterion leads to a very large pile length. Hence, the zero-toe-kick criterion is considered over-conservative for the monopiles with large diameter and was substituted by the vertical-tangent criterion in the new edition of the guideline in Germanischer Lloyd (2005). An embedded pile length conforms to the vertical-tangent criterion should be vertical at least at one location in the deflection curve (see Fig.1(a) and Fig. (b)).

The lateral deformation response of the monopile foundations with a large pile diameter under monotonic and cyclic loading are investigated with the finite element model in this study. The suitability of requirements regarding the different critical lengths $L_{c,const}$, $L_{c,vt}$ and $L_{c,ztk}$ for the minimum required pile length is discussed.

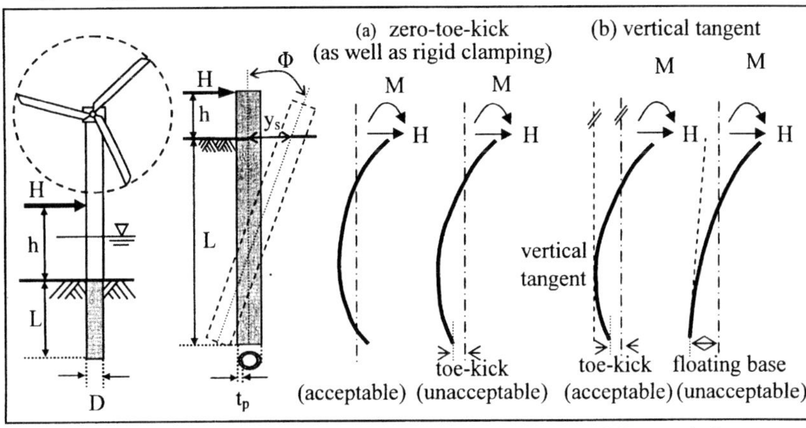

FIG. 1. Loading condition of monopile foundations for offshore wind energy converters and the design criteria for monopiles suggested by Germanischer Lloyd [Faber & Klose (2006)].

FINITE ELEMENT MODEL FOR MONOPILE FOUNDATIONS

The three-dimensional (3D) finite element model to investigate the lateral deformation response of a pile in sandy soil in this study was established by Abdel-Rahman & Achmus (2005). The computations were carried out using the finite element program system ABAQUS (Abaqus 2008). To account for the non-linear soil behavior, soil elements are simulated as elasto-plastic material with Mohr-Coulomb failure criterion and a stress dependency of the oedometer stiffness modulus was implemented as follows:

$$E_s = k\sigma_{at}\left(\sigma_m / \sigma_{at}\right)^{\lambda} \tag{8}$$

Herein σ_{at} is the reference (atmospheric) stress and σ_m is the current mean principal stress in the soil elements. The parameter κ determines the soil stiffness at the reference stress state and the parameter λ rules the stress dependency of the soil stiffness. To determine the stiffness parameters for large-diameter laterally loaded piles, the pile deflections of monopiles with small diameter are compared with the results of API p-y curves method [API (2000)]. The p-y curves method is known to give a suitable estimation of pile deflection for the piles with diameters upper limit to 2.5m. By fitting the deflection of piles with small diameter obtained from numerical method to the results of p-y curves method, the stiffness parameters are calibrated and are used in the finite element model of monopiles with large diameter [Achmus et al. (2008a)].

To consider the soil behavior under cyclic load, the degradation stiffness method incorporated with the cyclic triaxial test results is considered [Achmus et al. (2007), Kuo (2008)]. The secant soil stiffness E_s decreases with the number of cycles N as:

$$E_{sN} / E_{s1} = N^{-b_1 (X_c)^{b_2}} \tag{9}$$

Herein X_c is the characteristic cyclic stress ratio which depends on the static failure stress ratio and the cyclic stress level, and b_1, b_2 are regression parameters determined from cyclic triaxial tests [Huurman (1996)]. The regression parameters for dense sand based on the existing triaxial test results are documented by Kuo (2008). By calculating with a special numerical method, the degradation stiffness of each soil element can be determined and – by considering the degraded stiffnesses in a subsequent numerical simulation of the pile-soil system – the deformation of monopiles after a specified cycle number N can be obtained.

The model conditions of monopiles and material parameters used in this study are given in Table 1.

Table 1. Model conditions and parameters used in the finite element model

<table>
<tr><th colspan="9">Pile properties</th></tr>
<tr><th colspan="3">Pile geometry</th><th colspan="6">Pile parameters</th></tr>
<tr><td colspan="2">Pile diameter D (m)</td><td>5</td><td colspan="4">Elastic modulus of pile E_p (kPa)</td><td colspan="2">2.1E8</td></tr>
<tr><td colspan="2">Embedded length L (m)</td><td>20, 25, 30 40, 50, 60</td><td colspan="4">Submerged unit weight γ'_p (kN/m^3)</td><td colspan="2">68</td></tr>
<tr><td colspan="2">Pile thickness t_p (m)</td><td>0.06</td><td colspan="4">Poisson's ratio of pile ν_p</td><td colspan="2">0.2</td></tr>
<tr><th colspan="9">Soil properties</th></tr>
<tr><th rowspan="2">Soil type</th><th>Submerged unit weight (kN/m^3)</th><th colspan="2">Stiffness parameters</th><th>Friction angle (°)</th><th>Dilatancy angle (°)</th><th>Poisson's ratio</th><th colspan="2">Degradation parameters</th></tr>
<tr><th>γ'_s</th><th>κ</th><th>λ</th><th>φ'</th><th>ψ</th><th>ν_s</th><th>b_1</th><th>b_2</th></tr>
<tr><td>Dense sand</td><td>11</td><td>800</td><td>0.55</td><td>37.5</td><td>7.5</td><td>0.25</td><td>0.2</td><td>5.76</td></tr>
</table>

PILE BEHAVIOR UNDER MONOTONIC LOADING CONDITION

A lateral load H is applied on an offshore wind energy converter at a height h above soil surface (see Fig. 1) and results in a horizontal displacement y_s at soil surface and a secant rotation angle Φ on the monopile with a length L and pile diameter D. The load-deformation responses of monopiles with a diameter 5m are presented in Fig. 2. The lateral displacement of piles of course increases with the lateral load H and the moment arm h. The load-deformation curves for the monopiles with $L \geq 30$m are almost identical under a lateral load not exceeding 15MN. The deformation response could diverge by increasing the load level for the monopiles with L=30m under a large moment arm h=30m. Thus, the critical pile length $L_{c,const}$ determined from the finite element model is load-dependent.

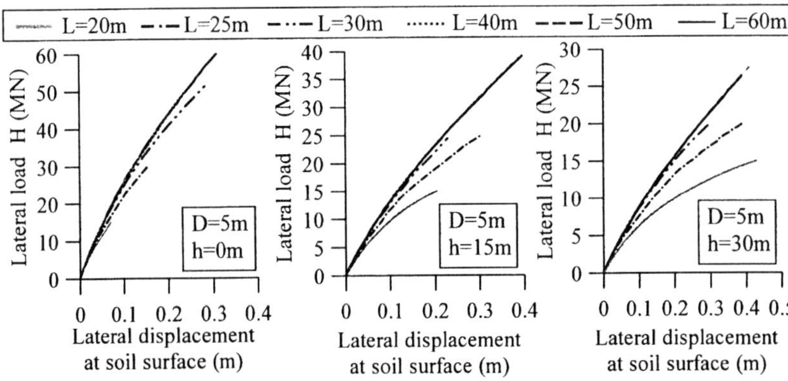

FIG. 2. Load-deformation response of monopile with diameter D=5m in dense sand.

Under a selected lateral load H=15MN, the monopiles with D=5m present different pile deflection responses with increasing embedded length (Fig. 3). This selected load level is high enough for the design of 5MW offshore wind turbine in German region based on the authors' experience. Although the lateral displacement at soil surface of monopiles of D=5m with L=30m to 60m are almost identical, the horizontal displacements at pile tip are quite different. For the pure lateral load condition (h=0), the horizontal displacements of the monopiles are not decreased by increasing the pile length when L≥40m under H=15MN. However, the monopile with L=40m is even not reaching the vertical-tangent criterion. Only the monopile with a very long embedded pile length as 60m reaches the zero-toe-kick criterion with a positive horizontal displacement at pile tip under a lateral load with a large moment arm h=30m. The monopile with L=20m has the largest deformation and a floating base rotation under all loading conditions. A monopile needs a larger embedded length to reach the vertical-tangent under a pure H condition than the monopile under a lateral load with a moment.

The pile length reached $L_{c,vt}$ for the monopiles of diameter 5m are presented in Fig. 4(a). For the monopiles with diameter of 5m, the vertical-tangent response happens only on the piles with length longer than 40m even under a very small lateral load. This pile length $L_{c,vt}$ depends on the lateral load H, moment arm h and pile stiffness. Under a given loading condition, for the monopiles with an embedded length shorter than $L_{c,vt}$ the floating base response applies. The monopiles subjected to a lateral load with a moment arm h need a shorter embedded length to reach the vertical tangent response than the monopiles subjected to an identical lateral load but without a moment arm.

The deformation responses of the pile top under the given loading conditions for different pile lengths are presented in Fig. 4(b). It is clear that the critical length $L_{c,const}$ depends significantly on the loading condition. Under a specified loading condition, the deformation of the pile top is not influenced by the pile length when the pile length exceeds $L_{c,const}$. The critical length of a monopile $L_{c,const}$ subjected to a lateral load with a moment arm is longer than a monopile subjected to a lateral load without a moment arm. This result is contrary to the results of $L_{c,vt}$ in Fig. 4(a).

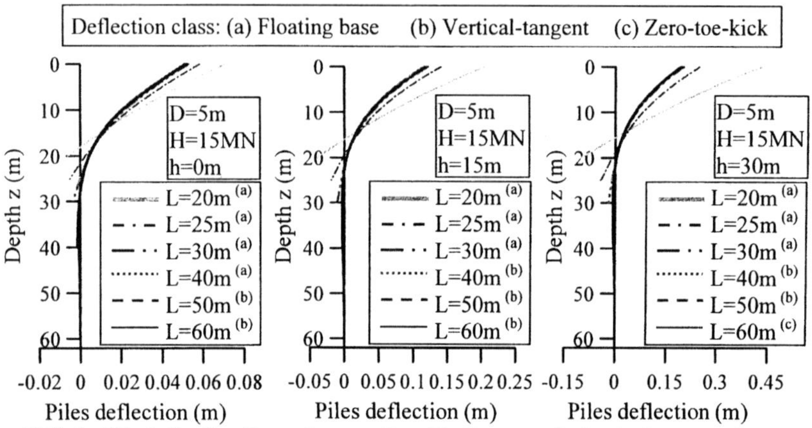

FIG. 3. Pile deflection lines of monopile with diameter D=5m in dense sand.

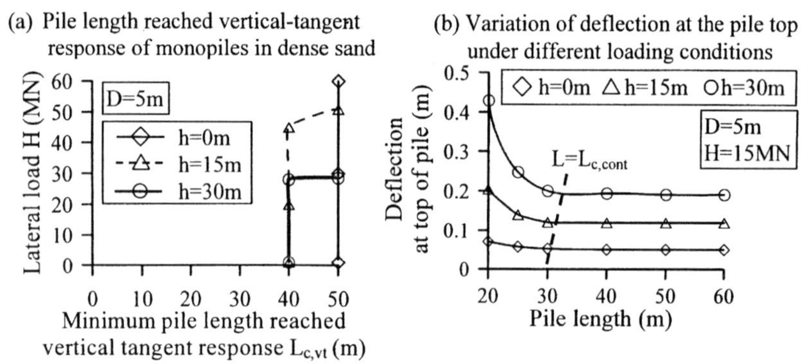

FIG. 4. Critical pile length $L_{c,cont}$ and $L_{c,vt}$ of monopile with diameter 5m in dense sand under different loading conditions calculated by FEM.

The critical pile length $L_{c,const}$ determined from the suggestion of Titze (1970) is 28.3m for the monopile with D=5m under the pure H loading condition, where the coefficient of subgrade modulus n_h=33900kN/m^3 for dense submerged sand is used based on the suggestion of API (2000). For the pure-moment loading condition, the critical length $L_{c,const}$ is 24.7m for the monopiles with D=5m.

To determine the critical length according to the suggestion of Randolph (1981), some iterative calculations are necessary. The shear modulus of soil is derived from the soil stiffness due to Eq. (8) and Poisson's ratio in Table 1. The value G_c is the average of G^* at depth of soil surface and active length of pile. The critical length $L_{c,const}$ determined from Eqs. (6) and (7) is 29.7m for the monopile with D=5m. In contrast to the critical lengths calculated by the finite element model with elasto-plastic soil materials in Fig. 4(b), the critical lengths $L_{c,const}$ determined from the method of Titze (1970) and Randolph (1991) are load-independent. The critical lengths $L_{c,const}$ determined from the methods of Titze (1970) and Randolph (1981) are close to the results of monopiles under low loading level in the finite element analysis. This is probably due to an elastic soil continuum considered in the methods of Titze (1970) and Randolph (1981).

For the monopile with D=5m subjected to a lateral load H=15MN with a moment arm h=0m, the critical length obtained from Fig. 4(b) is 30m. The pile deflection for a monopile with this length is still in a floating base response (Fig. 3). The pile length reached vertical-tangent exceeds 40m under the identical loading condition (Fig. 4(a)). Based on the calculation results, an identical deformation response of pile top can happen on the monopiles with floating base, vertical-tangent or zero-toe-kick response at pile tip when the loading condition is small enough. The deformation responses of monopiles are strongly dependent on the loading conditions and the critical length $L_{c,const}$ increases with the loading level H and moment arm h, as can be seen in Fig. 2 and Fig 4(b). Hence, the critical pile length based on the pile tip deformation response as $L_{c,vt}$ and $L_{c,ztk}$ are not suitable to determine the required pile length of the large-diameter monopile foundations in the consideration of limiting the pile deflection. The critical pile length $L_{c,const}$ is more relevant to the deformation response of pile top and is more suitable as a judgement of the required pile length for the design of offshore wind turbine foundations.

THE MINIMUM REQUIRED LENGTH UNDER CYCLIC LOADING CONDITION

Based on the laboratory test results of Hettler (1981) and field test results of Alizadeh & Davisson (1970), the horizontal displacements of monopiles increase with load cycles when the lateral load is high enough. For the offshore wind turbines cyclic loading results from wind and waves, and excess lateral deformation could happen on the monopile foundation and imperil the serviceability of wind energy converters. To ensure the serviceability of wind turbines, the tolerant deformation response of monopile foundations should be more suitable as a criterion to determine the minimum required pile length. In this section, the deformation responses of monopiles under cyclic loading are studied with the degradation stiffness method. The ability of monopiles with different pile length to limit the accumulation of lateral deformation is compared.

For the cyclic loading condition, the deformation responses of a monopile with D=5m are presented in the form of the accumulated rate of the horizontal displacement of pile top at soil surface in Fig. 5. The accumulated rate of horizontal displacement of the pile top is the ratio of the horizontal displacement of the Nth cycle y_{sN} to the horizontal displacement of the first cycle y_{s1}. For the monopiles with $L \geq 40$m, the accumulation rate of lateral displacement is identical under the given loading conditions. For the

monopiles subjected to the load H=15MN at h=0m, the accumulation of deformation can be reduced by increasing the pile length till the embedded length reaches the critical length $L_{c,const}$ =30m obtained from Fig. 4(b). The monopiles have an identical deformation response at pile top under cyclic loading when the pile length exceeds $L_{c,const}$ under the given loading condition based on the results of finite element analysis for monotonic loading condition.
For a monopile with L=30m and D=5m, an accumulated deformation response at pile tip with floating base is found under a cyclic lateral load H=15MN with h=15m in Fig. 6. For the monopiles with L=40m and 50m, although the horizontal displacements of the monopile at pile top are identical under H=15MN with h=15m in the monotonic and cyclic conditions in Fig. 3 and Fig. 5, different deformation responses at pile tip are presented. The pile tip deformation of a monopile with L=50m keeps the vertical-tangent response under the cyclic loading. A floating base response of the monopile with L=40m transfers from a vertical-tangent response after 10 cycles; even so this transformation of the pile tip deformation response shows no influence on the deformation of the pile top.

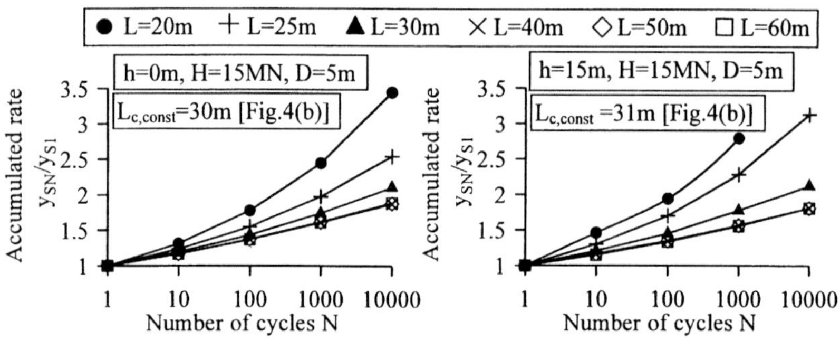

FIG. 5. Accumulation rate of horizontal displacement of monopiles at pile top with D=5m.

Based on the calculation results, no evidence is obtained that monopiles which reach zero-toe-kick or vertical-tangent criteria have always better deformation responses. All deformation responses are dependent on the loading condition. The suggestions for the deformation responses at pile tip as vertical-tangent and zero-toe-kick seem to be not suitable as the design criteria of the minimum required pile length under cyclic loading. The monopiles with a length longer than the $L_{c,const}$ in monotonic loading condition show the identical accumulation response of deformation in the cyclic loading condition. Hence, for determination of the minimum required length of monopiles under lateral loading, the tolerant deformation of wind energy converters under the cyclic loading in the whole lifetime is more relevant.

CONCLUSIONS

The minimum required embedded length for the monopiles with diameter 5 m embedded in dense sand is investigated by the finite element model under monotonic and cyclic lateral load. The calculation results are compared with the existing suggestions of the critical pile length as $L_{c,const}$, $L_{c,vt}$ and $L_{c,ztk}$.

Under the monotonic loading, the critical length $L_{c,const}$ increases with the loading level and moment arm. The deformation of pile top is not affected by the pile length when the pile length exceeds $L_{c,const}$. The suggestions of Titze (1970) and Randolph (1981) may underestimate the critical length $L_{c,const}$ under a high load level with a large moment arm. An identical deformation at pile top can be found on the monopile with different type of pile tip deformation under a specified lateral load. A floating base response at pile tip can even happen on the monopiles with a length which exceeds $L_{c,const}$. The critical length $L_{c,vt}$ and $L_{c,ztk}$ based on the deformations of pile tip are not suitable for determining the required pile length to limit the deformation response under monotonic loading.

Under the cyclic loading condition, a nearly identical accumulation rate of horizontal displacement at pile top is found when the pile length exceeds $L_{c,const}$ obtained from the finite element model. The type of deformation response at pile tip can be disregarded.

For an offshore wind turbine, the tolerant deformation of wind energy converters and the applied loading level are the most important factors to determine the minimum required pile length. For a given pile diameter, the critical pile length can be obtained under a design load level under monotonic condition. The accumulated deformation under cyclic loading of this monopile with the critical length should stay within the tolerance of wind energy converters. In case that the calculated deformation exceeds tolerant deformation of wind turbine, an alternative diameter should be considered.

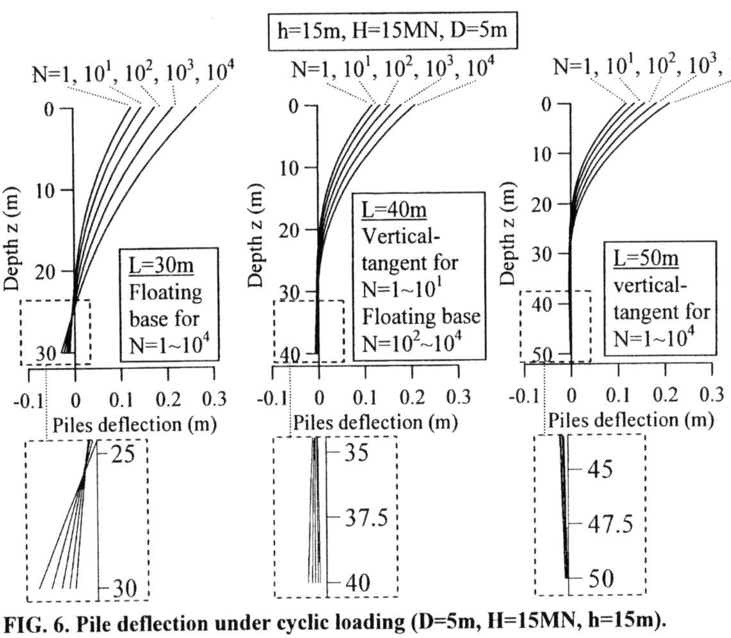

FIG. 6. Pile deflection under cyclic loading (D=5m, H=15MN, h=15m).

REFERENCES

ABAQUS User's Manual, Version 6.7, (2008). Simulia, Providence, RI, USA.

Abdel-Rahman, K. and Achmus, M. (2005). "Finite Element Modelling of horizontally loaded Monopile Foundations for Offshore Wind Energy Converters in Germany." *International Symposium on Frontiers in Offshore Geotechnics (ISFOG)*, Perth, Australia: 309-396.

Achmus, M., Abdel-Rahman, K. and Kuo, Y.-S. (2007). "Numerical Modelling of large Diameter Steel Piles under Monotonic and Cyclic Horizontal Loading." *Tenth International Symposium on Numerical Models in Geomechanics*, Greece: 453-459.

Achmus, M., Abdel-Rahman, K. and Kuo, Y-S. (2008a). "Design of Monopile Foundations for Offshore Wind Energy Plants." *11th Baltic Geotechnical Conference-Geotechnics in Maritime Engineering*, Gdansk, Poland, Vol. 1: 463-470.

Achmus, M.,. Kuo, Y-S. and Abdel-Rahman, K. (2008b). "Zur Bemessung von Monopiles für zyklische Lasten." *Bauingenieur*, Vol. 83 (8): 303-311 (in German).

Achmus, M.,. Kuo, Y-S. and Abdel-Rahman, K. (2009). "Behavior of monopile foundations under cyclic lateral load." *Computers and Geotechnics*, Vol.36 (5): 725-735.

Alizadeh, M., and Davisson, M. T. (1970). "Lateral load test on piles-Arkansas River project." *Journal of the Soil Mechanics and Foundations Division*, ASCE, Vol. 96(5): 1583-1604.

American Petroleum Institute (API) (2000). "Recommended Practice for planning, designing and constructing fixed offshore platforms- working stress design." *API Recommended Practice 2A-WSD (RP2A-WSD)*, 21st edition, Dallas.

Faber T. and Klose M. (2006) "Experiences with certification of offshore wind farms." *Proceedings of Sixteenth International Offshore and Polar Engineering Conference*, San Francisco: 375-382.

Germanischer Lloyd (2005). *Guideline for the Certification of Offshore Wind Turbines*, Hamburg, Germany.

Hetenyi, M. (1946). *Beams on elastic foundations*, Ann Arbor: University of Michigan Press.

Hettler, A. (1981). *Verschiebungen starrer und elastischer Gründungskörper in Sand bei monotoner und zyklischer Belastung*, Ph.D. thesis, Universität Fridericiana Karlsruhe, Heft 90.

Huurman M. (1996). "Development of traffic induced permanent strain in concrete block pavements." *Heron*, Vol. 41(1): 29-52.

Kuo, Y.-S. (2008). *On the behavior of large-diameter piles under cyclic lateral load*, Ph.D. thesis, Lebniz Universität Hannover, Hannover, Heft 65.

Lesny, K. & Wiemann, J. (2005). "Design aspects of monopiles in German offshore wind farms." *Proceedings of the first International Symposium on Frontiers in Offshore Geotechnics*, Perth, Australia: 383-389.

Randolph, M. F. (1981). "The response of flexible piles to lateral loading." *Géotechnique*,Vol.31(2): 247-259.

Titze, E. (1970). *Über den seitlichen Bodenwiderstand bei Pfahlgründungen*, (in German), Bauingenieur-Praxis, Hannover, Heft 77, Verlag Ernst & Sohn, Berlin.

Effects of PCC Pile Spacing on Expressway Embankment Settlement Control

Ting Zhang[1, 2, 3], MIEAust, Hanlong Liu[2, 3], Xiangtao Xu[4] and Ming Dai[2, 3]

[1] Sinclair Knight Merz Pty Ltd, 263 Adelaide Terrace, Perth, WA 6000, Australia; tzhang@skm.com.au
[2] Geotechnical Research Institute of Hohai University, Nanjing 210098, China
[3] Key Laboratory of Ministry of Education for Geomechanics and Embankment Engineering, Hohai University, Nanjing 210098, China
[4] Advanced Geomechanics Pty Ltd, 4 Leura St, Nedlands, WA 6009, Australia; xiangtaox@ag.com.au

ABSTRACT: The Large Diameter Cast-in-place Concrete Pipe Pile (referred as PCC pile) has been used primarily in soft ground improvement applications such as for bridge abutment support piling. To assist design of PCC pile group, this paper investigates the influence of pile spacing on the embankment settlement control based on a series of numerical analysis. The pile spacing was varied from 3D to 5D. The induced horizontal displacement on neighbouring piles upon loading was also presented. It was found that settlement reduces as pile spacing reduces. Pile spacing of 3D is reasonable in terms of settlement control and economic purposes.

INTRODUCTION

A new type of ground improvement method using large diameter cast-in-place concrete pipe pile (referred as PCC pile) has been patented in China (Liu et al 2003a, 2003b, 2004a, 2004b, 2009 and Xu et al 2006). The PCC pile is normally ~1m in diameter with a wall thickness of ~0.12m. The proprietary construction procedure for a PCC pile is: (i) vibro-driving of the temporary steel pipe pile, the base of which prevents soil entering the annulus during installation, but opens under application of a concrete pressure head (ii) pouring concrete into the annular void space of this pipe and (iii) withdrawing the steel pipe pile using the same driving equipment. The end result is an un-reinforced concrete pipe pile.

The PCC pile has been used primarily in soft ground improvement applications such as for bridge abutment support piling (BASP) and was developed to assist fast-track developments of expressway and other infrastructure in Jiangsu, Zhejiang, Hunan provinces and Tianjin city (Xu et al 2006, Liu et al 2005a and 2005b, Yang et al 2004). The PCC pile has a high shaft friction to concrete volume ratio (50% less in concrete volume compared to a solid pile with the same outer diameter) and avoids the need for reinforcement to deal with handling and installation stresses; it therefore offers a cost-effective solution to resist compression loads. The PCC's adaptability to various poor ground conditions has made it a popular choice since first being employed. Photos showing a group of completed PCC piles and finished expressway embankment are provided on Figures 1a,1b and 1c. Before construction of the expressway embankment, a cushion consisting of broken stones (~30-50cm in thickness) and one or two-layer of geotexile are placed on top of the piled foundation. This paper is aimed to study the response of the piled foundation under embankment loading, in particular, the influence of the pile spacing on embankment settlement control.

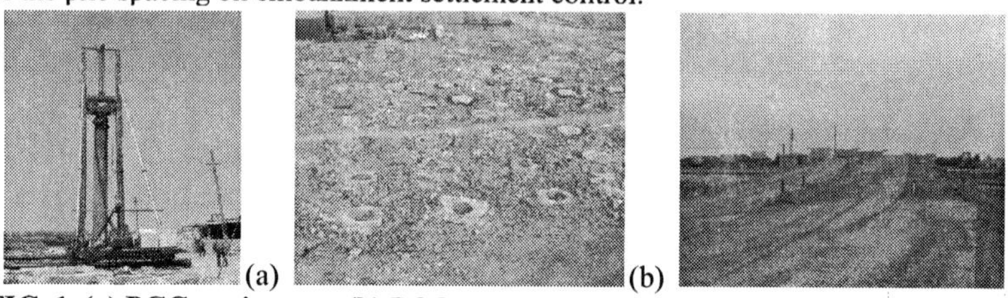

FIG. 1. (a) PCC equipment, (b) PCC pile group, and (c) Expressway embankment.

NUMERICAL ANALYSIS

The numerical analysis by ABAQUS in this study includes a group of PCC piles, the flexible cushion and expressway embankment. Considering the symmetry of the arrangement of PCC piles and importance of settlement at axis, a 3D finite element mesh (as shown in Figure 2a) was adopted for all analyses. The pile is 1m in diameter with a wall thickness of 0.12m and 18m in length. The thickness of the cushion is 0.3m. The height of mesh is 27m and the width is 36m (~3 times the bottom width of the embankment). The embankment was built in stages, each 0.4m up to a maximum height of 5m. The unit weight of the embankment fill is 20kN/m^3. The pile spacing investigated was 3.0D, 3.5D, 4.0D, 4.5D, and 5D. The details of the soil model and parameters are given by Wen et al (2005) which are shown in Table 1 and Table 2.

Table 1. Soil Properties

Soil Stratum	Name	Thickness	γ	Young's Modulus	c	φ
		m	kN/m3	MPa	kPa	°
1	Fills	0.7	18.9	4.8	20.3	15.4
2	Silty Clay	2.6	18.8	5.3	22.1	15.8
3	Sandy Silt	12.2	17.6	3.3	12.7	16.3
4	Silt	12.1	18.1	3.98	26.8	17.5
5	Silty Sand	17.3	19.1	14.2	1.8	31.2

Table 2. Fill and Pile Properties

Materials	Name	γ	Young's Modulus	Poisson's ratio
		kN/m3	MPa	-
1	PCC Pile	25	20000	0.15
2	Cushion	20	1000	0.15
3	Embankment	20	30	0.3

Linear elastic soil model was adopted for the embankment fill, cushion and the PCC pile. However, the soil response is highly nonlinear and can only be more realistically represented by Mohr-Colomb model with smoothing flowing potential functions. Thin layer element (Lu and Bao 2000) was adopted in the interface between the pile, soil and cushion. Standard fixity was adopted (i.e. total fixity in the bottom, and free in vertical direction in the left and right-hand boundaries).

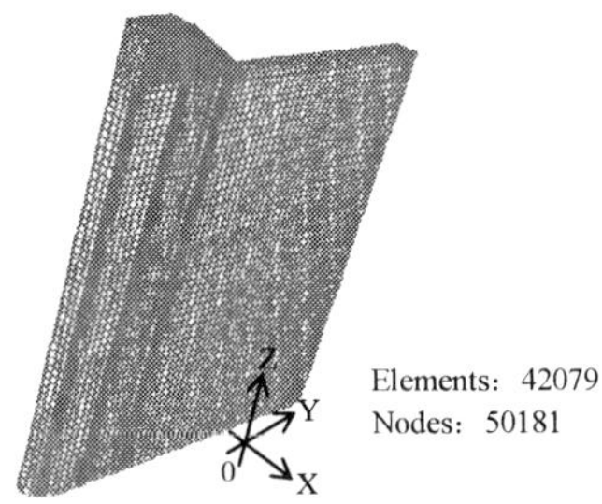

FIG. 2. 3D finite element mesh

ANALYSIS RESULTS

Soil Settlement

Under certain load, the maximum settlement of the foundation occurs in the soil in the middle of the embankment. Figure 3 shows the variation of the maximum settlement with fill height at different pile spacing. As illustrated, the maximum settlement is influenced by both the fill height and pile spacing. It increases as fill height increases and decreases as pile spacing reduces. For instance, at fill height of 4.8m, the maximum settlement reduces ~20mm increment with every 0.5m reduction in pile spacing. Therefore, for pile spacing of 3 to 5 times pile diameters (i.e. 3D to 5D), it appears reasonable to adjust the pile spacing to control the soil settlement.

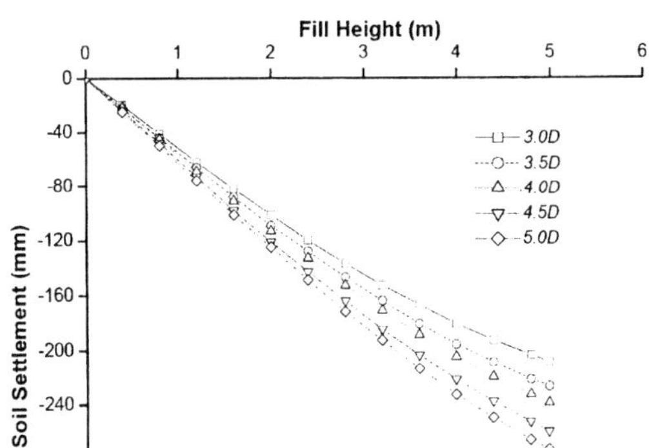

FIG. 3. Maximum soil settlement in the middle of the embankment.

Embankment Settlement

The settlement of the embankment depends on response of piled foundation and the cushion with geotextiles in between. To better quantify the results, the settlement of the embankment (w_{emb}) is normalized by the embankment fill height (h). The w_{emb}/h ratio is shown to decrease at the beginning of the loading (i.e. h<1.5m). At fill height greater than 1.5m, the w_{emb}/h ratio increases slightly with fill height. As shown in Figure 4, it also depends on the pile spacing (S). The relationship between w_{emb}/h, and S is further presented in Equation 1, which is derived based on linear regression of the data presented in Figure 4.

$$w/h = -0.03506 + 0.00198\,h - 0.00614\,S \quad (R^2=0.945) \tag{1}$$

where, w is the maximum displacement; S is pile spacing

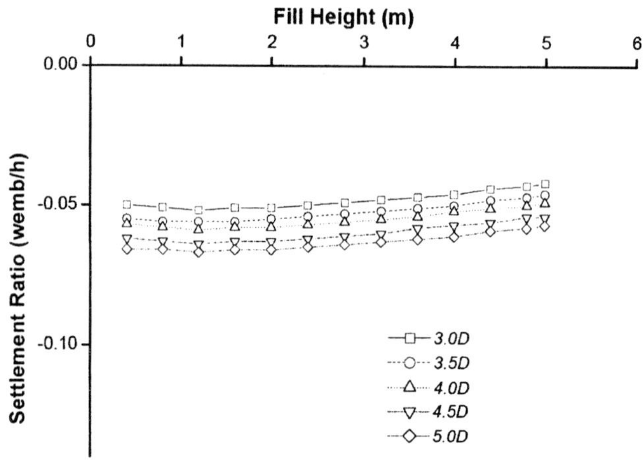

FIG. 4. Ratios of w_{emb}/h as a function of fill height (h) and pile spacing (S).

Pile Settlement

The settlement of the pile head upon loading is similar to that of the surrounding soil, but smaller in magnitude. The difference in settlement between the pile head and soil (Δw) increases with fill height. Values of Δw are significantly influenced by pile spacing. As illustrated in Figure 5, values of Δw reduce as pile spacing reduces. It is about 2mm for pile spacing of 3D, while 11mm for pile spacing of 5D under 5m embankment fills. It should be mentioned that the Δw may be underestimated since the consolidation of the soil upon loading is not considered in this numerical analysis.

Soil Settlement at Depths

Figure 5 shows the variation of the soil settlement with fill height and Figure 6 shows differential settlement in depths. It is apparent that at depth greater than pile length (i.e.>~18m), there is no difference between soil settlement under different pile spacing.

At shallow depth, the settlement at depths decreases with reduction in pile spacing. Therefore, it is more appropriate to adopt 3D pile spacing to minimize the settlement in pile group foundation design.

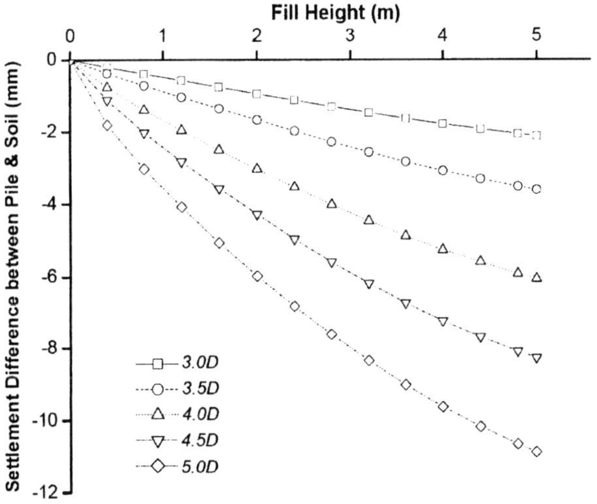

FIG. 5. Settlement difference (Δw) between pile head and the surrounding soil vs fill height.

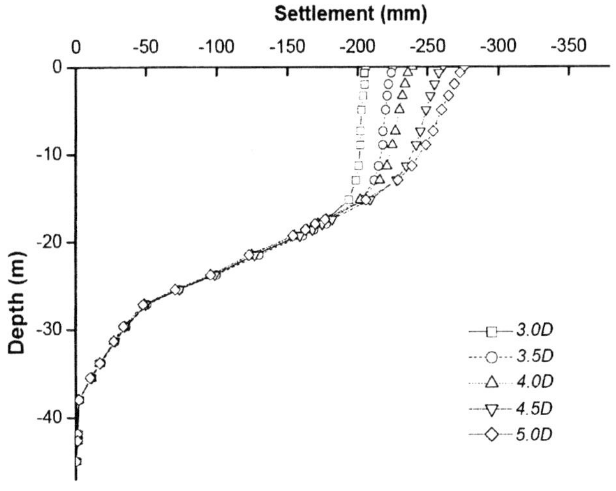

FIG. 6. Differential settlement in depth.

Horizontal Displacement

Figure 7 shows the horizontal displacement of the edge piles (i.e. under the toe of the expressway embankment) at different pile spacing under 4.8m high of embankment fill. The horizontal displacement (δ) increases with increase in depths. The δ value is much smaller at pile spacing of 3D than that at pile spacing of 5D. For example, at the pile tip level, δ is up to 35mm at pile spacing of 5D, while it is less than 25mm at pile spacing of 3D, which is about 30% less. This is because the with larger pile spacing, the surrounding soil is less confined laterally by the piles.

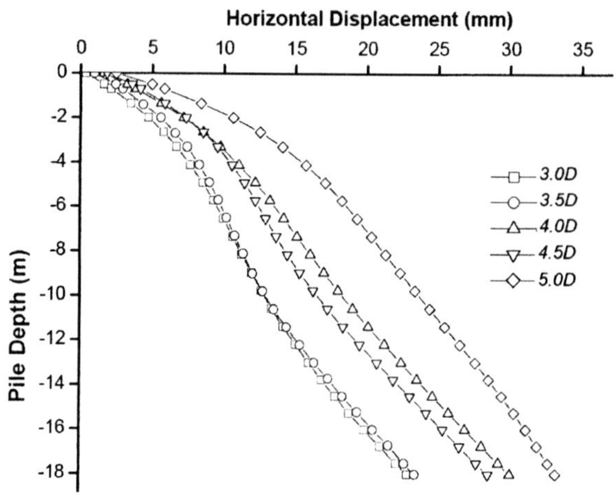

FIG. 7. Horizontal displacement of edge pile under 4.8m of fill

Load Distribution Ratio between Pile and Soil

The ration of load carried by the pile body and its surrounding soil is defined here as n. The values of n at various spacing (i.e. 3.5D, 4D, 4.5D & 5D) are normalized by that at 3D (referred as n_{3m}). Figure 8 presents the variation of n/n_{3m} with fill height and pile spacing. It is shown that n/n_{3m} reduces with increase in fill height, with the tendency to reach a constant value at certain pile spacing. For fill height greater than ~3m, the ratio of n/n_{3m} is relative constant. The n/n_{3m} ratio increase as pile spacing increases. In other words, it reduces as the replacement ratio increases.

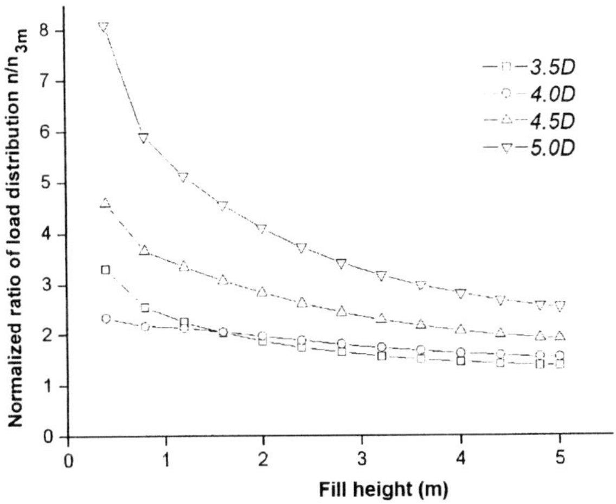

FIG. 8. The ratio of load distribution between the pile and surrounding soil

CONCLUSIONS

This paper illustrates the influence of the PCC pile spacing on embankment settlement control through a series of numerical analysis. The settlement increases as the pile spacing increases from 3D to 5D. This is also in consistent with previous field experimental results (Wen 2005). The ratio of load distribution between the soil and pile remains relatively independent of the pile spacing if the fill height is greater than 3D.

ACKNOWLEDGMENTS

The authors appreciate the support of the National Science Foundation of China NOs. 50679017 and 50778063.

REFERENCES

Liu, H. L., Fei, K., Ma, X. H. and Gao, Y. F. (2003) "Large-diameter driven cast-in-place concrete thin-wall pipe pile (I) Research and Development." *Journal of Rock and Soil Mechanics*, Vol. 24: 164-198.

Liu, H. L., Fei, K., Ma, X. H. and Gao, Y. F. (2003) "Large-diameter driven cast-in-place concrete thin-wall pipe pile (II) Application." *Journal of Rock and Soil Mechanics*, Vol. 24: 372-375.

Liu, H.L., Ma, X. H., Chu, H. Y., Chen, Y. H. and Gao, Y. F.(2004) "Cast-in-place pipe pile construction equipment for soft ground improvement." Patent Number: ZL 02219218.X.

Liu, H.L., Ma, X. H., Chu, H. Y., Chen, Y. H. and Gao, Y. F.(2004) "Cast-in-place pipe pile construction technology for soft ground improvement." Patent Number: ZL

021125384.4.6.

Liu, H.L., Fei, K. and Xu, X.T. (2005) "Development and application of the large-diameter driven cast-in-place concrete thin-wall pipe pile." *Proc. Of the 16th International Conference on Soil Mechanics and Geotechnical Engineering, Oosaka, Japan:* 2137-2140.

Liu, H.L., Fei, K. Deng, A. and Zhang T. (2005) "Erective Sea Embankment with PCC Pile." *China Ocean Engineering*, Vol. 19(2), 339-348.

Liu, H.L., Chu, J. and Deng, A. (2009) "Use of large diameter concrete cast-in-situ pipe pile for embankment over soft clay." *Can. Geotech J.*, Vol. 46: 915-927.

Lu, T.H. and Bao, F.B. (2000) "Coupled constitutive model of thin layer element of interface." *Journal of Water Resource*, Vol. 2: 71-75

Wen, X.J., Zhang, T. and Dai, M.(2005) "FEM analysis on settlement control of PCC pile composite foundation." Research report of GeoHohai, Nanjing.

Xu, X.T., Liu, H.L. and Lehane, B.M. (2006) "Pipe pile installation effects in soft clay." *Proceedings of the Institution of Civil Engineers, Geotechnical Engineering 159,* Issue GE4: 285-296.

Xu, X., Lehane, B.M., C. Gaudin, Zhang, T. and Liu, H.L. (2006) "Centrifuge studies of single and group displacement piles in clay." *Proc. of the 6th International Conference on Physical Modelling in Geotechnics, Hong Kong:* 895-900.

Yang, S.S., Liu, H.L., Zhou, Y.D. and Fei, K. (2004) "The application of PCC piles in ground improvement of expressway." *Chinese Journal of Geotechnical Engineering*, Vol. 26(6): 750-754.

Zhang, T., Liu, H.L., Hu, Y.X., Stewart, D. (2009) "Geotechnical Drum Centrifuge Technique and Its Engineering Application." *Journal of Rock and Soil Mechanics*, Vol. 30(4):1191-1196.

Strength and Deformation Characteristics of Lagoonal Clay Deposits in Lixia River Area by Seismic Piezocone Tests

G. J. Cai[1], S. Y. Liu[1] A. J. Puppala[2], L. Y. Tong[1] and G. Y. Du[1]

[1]Institute of Geotechnical Engineering, Southeast University, Nanjing 210096, China; email: focuscai@163.com
[2]Department of Civil Engineering, The University of Texas at Arlington, Arlington, Texas 76019; email: anand@uta.edu

ABSTRACT

The results of a site investigation on the soft, normally to slightly overconsolidated, Lixia River Lagoonal clay deposit underlying several expressway lines based on seismic piezocone penetration tests (SCPTUs), undisturbed soil sampling from an adjacent borehole, and laboratory tests, are reported here. SCPTU data were used both in the determination of soil profile and the interpretation of strength and deformation properties. The investigation demonstrated a close agreement between strength parameters derived from the field vane tests, and laboratory tests and interpreted SCPTU parameters with a cone factor of N_{kt}=16.5. Comparison of the results reveals the validity of SCPTU tests to interpret the strength and deformation properties of Lixia River Lagoonal clay deposits.

INTRODUCTION

The seismic piezocone (SCPTU), which provides nearly continuous measurement of tip resistance (q_t), sleeve friction (f_s), pore pressure (u_2), and shear wave velocity (V_s), appears to be an extremely promising tool to interpolate or extrapolate soil property parameters for preliminary design (Lunne et al., 1997). Springman et al. (1999) used the cone penetration or piezocone and triaxial test data to characterize the strength of soft Swiss lacustrine clay and to estimate an undrained shear strength profile. Simonini and Cola (2000) and Ricceri et al. (2002) characterized the soil profile and estimated the geotechnical properties of the Venice lagoonal soils based on piezocone and dilatometer tests. Therefore, the complementary nature of field and laboratory testing lead to correlations that can be used or applied on the basis of local experience.

In this study, a special test site had been selected at the Huaiyan expressway, where, in a limited area, several geotechnical investigations including deep boreholes together with cone penetration test, seismic piezocone, field vane test and down-hole tests have been performed. Some of the results of these investigations concerned with the seismic response of these soils measured in the laboratory and from SCPTU have been already published (Cai et al. 2008).

One of the aims of the field investigations on Lixia River lagoon is to evaluate the effectiveness and reliability of the seismic piezocone, to estimate the geotechnical strength properties, the latter approximately determined through the use of a series of semi-empirical correlative equations applied to the readings recorded throughout the test. Certain modifications are introduced to the correlations for interpreting strengths of the very soft clayey soils. Comparisons among results from various tests are also included.

SITE DESCRIPTION

The seismic piezocone test sites were selected in Huai-yan expressway at the city of Yancheng in eastern Jiangsu province. The ancient Lixia River lagoon is a transgression and regression integrated barrier-lagoon sedimentary system that was characterized based on the previous achievement on lagoon and through detailed investigation of regional geology of Jiangsu. The Lixia River lagoon soft soils developed under such sedimentary environment and influenced by the alluvium of the old Yellow River, therefore there formed the typical lagoonal soft soil with high water content, high organic matter content, high compressibility and inferior engineering properties. The location map of Lixia River area is shown in Figure 1.

The clayey soil samples were collected by means of a stationary piston sampler 76 mm in diameter before embankment construction at 1.0 m interval between 2.0 and 20 m. The lagoonal soft soils studied are slightly overconsolidated, with an OCR of about 1.8. Table 1 gives the typical values of the properties of the soil layers studied.

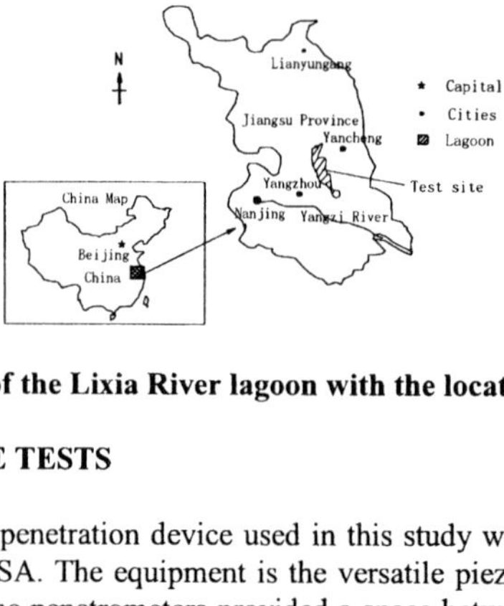

Figure 1. View of the Lixia River lagoon with the location of test sites

SEISMIC PIEZOCONE TESTS

The seismic piezocone penetration device used in this study was produced by Vertek-Hogentogler & Co. of USA. The equipment is the versatile piezocone system equipped with advanced digital cone penetrometers provided a space between 60° and tapered, 10 cm^2 tip area cone which provided measurements of q_c, f_s, and u_2 with a 5-mm-thick porous filter located just behind the cone tip. Water pressures can act over a part of the base area of the cone in a direction opposite to the resisting forces developed during pushing, thus the measured total stress q_c will be reduced. To obtain the actual total stress q_t, a correction must be applied to account for the design of the cone. This correction is done directly by the field computer data acquisition system during data processing for each set of readings, using the following relationship.

$$q_t = q_c + (1 - a) \times u_2 \tag{1}$$

Where q_c = measured cone resistance; u_2 = pore pressure measured at the cone base; q_t = corrected cone resistance, and a is net area ratio = 0.80.

The rate of penetration for all tests in this study was 20 mm/s. At this rate one set of readings could be obtained for every 50 mm of penetration. The shear wave velocities are measured at intervals of 1.0 m, corresponding to successive rod additions during advancement of the penetrometer. It should be noted that the seismic piezocone tests have been carried out close to the boreholes. The typical profiles of q_t, f_s, u_2 and V_s with depth for the test are presented in Figure 2. The profiles confirm the high non-homogeneity of the Lixia River lagoonal soils to the depths investigated.

Table 1. Typical geotechnical properties of the soil layers studied

Soil layer	γ (kN/m^3)	w (%)	w_p (%)	w_L (%)	e	c_v (10^{-3}cm^2/s)	M (MPa)
Fill	17.9-18.5	37.8-40.3	48.6-49.7	61.4-63.5	1.101-1.114	0.702-0.721	3.31-3.50
Clay	17.8-18.1	37.9-42.3	52.4-53.6	70.8-72.3	1.106-1.109	0.513-0.533	3.75-3.92
Muck	15.1-15.5	75.2-77.3	63.4-64.8	85.2-87.5	2.104-2.112	0.348-0.361	1.16-1.38
Mucky clay	17.2-17.9	38.2-43.1	54.3-56.4	77.5-79.4	1.154-1.167	1.711-1.732	5.75-6.23
Silty clay	18.3-18.9	22.3-24.7	40.3-42.6	52.1-54.8	0.858-0.864	4.510-4.532	6.14-6.57

Note: γ, unit weight; w, water content; w_p, plastic limit; w_L, liquid limit; e, void ratio; c_v, coefficient of consolidation; M, constrained modulus.

STRENGTH AND DEFORMATION CHARACTERISTICS

Undrained Shear Strength

The undrained shear strength is an important parameter for evaluation of strength properties of clays. A large amount of work has been reported in literature on the interpretation of undrained shear strength of clays with CPTU results since the advent of piezocone penetrometers (Lunne et al. 1997). Konrad and Law (1987) reviewed the primary approaches and proposed a new interpretation method which takes the measured pore pressures into account. There are mainly two interpretation approaches: one is based on theoretical solutions, and the other is based on empirical correlations. On the basis of bearing capacity theory the estimation of S_u from CPTU using cone resistance is made from the following equation:

$$S_u = \frac{q_t - \sigma_{v0}}{N_{kt}} \tag{2}$$

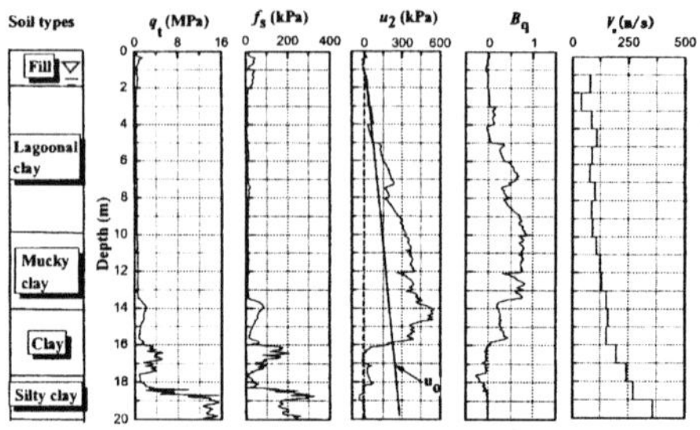

Figure 2. SCPTU profiles at Huaiyan expressway site

Where N_{kt} is an empirical cone bearing capacity factor, and σ_{vo} is the total in situ vertical stress. Several approaches are available for N_{kt} (Yu and Mitchell 1998), these being dependent on the type of analytical solution assumed to describe the undrained penetration process (the limit equilibrium method, the cavity expansion theory, and the strain path method). The bearing factor N_{kt} is characterized by a large range of variation, between 11 and 19 for normally consolidated clays, approaching 30 for heavily over-consolidated soils. In general, a representative value of N_{kt} =15 is adopted to obtain the average S_u in intact clays.

On the basis of undrained shear strength measured in the field vane shear tests (FVT), values of N_{kt} were calculated for soft clay layers. It can be seen from Figure 3 that there is a correlation between the S_u measured from FVT results and $q_t - \sigma_{vo}$ (r^2=0.75). The linear regression analysis result is also shown in Figure 3.

$$S_u = \frac{q_t - \sigma_{v0}}{16.5} \qquad (r^2=0.75) \tag{3}$$

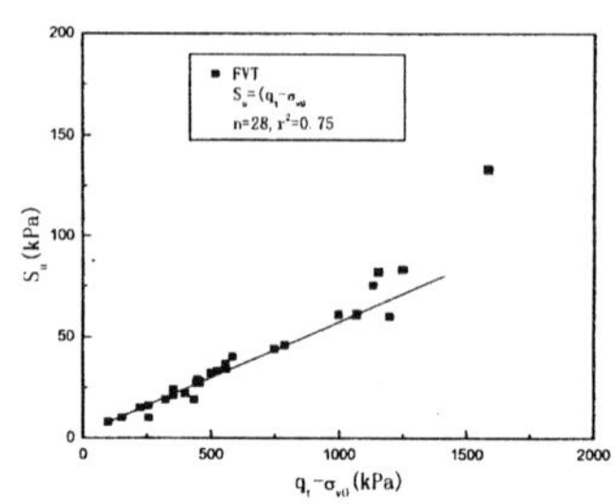

Figure 3. Undrained shear strength from field vane tests versus cone resistance

Therefore Eq. (3) can be used for a preliminary and approximate estimate of undrained strength properties. With theoretical and semi-theoretical approaches based on cavity expansion theory, a number of relationships have been proposed between excess pore pressure Δu and S_u. One such relationship is expressed in the form of following equation:

$$S_u = \frac{\Delta u}{N_{\Delta u}} = \frac{u_2 - u_0}{N_{\Delta u}} \tag{4}$$

Where $N_{\Delta u}$ is an empirical cone pore pressure factor, u_2 is pore pressure measured between the cone and the friction sleeve, and u_o is the equilibrium pore pressure. Based on cavity expansion theory, $N_{\Delta u}$ is theoretically shown to vary between 2 and 20. The pore pressure at failure is related to a change in octahedral normal stresses and shear stresses, and also to OCR. Senneset at al. (1982) defined the pore pressure ratio $(u_2-u_o)/(q_t-\sigma_{vo})$ denoted by B_q to correlate with the OCR. Due to the analogy, Lunne and Lacasse (1985) found that $N_{\Delta u}$ correlates well with B_q for North Sea clays taking the anisotropic consolidated undrained triaxial test sheared in compression (CAUC) strength as the reference values. For the Lixia River lagoonal clays, using S_u values from CAUC tests on undisturbed samples, a better correlation between cone pore pressure factor and the pore pressure ratio B_q is obtained as shown in Figure 4.

$$N_{\Delta u} = 10.2 \times B_q \qquad (r^2=0.88) \tag{5}$$

The above correlations for $N_{\Delta u}$ were derived for normally to slightly overconsolidated clays. The pore pressure ratio can bc obtained using the field measured CPTU data and this leads to the determination of B_q parameter, which in turn can be used to estimate the undrained shear strength using the Equation 4.

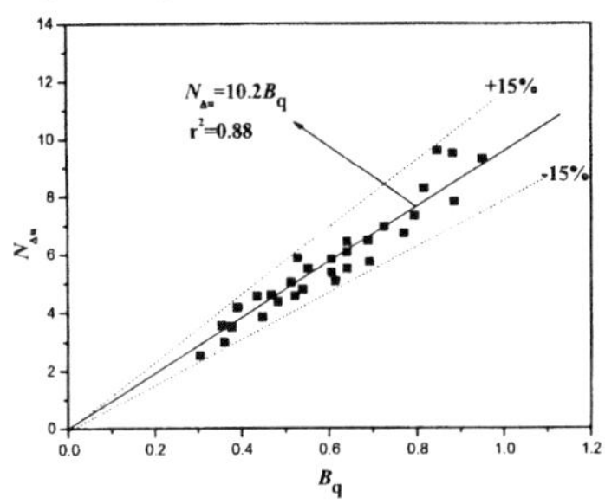

Figure 4. Cone Pore Pressure factor $N_{\Delta u}$ verse pore pressure ratio B_q

Constrained Modulus

The one-dimensional constrained modulus, M, as measured in an oedometer test, has been expressed in terms of a coefficient, α_m, and cone resistance:

$$M = \alpha_m q_c \tag{6}$$

Where α_m is a correlation factor.

In practice, it has been usual to correlate the modulus M to a penetration resistance (Mitchell and Gardner 1975; Schmertmann 1978; Jamiolkowski et al. 1985; Senneset et al. 1989; Kulhawy and Mayne, 1990).

To estimate one-dimensional constrained modulus M, the correlation with net cone resistance $(q_t - \sigma_{v0})$ is used in the form (Kulhawy and Mayne, 1990):

$$M = 8.25 \times (q_t - \sigma_{vo}) \tag{7}$$

Comparisons between M from CPTU with laboratory oedometric modulus for various types of soil proposed by Kulhawy and Mayne (1990) showed that the ratio $M_{\text{-CPTU}}/M_{\text{-lab}}$

could equal to 2.21 for high-plasticity clays and silty soils. In Figure 5 the constrained modulus estimated with relationship (7) is plotted against that determined by laboratory oedometric tests, carried out on all the types of Lixia River lagoonal soils. In our case, the ratio $M_{\text{-CPTU}}/M_{\text{-lab}}$ is always greater than the unity and is not influenced by the type of soil or by its cone resistance value.

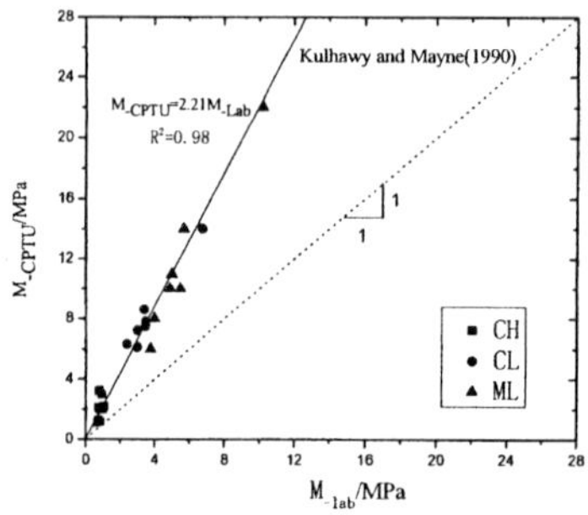

Figure 5. Measured versus predicted constrained modulus values

To examine the possibility for better correlations to estimate the constrained modulus from CPTU data, the corrected cone tip resistance (q_t) and the net cone resistance (q_t -σ_{vo}) were plotted against the laboratory measured constrained modulus as shown in Figures 6(a) and 6(b). A linear correlation was obtained between $M_{\text{-Lab}}$ and q_t as follows:

$$M_{-Lab} = 3.36 q_t \qquad R^2=0.78 \tag{8}$$

And the following linear correlation was also obtained between M and (q_t -σ_{vo}) given as follows:

$$M_{-Lab} = 3.73(q_t - \sigma_{v0}) \qquad R^2=0.92 \tag{9}$$

The arithmetic mean and standard deviation of ($M_{\text{-CPTU}}/M_{\text{-lab}}$) are 2.12 and 0.23 for the first correlation ($M_{\text{-Lab}}$ =3.36q_t), whereas 2.31 and 0.29 for the second correlation [$M_{\text{-Lab}}$ =3.73(q_t -σ_{vo})].

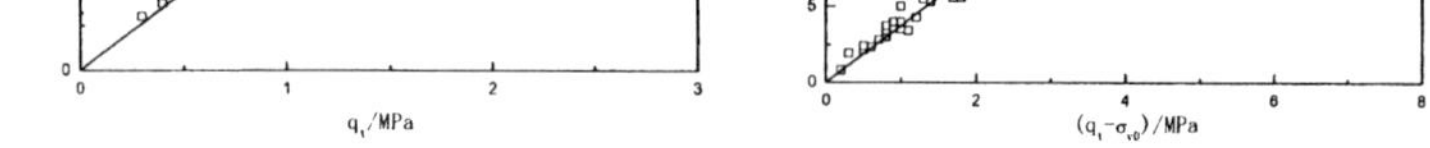

(a) (b)

Figure 6. (a) $q_t \sim M_{\text{-Lab}}$; (b) $(q_t - \sigma_{vo}) \sim M_{\text{-Lab}}$

The profile of predicted constrained modulus using the proposed correlations was compared with the profile of laboratory measured constrained modulus in Lixia River area (Figure 7). The comparison clearly showed that the predicted constrained moduli using the proposed two correlations agree well with the measured constrained moduli.

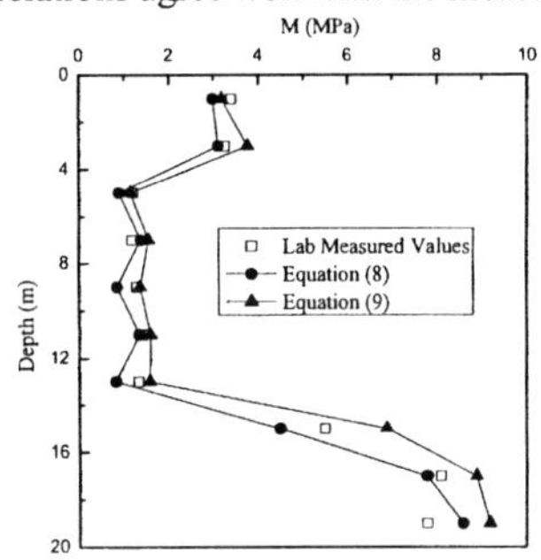

Figure 7. Measured versus predicted *M* values

SUMMARY AND CONCLUSIONS

This paper presents a SCPTU method used to characterize both the soil profile and to estimate the main geotechnical properties of soft, normally consolidated to slightly overconsolidated clays interbedded with silts or low plasticity clays. From the experimental investigation and the comparison between in situ and laboratory results, the following conclusions can be drawn:

(1) The investigation demonstrated a close agreement between strength parameters derived from the field vane tests and laboratory testing when a piezocone factor of N_{kt} =16.5 was applied.

(2) The predicted *M* values obtained from the interpretation method by Kulhawy and Mayne (1990) were compared with the laboratory reference *M* values from laboratory oedometer tests and a few new correlations for *M* were developed for the local soft clayey soils.

(3) The profile of predicted constrained modulus using the proposed correlations was compared with the profile of laboratory measured constrained modulus for the different sites in Lixia River area. The comparison showed that the predicted constrained moduli using the proposed two correlations agree well with the measured constrained moduli.

ACKNOWLEDGEMENTS

This research was supported by the National Natural Science Foundation of China (NSFC) (Grant No. 40702047), the Excellent Doctoral Dissertation Foundation of Southeast University, China (Grant No. YBJJ0714) and the Jiangsu Transportation Research Foundation (Grant No. 8821006021). The first author received financial support from the China Scholarship Council to attend the University of Texas at Arlington as a visiting research scholar under the guidance of Dr. Anand J. Puppala.

REFERENCES

Cai, G. J., Liu, S. Y., Tong, L. Y., and Du, G. Y. (2008). Evaluation of maximum shear modulus of soft clay from seismic piezocone tests (SCPTU). Rock and Soil Mechanics, 29(9): 2556-2560.

Jamiolkowski, M., Ladd, C.C., Germaine, J.T., and Lancellotta, R. (1985). New developments in the field and laboratory testing of soils: State of the art. Proceedings of the 11th International Conference on Soil Mechanics and Foundation Engineering, San Francisco, 12–16 August 1985. A.A. Balkema, Rotterdam. Vol. 1, pp. 57–153.

Konrad, J. M., and Law, K. T. (1987). Undrained shear strength from piezocone tests. Canadian Geotechnical Journal 24, 392-405.

Kulhawy, F.H., and Mayne, P.W. (1990). Manual on estimating soil properties for foundation design. Report No. EL-68000, Electric Power Research Institute, EPRI, August 1990. Palo Alto. 306 pp.

Lunne, T., and Lacasse, S. (1985). Use of in situ tests in North Sea soil investigations. Proceedings of the Symposium: From theory to practice in deep foundations. Porto Allegre, Brazil, Oct. 1985, Published in: Norwegian Geotechnical Institute, Oslo, Publication, 169.

Lunne, T., Robertson, P. K., and Powell, J. J. M. (1997). Cone penetration testing in geotechnical practice. Blackie Academic and Professional.

Mitchell, J. K., and Gardner, W. S. (1975). In situ measurement of volume change characteristics. Proceedings of the ASCE Specialty Conference on In Situ Measurement of soil Properties, Raleigh, North Carolina, 2, 279-345, ASCE.

Ricceri, G., Simonini, P., and Cola, S. (2002). Applicability of piezocone and dilatometer to characterize the soils of the Venice Lagoon. Geotechnical and Geological Engineering, 20: 89-121.

Schmertmann, J. H. (1978). Guidelines for Cone Penetration Test, United States Department of Transport, Report FHWA TS-78-209, 145 pp.

Senneset, K., Janbu, N., and Svano, G. (1982). Strength and deformation parameters for CPT. Proceedings of the 2nd European Symposium on Penetration Testing, ESOPT II , Amsterdam, The Netherlands, 2, 863-870.

Senneset, K., Sandven, R., and Janbu, N. (1989). The evaluation of soil parameters from piezocone tests. Transportation Research Record, No. 1235, 24-37.

Springman, S. S., Trausch, J. G., Heil, H. M., and Heim, R. (1999). Strength of soft Swiss lacustrine clay: cone penetration and triaxial test data. Journal of the Transportation Research Board, No. 1675: 1-9.

Yu, H. S., and Mitchell, J. K. (1998). Analysis of cone resistance: review of methods. Journal of Geotechnical and Geoenvironmental Engineering, 124(2): 140-149.

Monitoring In-Situ Soil Moisture Variations of Expansive Clay using Neutron Probes

Jie Li[1] and Gang Ren[1]

[1]School of Civil, Environmental and Chemical Engineering, RMIT University, GPO Box 2476V, Melbourne, VIC 3001 Australia; jie.li@rmit.edu.au
[1]School of Civil, Environmental and Chemical Engineering, RMIT University, GPO Box 2476V, Melbourne, VIC 3001 Australia; gang.ren@rmit.edu.au

ABSTRACT: The neutron probe has proven to be an effective means for long-term monitoring of *in-situ* soil moisture variations. However, correlation of neutron probe data (i.e. neutron counts) with moisture contents to give accurate absolute values is a very difficult task, particularly for expansive soils due to problems associated with soil volume changing and shrinkage cracks. In this study, a numerical model of neutron diffusion in soils is used to calibrate neutron-probe data collected in a field site over a period of 5 years. The envelopes of soil moisture content derived from the neutron probe, representing the extreme observation at various depths are reported. The effects of tress and covers on soil moisture changes are also examined.

INTRODUCTION

Expansive soils swell or shrink as the moisture content gets higher or lower because of their mineralogical composition. Lightly loaded structures built on expansive soils are frequently subjected to differential movement as a result of volume changes within the foundation soil. The extent of the movement depends mainly on the extent of soil moisture or suction change under the footing. These moisture changes are often induced by seasonal changes in rainfall and evaporation, watering of gardens, leakage from waterpipes, or extraction of water by trees and shrubs. Therefore, accurate measurement of soil moisture changes in the field is essential for foundation-expansive soil-water interaction research. Such measurement can be obtained by the neutron probe technology which has gained wide acceptance as a tool for measuring soil moisture contents in-situ.

As part of a long term study into the behaviour of unsaturated, expansive soils, a field site was established in 1993, in Newcastle, Australia. The Maryland site was selected for this study because it is typical of many new and existing residential housing estates on the eastern seaboard of Australia, where seasonally induced ground movements of between 40 and 70 mm cause significant distress to old and poorly engineered residential structures. The site has been extensively instrumented to allow soil moisture conditions and ground movements to be closely monitored. The primary objective of the Maryland field study is to collect high quality field data that can be used to check current design methods for footings on expansive soils.

Selected results from the Maryland site study have been published by Allman *et al.* (1994) and Fityus *et al* (2004). The present paper is focused on the neutron probe calibration and the field monitoring soil moisture variations.

THE SITE DESCRIPTION

The details of the Maryland site arrangement is shown in Fig 1. The test site, approximately 80m by 25m, is divided into four distinct zones for study: (1) A naturally grassed area, (2) A tree affected area, (3) A 10m x 10m area with an unloaded flexible cover, and (4) A 10m x 10m area covered by a loaded reinforced concrete slab. The soil profile across the site is relatively uniform and can be generally described as 0.25m silty clay topsoil underlain by high plasticity clay to a depth of approximately 1.2m, then medium plastic silty clay to approximately 2m where highly to extremely weathered siltstone is encountered. Soil type changes are gradual with no distinct layer boundaries evident below the base of the topsoil. Clay soils on the site are expansive, realising 5-10% volumetric strain when subjected to a water content change equal to that corresponding to the change from a dry season to a wet season at the field site. Detailed description of the test site can be found in Fityus *et al* (2004).

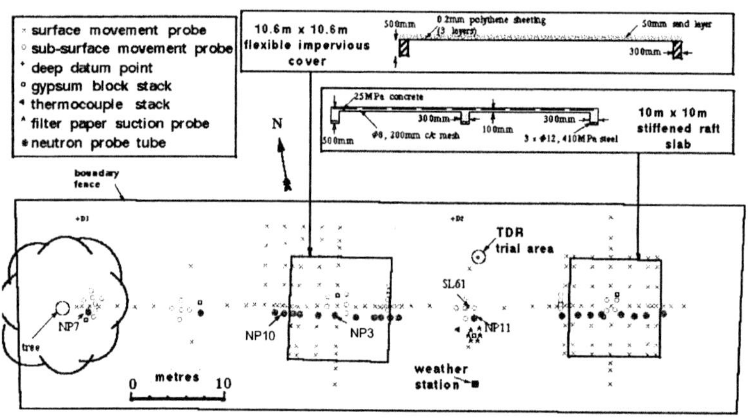

FIG. 1. The Maryland site arrangement (after Fityus *et al*, 2004).

Since 1994, the *in situ* soil moisture content has been monitored using a neutron moisture gauge that consists of a source of fast (high-energy) neutrons, a thermal neutron detector, and the associated electronic equipment necessary to power the detector and to display the results (see Fig. 2). Soil water content is estimated by lowering the neutron source into the ground through the access tube, and counting the number of thermalised neutrons that find their way back to the detector. The main advantages of the neutron method compared to the gravimetric method are: (a) it is non-destructive (after initial installation), (b) it is fast, and (c) repeated measurements can be carried out *in situ*. However, correlation of neutron probe data (i.e. neutron counts) with moisture contents to give accurate absolute values is a very difficult task. It has been shown that the factory calibrations are often inaccurate, particularly for clay soils (Chanasyk *et al.* 1996). The calibration relationship is influenced by the strength of the neutron source, the size and type of the neutron detector, the position of the detector relative to the source, the position of the detector relative to the ground surface (or water table), the size and composition of the access tube, the physical and chemical properties of the soil, the dry density and the water content of the soil (Chanasyk *et al.* 1996; Li *et al* 2003). The problem is more complicated for expansive soils due to the fact that the soil volume and bulk density change as in situ soil moisture changes. In order to get a satisfactory calibration for expansive soils, a generalized calibration has been developed based on the multigroup neutron diffusion theory and finite element method.

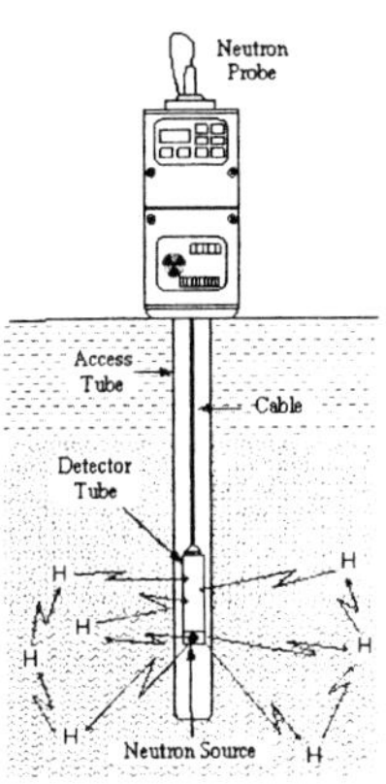

FIG. 2. Schematic drawing of a neutron probe in use.

THE NUMERCAL MODEL FOR NEUTRON PROBE CALIBRATION

The numerical model used in this study was developed by the first author and has been described in detail in Li *et al* (2003). In the finite element model, the complete energy spectrum of neutrons, from initial fast state to thermalised state, is subdivided into a number of discrete energy groups so that each energy group can, to reasonable approximation, be treated as monoenergetic with constant parameters. For a given neutron energy group, the neutron balance (or conservation) equation under steady-state conditions is expressed as:

$$\overbrace{\left(-\nabla \boldsymbol{J}\right)}^{\text{Leakage}} + \overbrace{\left(\Sigma_{sl_i}\phi_i + \Sigma_{a_i}\phi_i\right)}^{\text{Sink}} = \overbrace{\left(S \text{ or } \Sigma_{sl_{i-1}}\phi_{i-1}\right)}^{\text{Source}} \tag{1}$$

From this neutron balance, the neutron diffusion equation for an n-group diffusion model can be written in the form:

$$\left.\begin{array}{ll}
D_1\nabla^2\phi_1 - \Sigma_{sl_1}\phi_1 - \Sigma_{a_1}\phi_1 + S = 0 & E_1 \geq E > E_2 \\
D_2\nabla^2\phi_2 + \Sigma_{sl_1}\phi_1 - \Sigma_{sl_2}\phi_2 - \Sigma_{a_2}\phi_2 = 0 & E_2 \geq E > E_3 \\
D_3\nabla^2\phi_3 + \Sigma_{sl_2}\phi_2 - \Sigma_{sl_3}\phi_3 - \Sigma_{a_3}\phi_3 = 0 & E_3 \geq E > E_4 \\
\cdots\ \cdots \qquad \cdots\cdots & \\
D_n\nabla^2\phi_n + \Sigma_{sl_{n-1}}\phi_{n-1} - \Sigma_{a_n}\phi_n = 0 & E_n \geq E
\end{array}\right\} \tag{2}$$

where $\boldsymbol{J}$ is the neutron current vector, S is the high energy neutron source term, and D_i, Σ_{sl_i}, Σ_{a_i} are the diffusion coefficient, the slow-down cross-section, and the macroscopic absorption cross-section respectively for the ith energy group, E_i. The unknown quantity ϕ_i is the neutron flux distribution, which is defined as the product of the neutron density and the velocity. D_i, Σ_{sl_i}, and Σ_{a_i} are calculated from the chemical composition of the soil (Li *et al.* 2003).

Once the distribution of the thermal neutron flux, ϕ_{th}, in the neutron detector is calculated, the neutron counts detected in a given time T can be estimated by the following integration over the spatial domain of the detector:

$$CR = \int_v \phi_{th} \cdot \Sigma_{a,D} \cdot T \cdot dv \tag{3}$$

where CR is the count rate and $\Sigma_{a,D}$ represents the thermal absorption cross-section of the neutron detector.

In order to apply the numerical analysis, it is necessary to first know the elemental composition of the soil. A total of fourteen chemical analyses of the Maryland soil at different depths were carried out (Li *et al.* 2003). A total of eight theoretical calibration equations have been developed for the Maryland soil at different depths. The calibration equation generally takes the form:

$$m_v = \text{A} + \text{B}\,(CR) \tag{4}$$

where A, B = calibration constants (depend on the chemical composition of the soil),
CR = count rate.

RESULTS

The *in situ* soil moisture content was measured by using a CPN 503 Hydroprobe (neutron probe) at 25 locations as shown in Fig. 1 on a monthly base. The complete presentation of the neutron probe soil moisture profiles for all locations is beyond scope of this paper due to the space limitation. Figure 3 shows the extremes of moisture content measured with the neutron probe at four different locations (NP3, NP7, NP10 and NP11). It should be pointed out that aluminum access tubes were installed to a depth of 1.5m only at locations of NP3 ad NP10. The profiles shown represent the upper and lower envelopes of all readings taken at each depth during the project. Intermediate values were omitted for clarity.

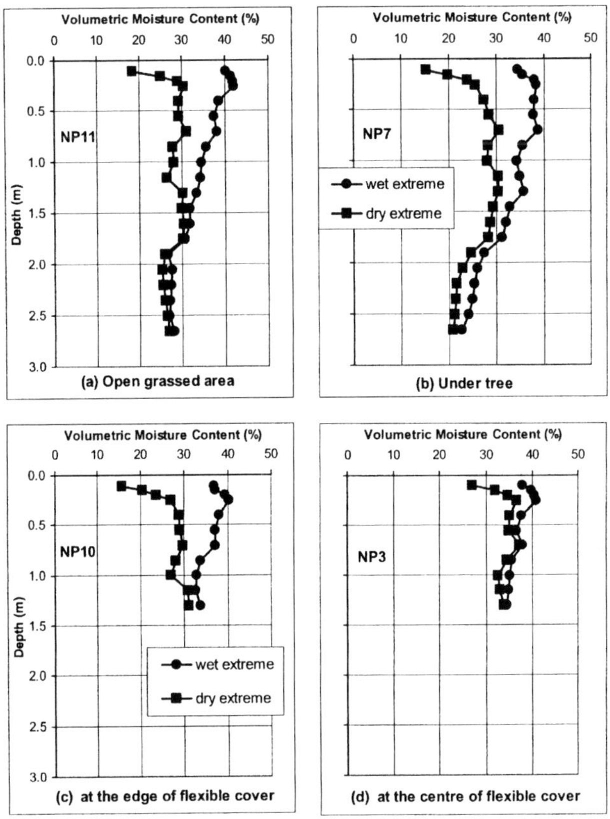

FIG. 3. Volumetric moisture content change profiles

From Fig. 3, a number of observations can be made. These include:

- The depth of seasonal moisture variation for the Maryland site can be taken as 1.5m as no significant seasonal moisture changes (hence no volume changes occurs in response to climate) below 1.5m at the open grassed area (Fig. 3a).
- The effect of tree is evident at NP7 (Fig. 3b) where a large eucalypt tree is located about 2m away as the active depth is extended below 2.6m.
- Only a small change in soil moisture occurred under the flexible cover and this change was confined to the top 0.5m. The maximum change in moisture content at the centre of the flexible cover (Fig. 3d) was less than half that occurred at the edge of flexible cover (Fig. 3c).
- For all four envelopes, the values of observed moisture content are much lower above a depth of about 0.35m. This can be attributed to the equipment constraints. The neutron moisture probe measures the average moisture content in a 'sphere' of soil of variable radius between 0.1m and 0.3m (decreasing with increasing moisture content) (Li *et al.* 2003). For reading taken at depths of less than 0.35m, some of this volume extends above the ground surface where there is only air and, hence, the values will be low.

CONCLUSIONS

A field site for study of expansive soil behaviour was established near Newcastle, Australia. The *in situ* soil moisture has been monitored using a neutron moisture probe. In this study, a numerical model based on multi-group diffusion theory has been used to develop the calibration equations for neutron probe at different depths. Typical volumetric moisture content change profiles derived from the neutron probe readings, representing the extreme observation at four different locations are presented. The results indicate that the depth of seasonal moisture variation at the field site can be taken as 1.5m. This research has shown that a calibration relationship for expansive soils can be accurately estimated using the neutron diffusion model if the elemental composition of the soil is known.

REFERENCES

Allman, M. A., Smith, D. W. and Sloan, S. W. (1994). "Establishment of a reactive soil test site near Newcastle." *Australian Geomechanics*, Vol. 26, pp 46-56

Chanasyk, D.S. and Naeth, M.A. (1996). "Field measurement of soil moisture using neutron probes." *Can. J. Soil Sci.*, Vol. 76: 317–323.

Fityus, S.G., Smith, D.W. and Allman, M.A. (2004). "An expansive soil test site near Newcastle." *J. Geotechnical & Geoenv. Engrg.*, Vol . 130: pp.686-695.

Li, J., Smith, D. W., Fityus, S.G. and Sheng, D. C. (2003). "The Numerical Analysis of Neutron Moisture Probe Measurements", *ASCE International Journal of Geomechanics,* Vol. 3 (1): pp. 11-20.

Correlating Vane Shear Results with Undrained Strength Parameters for Slope Stability Analysis

Shuwang Yan[1], Xiaowei Feng[1] and Jinfang Hou[1]

[1]School of the civil engineering, Tianjin University, Tianjin, China, 300072; yanshuwang@tju.edu.cn

ABSTRACT: When using vane strength to evaluate the stability of soft clay slopes in Tianjin Harbor area, the calculated safety factors are often much less than 1.0 for stable slopes. A reasonable explanation is given to this phenomenon and a new way of using vane strength is presented for slope stability analysis. A correlation between the in-situ vane shear strength and undrained shear strength parameters is proposed based on the considerations that the vane strength is anisotropic and that the vane strength increases with depth linearly. Stability analysis was performed for several existing slopes. The results indicate that the safety factors calculated using the correlated shear strength parameters are more reasonable than those calculated with the vane strength directly.

INTRODUCTION

According to the design guidelines, such as the British Standard of Code for Earthworks, undrained shear strength should be adopted for short term stability analysis of soft clay slope (BRITISH STANDAR, 1981). Tianjin Harbor area is covered with a soft clay layer of about 18 m thick. The soft clay layer is a sedimentary deposit of recent geological ages and consists of very fine clay particles. This soil layer is highly compressible with large void ratio and low shear strength. Meanwhile, the soil has a sensitivity of 3 to 4 (Hou and Chen, 1987). Since Tianjin Harbor area is constructed on such soil conditions, several serious slope failures have taken place. Therefore, to accurately assess the stability of slopes in the port construction becomes an important subject (Hou and Chen, 1987; Wei, 1987).

The highly sensitive nature of the soft deposit in Tianjin Harbor area makes it difficult to retrieve undisturbed soil samples to obtain the undrained strength at laboratory . Soil sampling, handling, transportation, and testing will inevitably cause disturbance and reduce the undrained shear strength of the soil (Hou and Chen, 1987; and Wei, 1987). Using of these strength data in the analysis and design of soil slopes often results in a safety factor unrealistically lower than 1.0 even for stable slopes (Li , et al ., 2005; Hou and Chen, 1987; and Wei, 1987).

To minimize possible sample disturbance, the in-situ vane shear test method has been commonly adopted to determine the undrained shear strength for design and analysis of soil slopes in Tianjin Harbor area. However, even with the use of in-situ vane shear strength, the safety factors of stable slopes were estimated to be less than 1.0. This is because the vane shear strength mainly reflects the undrained shear strength of the vertical shear surface, which provides smaller strength than other shear surfaces do (Craig, 1992; Qian and Yin, 1996). As a consequence, the safety factors determined from vane shear strength are far below 1.0 for stable slopes in soft clay areas, and may not realistically reflect the stability of the actual slopes (Wei, 1987). Therefore, the vane shear strength can not directly be used to assess the stability of the slopes in this area.

CHARACTERISTICS OF VANE SHEAR STRENGTH

The main advantage of vane shear test is that the test itself causes little disturbance of the soil and is carried out below the bottom of the borehole in virtually undisturbed soil. This is particularly important in sensitive clays, because higher shear strength tends to be resulted from the in-situ vane shear test than from laboratory tests on samples obtained using a general purpose sampler (such as a Shelby tube sampler).

For normally consolidated clays with certain plasticity index, I_P, the undrained shear strength is increased linearly with depth (Hou, 2007):

$$\frac{c_u}{\sigma_v'}=0.11+0.0037I_p \tag{1}$$

in which c_u is the undrained shear strength, σ_v' is the effective overburden pressure.

Like undrained strengths obtained using other methods, the undrained shear strength of clay from in-situ vane shear test varies linearly with depth. FIG.1 gives some typical vane test results in Singapore and Tianjin, China (Hou, 2007).

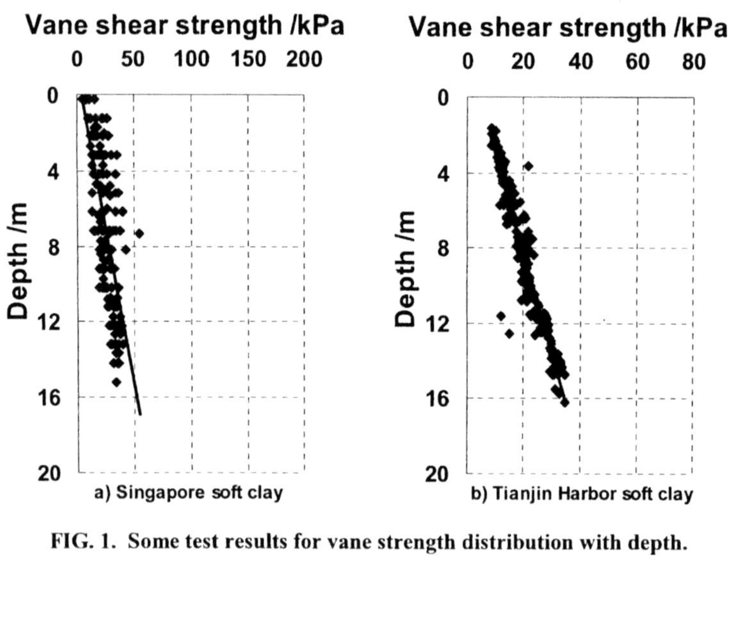

FIG. 1. Some test results for vane strength distribution with depth.

The undrained shear strength obtained using the conventional test methods can only reflect the strength at certain shear surface. The differences of undrained shear strength at different shear surface are ignored. The vane shear strength is then assumed to be equal in all directions, which makes many tough problems easily be solved. As a result, some inaccuracy may arise.

CORRELATING VANE SHEAR STRENGTH WITH SHEAR STRENGTH PARAMETERS

Skempton's Formula

Researches have been performed to relate the vane shear strength to other soil strength parameters. For example, Skempton and Sowa (1963) presented the following equation to correlate the vane shear strength with the consolidated undrained friction angle:

$$\sin\varphi_{cu} = \frac{S_u}{S_u + p} \tag{2}$$

Where,

φ_{cu} = the consolidated undrained strength parameter, the inner friction angle;

S_u = the vane shear strength;

p = overburden.

The above correlation did not consider the strength anisotropy.

Correlating the Vane Shear Strength with Shear Strength Parameters Considering Strength Anisotropy

The vane device and the surfaces on which the resistance will balance the applied torque are shown in FIG.2.

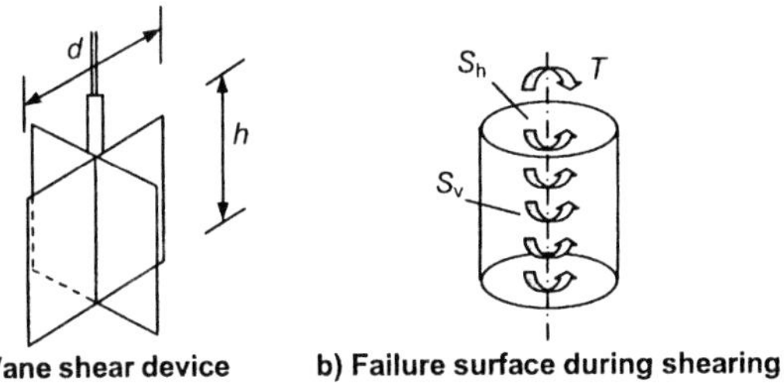

FIG. 2. Vane shear device and failed surfaces during shearing

Since the effective normal stress varies with plane direction as well as depth, the shearing resistance on the cylindrical vertical surface and horizontal shear surface in the vane shear testing will be different as shown below:

$$S_v = K_0\gamma' z\tan\varphi + c \tag{3}$$

$$S_h = \gamma' z\tan\varphi + c \tag{4}$$

Where,

S_v and S_h = shear strengths on vertical and horizontal surfaces, respectively;
K_0 = coefficient of earth pressure at-rest;
γ' =submerged unit weight of soil;
z =depth below ground surface;
φ = internal friction angle of soil;
c = cohesion of soil.

At failure of vane shear test, the limit equilibrium equation can be expressed as follows, assuming a uniform shear stress distribution at perimeter:

$$T=\frac{1}{2}\pi d^2 h S_v+\frac{1}{6}\pi d^3 S_h \tag{5}$$

In which
T = vane shear torque at failure;
d = vane width;
h = vane length.

Substituting Eq.(3) and Eq.(4) into Eq.(6), the average vane shear strength can be expressed as

$$S_{vane}=\frac{T}{\pi(\frac{d^2 h}{2}+\frac{d^3}{6})}=\frac{\frac{1}{2}d^2 h K_0+\frac{1}{6}d^3}{\frac{1}{2}d^2 h+\frac{1}{6}d^3}\gamma' z\tan\varphi+c \tag{6}$$

Denote Equation (6) as the following linear regression equation:

$$\overline{Y}=a\overline{X}+b \tag{7}$$

In which

$$\overline{Y}=\frac{T}{\pi(\frac{d^2 h}{2}+\frac{d^3}{6})}=S_{vane} \tag{8}$$

$$\overline{X}=\frac{\frac{1}{2}d^2 h K_0+\frac{1}{6}d^3}{\frac{1}{2}d^2 h+\frac{1}{6}d^3}\gamma' z \tag{9}$$

$$a=\tan\varphi\;;b=c \tag{10}$$

For uniform soil stratums with adequate vane test data, the constant a and b in Eq.(9) can be obtained using regression method, such as that by Brockett and Levine (1984).

With the rapid development of Tianjin Harbor area, large number of vane shear tests has been performed and tested data are accumulated in a database (Hou, 2007). Other soil parameters, such as K_0 and γ', are also tested and are acceptable (Hou and Chen, 1987). Therefore, it is possible to convert the vane shear strength into a set of undrained strength parameters of the soil in this area, which could be used to calculate slope stability, bearing capacity and earth pressure with more convenience and accuracy.

The Converted Shear Strength Parameters of Tianjin Soft Clay

The soil profile at a site of Tianjin Harbor is described in Table 1. The profile includes a 1-m thick hard crust at the top, followed successively by soft clay, silty clay, clayey silt, and dense sand. The soft clay stratum, nearly 18-m thick, consists of varying thin layers of organic matters, silty sand and sandy soil pockets. Most of the cut slopes are located in this stratum. The soil in this stratum is in liquid to plastic state of consistency, and is highly sensitive and compressible with low strength.

Figure.1 (b) gives vane test results in 20 boreholes in Dongjiang Port area of Tianjin Harbor. It can be seen that the vane shear strength increases linearly with depth without much scatter. According to Eq.(7), we can get

$$a=0.281 \text{ and } b=6.60$$

Then the strength parameters can be determined with Eq.(10):

$$\varphi=15.7° \text{ and } c=6.60 \text{ kPa}$$

SLOPE STABILITY ANALYSIS

The Fellenius method of slice is adopted in the slope stability analysis. The safety factor against slope failure is expressed in terms of effective stress as follows (Morgenstern and Price, 1967):

$$SF=\frac{cL_a+\tan\varphi\Sigma W\cos\alpha}{\Sigma W\sin\alpha} \tag{11}$$

Where,

L_a = base length of sliding surface;

W = weight of each slice;

α = base angle from horizontal of each slice;

SF = safety factor;

Other factors have been defined earlier.

Five slopes were analyzed. The slope and surcharge information are presented in Table 1. Stability analyses of these five slopes were performed using shear strength parameters obtained from unconsolidated undrained (UU) triaxial and in-situ vane shear tests. The eatimated safety factors are summarized in Table 2.

Table 1. Stability Analysis Results of Some Slopes in Tianjin Binhai New Area

Project	Bottom Elevation (m)	Top Elevation (m)	Water Level (m)	Slope Ratio	Surcharge (kPa)
No.I dock	-9.0	+4.3	-1.76	1:3.01	0
No.III dock	-11.0	+4.8	-2.00	1:2.66	30
Steel dock	-11.0	+4.9	-2.00	1:4.09	65
Wood dock	-10.5	+6.1	+0.50	1:2.65	30
Nonmetal dock	-11.8	+4.5	+0.50	1:2.50	30

The data in Table 2 show that the safety factor estimated using UU strength are unrealistically low due to sample distribution. Additionally, the safety factors determined from vane shear strengths were underestimated. The safety factors obtained from correlated shear strength parameters appear to match the state of slope well.

Table 2. Safety Factors and Status of the Analyzed Slopes

Project	Safety Factors with Different Type of Strength			Slipping Depth (m)	Status of Specified Slope
	UU	Vane Shear Strength	Correlated Strength Parameters		
No. I dock	0.77	0.836	1.17	-12.1	Stable
No.III dock	0.78	0.813	1.07	-12.9	Stable
Steel dock	0.75	0.953	1.22	-8.80	Stable
Wood dock	0.72	0.848	1.03	-8.20	Stable
Nonmetal dock	0.67	0.802	0.93	-12.3	Failed

SUMMARY AND CONCLUSIONS

The shear strength parameters (c and φ) determined for the in-situ vane shear strengths were subjected to statistical regression analysis considering the effect of anisotropy of the shear strength at different orientations. Using the correlated shear strength parameters, slope stability analyses were performed. The same slopes were also analyzed using in-situ vane shear strengths, as well as unconsolidated-undrained shear strength parameters obtained from laboratory testing on so called undisturbed samples. The results indicate that the shear strength parameters converted from in-situ vane shear strength are able to estimate reasonable safety factors.

REFERENCES

British Standard. (1981). "Code of Practice for Earthworks " BS6031.

Hou, Z. and Chen, H. (1987). "Tianjin soft soil foundation." *Tianjin Science and Technology Publishing House*. Tianjin, China. (in Chinese).

Wei, R. L.(1987). "The strength and displacement of soft clay." *People's Transportation Publishing House.*, Beijing, China. (in Chinese).

Li, S. , Yue, Z. Q., Tham, L. G., Lee, C. F. and Yan, S. W. (2005). "Slope failure in unconsolidated soft soils during the development of a port in Tianjin China." *Canadian Geotechnical Journal*, Vol. 42 (1): 147-165.

Criag, R. F., (1992). "Soil Mechanics." *Chapman & Hall*, UK.

Qian, J. H. and Yin, Z. Z. (1996). "Geotechnical principles and calculations (Second Edition)" *Chinese Hydraulic Publishing House*, Beijing China. (in Chinese)

Skempton, A. W. , Sowa, V. A. (1963). "The behavior of saturated clays during sampling and testing." *Geotechnique*, Vol. 13: 269-290.

Hou, J. F. (2007). "Research on stability of soft clay slopes." *Doctorial dissertation, Tianjin University*, China. (in Chinese).

Brockett, P. and Levine, A. (1984). "Statistics and probability and their applications." CBS College Publishing, USA.

Morgenstern, N. R. and Price, V. E. (1967). "A numerical method for solving the equations of stability of general slip surfaces." *Computer Journal*, Vol. 9: 388-393.

Soft Clay Erosion and Its Undrained Shear Strength Measurements

Hang Yin and Guoping Zhang

Department of Civil and Environmental Engineering, Louisiana State University, Baton Rouge, LA 70803, USA
Email: hyin5@lsu.edu; gzhang@lsu.edu

ABSTRACT: Characterization of the in-situ critical shear stress of cohesive soils plays a vital role in bridge scour design. This stress is closely related to the undrained shear strength, which is conventionally obtained using a field shear vane (FSV) in situ or an erosion function apparatus (EFA) in laboratory. However, neither method is appropriate for assessing the erosional resistance of top surface layer of soft clays in streambeds, because: (1) FSV lacks sufficient sensitivity for the strength of very soft clays; (2) it is very difficult to obtain high-quality samples from such soft clays to be tested in laboratory; and (3) the properties of cohesive soils vary significantly with consolidation, organic matter content, microorganisms and biofilms, and seasonal changes. This paper describes the development of a new Louisiana Scour Vane (LSV) for in-situ measuring the undrained shear strength of surficial soft clay layer. Laboratory testing on a reconstituted clay was conducted to validate the LSV's capability of measuring shear resistance changes at small depth intervals. This new portable and easy-to-operate apparatus will act as a suitable supplement to the FSV for geotechnical engineers.

INTRODUCTION

Evaluation of the erodibility of bridge foundation soils is a critical process prior to the bridge scour design. This assessment is usually achieved by determining the erosion rate ($\dot{z}$ = *dz/dt*) (Fig. 1) using an erosion function apparatus (EFA) (Richardson and Davis 2001). Initially the erosion rate is typically zero before a critical shear stress (τ_c) of the soil is reached (Tolhurst et al. 1999). For cohesive soils abundant in Louisiana's streams and wetlands, this value is closely related to the soil's undrained shear strength (s_u), an indicator of erosion resistance (Zreik et al. 1998, Sheppard and Bloomquist 2003, Zimbone et al. 1996, Watts et al. 2003).

The undrained shear strength of cohesive soils in streambeds varies significantly with depth. For instance, the top layer can be roughly divided into three zones, as shown in Fig. 2 (Parchure and Mehta 1985). Zone 1 defines a critical depth from the bed surface (soil-water interface) to a depth z_c, in which the shear resistance is very low but increases exponentially with depth; Zone 2 defines a soil layer beneath Zone 1 where the shear resistance increases with depth but has a much smoother (or smaller) gradient than in Zone 1; Beneath Zone 2 is Zone 3 where the bed shear strength remains nearly constant. Zones 1 and 2 can reach a thickness of up to 10 cm.

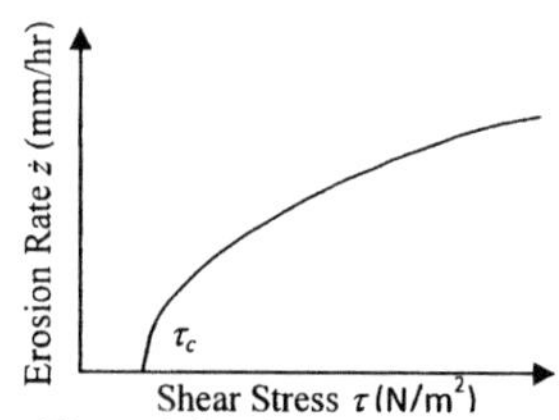

Figure 1. Erosion rate vs. shear stress obtained bv an EFA

According to Fig. 2, the small thickness and very low shear strength of Zones 1 and 2 present a significant challenge for characterizing their shear strength using conventional geotechnical in-situ testing methods or devices. For instance, the field shear vane (FSV) with blades of >10 cm in height and large self-weight lacks sufficient resolution and sensitivity for the strength characterization of a thin, soft layer. Other more sophisticated instruments, such as benthic flumes and vertical jet systems (e.g., a cohesive strength meter (CSM)) (Black and Paterson 1997), are mainly used in hydraulic practices or not commercially available to geotechnical engineers.

This paper describes the design and validation of a new Louisiana Scour Vane (LSV) for in-situ measurements of the undrained shear strength of surficial layer of soft cohesive soils. This new vane consists of eight blades of only 1 cm in height fixed in a large ring. Its configuration is totally different from the traditional FSV. A similar device was also developed for the characterization of intertidal flat sediments (Bassoullet and Le Hir 2007). It can be connected to the conventional FSV's torque instrument to maintain interchangeability between these two devices. Compared with the erosion function apparatus (EFA), this new apparatus can measure the strength in-situ to eliminate the effect of sample disturbance. It can also conveniently and rapidly characterize a large area for spatial and temporal variability in strength (Grant and Gust 1987, Paterson 1989, Yallop et al. 2000). Additionally, the Hydraulic Engineering Circulation No. 18 (HEC-18) (Richardson and Davis 2001) particularly emphasizes the need for research on "field measurements of scour".

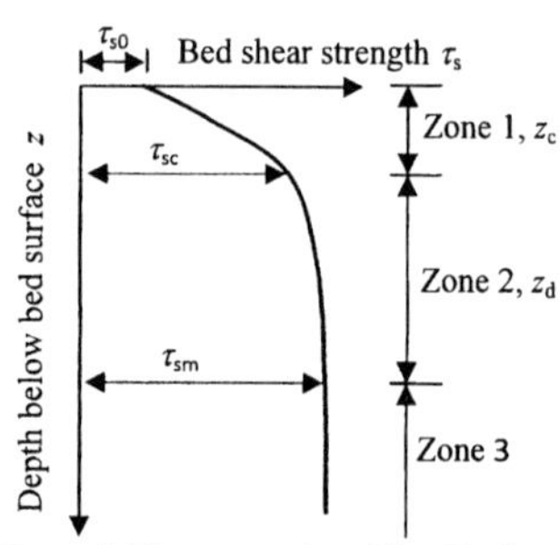

Figure 2. Three-zoned profile of bed erosional shear strength of cohesive soils.

DESIGN OF THE LSV

A significant concern with the FSV's capability to measure the undrained shear strength of top soft clay layer is the low sensitivity caused by the long vane blades and relatively small diameter. This traditional device (e.g., Geonor vane H-10, H-60, or H-70) functions by measuring the torque (T) that fails the soil through a cylindrical surfaces (both circumferential and end surfaces) (Fig. 3).

According to ASTM D 2573-08 (ASTM International 2008), the shear strength (τ_s) can be determined based on the measured torque (T):

$$T = \pi\left(2r^2h + \frac{4}{3}r^3\right)\tau_s \text{, if the vane is inserted and submerged into the soil} \quad (1)$$

$$T = \pi\left(2r^2h + \frac{2}{3}r^3\right)\tau_s \text{, if the vane is maintained on the surface of the soil} \quad (2)$$

where, r and h are the radius and height of the vane, respectively; τ_s is the undrained shear strength of the soil; and T is the measured torque.

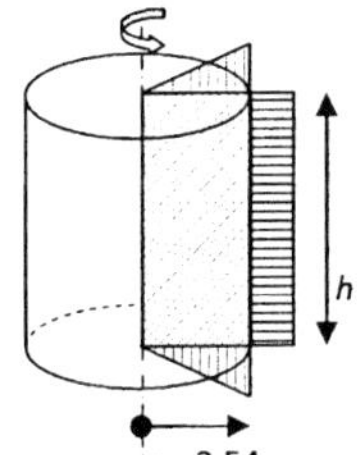

Figure 3. FSV cylindrical failure surfaces (including top & bottom surfaces) and shear stress distribution.

The new LVS is designed based on a similar failure mechanism, but is targeted for characterizing the surficial layer of cohesive soils whose shear strength varies significantly with depth. As such, the height of the new LSV blades was reduced to only 1 cm in order to meet the requirement for depth sensitivity. However, the reduction in height has a negative effect on the torque resistance from soft soils, as the shorter blades actually decrease the measured torque (see Eqs. 1-2). Therefore, to maintain both the strength sensitivity and depth sensitivity, extending the failure surface diameter was the second improvement. Because the torque is proportional to r^2 and r^3 but h, it can be augmented nearly 30 times if r is increase to 12 cm compared to the traditional vane.

However, this design could not be used in-situ for its extremely long arms that are prone to bending or damage. To avoid this problem, the final design uses a ring failure surface instead of the cylinder, together with eight blades. Moreover, a supporting ring is added to the top edge of the blades to increase the device's stability (Fig. 4).

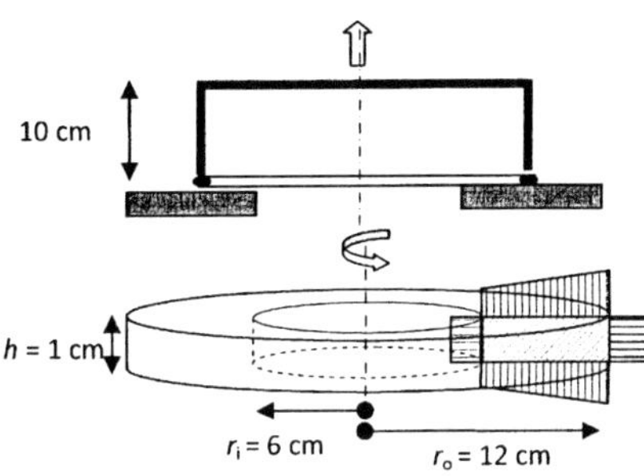

Figure 4. The new LSV and soil failure surface

The theoretical solution to the shear strength (τ_s) based on measured torque (T) for this new LSV is given as follows:

$$T = \pi\left[2\left(r_i^2 + r_o^2\right)h + \tfrac{4}{3}\left(r_o^3 - r_i^3\right)\right]\tau_s ; \quad (3)$$

or

$$T = \pi\left[2\left(r_i^2 + r_o^2\right)h + \tfrac{2}{3}\left(r_o^3 - r_i^3\right)\right]\tau_s , \quad (4)$$

Equations (3) and (4) refer to two cases where the vane is entirely inserted below the soil surface and the vane is just maintained on the surface of the soil, respectively.

LSV STRUCTURAL DESIGN AND FABRICATION

The new LSV has to meet another requirement – the structural stability to transfer the soil's resistance to the shaft which then transfers the resistance to the torque instrument device. The proposed long arms, aimed to increase the increase the torque sensitivity, require significant large bending and shearing resistance, which influence the selection of materials used for the arms and rings. This structural design plays an important role in ensuring the proper function of the new LSV.

A soft clay with an undrained shear strength of 0.6 kPa at a depth of 10 cm is considered in the structural design. The LSV measured torque T should be:

$$T = \pi\left[2\left(r_i^2 + r_o^2\right)h + \tfrac{4}{3}\left(r_o^3 - r_i^3\right)\right]\tau_s = 4.5 N\cdot m\,; \tag{5}$$

Shear force F for each horizontal arm of the supporting frame should be:

$$F = T/\left(4R_{shaft}\right) = 124.4N\,; \tag{6}$$

Bending modulus E is subjected to:

$$E = \frac{FL^3}{3I\Delta} = \frac{FL^3}{3\times\left(\pi D^4/64\right)\Delta} = 2\cdot 10^{11}\,Pa \text{ (assume } \Delta \approx 5mm\text{)}; \tag{7}$$

Shear modulus G is determined as:

$$G = M_N L/\left(I_P\varphi\right) = 2\cdot 10^{11}\,Pa \text{ (assume } \varphi \approx 1^\circ\text{)}. \tag{8}$$

where R_{shaft} is the radius of the connecting rod; L stands for the radius of the supporting frames; I is the area moment of inertia of the horizontal frames; Δ is the maximum tolerable bending deformation of the horizontal frames; φ is the maximum affordable angle of twist of the supporting frames; I_p is the polar moment of inertia; and M_N stands for the resisting torque; D is the diameter of the stainless steel ring; GI_P is the torsional stiffness of the material.

Based on the above possible failure conditions, the bending resistance strength is the most critical factor for the determination of what kind of material should be chosen for the device. In this research, stainless steel has been selected for the FSV fabrication, for two concerns: (1) to meet the structural stability requirements; and (2) to prevent corrosion of the vane and increase its durability for long-term use. The blades and fixing wiring ring will also be made of stainless steel. The connection between the vane and the rod will be the same with the conventional vane to maintain interchangeability between the conventional H-70 FSV and the newly designed LSV (Fig. 5).

Figure 5. The new vane connected to the H-60 torque instrument.

SAMPLE PREPARATION

Laboratory reconstituted soft clays were used to test the new LSV and validate its capability. The method proposed by Burland (Burland 1990) was adopted to prepare a soft clay sample for the LSV test. This method basically prepares a homogeneous saturated clay sample by consolidating soil slurry with an initial water content of 1.0-1.5 times the liquid limit of the soil.

As described above, the diameter of the new LSV is 24 cm, indicating that the tested clay sample must have a diameter >24 cm. Also, to check the vertical strength sensitivity, a sample height of ~15 cm is preferred. Therefore, the cylindrical consolidation mold is chosen to be 30.5 x 30.5 cm (height x diameter), in order to be able to accommodate the new LSV and to provide enough space between the inner wall of the mold and the outer surface of the vane (for minimizing the boundary influence). However, a problematic issue arises with the height of the sample. The consolidation process for such a thick sample may take up to several months. Therefore, in order to expedite the consolidation process or reducing the time required for the consolidation, a fine-grained soil consisting of 50 wt.% pure kaolinite clay (with 90 wt.% of particles < 2 μm) and 50 wt.% of river silt (with >85 wt.% of particles of 2-75 μm in size) are mixed together. The total dry weight of the kaolinite and silt mixture used for the consolidation was 8.83 kg. Since the liquid limit of pure kaolinite clay is 65%. As such, the water content of the prepared clay and silt slurry was 74%, with an initial void ratio of 2.0.

VANE SHEAR TESTING AND CALIBRATION

Table 1 and Fig. 6 show the tests results with traditional FSV H-60 and new LSV connected with different torque instruments. The calibrated τ_s is obtained by timing the measured τ_s with the calibration factor. For the conventional Geonor® FSV H-60 vane, this factor is given in the user manual as 0.5 for the largest H-60 vane (i.e., the dimension used for this test). For LSV, this factor is determined according to the equations recommended by ASTM D 2573 (ASTM International 2008).

Table 1. Test results for the clay-silt mixture

Depth (cm)	Vane Type	Torque Instrument	Measured τ_s (kPa)	Calibrated τ_s (kPa)
4.5	FSV H-60	H-60	2	1
9.0	FSV H-60	H-60	2	1
1.0	LSV	H-60	7.5	0.05
3.0	LSV	H-60	140	0.55
4.0	LSV	H-70	10	0.70
6.0	LSV	H-70	12	0.84

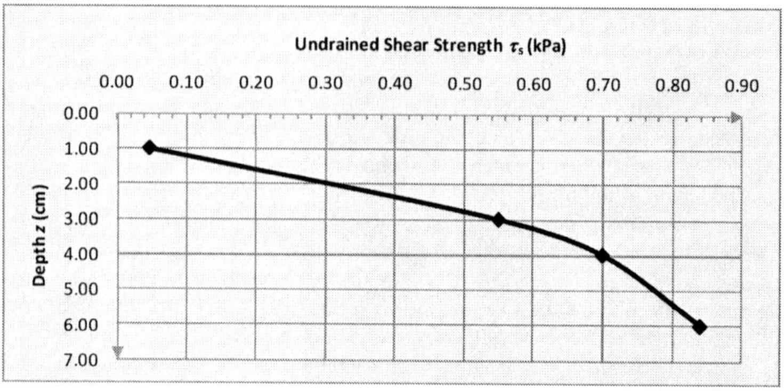

Figure 6. Undrained shear strength profile of the reconstituted sample

No matter which vane is used, the real τ_s should be the same. Thus, this value times the area of real failure surface will be the real torque measured by the instrument. The area of H-60 vane failure surface should be 2.93×10^{-5} m^2 and 6.01×10^{-5} m^2 (regular vane and big vane separately), while the area of the LSV failure surface should be 7.46×10^{-3} m^2. If the shear strength change on the top and bottom surface of the blade was assumed to be linear (which would underestimate the actual value (Perez-Foguet et al. 1999)), while on the inner and outer cylindrical surfaces, it remains the same (ignore the influence of depth change), the calibration factor for this new vane to H-60 vane torque instrument should be 0.004. However, if the vane is not inserted into the soil, this factor should be adjusted to 0.006 since the top surface of the blade does not provide shear resistance. Using this similar calculation, the calibration factor for converting LSV strength to the H-70 torque instrument is also obtained as 0.070.

From this plot, the newly developed LSV can accurately obtain the undrained shear strength profile via simple testing and calculation. This vertical strength profile is very consistent with that of the widely accepted three-zoned subdivision shown in Fig. 2.

DISCUSSION

The testing results demonstrate that this new device provide an easy-to-operate vane shear apparatus for the in-situ measurement of undrained shear strength and other geotechnical field testing by geotechnical engineers. Its extended shear arms and exchangeability with the traditional FSV torque instrument make it feasible for field testing of extremely soft cohesive soils, particularly for the surficial shallow layer of sediments in streambed or seabed.

An additional advantage of this design is that the large diameter connecting ring that holds all the blades together can also serve as a supporting platform to prevent the device from undesired movement under its self weight in soft soils. The conventional FSV used in this study could not be stabilized over the soft soil sample and continuously sank into the soil under the self-weight of the vane. While for the new LSV, little movement happened. A stabilized device is of significant importance for measuring accurately and reliably the undrained shear strength of top surficial layer of soils.

These long shear arms do have some disadvantages. One problem is the high material strength and stiffness required to resist the bending of arms. This issue makes infeasible to measure the undrained shear strength of a medium to stiff soil at greater depth where the shear strength is high enough to cause a shearing failure of the device itself, particularly the shear failure of the extended shear arms and the connection joints between the arms and the central shaft (details can be found from the LSV design section, where an undrained shear strength of 0.6 kPa is assumed for the design). In other words, this design is a special device for measuring the undrained shear strength in extremely soft top layers of cohesive soils, where a traditional FSV cannot provide this capability. However, comparison and verification from the other available strength test results are still expected to further validate this device.

CONCLUSIONS

A new shear vane – Louisiana Scour Vane (LSV) was designed and fabricated to provide geotechnical engineers a portable and easy-to-operate device for in-situ characterization of the undrained shear strength of the surficial shallow layer (e.g., ~10 cm) of soft sedimentary cohesive soils on streambeds. This field testing instrument is of great importance to assess the scour potential of bridge foundations as the critical erosion resistance of a cohesive soil is closely related to the undrained shear strength.

Laboratory testing was also conducted to calibrate the new LSV and assess its capability. Results demonstrate that the new LSV possesses both the strength sensitivity (as it can measure the strength as low as 0.05 kPa) and vertical profile sensitivity (as it can measure the strength at intervals as small as ~1-2 cm). Furthermore, compared with the conventional FSV, the new LSV can have better stability (i.e., not sinking under its self-weight) due to the extended arms, a large-diameter supporting ring, and sometimes higher frictional resistance.

With this new device, geotechnical engineers can easily obtain the streambed shear strength profile in a particular site via in-situ testing. It is also expected to be applicable to a wide variety of other geotechnical engineering practices, such as the characterization of the coastal and intertidal flat sediments. By correlating the LSV measurements with other hydraulic devices such as vertical jet systems or benthic flumes, it could dramatically reduce the field exploration cost. Meanwhile, it can be simply used for in-situ testing of large areas to map regional, large scare variations of shear strength, which cannot be easily achieved by other similar techniques.

ACKNOWLEDGMENTS

The work was supported by the LTRC's TIRE Program. The continuous support from the Project Managers, Mr. Walid R. Alaywan and previously Dr. Doc Zhang, are gratefully acknowledged. Thanks also go to Dr. Zhongxin Wei who provided technical help to the experiments. Generous sponsorship from GeoEngineers, Inc., who kindly provided the GEONOR H-60 FSV to the study is also greatly acknowledged.

REFERENCES

ASTM International, West Conshohochen, PA. *ASTM Standard D2573 -08, "Standard Test Method for Field Vane Shear Test in Cohesive Soil"*. 2008. DOI: 10.1520/D2573-01R07, www.astm.org.

Bassoullet, P., and P. Le Hir. "In situ measurements of surficial mud strength: A new vane tester suitable for soft intertidal muds." *Continental Shelf Research 27*, 2007: 1200-1205.

Black, K. S., and D. M. Paterson. "Measurement of the erosion potential of cohesive marine sediments: a review of current in-situ technology." *Journal of Marine Environmental Engineering 4*, 1997: 43-83.

Burland, J. B. "30th Rankine Lecture - On the Compressibility and Shear-Strength of Natural Clays." *Geotechnique*, 1990: 329-378.

Grant, J., and G. Gust. "Prediction of coastal sediment stability from photopigment content of mats of purple sulphur bacteria." *Nature 330*, 1987: 244-246.

Parchure, T. M., and A. J. Mehta. "Erosion of Soft Cohesive Sediment Deposits." *ASCE Journal of Hydraulic Engineering 111*, 1985: 1308-1326.

Paterson, D. M. "Short-term changes in the erodibility of intertidal cohesive sediments related to the migratory behavior of epipelic diatoms." *Limnology and Oceanography 34*, 1989: 223-234.

Perez-Foguet, A., A. Ledesma, and A. Huerta. "Analysis of the vane test considering size and time effects." *International Journal for Numerical and Analytical Methods in Geomechanics 23*, 1999: 383-412.

Richardson, E. V., and S. R. Davis. "Evaluating Scour at Bridges." *Hydraulic Engineering Circular No. 18*, 2001, 4th ed.

Sheppard, D. M., and D. Bloomquist. "Rate of Erosion Properties of Rock and Clay." *Florida Department of Transportation*, 2003.

Tolhurst, T. J., et al. "Measuring the in situ Erosion Shear Stress of Intertidal Sediments with the Cohesive Strength Meter (CSM)." *Estuarine, Coastal and Shelf Science 49*, 1999: 281-294.

Watts, C. W., T. J. Tolhurst, K. S. Black, and A. P. Whitmore. "In situ measurements of erosion shear stress and geotechnical sheaer strength of the intertidal sediments of the experimental managed realignment scheme at Tollesbury, Essex, UK." *Estuarine, Coastal and Shelf Science 58*, 2003: 611-620.

Yallop, M. L., D. M. Paterson, and P. Wellsbury. "Interrelationships between rates of microbial production, exopolymer production, microbial biomass, and sediment stability in biofilms of intertidal sediments." *Microbial Ecology 391*, 2000: 116-127.

Zimbone, S. M., A. Vickers, R. P. C. Morgan, and P. Vella. "Field investigations of different techniques for measuring surface soil sheaer strength." *Soil Technology 9*, 1996: 101-111.

Zreik, D. A., B. G. Krishnappan, J. T. Germaine, O. S. Madsen, and C. C. Ladd. "Erosional and mechanical strengths of deposited cohesive sediments." *ASCE Journal of Hydraulic Engineering 124*, 1998: 1076-1085.

Estimation of In-situ Strength for Sandy Soils using CPT Cone Resistance

Junhwan Lee[1], Doohyun Kyung[2], and Jungmoo Hong[2]

[1]Associate professor, School of Civil & Environmental Eng., Yonsei University, 134 Shinchon-dong, Seodaemun-gu, Seoul 120-749, Korea; junlee@yonsei.ac.kr
[2]Graduate research assistant, School of Civil & Environmental Eng., Yonsei University, 134 Shinchon-dong, Seodaemun-gu, Seoul 120-749, Korea

ABSTRACT: The shear strength of soils is key design property that governs the stability of geotechnical structures. For clays, the undrained shear strength is commonly adopted in design, while the friction angle is a soil property that represents the shear strength of sands. Estimation of these properties is a challenging task, and the challenge is even greater for sandy soils since the strength is highly state-dependent and undisturbed soil sampling is difficult. As a result, various empirical correlations based on in-situ test results, such as SPT and CPT, have been proposed. However, further investigation is still necessary for the consideration of the state-dependent dilatancy and soil constitution. In the present study, a CPT-based method for the estimation of the in-situ strength for granular soils is investigated. Various field variables such as fines content, soil density and confining stress are considered. A series of laboratory test results obtained for various soil conditions are used in the analysis and investigation.

INTRODUCTION

For stability problems in clays, the undrained shear strength (s_u) is commonly adopted in the design and analysis, while the friction angle (ϕ') is key strength parameter for sandy soils. Estimation of these properties is a challenging task, primarily due to complex soil constitutions, non-homogeneity, and non-linear soil behavior. For sandy soils, the challenge is even greater, as the strength is highly state-dependent and undisturbed soil sampling is not a practical option. As a result, various empirical correlations based on in-situ test results, such as SPT blow count, N_{SPT} or CPT cone resistance q_c, have been proposed (Durgunoglu and Mitchell1975, Robertson and Campanella 1983, Chen and Juang 1996). While these have provided useful tools for the interpretation of CPT measurements, further investigation is still necessary and under way for the consideration of the effect of the state-dependent dilatancy and soil constitution.

In this paper, the estimation of the in-situ strength for sandy soils based on the CPT cone resistance q_c is investigated. Soil variables considered in this analysis include silt content, relative density and confining stress. Results from laboratory tests and penetration analysis for various soil conditions are used in the analysis and investigation. Results from calibration chamber tests are adopted and used in the comparison between measured and predicted results.

EXPERIMENTS

Triaxial test results from the literature (Salgado et al. 2000, Lee et al. 2004, Lee et al. 2008) were adopted for investigating a CPT-based approach for the estimation of in-situ strength for sandy soils. The tests were conducted using sand samples at various conditions of different silt contents (s_{co}) of 0, 2, 5, 10, 15 and 20%, relative densities (D_R), and confining stresses. Detailed soil and test conditions can be found from Salgado et al. (2000) and Lee et al. (2004).

Calibration chamber cone penetration test results (Lee et al. 2008) were adopted for the comparison between measured and calculated results. Since different stress states and relative densities can be used in calibration chamber tests, the comparison between measured and calculated results is readily available under various soil conditions. Two relative densities of 55 and 86% were adopted in this study. The cone penetrometer used in the calibration chamber tests in this study consists of the cone probe with extension rods, pushing device, depth encoder, and data acquisition system. The cone was of a miniature type with a diameter of 1.6 cm.

STATE-DEPENDENT DILATANCY RELATIONSHIP FOR SANDS

The peak friction angle ϕ'_p consists of the critical-state friction and dilatancy angles. The critical-state friction angle ϕ'_c is an intrinsic soil variable constant for a given soil type, while the dilatancy angle varies with both relative density and confining stress. In order to quantify the dilatancy of sands, Bolton (1986) proposed the following well-known dilatancy relationship:

$$\phi'_p = \phi'_c + R_D \cdot I_R \tag{1}$$

where $R_D = 3$ and 5 for triaxial and plane-strain conditions, respectively. The dilatancy index I_R is given by:

$$I_R = I_D\left[Q - \ln\left(\frac{100\sigma'_{mp}}{p_A}\right)\right] - R \tag{2}$$

where I_D = relative density = $D_R(\%)/100$; p_A = reference stress = 100 kPa; σ'_{mp} = mean effective stress at peak strength; and Q and R = intrinsic soil variables = 10 and 1 for clean quartz sands, respectively. While Eqs. (1) and (2) have been widely used, field application is still limited since σ'_{mp} is difficult to obtain due to unknown stress states mobilized upon loading.

For each case of triaxial tests adopted in this study, values of the cone resistance q_c were obtained from the penetration resistance analysis using the program CONPOINT

(Salgado and Randolph 2001). Since the confining stress (σ'_c) in TX tests is commonly determined from in-situ vertical (σ'_{v0}) and horizontal (σ'_{h0}) stresses, values of σ'_{v0} and σ'_{h0} at different K_0 values, equivalent to σ'_c values, were assumed and used in the calculation. Fig. 1 shows values of the mean effective stress at peak σ'_{mp}, measured from TX tests, versus the cone resistance q_c normalized with σ'_{h0}.

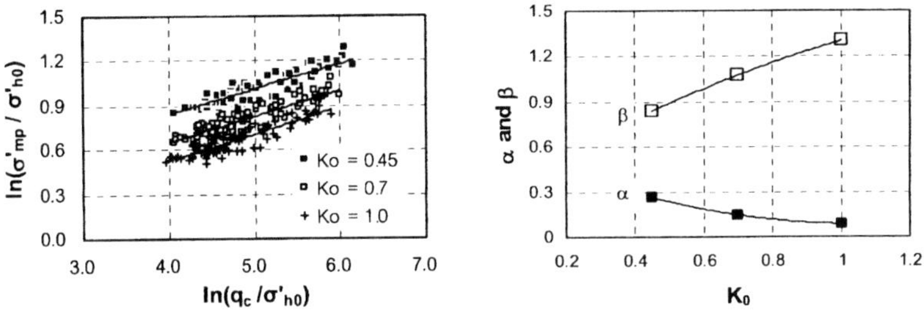

FIG. 1. Normalized σ'_{mp}-q_c relationship and values of α and β.

As can be seen in Fig. 1, quite unique correlations between $\sigma'_{mp}/\sigma'_{h0}$ and q_c/σ'_{h0} are observed for all the soil conditions considered in this study. Correlations in Fig. 1 can be given by:

$$\ln\frac{\sigma'_{mp}}{\sigma'_{h0}} = \alpha \cdot \left(\ln\frac{q_c}{\sigma'_{h0}}\right)^{\beta} \tag{3}$$

where α and β = correlation parameters. Values of α and β varied as a function of K_0, and are given in Fig. 1. If the correlation of Eq. (3) is adopted, Eq. (2) for I_R and ϕ'_p can be rewritten as the following equation:

$$I_{R,CPT} = I_D\left[Q - \alpha \cdot \left(\ln\frac{q_c}{\sigma'_{h0}}\right)^{\beta} - \ln\left(\frac{100\sigma'_{h0}}{p_A}\right)\right] - R \tag{4}$$

$$\phi'_p = \phi'_c + R_D \cdot I_{R,CPT} \tag{5}$$

where $I_{R,CPT}$ = modified dilatancy index in terms of q_c. While Eq. (4) requires the estimation of in-situ horizontal stress σ'_{h0}, the $I_{R,CPT}$ equation of Eq. (4) may be more effective and straightforward because it is based on an in-situ measurement (i.e., q_c), instead of a certain empirical correlation to estimate σ'_{mp}.

Since both q_c and σ'_{mp} are primarily governed by D_R and the confining stress, Eqs. (2) and (1) may further be modified using q_c as follows:

$$I_{R,CPT*} = I_D\left[Q_{CPT} - \ln\left(\frac{100q_c}{p_A}\right)\right] - R_{CPT} \tag{6}$$

$$\phi'_p = \phi'_c + R_D \cdot I_{R,CPT*} \tag{7}$$

where $I_{R,CPT*}$ = simplified dilatancy index in terms of q_c; Q_{CPT} and R_{CPT} = intrinsic soil variables corresponding to Q and R in Eq. (2). In order to obtain values of Q_{CPT} and R_{CPT}, a regression analysis was performed using TX test results and values of q_c from CONPOINT. Fig. 2 shows values of Q_{CPT} and R_{CPT}, as a function of silt content, obtained from the regression analysis. In addition, from the comparison with Q and R of Eq. (2), it was found that the differences of the parameters (i.e., Q vs. Q_{CPT} and R vs. R_{CPT}) are fairly consistent for all the silt contents considered herein, indicating that the differences are attributed mainly to the regression analysis using cone data.

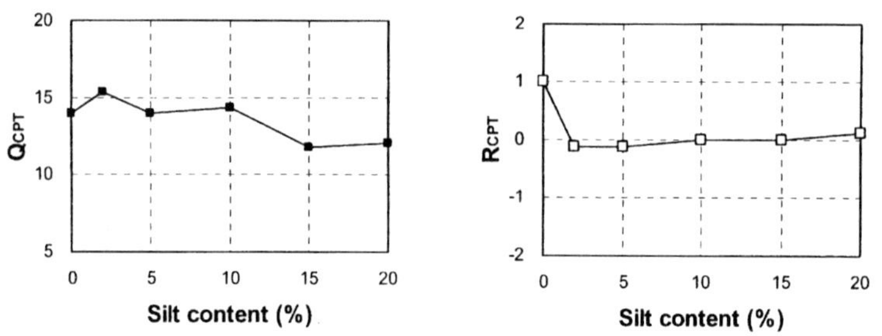

FIG. 2. Values of Q_{CPT} and R_{CPT} with silt content.

COMPARISON WITH CALIBRATION CHAMBER TEST RESULTS

In order to check the validity of the proposed methods, results from calibration chamber cone penetration tests were adopted and used in comparison and verification (Lee et al 2008). A total of 11 calibration chamber CPTs were performed at different relative densities and stress states. Calibration chamber test results obtained by Houlsby and Hitchman (1988) were also included in the comparison. The chamber used in the tests by Houlsby and Hitchman (1988) was of height and diameter equal to 1.0 and 0.9 m, respectively, whereas CPTs were conducted using the standard 36-mm cone. More detailed test conditions can be found from Houlsby and Hitchman (1988). A total of 15 calibration chamber CPT results were adopted in this comparison.

Fig. 3 shows measured and calculated values of ϕ'_p for calibration chamber test results from Lee et al. (2008) and Houlsby and Hitchman (1988). For Lee et al. (2008), measured values of ϕ'_p were obtained from the triaxial tests. For Houlsby and Hitchman (1988), since values of measured ϕ'_p were not reported, values of ϕ'_p obtained from Bolton's dilatancy relationship were adopted as reference measured values as Bolton's relationship is known to reasonably well reproduce strength of sands close to measured values. For calculated values of ϕ'_p, Eqs. (5) and (7) using CPT-based modified and simplified dilatancy relationship were used.

From Fig. 3, it is observed that predicted ϕ'_p values from Eqs. (5) and (7) show reasonable agreement with measured values. It is observed that the modified dilatancy equation based on $I_{R,CPT}$ of Eq. (5) produces some results quite close to measured values. The simplified dilatancy equation of Eq. (7) tends to underestimate values of ϕ'_p as soil becomes more dilatant. However, when typical design range of ϕ'_p values in 30° – 40° is considered, it is observed that overestimated ϕ'_p values could result in from both Eqs. (5) and (7).

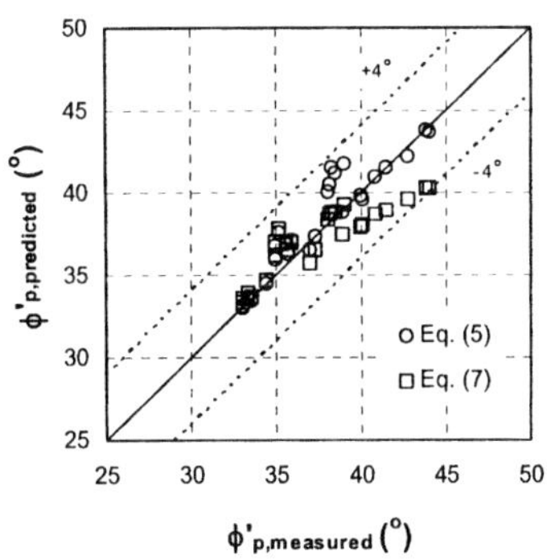

FIG. 3. Measured versus predicted ϕ'_p values from calibration chamber tests.

CONCLUSIONS

In this study, the field estimation of the shear strength for sandy soils based on CPT cone resistance was investigated. Results from laboratory test and calibration chamber cone penetration test results were used for this investigation. Based on correlations between q_c and σ'_{mp} for a given TX and equivalent field stress state, a modified dilatancy index $I_{R,CPT}$ in terms of q_c was presented. Further simplified dilatancy index $I_{R,CPT*}$ in terms of q_c was also presented. Values of intrinsic variables Q_{CPT} and R_{CPT} for $I_{R,CPT*}$ were evaluated and given as a function of silt content.

In order to check the validity of the CPT-based approach for in-situ evaluation of shear strength and dilatancy for sandy soils, calibration chamber cone penetration test results were adopted and used in the comparison. Various soil conditions were considered in the tests. It was observed that predicted ϕ'_p values from the CPT-based methods based on the modified and simplified dilatancy equations showed reasonable agreement with measured values. Overall, better matches to measured ϕ'_p values were observed from the modified dilatancy equation while the simplified dilatancy equation tends to underestimate values of ϕ'_p as soil becomes more dilatant. However, when typical design range of ϕ'_p values is considered, it was observed that overestimated ϕ'_p values could result in from both modified and simplified dilatancy equations.

REFERENCES

Bolton, M.D. (1986). "The strength and dilatancy of sands." *Geotechnique*, Vol. 36(1): 65-78.

Chen, J.W. and Juang, C.H. (1996). "Determination of drained friction angle of sands from CPT." *J. Geotechnical Engineering*, Vol. 122(5): 374 – 380.

Durgunoglu, H.T. and Mitchell, J.K. (1975). "Static penetration resistance of soils I: analysis." *Proc. of ASCE Conference on In Situ Measurement of Soil Properties*, ASCE, New York, Vol. 1: 151 – 171.

Houlsby, G. T., and Hitchman, R. (1988). "Calibration chamber tests of a cone penetrometer in sand." *Geotechnique*, Vol. 38(1): 39 – 44.

Lee, J., Salgado, R., and Carraro, A. (2004). "Stiffness degradation and shear strength of silty sands." *Canadian Geotechnical Journal*, Vol. 41(5): 831 – 843.

Lee, J., Eun, J., Lee, K., Park, Y. and Kim, M. (2008). "In-situ evaluation of strength and dilatancy of sands based on CPT results." *Soils and Foundation*, Vol. 48(2): 262 – 271.

Robertson, P. K. and Campanella, R.G. (1983). "Interpretation of cone penetration tests I: sand." *Canadian Geotechnical Journal*, Vol. 109(11): 1449 – 1459.

Salgado, R., Bandini, P., and Karim, A. (2000). "Stiffness and strength of silty sand." *J. Geotechnical and Geoenvironmental Engineering*, Vol. 126(5): 451 – 462.

Salgado, R. and Randolph, M. F. (2001). "Analysis of cavity expansion in sands." *International Journal of Geomechanics*, Vol. 1(2): 175-192.

Interpretation of Piezocone Tests in Venetian Silty Soils and the Issue of Partial Drainage

Laura Tonni[1] and Guido Gottardi[2]

[1]Lecturer, Dept. DISTART, University of Bologna, Viale Risorgimento 2, 40136 Bologna, Italy; laura.tonni@unibo.it
[2]Professor, Dept. DISTART, University of Bologna, Viale Risorgimento 2, 40136 Bologna, Italy; guido.gottardi2@unibo.it

ABSTRACT: The paper reports results of an extensive piezocone testing campaign performed at the Treporti Test Site (Venice, Italy) within a long-lasting research project aimed at developing a reliable geotechnical model of the heterogeneous and highly stratified Venetian lagoon subsoil. The analysis and the interpretation of the large amount of piezocone data in terms of compressibility and consolidation properties of the Venetian silty sediments reveal limitations of standard approaches for the characterization of such intermediate soils. Potentially dilative behaviour as well as partial drainage effects during cone penetration in silts must be taken in careful consideration in order to derive reliable soil parameters from *in-situ* measurements. According to recently proposed procedures, a preliminary attempt to assess drainage degree during piezocone tests is finally proposed.

INTRODUCTION

The paper presents results of a field study based on a consistent set of piezocone tests performed within a rather ambitious research programme carried out by the Italian Universities of Padova, Bologna and l'Aquila at the Treporti Test Site (*TTS*, Venice, Italy), with the aim of better understanding the behaviour of the heterogeneous and highly stratified silty sediments forming the upper 100 m of the Venetian lagoon basin. The whole research programme, consisting of a detailed geotechnical characterization through *in-situ* tests, continuous coring boreholes and high quality laboratory tests, also included the construction of a full scale 6.7 m high, 40 m diameter geogrid-reinforced vertical-walled cylindrical test bank. Details about the project as well as the monitoring system of settlements, local vertical strains and pore pressures can be found in Simonini (2004) and Gottardi & Tonni (2004).

As a consequence of such extensive site investigation, a complete database of piezocone measurements has become available for interpretation in terms of soil compressibility and consolidation parameters.

However, the application of existing and well-established empirical correlations to

Venetian soils revealed evident limitations of classical approaches, thus suggesting that the experience gained for clays and sands cannot be simply extended to such highly heterogeneous silty sediments. In particular, difficulties can be related to soil consolidation characteristics and partial drainage effects. Indeed, due to the essentially silty nature and the complex macrofabric of Venetian subsoil, different degrees of partial drainage are very likely to occur under a standard rate of penetration and the preliminary evaluation of such condition turns out to be of crucial importance for the assessment of representative mechanical parameters.

This paper presents adaptations of already established relationships and propose an application to Treporti piezocone measurements of recent interpretation procedures (Schnaid *et al.*, 2004; Schneider *et al.*, 2007), intended to recognize partial drainage conditions during penetration.

THE PIEZOCONE TESTING CAMPAIGN AT *TTS*

Sediments forming the upper 100 m of the Venetian lagoon basin consist of a chaotic assortment of interbedded NC or slightly OC silts, silty sands and silty clays, as a result of the complex depositional history over the last glacial Pleistocene period. According to Cola & Simonini (2002), the 95% of such sediments can be grouped into 3 classes: medium-fine sands (SP-SM) with sub-angular grains, silts (ML) and very silty clays (CL). The remaining 5% may be classified as organic clay and peat.

A representative test site of such heterogeneous and stratified subsoil was located outside Treporti, an old fishermen village facing the North Eastern lagoon (Fig. 1a). As shown in Figure 1b, the extensive piezocone testing programme was performed in four different phases. The *first Phase*, including tests located throughout the selected area, was intended to give an initial description of the site subsoil, with particular reference to the detection of the most compressible soil layers. Afterwards, a second series of CPTUs, solely located beneath the circular area of the loading bank to be built,

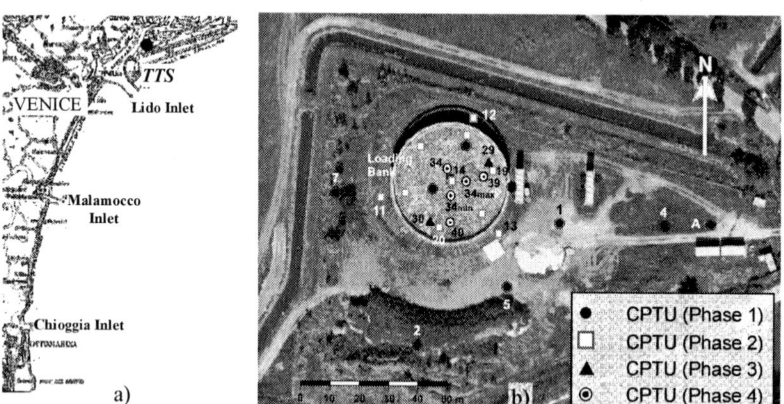

FIG. 1. a) View of the Venetian lagoon and location of the Treporti Test Site (*TTS*); b) Location of piezocone tests.

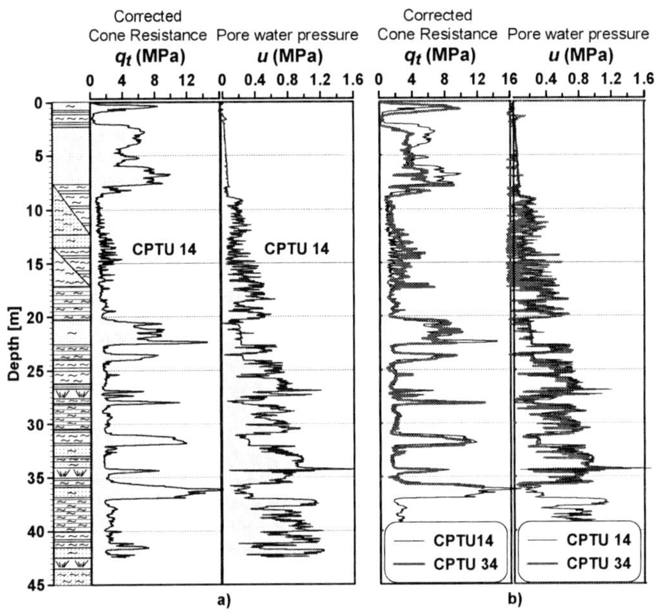

FIG. 2. a) CPTU 14 log profiles; b) Data comparison of second and fourth phase piezocone tests.

was carried out. At the end of the bank construction, a limited number of tests, belonging to the so called *Phase 3*, was performed from the top of the bank itself.

The final campaign was launched four years later, after completion of the gradual removal of the test embankment, and included tests (CPTU 34, CPTU 39, CPTU 40 in Figure 1b) located close to selected verticals already investigated in both the second and the third phases, as well as two further tests (CPTU 34_{min}, CPTU 34_{max}) performed near the centre of the loaded area at non-standard penetration rates, i.e. v_{min} = 0.15 cm/s and v_{max} = 4.0 cm/s (Tonni & Gottardi, 2009). Its main purpose was to examine the stress-history effects on piezocone measurements, as well as to detect drainage conditions during penetration in silty layers.

A typical CPTU profile of Venetian subsoil is reported in Figure 2a, showing the corrected cone resistance q_t and the pore water pressure u of the second phase test CPTU 14 (located at the centre of the loaded area), in conjunction with the log of a nearby borehole. Such profile provides immediate evidence of a highly stratified subsoil, with a well-defined top layer of sand, 6-7 m thick, followed by a silty unit from 7.5 m to 20 m depth and then by a dense alternation of thin layers of sandy and clayey silts.

According to grain size analyses referred to the upper 60 m of Treporti subsoil (Simonini, 2004), the typical Venetian soil classes SP-SM, SM and CL are here represented in the following proportions: 22 %, 32% and 37 % respectively, whilst organic clays and peats are 9%.

Profiles of adjacent tests CPTU 14 and CPTU 34, performed in the second and fourth phases respectively, are compared in Figure 2b. As a consequence of overconsolidation, a general increase in the corrected cone resistance q_t, in conjunction with a more pronounced dilative behaviour, can be detected within the silty unit. The unexpected decrease of q_t within the shallow sandy layer is likely to be ascribed to the high horizontal spatial variability of the local subsoil.

COMPRESSIBILITY

Compressibility of the upper 20 m sediments, expressed in terms of the *1D* modulus *M*, has been evaluated from field data, using the soil vertical strain measurements provided at 1 m intervals by the sliding deformeter installed beneath the centre of the test bank.

The construction of the earth-reinforced cylindrical sand bank, 6.7 m high, was carried out in 13 steps and required almost six months of work. Once completed, a total surcharge of approximately 106 kPa was applied to the ground surface.

Pore pressure monitoring, in conjunction with the analysis of vertical settlement with time, revealed that consolidation took place almost entirely during the bank construction. Therefore, field drained moduli *M* were determined from the vertical strains measured soon after bank completion. Furthermore, it was observed that the maximum horizontal displacement provided by the inclinometers located just outside the test bank was one order of magnitude smaller than the total vertical displacement recorded by the adjacent sliding deformeters, thus confirming that the deformation process prevalently developed in the vertical direction and that the assumption of one dimensional conditions in the centre of the loaded area can be reasonably accepted.

Moduli *M* were also estimated from CPTU 14 data, using common empirical correlations (Lunne & Christophersen, 1983; Kulhawy & Mayne, 1990), based on a clear distinction between drained and undrained conditions. However, the relevant predicted displacements (Fig. 3) are quite far from field values.

A better agreement can be reached either using the following semi-empirical relationship (Senneset *et al.*, 1988), based on experimental evidence from silty deposits of the North Sea:

$$M = 2q_t \cdot \sqrt{\frac{\sigma'_{v0}+\Delta\sigma'_v/2}{\sigma'_{v0}}}\text{, for } q_t < 2.5\text{ MPa} \quad (1a)$$

$$M = (4q_t - 5) \cdot \sqrt{\frac{\sigma'_{v0}+\Delta\sigma'_v/2}{\sigma'_{v0}}}\text{, for } 2.5\text{ MPa} < q_t < 5\text{ MPa} \quad (1b)$$

or considering a simple modification of the Kulhawy and Mayne's formula, given by:

$$M = \beta \cdot (q_t - \sigma_{vo}) \quad (2)$$

where constant $\beta = 3.5$ was back-calculated from sliding deformeter measurements.

Although approaches described by eqs. (1a-1b) and (2) result in better descriptions of the overall subsoil compressibility (see Fig. 3), it is worth remarking that both of them fully disregard the complex mechanics of cone penetration in silts. Indeed, permea-

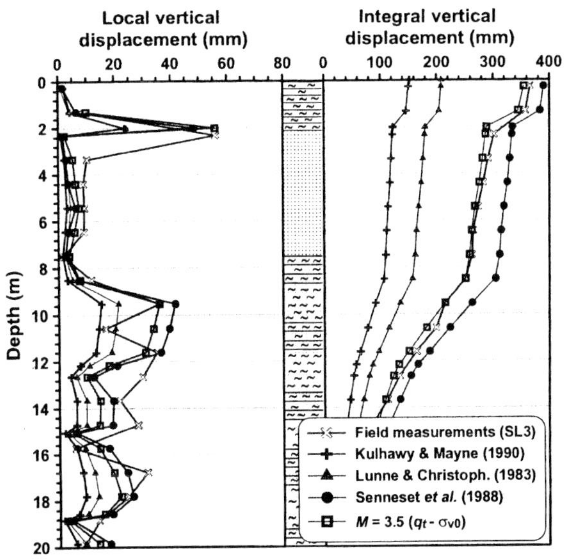

FIG. 3. Local and integral subsoil vertical displacements from 0 to 20 m depth.

bility and consolidation characteristics of natural silty deposits often fall within the range in which partial drainage is likely to occur, hence the simplest accepted approach of a broad distinction between drained and undrained conditions cannot be extended without a great deal of uncertainty (Schnaid *et al.*, 2004).

PARTIAL DRAINAGE CONDITIONS

The issue of partial drainage detection during cone penetration in Venetian sediments can be analysed from the perspective of two interpretation procedures, proposed by Schnaid *et al.* (2004) and Randolph & Hope (2004) respectively.

Figure 4 shows the application of the first approach, based on plotting the normalized cone resistance Q_t vs. the pore pressure parameter B_q in combination with the undrained strength ratio s_u/σ'_{v0}, to CPTU 14 data from 7.7 to 20 m depth, related to the predominantly silty unit. A cone factor N_{kt} equal to 15 was assumed to convert the cone resistance to undrained shear strength.

Basic properties of the relevant sediments are: average liquid limit LL = 32±10 %, average plasticity index PI = 10±10 %, coefficient of horizontal consolidation, as obtained from the interpretation of piezocone dissipation tests, $c_h = 10^{-5} \div 10^{-6}$ m^2/s. Most points of Figure 4 fall in the range where $B_q < 0.3$, corresponding to the domain in which partial drainage is prevailing when testing normally consolidated soils at a standard rate of penetration (v = 2 cm/s). It is worth noting that the calculated undrained strength ratio is in general significantly higher than values commonly accepted for NC or slightly OC soils and exhibits considerable scatter, thus suggesting

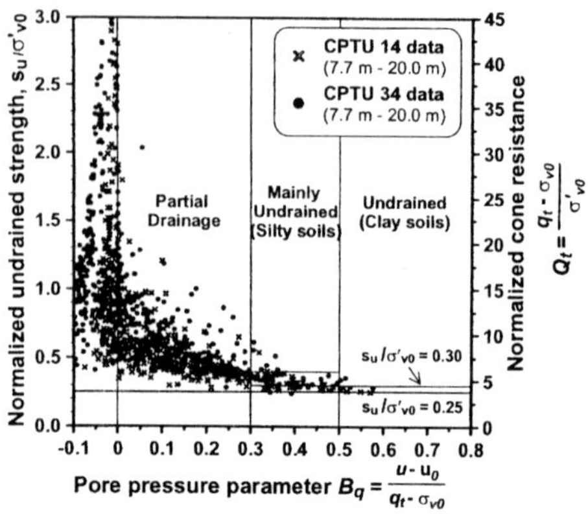

FIG. 4. Drainage conditions on Venetian silts

that deviation from this pattern in the case of Venetian sediments is essentially related to partial drainage phenomena. Data from the adjacent CPTU 34 test are also plotted in Figure 4. Due to the undergone compression and overconsolidation, silty sediments show a more pronounced dilative behaviour, with low or even negative excess pore pressures ($B_q < 0$).

Therefore, such approach for the evaluation of partial drainage should also consider the possible dilative behaviour of soils, as further discussed in the following section with reference to the piezocone tests performed at non-standard penetration rates.

DRAINAGE ASSESSMENT

Following Randolph & Hope (2004), tests at different penetration rates have been considered as a simple way to detect the transition point from undrained to partially drained response. Figure 5 shows cone resistance and pore pressure profiles of adjacent tests CPTU 34_{min} and CPTU 34_{max}, carried out at v = 0.15 cm/s and 4 cm/s respectively. It can be easily observed that a reduction in the rate of penetration results in a general increase in corrected cone resistance q_t and a decrease in pore water pressures u within the predominantly silty unit from 7.5 to 20 m depth.

In Figure 6 data from tests CPTU 34, CPTU 34_{min} and CPTU 34_{max} are represented in terms of the pore pressure ratio B_q vs. the normalized penetration rate $V = v \cdot d/c_v$, where d is the probe diameter and c_v is the coefficient of vertical consolidation. The figure also includes the curves provided by Schneider *et al.* (2007), obtained by fitting model piezocone data from centrifuge tests in normally consolidated and overconsolidated specimens of kaolin clay and silty clay.

The depicted trend lines clearly show that pore pressure responses of silty soils and kaolin differ significantly. Limits for drained and undrained conditions seem to correspond to non-dimensional velocities V of about 0.1 and 100 respectively, although the complex interaction of viscosity effects, silt content and stress history makes difficult detecting the transition from undrained to partially drained penetration. In Figure 6 data from Treporti have been also superimposed. Unfortunately, only limited

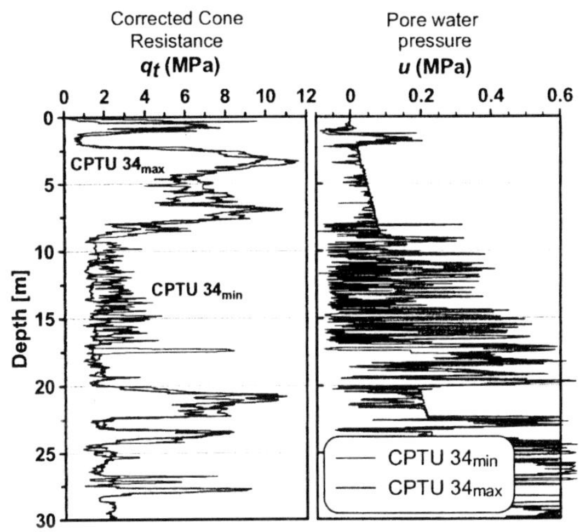

FIG. 5. Piezocone profiles of adjacent tests at different penetration rates.

data for a detailed characterization of such heterogeneous natural deposits are available. Discrepancies with respect to trend lines could be partly due to uncertainties about the evaluation of the relevant coefficient of consolidation, which was estimated using well-established interpretation procedures of dissipation tests based on the assumption of undrained behaviour during penetration, which can be questionable in such intermediate soils.

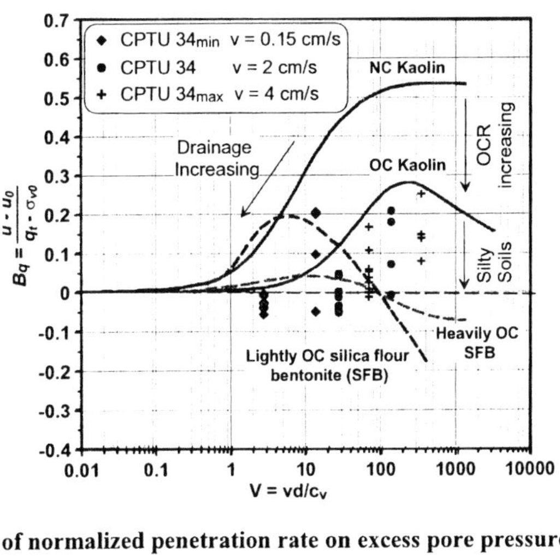

FIG. 6. Effect of normalized penetration rate on excess pore pressure ratio.

CONCLUSIONS

The paper has outlined the issues related to the interpretation of piezocone tests in the heterogeneous and highly stratified silty sediments forming the shallowest part of the Venetian lagoon basin. Difficulties in interpreting *in-situ* testing by using standard approaches was related to soil consolidation characteristics and associated partial drainage effects.

A preliminary attempt to assess drainage conditions has been performed through the application of two existing approaches to specific piezocone tests, conducted at different penetration rates. Although the occurrence of partial drainage within the predominantly silty unit was confirmed, additional testing at variable rates is required in order to get a better insight into the complex mechanics of cone penetration and pore pressure response.

REFERENCES

Cola, S. and Simonini, P. (2002). "Mechanical behaviour of silty soils of the Venice lagoon as a function of their grading properties." *Can. Geotech. J.*, Vol. 39: 879-893.

Gottardi, G. and Tonni, L. (2004). "Use of piezocone tests to characterize the silty soils of the Venetian lagoon (Treporti test site)." *Proc. ISC-2 on Geotech. and Geophysical Site Characterization*, Porto: 1643-1650.

Kulhawy, F.H. and Mayne, P.H. (1990). "Manual on estimating soil properties for foundation design." *Electric Power Research Institute*, EPRI, August 1990.

Lunne, T. and Christophersen, H.P. (1983). "Interpretation of cone penetrometer data for offshore sands." *Proc. of the Offshore Technology Conference*, Richardson, Texas, Paper No. 4464.

Randolph, M.F. and Hope, S. (2004). "Effect of cone velocity on cone resistance and excess pore pressures." *Proc. Int. Symp. on Eng. Practice and Performance of Soft Deposits*, Osaka: 147-152.

Schnaid, F., Lehane, B.M. and Fahey, M. (2004). "Characterisation of unusual geomaterials." *Proc. ISC-2 on Geotechnical and Geophysical Site Charact.*, Porto: 49-73.

Schneider, J.A., Lehane, B.M. and Schnaid, F. (2007). "Velocity effects on piezocone measurements in normally and over consolidated clays." *Int. J. Physical Modelling in Geotechnics*, Vol. 2: 23-34.

Senneset, K., Sandven, R., Lunne, T., By, T. and Amundsen, T. (1988). "Piezocone tests in silty soils." *Proc. Int. Symp. on Penetration Testing, ISOPT-1*, Orlando, Vol. 2: 955-966.

Simonini, P. (2004). "Characterization of the Venice lagoon silts from in-situ tests and the performance of a test embankment." *Proc. ISC-2 on Geotech. and Geophysical Site Characterization*, Porto: 187-207.

Tonni, L. and Gottardi, G. (2009). "Partial drainage effects in the interpretation of piezocone tests in Venetian silty soils." *Proc. 17th Int. Conf. Soil Mech. Geot. Eng.*, Alexandria (Egypt), Vol. 2: 1004-1007.

Cylindrical Cavity Expansion from a Finite Radius

Haiying Huang[1] and Emmanuel Detournay[2]

[1]Assistant Professor, Georgia Institute of Technology, Atlanta, GA, haiying.huang@ce.gatech.edu
[2]Professor, University of Minnesota, Minneapolis, MN, detou001@umn.edu

ABSTRACT: Solutions of cavity expansion have many practical applications in geotechnical engineering. In the large body of literature on the cavity expansion models for cohesive frictional materials, the solutions have mostly been derived by adopting a finite logarithmic strain definition in the plastic zone. A more rigorous alternative approach using the rate formulation for cylindrical cavity expansion from a finite radius is presented in this paper. The material is assumed to behave as elasto-perfectly plastic, obeying a Mohr-Coulomb yield criterion and the associated or nonassociated flow rule. The "time" (or evolution) variable is chosen to be the current cavity radius. The solution for the onset of plasticity serve as both the continuity condition for the elastoplastic interface and the initial condition for cavity expansion. The pressure-expansion relationship obtained is a first order ordinary differential equation and converges to a self-similar solution for expansion at a large radius.

INTRODUCTION

The solutions of cavity expansion have many practical applications in geotechnical engineering, among them, analysis of the bearing capacity of driven piles and interpretation of the pressuremeter and cone penetrometer testing results. In the large body of literature on the cavity expansion models for cohesive frictional materials (see reviews in Yu (2000) and Yu and Carter (2002)), the solutions have mostly been derived by adopting a finite logarithmic strain definition in the plastic zone, following Chadwick (1959). A general solution for both cylindrical and spherical cavity expansion from a finite radius using the total strain formulation was given in Yu and Houlsby (1991). A more rigorous alternative approach using the rate formulation was first carried out in Hill (1950) for Tresca materials and was adopted in Carter *et al.* (1986) for cohesive frictional materials. However, the cavity expansion solution in Carter *et al.* (1986) was obtained by assuming incompressibility in the elastic zone and by neglecting the convective rate of change in the time derivative (a small plastic strain assumption effectively). As a result, the pressure-expansion relationship obtained in Carter *et al.* (1986) is valid for small strain only and will not asymptote to the corresponding self-similar solution, i.e., the limit solution for expansion from a zero radius. Rate formulation of the cavity expansion solution with these two assumptions removed was presented in Yu and Carter (2002) for self-similar expansion. A non-self-similar expansion (i.e., from a finite radius) for the application of rock indentation was derived in Alehossein *et al.* (2000), where the kinematic conditions at the cavity wall, however, differ from those of the pressurized cavity expansion problem. A non-self-similar solution was also derived in Davis and Selvadurai (2002) for frictional materials using the rate formulation. However, the time integration to obtain the radial displacement from the velocity field in Davis and Selvadurai (2002) is incorrect.

A characteristic of cavity expansion from a zero radius is that the problem does not contain a length scale and is therefore geometrically self-similar. Consequently, field quantities such as stresses remain unchanged in a transformed coordinate system $(x/a, y/a)$, where a is the current cavity radius; and quantities such as stress gradients will be inversely proportional to the cavity radius a. As a result, regardless of the cavity size, the expansion pressure will be a constant and is a function of the material properties and the far field stress only. The characteristic of self-similarity no longer exists if the cavity expands from a finite radius. The non-self-similar solution will depend on the cavity size and should converge to the self-similar solution for expansion at a large radius.

In this paper, the pressure-expansion relationship for pressurized cylindrical cavity expansion from a finite radius is derived using the rate formulation. The material is assumed to behave as elasto-perfectly plastic, obeying a Mohr-Coulomb yield criterion and the associated or nonassociated flow rule. In this formulation, the "time" (or evolution) variable is chosen to be the current cavity radius. The solution for the onset of plasticity serves as both the continuity condition for the elastoplastic interface and the initial condition for cavity expansion. The pressure-expansion relationship can be obtained by solving a first order ordinary differential equation.

STRESS FIELD

Assume a cylindrical cavity with an initial radius a_0 is subjected to uniform internal pressure p and far field stress p_0, i.e.,

$$\sigma_r\big|_{r=a} = -p \quad \sigma_r\big|_{r\to\infty} = -p_0$$

The sign convention is assumed to be tension positive here. Axisymmetric expansion of the cavity due to pressure p will lead to development of a plastic zone of size $r = r_*$ (see FIG. 1). Denote the radial stress at the elasto-plastic interface as σ_r^*. The stress field in the elastic zone can be obtained from the Lamé solution,

$$\sigma_r = (\sigma_r^* + p_0)\left(\frac{\xi_*}{\xi}\right)^2 - p_0 \quad \sigma_\theta = -(\sigma_r^* + p_0)\left(\frac{\xi_*}{\xi}\right)^2 - p_0 \tag{1}$$

where σ_r and σ_θ are the radial and tangential stresses; $\xi = r/a$ is the radial coordinate scaled with respect to the current cavity size a and $\xi_* = r_*/a$ is the scaled plastic zone size.

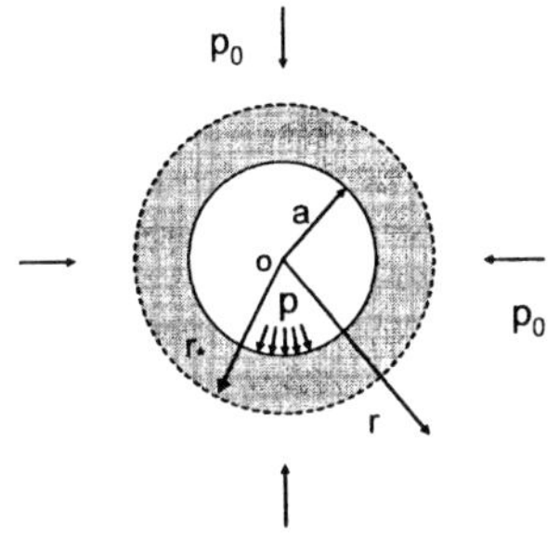

FIG. 1. Schematic of a cylindrical cavity subjected to an internal pressure p and far field stress p_0 (plastic zone in shade).

If the material fails according to a Mohr-Coulomb failure criterion, i.e.,

$$\sigma_r = k_p \sigma_\theta - \sigma_c \tag{2}$$

where $k_p = (1+\sin\phi)/(1-\sin\phi)$ is the passive failure coefficient defined in terms of the internal friction angle ϕ and σ_c is the uniaxial compressive strength, the stress field in the plastic zone can be determined from the equilibrium equation and the failure criterion as

$$\sigma_r = h - (p+h)\xi^{(1-k_p)/k_p} \qquad \sigma_\theta = h - \frac{1}{k_p}(p+h)\xi^{(1-k_p)/k_p} \tag{3}$$

where $h = \sigma_c/(k_p - 1)$. By applying the stress continuity conditions at the elasto-plastic interface, the pressure at the cavity wall can be related to the scaled plastic zone size as follows,

$$p = -h + \frac{2k_p}{k_p+1}(h+p_0)\xi_*^{(k_p-1)/k_p} \tag{4}$$

And at the elasto-plastic interface,

$$\sigma_r^* = h - \frac{2k_p}{k_p+1}(h+p_0) \tag{5}$$

Substituting Eq. (4) into Eq. (3) yields the stresses in the plastic zone as

$$\begin{aligned} \sigma_r &= h - \frac{2k_p}{k_p+1}(h+p_0)\left(\frac{\xi_*}{\xi}\right)^{(k_p-1)/k_p} \\ \sigma_\theta &= h - \frac{2}{k_p+1}(h+p_0)\left(\frac{\xi_*}{\xi}\right)^{(k_p-1)/k_p} \end{aligned} \tag{6}$$

The stress field above is statically determined up to one variable, namely, the size of the plastic zone ξ_*, which can only be obtained from kinematic considerations.

Note since the total strain formulation and the rate formulation concern only the kinematic aspect of the problem, derivation of the stress field is the same for the two approaches.

VELOCITY FIELD

A general non-associative flow rule for cohesive frictional materials can be expressed as

$$\dot{\varepsilon}_\theta^p = -k_d \dot{\varepsilon}_r^p \tag{7}$$

where $k_d = (1+\sin\psi)/(1-\sin\psi)$ is the dilatancy coefficient defined in terms of the dilatancy angle ψ ($0 \le \psi \le \phi$). Expressing the total strain rate in terms of the radial velocity v,

$$\dot{\varepsilon}_r = \frac{dv}{dr} \quad \text{and} \quad \dot{\varepsilon}_\theta = \frac{v}{r} \tag{8}$$

and decomposing the total strain rate into the sum of the elastic and plastic strain rates yield

$$k_d \frac{dv}{dr} + \frac{v}{r} = \dot{\varepsilon}_\theta^e + k_d \dot{\varepsilon}_r^e \tag{9}$$

The velocity here is interpreted as the rate of change of a material point position or its displacement with respect to the current cavity radius a. Under the coordinate transformation, $\xi = r/a$, the above equation becomes

$$k_d \frac{dv}{d\xi} + \frac{v}{\xi} = a(\dot{\varepsilon}_\theta^e + k_d \dot{\varepsilon}_r^e) \tag{10}$$

Furthermore, by invoking Hooke's law and the yield condition, it can be shown that,

$$\dot{\varepsilon}_\theta^e + k_d \dot{\varepsilon}_r^e = \frac{\omega \dot{\sigma}_r}{2G} \tag{11}$$

where

$$\omega = \frac{k_p k_d + 1 - \nu(k_p+1)(k_d+1)}{k_p} \tag{12}$$

and G and ν are the shear modulus and the Poisson's ratio.

The material time derivative for an Eulerian coordinate system (r,a) with a being the evolution parameter can be expressed as

$$\dot{(\;)} = \frac{\partial}{\partial a} + v\frac{\partial}{\partial r} = \frac{\partial}{\partial a} + \frac{v}{a}\frac{\partial}{\partial \xi} \tag{13}$$

Alternatively, the scaled coordinate ξ and the cavity radius a can also be used to describe the kinematics of the material points. A Lagrangian time derivative D/Da at a fixed ξ for the coordinate system $(a\xi, a)$ (note $r = a\xi$) can therefore be defined as,

$$\left.\frac{D}{Da}\right|_{\xi} = \left.\frac{d}{da}\right|_{\xi} = \frac{\partial}{\partial a} + \frac{\xi}{a}\frac{\partial}{\partial \xi} \tag{14}$$

The material time derivative can then be expressed in terms of the special operator D/Da as

$$\dot{(\,)} = \frac{D}{Da} + \frac{1}{a}(v - \xi)\frac{\partial}{\partial \xi} \tag{15}$$

Note that in self-similar problems, $D/Da = 0$.

Using the differentiation rule (15), the radial stress rate in the plastic region can be readily derived from Eq. (3),

$$\dot{\sigma}_r = 2(h + p_0)\left(\frac{k_p - 1}{k_p + 1}\right)\left[\frac{1}{a}\left(\frac{v}{\xi} - 1\right) - \frac{1}{\xi_*}\frac{d\xi_*}{da}\right]\left(\frac{\xi_*}{\xi}\right)^{(k_p-1)/k_p} \tag{16}$$

The governing equation (10) for the velocity in the plastic region thus becomes

$$k_d\frac{dv}{d\xi} + \frac{v}{\xi}\left[1 - \omega\kappa\left(\frac{\xi_*}{\xi}\right)^{(k_p-1)/k_p}\right] = -\omega\kappa\left(1 + \frac{a}{\xi_*}\frac{d\xi_*}{da}\right)\left(\frac{\xi_*}{\xi}\right)^{(k_p-1)/k_p} \tag{17}$$

where

$$\kappa = \frac{h + p_0}{G}\left(\frac{k_p - 1}{k_p + 1}\right) \tag{18}$$

Typically $\kappa << 1$ for both soils and rocks. Neglecting the nonlinear term involving the small number κ yields,

$$v = A\xi^{-1/k_d} - \mu\kappa\left(\xi_* + a\frac{d\xi_*}{da}\right)\left(\frac{\xi_*}{\xi}\right)^{-1/k_p} \tag{19}$$

where A is a constant of integration and

$$\mu = \frac{\omega k_p}{k_p + k_d} \tag{20}$$

Note that the velocity field obtained above is the first order approximation in terms of κ. The full solution to Eq. (17) can also be solved analytically, however, involving the incomplete Gamma function (or series expansion).

The constant A can be determined by invoking the velocity condition at the elasto-plastic boundary. Utilizing Eq. (13) that relates the Lagrangian to the Eulerian time derivative, we have,

$$v = \frac{\frac{\partial u}{\partial a}}{1 - \frac{\partial u}{\partial r}} \tag{21}$$

where the displacement u is seen here as a function of the position r and the "time" parameter a. In the elastic region, the radial displacement field is given by the Lamé solution,

$$u = \frac{\kappa r}{2}\left(\frac{a\xi_*}{r}\right)^2 = \frac{a\kappa\xi}{2}\left(\frac{\xi_*}{\xi}\right)^2 \tag{22}$$

The radial strain at the elasto-plastic interface is $\varepsilon_r^* = \kappa/2$. Therefore, ignoring the nonlinear term associated with κ is equivalent to assuming small strain in the elastic zone, which differs from the assumption of small strain everywhere in Carter *et al.* (1986).

It follows from Eqs. (21) and (22) that the radial velocity at the elasto-plastic interface reads

$$v(\xi_*) = \frac{2\kappa}{\kappa+2}\left(\xi_* + a\frac{d\xi_*}{da}\right) \approx \kappa\left(\xi_* + a\frac{d\xi_*}{da}\right) \tag{23}$$

Substituting Eq. (23) into Eq. (19) yields the expression for the constant A,

$$A = (1+\mu)\kappa\left(\xi_* + a\frac{d\xi_*}{da}\right)\xi_*^{1/k_d} \tag{24}$$

and the velocity field in the plastic region,

$$v(\xi) = \left[(1+\mu)\left(\frac{\xi_*}{\xi}\right)^{1/k_d} - \mu\left(\frac{\xi_*}{\xi}\right)^{-1/k_p}\right]\kappa\left(\xi_* + a\frac{d\xi_*}{da}\right) \tag{25}$$

PRESSURE-EXPANSION RELATIONSHIP

After the velocity is known, the scaled size of the elasto-plastic boundary ξ_* can be determined readily by considering the velocity condition at the cavity wall. Since the velocity is defined with the cavity size a as the "time" parameter, the velocity at $\xi = 1$ is simply,

$$v(1) = 1 \tag{26}$$

Denote $\delta = a/a_0$ as the cavity size ratio. Substituting Eq. (26) into Eq. (25) yields a first order differential equation for the relationship between the scaled plastic zone size ξ_* and the cavity size ratio δ,

$$\delta\frac{d\xi_*}{d\delta} = -\xi_* + \frac{1}{\kappa}\frac{1}{(1+\mu)\xi_*^{1/k_d} - \mu\xi_*^{-1/k_p}} \tag{27}$$

Since the Poisson's ratio varies only slightly for soils and rocks, the set of governing parameters now reduces to $\{k_p, k_d, \kappa\}$ or $\{\phi, \psi, (h+p_0)/G\}$ only.

The initial condition for this first order ODE can be found from the condition for the onset of plasticity, i.e., when $\xi_* = 1$. Denote the cavity radius at the onset of plasticity as $\bar{a}$. According to the Lamé solution, the critical cavity displacement is then $\bar{u} = \kappa\bar{a}/2$. Since at the cavity wall, the radial displacement $\bar{u}$ is simply the difference between the current and the initial cavity radius, i.e., $\bar{u} = \bar{a} - a_0$, we have

$$\delta=\frac{\bar{a}}{a_0}=\frac{1}{1-\frac{\kappa}{2}} \tag{28}$$

The ODE, Eq. (27), with the above initial condition can also be combined into an integral equation,

$$\delta=\frac{1}{1-\frac{\kappa}{2}}\exp\left\{\int_1^{\xi_*}\left[-\xi_*+\frac{1}{\kappa}\frac{1}{(1+\mu)\xi_*^{1/k_d}-\mu\xi_*^{-1/k_p}}\right]^{-1}d\xi_*\right\} \tag{29}$$

Once the relationship between the scaled plastic zone size ξ_* and the cavity size ratio δ is obtained numerically by solving either Eq. (27) with Eq. (28) or Eq. (29) alone, the results with the aid of Eq. (4) can then be expressed in form of pressure p versus expansion δ.

Eq. (27) reduces to the self-similar solution after setting $d\xi_*/d\delta=0$, i.e.,

$$(1+\mu)\xi_*^{1/k_d+1}-\mu\xi_*^{-1/k_p+1}=\frac{1}{\kappa} \tag{30}$$

It can be seen that the scaled plastic zone size ξ_* and therefore the limiting pressure for the self-similar case remains constant. The limiting solution derived here is identical to the solution in Carter *et al.* (1986).

Using the notation in this paper, the non-self-similar solution derived using the total strain formulation in Yu and Houlsby (1991) can be rewritten as

$$\frac{\eta}{\gamma}\left[\left(1-\frac{\kappa}{2}\right)^{(k_d+1)/k_d}-(\delta\xi_*)^{-(k_d+1)/k_d}\right]=\int_1^{\xi_*^{(k_p-1)/k_p}}\exp(\zeta\rho)\rho^{-\gamma-1}d\rho \tag{31}$$

where

$$\eta=\exp\left[\frac{(k_d+1)(1-2\nu)(h+p_0)}{2Gk_d}\right]$$
$$\gamma=\frac{k_p(k_d+1)}{k_d(k_p-1)} \qquad \zeta=\frac{\kappa(1-\nu)}{(k_p-1)k_d}\left[k_dk_p+1-\frac{\nu(k_p+k_d)}{1-\nu}\right] \tag{32}$$

The equation can be rearranged into

$$\delta=\frac{1}{\xi_*}\left[-\frac{\gamma}{\eta}\int_1^{\xi_*^{(k_p-1)/k_p}}\exp(\zeta\rho)\rho^{-\gamma-1}d\rho+\left(1-\frac{\kappa}{2}\right)^{(k_d+1)/k_d}\right]^{-k_d/(k_d+1)} \tag{33}$$

The solution from this analysis is compared with those from Yu and Houlsby (1991) and Carter *et al.* (1986) in FIG. 2. The results were obtained with $\phi=\psi=30^\circ$ and $\kappa=0.001$ (which corresponds to a representative set of data of $E=1$ GPa, $\nu=0.25$, $\sigma_c=1$ MPa, $p_0=0.3$ MPa). As can be see from FIG. 2, since both the solution from this analysis and that in Yu and Houlsby (1991) consider large strain in the plastic zone, the two solutions agree well with each other and asymptote to the self-similar solution. Meanwhile, since the solution in Carter *et al.* (1986) is derived with the assumption of small strain everywhere, at $\delta=1.1$, the error is about 9.5%.

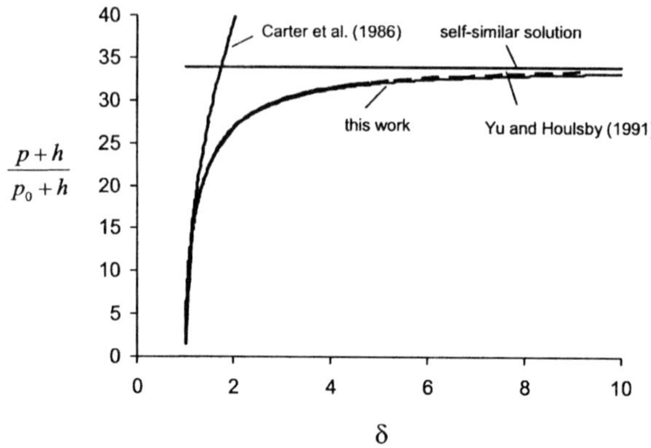

FIG. 2. Scaled pressure-expansion relationship for $\phi = \psi = 30^{\circ}$ and $\kappa = 0.001$.

CONCLUSIONS

A non-self-similar solution for cylindrical cavity expansion from a finite radius is derived using the rate formulation for cohesive frictional materials. The pressure-expansion relationship can be obtained numerically by solving a first order ordinary differential equation. The solution agrees well with that in Yu and Houlsby (1991) derived with a finite logarithmic strain definition in the plastic zone. Both solutions asymptote to a self-similar solution for expansion from zero radius. As expected, the solution in Carter *et al.* (1986), due to its assumption of small total strain, is valid only for the early stage of expansion.

ACKNOWLEDGEMENTS

Financial support from Schlumberger to HH is gratefully acknowledged.

REFERENCES

Alehossein, H., Detournay, E., and Huang, H. (2000). "An analytic model for the indentation of rocks by blunt tools." *Rock Mechanics and Rock Engineering*, 33(4), 267 - 284.

Carter, J., Booker, J., and Yeung, S. (1986). "Cavity expansion in cohesive frictional soils." *Géotechnique*, 36(3), 349 - 358.

Chadwick, P. (1959). "The quasi-static expansion of a spherical cavity in metals and ideal soils." *Quart. Journ. Mech. and Applied Math.*, XII, 52 - 71.

Davis, R. and Selvadurai, A. (2002). *Plasticity and geomechanics*. Cambridge University Press.

Hill, R. (1950). *The Mathematical Theory of Plasticity*. The Oxford Engineering Science Series. Oxford University Press.

Yu, H. (2000). *Cavity Expansion Methods in Geomechanics*. Kluwer Academic

Publishers.

Yu, H. and Carter, J. (2002). "Rigorous similarity solutions for cavity expansion in cohesive-frictional soils." *Int. J. Geomech.*, 2(2), 233 - 258.

Yu, H. and Houlsby, G. (1991). "Finite cavity expansion in dilatant soils: loading analysis." *Géotechnique*, 41(2), 173 - 183.

Identification of Soil Stratigraphy of Soft Deposit in Shanghai from CPTU Test

Shui-Long Shen[1], Jun-Peng Wang[2,3], and Lei Ma[2]

[1]Professor, Department of Civil Engineering, Shanghai Jiao Tong University, 800 Dongchuan Road, Shanghai, 200240; PH (86) 21-34204301; e-mail: slshen@sjtu.edu.cn
[2]PhD. Student, ditto;
[3]Lecturer, Department of Management, Qingdao Technological University, 2 Changjiang Road, Qingdao, 266520; PH (86) 532-86875755; e-mail: wangjunpeng79@sina.com

ABSTRACT: This paper presents a well recorded case history of cone penetration test with pore water pressure measurement (CPTU) in the soft deposit of Shanghai. In China, stratigraphy for soft deposit is obtained through physical properties (e.g. grain size and color) and mechanical properties (e.g. cohesion and inner friction angle from direct shear test) from borehole samples. Identification of soil stratigraphy with CPTU was developed 30 years ago. The field data at test site of Yan'an Road Tunnel in Shanghai is used in the analysis. The soil classification chart proposed by Robertson is employed to identify soil strata. The results are compared with that from plasticity index and grain size distribution of borehole samples (refer to code method). The results show that Robertson's chart can identify the stratigraphy of the soft deposit in Shanghai. However, there are some discrepancies of the results from Robertson's chart and the code method. The reason of the discrepancies is also discussed in the paper.

INTRODUCTION

Cone penetration test with pore water pressure measurement (CPTU) is a useful ground investigation technology employed in many countries around the world for providing a quantitative measurement of soil properties in its natural environment. Identification of in-situ stratigraphic profiles is one of main applications. This identification is based on the field observed results, of which there are dependencies between soil type and cone resistance as well as skin friction ratio, for example, sandy soils tend to produce high cone resistance and low friction ratio, whereas soft clay soils tend to produce low cone resistance and high friction ratio (Begemann, 1965; Douglas and Olsen, 1981; Robertson et al. 1986; Lunne et al., 2002).

In Shanghai, engineering practice shows that soil type is identified through grain size distribution and plasticity index (I_P) of borehole samples, which is called code method hereafter (CCES, 1999 and CCEMS, 2002). Except for rock existing in few hilly areas, the soft deposit in Shanghai belongs to the Quaternary sediments and is mainly composed of cohesive soil (clay and silt) and sand.

In this research, field data of CPTU tests and adjacent borehole data at the site of Yan'an Road Tunnel in Shanghai are collected; possibility of application of stratigraphy chart proposed by Robertson et al. (1986) to identify the stratigraphy of the soft deposit of Shanghai is presented.

STRATA IDENTIFICATION AND SOIL PROPERTIES FROM BOREHOLE

In Shanghai, the soft deposit is composed of six aquifers and five aquitards, overburdened one another, called multi-aquifer-aquitard system (MAAS) (Xu et al., 2009). This deposit was formed mainly within the warm geological era updated to 2.6 million years ago (Xu et al., 2009). The geotechnical investigation is generally to a depth of 120 m, to which the soft deposit is divided into 12 layers with the soil type of clay (particle size: d_c<0.005 mm), silt (particle size: 0.005 mm <d_c<0.074 mm), and sand (particle size: d_c>0.074 mm). Soil types of layer are defined into two categories with six soil types according to the grain size distribution and plasticity index (I_P) as established in the geotechnical investigation standard of Shanghai (CCES, 1999; CCEMS, 2002), that is cohesive soil with plasticity (clay, silty clay, and clayey silt) and sandy soil with non-plasticity (sandy silt, silty sand, and fine sand). Figure 1 illustrates this definition. Clay layer is defined based on plasticity index I_P >17; silty clay is defined according to 10<I_P≤17; clayey silt is defined according to I_P≤10 and clay content (M_c): 10%≤M_c≤15%. However, sandy soil is defined according to the grain size; sandy silt: clay content M_c<10%; silt sand: sand content (M_s) 50%≤M_s≤85%; fine sand: clay content M_s>85%. Most of the cohesive soil has silt content over 50%.

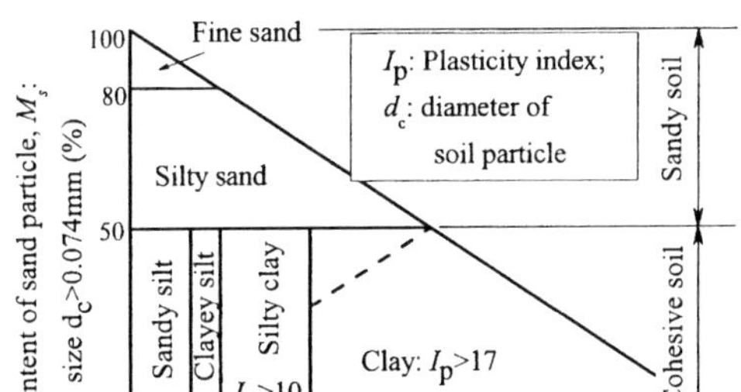

FIG. 1. Definition of soil layer type in Shanghai Code (CCEMS, 2002).

The site is located at Yan'an Road Tunnel, Shanghai, China. The tunnel passed under the Huangpu River. Figure 2 shows the soil profile and geotechnical properties obtained by borehole at this site based on the method as established in Shanghai (CCEMS, 2002). A gray mucky clay layer (MC) exists near the ground surface (from 0 to 5.5 m deep). The underlying layer from a depth of 5.5 m to 15 m is gray mucky silty clay (MSC). Below the MSC layer there is a silty clay (SC) layer, but its color is gray (from 15 m to 29.5 m deep) and dark green (from 29.5 m to 32.3 m deep), respectively.

As shown in Fig. 2, distribution of liquid limit (w_L) and natural water content (w) with depth are almost the same. The changes of plastic limit (w_P) are very small. Large scatter is noticed in the distribution of unit weight, void ratio, and coefficient of volume compressibility. Generally in Fig. 2, w, w_P and e decrease with depth while γ_t, m_v increases with the depth. However, under 25 m, m_v becomes also small.

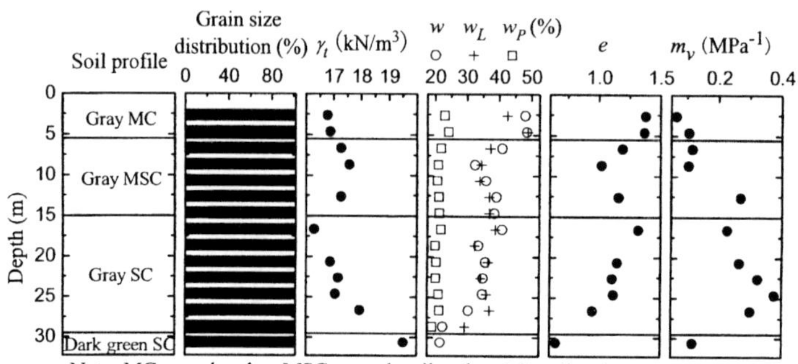

Note: MC = mucky clay, MSC = mucky silty clay, SC = silty clay, γ_t = unit weight, w_p = plastic limit, w = natural water content, w_L = liquid limit, e = void ratio, m_v = coefficient of volume compressibility, ■ clay ■ silt ■ sand

FIG. 2. Soil profile and geotechnical properties of borehole samples.

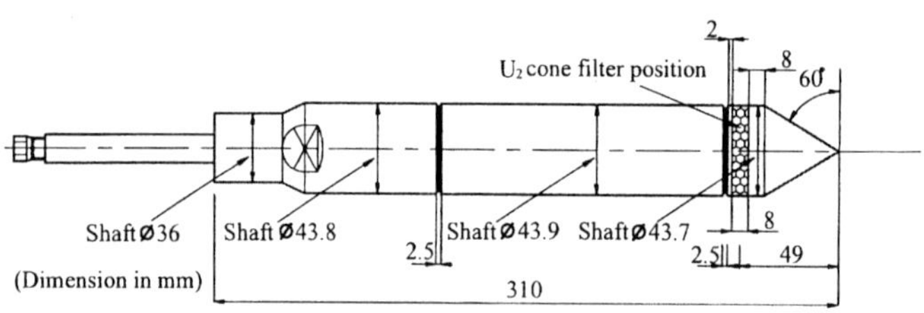

FIG. 3. Sectional view of the CPTU cone used in the field test.

CPTU TEST RESULTS

Description of the Cone

Figure 3 depicts the cone employed in this study, which has an equal end area to friction sleeve and an interchangeable tip to allocate the porous filter. The cone has a 60° apex angle, 1,500 mm^2 base area, 20,000 mm^2 friction sleeve f_s and a conical tip with an end resistance q_c. A ceramic filter is located 2 mm behind the conical tip. An electric cable usually connects the cone with the recording equipment at ground surface. The cone offers obvious advantages such as rapid testing procedures, continuous recording, high accuracy and repeatability (Robertson et al., 1983b). The device was pushed down into the ground at a rate of 0.02 m/s as specified in the European standard and also in the International Standard for Cone Penetration Testing (ISSMFE, 1989; Lunne et al., 2002). Data was measured in the interval of 20 mm of this field case.

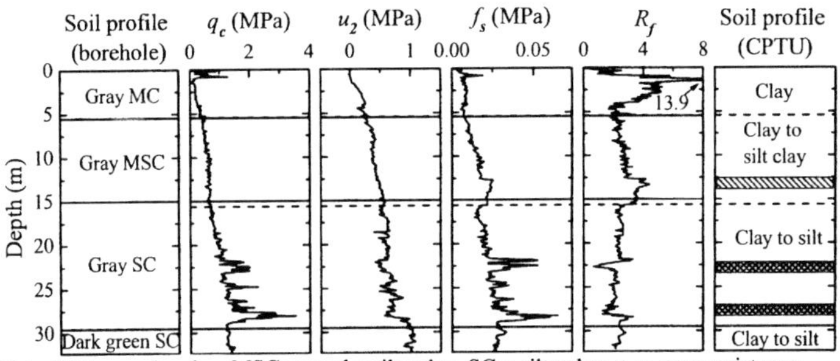

Note: MC = mucky clay, MSC = mucky silty clay, SC = silty clay, q_c = cone resistance, u_2 = pore pressure, f_s = sleeve friction, R_f = friction ratio; clay lens, sand lens.

FIG. 4. Geotechnical profile and soil properties from CPTU.

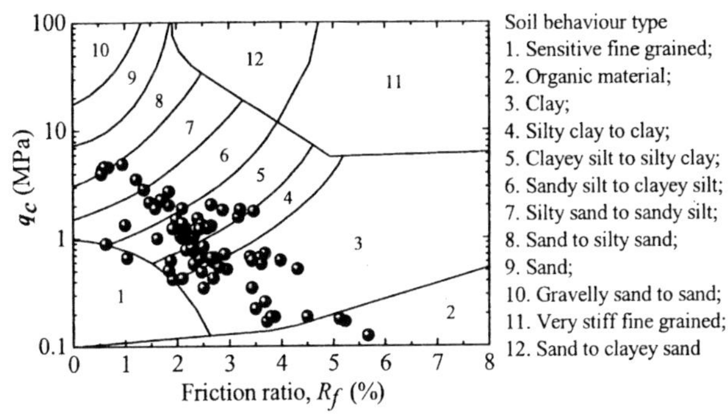

FIG. 5. Plot of all data from CPTU at the site in Robertson's chart.

Description of CPTU Test Profile

The CPTU with controlled rate (0.02 m/s) of penetration were carried out at this test site. The test depth was 32.30 m. The continuous CPTU profile of the soft deposit is shown in Fig. 4. The data include the cone resistance q_c, pore pressure measured behind the cone u_2, sleeve friction f_s, and friction ratio R_f. At this site, there is a general increase in q_c, u_2, and f_s along with depth while no obvious tendency is noticed in R_f along with depth except that there is a sudden increase in the depth of 1.4 m, at which it is thought of as the effect of drying (evaporation) of the top layer.

Soil Identification from Classification Chart

Robertson and Campanella (1983a) proposed a soil classification chart based on cone resistance, q_c and skin friction ratio, R_f. In later engineering practice, this classification charts have been adapted and improved based on an expanded database (Robertson et al., 1986; Olsen and Farr, 1986). In this research, a set of the data was recorded every 20 mm in depth, as shown in Fig. 4. The cone resistance q_c and friction ratio R_f in each set were redrawn in the Robertson's chart (Robertson et al., 1986) as shown in Fig. 5. The layer profile based on chart is illustrated in Fig. 4. From Fig. 4, soil profile from CPTU tests is basically consistent with that from borehole data except for some discrepancies in the depth with clay lens and sandy lens as shown in Fig. 4. In these layers, there are sudden changes of R_f, f_s and q_c. It is difficult to identify such thin layers from plasticity index and grain size based on borehole sample.

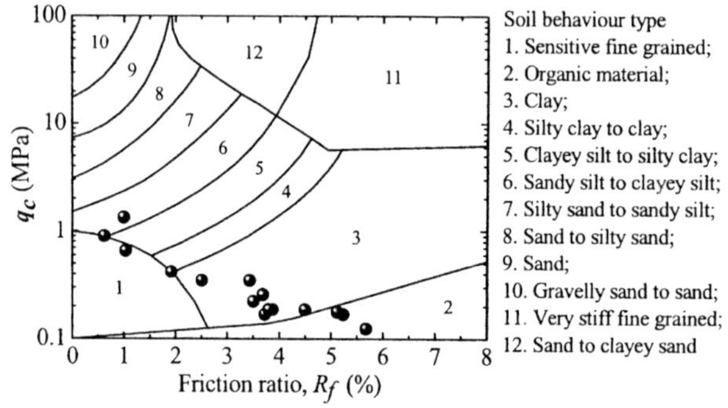

FIG. 6. Plot of CPTU data of MC layer from borehole in Robertson's chart.

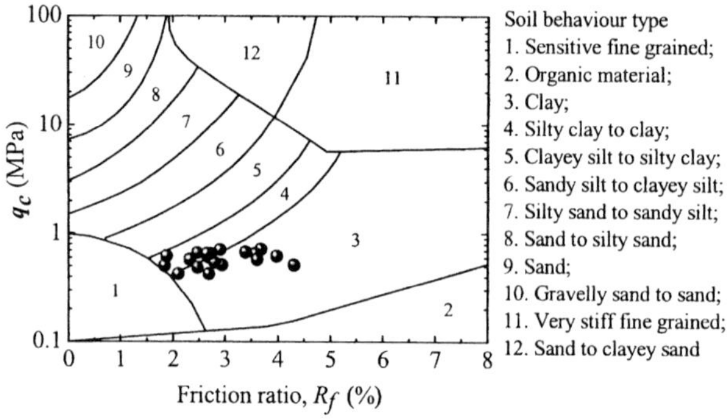

FIG. 7. Plot of CPTU data of MSC layer from borehole in Robertson's chart.

DISCUSSION

In order to investigate the discrepancies on soil identification between borehole results and CPTU tests, cone resistance q_c and friction ratio R_f of each soil layer of the soft deposit identified by borehole are redrawn in Robertson's chart as shown in Fig. 6 to Fig. 9. Figure 6 shows the CPTU data of gray mucky clay identified from borehole samples. Most of the CPTU data of the MC layer is located in zone 3, which represents that the soil type is clay while a few data fall in zones 1, 2, 4, and 6, which represent that this layer also includes the sensitive fine grained soil, organic material, silty clay, and clayey silt to sandy silt. These characteristics of soil are difficult to be identified from borehole samples.

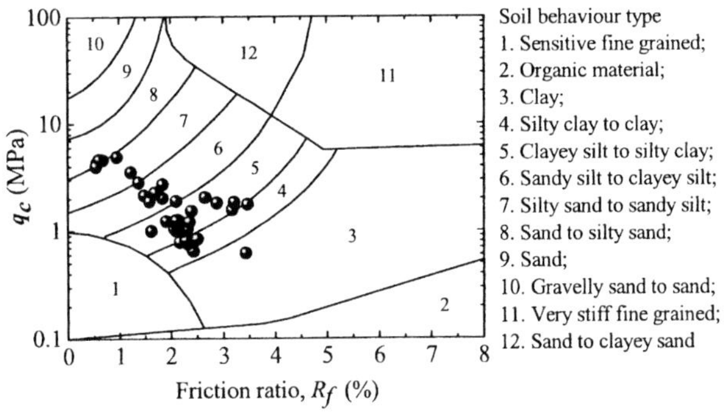

FIG. 8. Plot of CPTU data of SC layer from borehole in Robertson's chart.

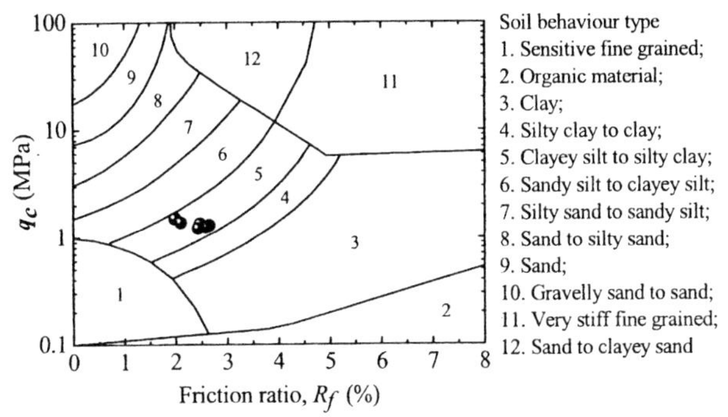

FIG. 9. Plot of CPTU data of dark green SC layer from borehole in Robertson's chart.

Figure 7 plots the CPTU data of gray mucky silty clay (MSC) identified from borehole samples. The CPTU data of the MSC layer is located in zones 3 and 4, which represents that the soil type is from clay to silty clay. From CPTU data, it can be also found that the MSC layer also contains a thin layer of clay lens as shown in Fig. 4. Figure 8 shows the CPTU data of gray silty clay layer identified from borehole samples. According to Fig. 8, most of the CPTU data of the SC layer is located in zones 5 and 6, which represent that the soil types of this layer are clayey silt to sandy silt, while a few data fall in zones 3, 4, 7, and 8, which means that this layer also includes clay, silty clay, and silty sand. From CPTU data, it can be found that there also exist two thin layers of sand lens as shown in Fig. 4. Figure 9 shows the CPTU data of dark green SC layer identified from borehole samples. All CPTU data of dark green silty clay layer are located in zone 5, which represent that the soil type of this layer is clayey silt to silty clay. This result is the same as that identified from borehole samples.

Table 1. Comparison of each soil layer identified both from borehole samples and CPTU tests.

Soil profile	Depth (m)	Number of same results	Number of different results	Percentage of different results
Gray MC	0-5.5	9	7	43.8%
Gray MSC	5.5-15	10	10	50%
Gray SC	15-29.5	26	10	27.8%
Dark green SC	29.5-32.3	6	0	0

Table 1 tabulates the comparison of each soil layer identified both from borehole and CPTU test. From Table 1, both of borehole results and CPTU test results is similar (>50%) in identification of soil stratigraphy, however, there are some discrepancies at the depth from 12.5 m to 14.9 m, where a thin clay lens exists, and from 22 m to 23.5 m and 26.4 m to 28.3 m, where two thin layers of sand lens exist. For these thin lens layers, the borehole sample can not identify. This is because in CPTU test data is measured continuously with a very small time interval or vertical distance, e.g. one data within 20 mm. However, for borehole sample method, the sample must be used to do following tests: physical tests (grain size, water content, unit weight) and mechanical tests (oedometer, shear test, and triangle test) so that the data interval at least 500 mm, in this field case, the borehole data was obtained with the depth interval of 1 m. As shown in Fig. 4, obvious changes occur in the curves of CPTU data when the cone passes through these thin layers: cone resistance q_c and friction ratio R_f suddenly increase compared with that at other depths. Therefore, these thin layers can be identified from CPTU test. In this sense, CPTU test becomes very useful and significant for geotechnical investigation.

CONCLUSIONS

1) Based on the CPTU data and combined with Robertson's chart, the stratigraphy of soft deposit in Shanghai was identified. The result is similar to those from borehole samples as established in the code method of Shanghai, which is based on the grain size distribution and plasticity index of soil sample.
2) Based on the CPTU data, thin layers of clay lens and sand lens can be identified through skin friction ratio. However, this trait is very difficult to be found by the borehole sample as the code method in Shanghai.
3) Based on large number of the researches on CPTU, CPTU test has its strength merit compared to traditional code method. However, the CPTU method with Robertson's chart has not been universally accepted in China and the reasons are i) the engineers has already used to the traditional code method; ii) the CPTU method was introduced in China in a short period.
4) More detailed researches on the CPTU method in different location and soil types must to be carried out further to establish the relationship between the CPTU

method and traditional code method so that the Chinese geotechnical engineers can be familiar with this method. Finally, based on the universal research results, the CPTU method with Robertson chart can be accepted by the Chinese geotechnical society and to be one of the code designated methods.

ACKNOWLEDGMENTS

The research work described herein was funded by the National Nature Science Foundation of China (NSFC) (Grant No.50779035) and the joint research project between NSFC and JSPS (Grant No.50911140105). It is also supported by Shanghai Leading Academic Discipline Project (Project Number: B208). These financial supports are gratefully acknowledged.

REFERENCES

Begemann, H.K.S. (1965). "The friction jacket cone as an aid in determining the soil profile." *Proc. 6th International Conference on Soil Mechanics and Foundation Engineering, Montreal,* Vol. 1: 17-20.

Committee of Civil Engineering in Shanghai (CCES) (1999). *Shanghai Standard Codes-Foundation Design Code* (DGJ08-11-1999), Shanghai. (in Chinese)

Committee of Civil Engineeing and Management in Shanghai (CCEMS) (2002). *Code for Investigation of Geotechnical Engineering* (DGJ08-37-2002), Shanghai. (in Chinese)

Douglas, B.J. and Olsen, R.S. (1981). "Soil classification using electric cone penetrometer: Cone Penetration Testing and Experience." *Proc. ASCE National Convention*, St. Louis: 209-227.

ISSMFE (1989). "International reference test procedure for cone penetration test (CPT)". *Report of the ISSMFE Technical Committee on Penetration Testing of Soils*-TC 16, with Reference to Test Procedures, Swedish Geotechnical Institute, Linköping, Information, 7: 6-16.

Lunne, T., Robertson, P. K., and Powell, J. M. (2002). *Cone Penetration Testing in Geotechnical Practice*. Spon Press, Taylor & Francis Group, London and New York.

Olsen, R.S. and Farr, J.V. (1986). "Site characterization using the cone penetration test." *Proc. ASCE Special Conference* - In Situ '86: Use of In Situ Tests in Geotechnical Engineering, Blacksburg: 854-868.

Robertson, P.K. and Campanella, R.G. (1983a). "Interpretation of Cone Penetration Tests: Part II–Clay." *Canadian Geotechnical Journal*, 20: 734-745.

Robertson, P.K., Campanella, R.G., and Wightman A. (1983b). "SPT-CPT Correlations." *Journal of Geotechnical Engineering*, ASCE, Vol. 109 (11): 1449-1459.

Robertson, P.K., Campanella, R.G., Gillespie, D., and Greig, J. (1986). "Use of piezometer cone data." *Proc. ASCE Special Conference*- In Situ '86: Use of In Situ Tests in Geotechnical Engineering, Blacksburg: 1263-80.

Xu, Y.-S., Shen, S.-L., and Du, Y.-J. (2009). "Geological and hydrogeological environment in Shanghai with geohazards to construction and maintenance of infrastructures." *Engineering Geology*, Vol. 109 (3-4): 241-254.

Author Index

Page number refers to the first page of paper

Subject Index

Page number refers to the first page of paper